P9-CLU-546

Mechanics, Heat,
and Sound

UNIVERSITY PHYSICS

Fourth Edition

Francis Weston Sears is Appleton Professor of Natural Philosophy, Emeritus, Dartmouth College. Before joining the Dartmouth faculty, he was Professor of Physics at Massachusetts Institute of Technology. During 1959 he was President of the American Association of Physics Teachers, and in 1962 was awarded the Oersted Medal by that organization for distinguished contributions to the teaching of physics. Professor Sears is author or co-author of many best selling texts in physics published by Addison-Wesley.

Mark W. Zemansky is Professor Emeritus of Physics at the City College of the City University of New York, where he taught for forty-five years. He received the Ph.D. degree in physics from Columbia University and has done research at Columbia University, Princeton University, and in Europe. He has served as President and recently as Executive Secretary of the American Association of Physics Teachers, and was a recipient of their Oersted Medal as well as the Townsend Harris Medal of the City College Associate Alumni. With Professor Sears he has co-authored several other Addison-Wesley texts.

Mechanics, Heat, and Sound

UNIVERSITY PHYSICS

Fourth Edition

FRANCIS WESTON SEARS
Professor Emeritus,
Dartmouth College

MARK W. ZEMANSKY
Professor Emeritus,
The City College of the City University of New York

ADDISON-WESLEY PUBLISHING COMPANY, INC.
Reading, Massachusetts
Menlo Park, California · London · Don Mills, Ontario

This book is in the
ADDISON-WESLEY SERIES IN PHYSICS

 Printed in the United States of America. Published simultaneously in Canada. Library of Congress Catalog Card No. 70–93991.

Preface

University Physics is available in one complete volume or as two separate parts: Part I covers the subjects of mechanics, heat, and sound; Part II includes electricity and magnetism, light, and atomic physics. The total number of topics is small enough so that the complete text may be taught in two semesters.

The text is intended for students of science and engineering who are taking a course in calculus concurrently and to whom calculus is still a new tool. The emphasis is on physical principles; historical background and practical applications have been given a place of secondary importance.

Three systems of units are used: the British gravitational system because it is the one used in engineering work throughout the country; the cgs system because some familiarity with it is essential for any intelligent reading of the literature of physics; and the mks system because of its increasing use in electricity and magnetism, as well as because it seems destined to eventually supplant the cgs system. The symbols and terminology, with few exceptions, are those recommended by the Committee on Letter Symbols and Abbreviations of the American Association of Physics Teachers as listed in the American Standard, ASA–Z10, published in 1947.

In this fourth edition, many of the features of the third edition are retained. Numerous illustrative problems are worked out in the body of the text; in each of them, all physical quantities are expressed by numerics along with the appropriate units. The problems at the end of each chapter deal with physical topics in the same order as in the body of the text; they are graded in difficulty and are designed to provide opportunities for routine drill as well as for independent thinking.

The level of mathematical and physical sophistication has been slightly reduced. The concept of the vector product has been removed from the treatment of elementary statics and postponed until the subject of angular momentum is encountered in Chapter 9. The relativistic treatment of relative velocity and the detailed diagrammatic analysis of length contraction and time dilation have been removed from elementary kinematics. Only the variation of mass with velocity has been retained in order to illuminate the Einstein relation between mass and energy. The concept of a magnetic field is introduced in the same manner as that of the second edition, without using relativistic principles.

Some of the more advanced topics that the students will meet in their later intermediate courses have been removed. Chief among these are the concept of circulation in a velocity field, details concerned with radiative heat transfer, and treatments of Maxwellian velocity distribution, mean free path, and viscosity, in the kinetic theory of gases. The mathematical treatment of the propagation of electromagnetic waves has been simplified and shortened.

In the early 1960's the level of difficulty of first-year physics for colleges and universities underwent a notable rise. Sparked by activities of the Commission on College Physics, by increased use of the materials prepared by the Physical Science Study Committee, and by wishful thinking, new courses were developed which went beyond the mathematical and physical limits imposed by the previous preparation of the average entering freshman. Unfortunately this preparation on the part of many students has not increased to the level demanded by the more difficult texts. As a result, some readjustment is necessary, and this fourth edition of *University Physics* may be regarded as a slight move in an attempt to reach the most realistic level.

It is with a feeling of sincere gratitude that we acknowledge the help we have received from the following teachers of physics: J. G. Anderson, J. D. Barnett, H. H. Barschall, A. C. Braden, P. Catranides, D. S. Duncan, D. Frantszog, J. L. Glathart, R. A. Kromhout, Robert J. Lee, Gerald P. Leitz, James A. Richards, Jr., T. L. Rokoske, M. Russel Wehr, L. W. Seagondollar, Lester V. Whitney, Jack Willis, and R. E. Worley.

Hanover and New York F.W.S.
January 1969 M.W.Z.

Contents

1 Composition and Resolution of Vectors

1-1 THE FUNDAMENTAL INDEFINABLES OF MECHANICS

Physics has been called the science of measurement. To quote from Lord Kelvin (1824–1907), "I often say that when you can measure what you are speaking about, and express it in numbers, you know something about it; but when you cannot express it in numbers, your knowledge is of a meagre and unsatisfactory kind; it may be the beginning of knowledge, but you have scarcely, in your thoughts, advanced to the stage of *Science*, whatever the matter may be."

A definition of a quantity in physics must provide a set of rules for calculating it in terms of other quantities that can be measured. Thus, when momentum is defined as the product of "mass" and "velocity," the rule for calculating momentum is contained within the definition, and all that is necessary is to know how to measure mass and velocity. The definition of velocity is given in terms of length and time, but there are no simpler or more fundamental quantities in terms of which length and time may be expressed. *Length and time are two of the indefinables of mechanics.* It has been found possible to express all the quantities of mechanics in terms of only three indefinables. The third may be taken to be "mass" or "force" with equal justification. *We shall choose mass as the third indefinable of mechanics.*

In geometry, the fundamental indefinable is the "point." The geometer asks his disciple to build any picture of a point in his mind, provided the picture is consistent with what the geometer *says* about the point. In physics, the situation is not so subtle. Physicists from all over the world have international committees at whose meetings the rules of measurement of the indefinables are adopted. The rule for measuring an indefinable takes the place of a definition.

1-2 STANDARDS AND UNITS

The set of rules for measuring the indefinables of mechanics is determined by an international committee called the *General Conference on Weights and Measures*, to which all the major countries send delegates. One of the chief functions of the Conference is to decide on a standard for each indefinable. A standard may be an actual object, in which case its main characteristic must be *durability*. Thus in 1889 when the meter bar of platinum-iridium alloy was chosen as the standard of length, it was felt that this alloy was particularly stable in its chemical structure. If, instead of platinum-iridium, a glass bar had been chosen, its length would have changed throughout the years because of the unavoidable crystallization that glass undergoes as it ages. Although platinum-iridium is an unusually stable alloy, the preservation of a bar of this material as a world standard entails a number of cumbersome provisions, such as making a large number of replicas for all the major countries and comparing these replicas with the world standard at periodic intervals. On October 14, 1960, the General Conference changed the standard of length to an *atomic constant*, namely, the *wavelength of the orange-red light emitted by the individual atoms of krypton-86* in a tube filled with krypton gas in which an electrical discharge is maintained.

The standard of mass is the mass of a *cylinder of platinum-iridium*, designated as *one kilogram*, and kept

TABLE 1-1 STANDARDS AND UNITS AS OF 1969

	Standard	Measuring device	Unit
Length	Wavelength of orange-red light from krypton-86	Optical interferometer	1 meter = 1,650,763.73 wavelengths
Mass	Platinum-iridium cylinder, 1 kilogram	Equal-arm balance	1 kilogram
Time	Periodic time associated with a transition between two energy levels of cesium-133 atom	Atomic clock	1 second = 9,192,631,770 cesium periods

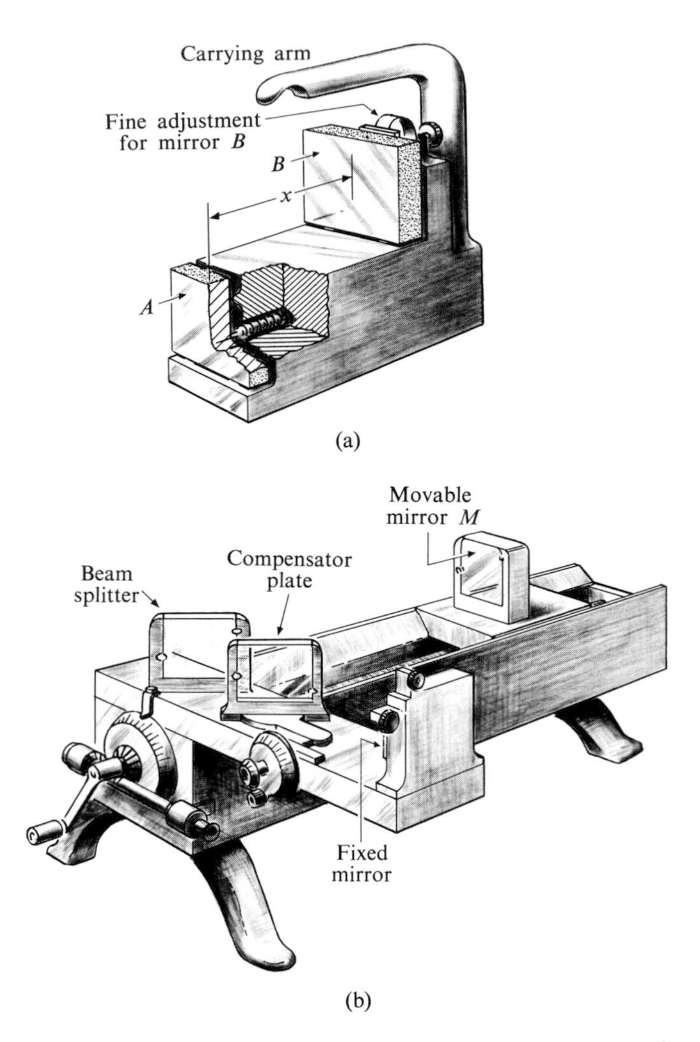

1-1 (a) Etalon and (b) Michelson interferometer for use in measuring the distance x in terms of the wavelength of light.

at the International Bureau of Weights and Measures at Sèvres, near Paris.

Before 1960, the standard of time was the interval of time between successive appearances of the sun overhead, averaged over a year, and called the *mean solar day*. Between 1960 and 1967 it was changed to the *tropical year 1900*, that is, the time it took the sun to move from a certain point in the heavens, known as the *vernal equinox*, back to the same point in 1900. In October 1967, the standard was changed again to the *periodic time of the radiation corresponding to the transition between the two hyperfine levels of the fundamental state of the atom of cesium-133.*

The three standards are listed in Table 1-1.

After the choice of a standard, the next step is to decide upon an instrument and a technique for comparing the standard with an unknown. Consider, for example, the distance x between two mirrors, A and B, of the device called an *etalon*, shown in Fig. 1-1(a). To find the number of wavelengths of orange-red light of krypton-86 in the distance x requires the use of an *optical interferometer*, one type of which, due to Michelson, is shown in Fig. 1-1(b). A movable mirror M on the Michelson interferometer is first made to coincide in position with A on the etalon. Then the mirror is moved slowly to coincide with B, during which time gradations of orange and black, known as *interference fringes*, move past the cross hair in the field of view of a telescope and are counted. The motion of one complete fringe corresponds to a motion of mirror M of exactly one-half a wavelength. A length known as *one meter* is defined in this way as

1 meter = 1,650,763.73 wavelengths of orange-red light of krypton-86.

Other units of length frequently used in pure science are:

1 angstrom unit = 1 Å = 10^{-10} m (used by spectroscopists),

1 nanometer = 1 nm = 10^{-9} m (used by optical designers),

TABLE 1-2 PREFIXES FOR POWERS OF TEN

Power of ten	10^{-12}	10^{-9}	10^{-6}	10^{-3}	10^{-2}	10^{3}	10^{6}	10^{9}	10^{12}
Prefix	pico-	nano-	micro-	milli-	centi-	kilo-	mega-	giga-	tera-
Abbreviation	p	n	μ	m	c	k	M	G	T

1 micrometer $= 1\ \mu\text{m} = 10^{-6}$ m (used commonly in biology),
1 millimeter $= 1$ mm $= 10^{-3}$ m
1 centimeter $= 1$ cm $= 10^{-2}$ m (used most often),
1 kilometer $= 1$ km $= 10^{3}$ m (a common European unit of distance).

The words "nanometer," "micrometer," and "kilometer" are all accented on the *first* syllable, *not* the second, just like the words "millimeter" and "centimeter." The prefix "nano" is pronounced "nanno." Units of length used in everyday life and in engineering in both the United States and the United Kingdom are defined as follows:

1 inch = 1 in. = {41,929.399 wavelengths of Kr light, 2.54 cm,
1 foot = 1 ft = 12 in.,
1 yard = 1 yd = 3 ft,
1 mile = 1 mi = 5280 ft.

The device used to subdivide the standard of mass, the kilogram, into equal submasses is the *equal-arm balance*, which will be discussed in Chapter 5. Frequently used units of mass are:

1 microgram $= 1\ \mu\text{g} = 10^{-9}$ kg,
1 milligram $= 1$ mg $= 10^{-6}$ kg,
1 gram $= 1$ g $= 10^{-3}$ kg,
1 pound mass = 1 lbm = 0.45359237 kg.

The clock making use of the standard time interval is the *cesium clock*, a large, complex, and expensive laboratory instrument. It is extraordinarily precise and maintains its frequency constant to one part in one hundred billion (10^{11}) or better. Furthermore, it may be compared with other high-precision clocks in an hour or so as against the years required for comparison with the old astronomical standard. In the atomic clock, a beam of cesium-133 atoms passes through a long metal cylinder and interacts with microwaves brought in by a wave guide from a generator controlled by a quartz oscillator. The *unit* of time used throughout the world is called the *second* and is defined to be

1 second = 1 s = 9,192,631,770 Cs periods.

Other common units of time are:

1 nanosecond = 1 ns $= 10^{-9}$ s,
1 microsecond $= 1\ \mu\text{s} = 10^{-6}$ s,
1 millisecond = 1 ms $= 10^{-3}$ s,
1 minute = 1 min = 60 s,
1 hour = 1 hr = 3600 s,
1 day = 1 day = 86,400 s.

It will be found useful to memorize the prefixes and abbreviations for the various powers of ten which are collected in Table 1-2. (The prefix "giga" is pronounced "jeega.")

1-3 SYMBOLS FOR PHYSICAL QUANTITIES

We shall adopt the convention that an algebraic symbol representing a physical quantity, such as F, p, or v, stands for both a *number* and a *unit*. For example, F might represent a force of 10 lb, p a pressure of 15 lb ft^{-2}, and v a velocity of 15 ft s^{-1}.

When we write

$$x = v_0 t + \tfrac{1}{2}at^2,$$

if x is in feet, then the terms $v_0 t$ and $\frac{1}{2}at^2$ must be in feet also. Suppose t is in seconds. Then the units of v_0 must be ft s^{-1} and those of a must be ft s^{-2}. (The factor $\frac{1}{2}$ is a *pure number*, without units.) As a numerical example, let $v_0 = 10$ ft s^{-1}, $a = 4$ ft s^{-2}, $t = 10$ s. Then the preceding equation would be written

$$x = 10\ \text{ft}\ \not{s}^{-1} \times 10\ \not{s} + \tfrac{1}{2} \times 4\ \text{ft}\ \not{s}^{-2} \times 10\ \not{s}^2.$$

The units are treated like algebraic symbols. The s's cancel in the first term and the s^2's in the second, and

$$x = 100 \text{ ft} + 200 \text{ ft} = 300 \text{ ft}.$$

The beginning student will do well to include the units of all physical quantities, as well as their magnitudes, in all his calculations. This will be done consistently in the numerical examples throughout the book.

1-4 FORCE

Mechanics is the branch of physics which deals with the motion of material bodies and with the forces that bring about the motion. Since motion is best described by the methods of calculus and many readers of this book are just beginning their study of this subject, we shall postpone a discussion of motion until Chapter 4, and start with a study of forces.

When we push or pull on a body, we are said to exert a *force* on it. Forces can also be exerted by inanimate objects: a stretched spring exerts forces on the bodies to which its ends are attached; compressed air exerts a force on the walls of its container; a locomotive exerts a force on the train it is drawing. The force of which we are most aware in our daily lives is the force of gravitational attraction exerted on every body by the earth, and is called the *weight* of the body. Gravitational forces (and electrical and magnetic forces also) can act through empty space without contact. In this respect they differ from the forces mentioned above, where the body doing the pushing or pulling must make contact with the body being pushed or pulled.

We are not yet in a position to show how a unit of force can be defined in terms of the units of mass, length, and time. This will be done in Chapter 5. For the present, a unit of force can be defined as follows. We select as a standard body the standard pound, defined in Section 1-2 as a certain fraction (approximately 0.454) of a standard kilogram. The force with which the earth attracts this body, at some specified point on the earth's surface, is then a perfectly definite, reproducible force and is called a force of *one pound* (avoirdupois). A particular point on the earth's surface must be specified, since the attraction of the earth for a given body varies slightly from one point to another. If great precision is not required, it suffices to take any point at sea level and 45° latitude.

In order that an unknown force can be compared with the force unit, and thereby measured, some measurable effect produced by a force must be used. One such effect is to alter the dimensions or shape of a body on which the force is exerted; another is to alter the state of motion of the body. Both of these effects are used in the measurement of forces. In this chapter we shall consider only the former; the latter will be discussed in Chapter 5.

The instrument most commonly used to measure forces is the spring balance, which consists of a coil spring enclosed in a case for protection and carrying at one end a pointer that moves over a scale. A force exerted on the balance changes the length of the spring. The balance can be calibrated as follows. The standard pound is first suspended from the balance at sea level and 45° latitude and the position of the pointer is marked 1 lb. Any number of duplicates of the standard can then be prepared by suspending a body from the balance and adding or removing material until the index again stands at 1 lb. Then when two, three, or more of these are suspended simultaneously from the balance, the force stretching it is 2 lb, 3 lb, etc., and the corresponding positions of the pointer can be labeled 2 lb, 3 lb, etc. This procedure makes no assumption about the elastic properties of the spring except that the force exerted on it is always the same when the pointer stands at the same position. The calibrated balance can then be used to measure an unknown force.

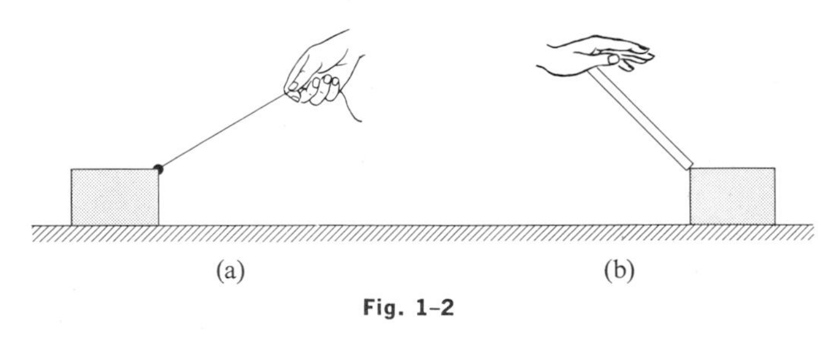

Fig. 1-2

1-5 GRAPHICAL REPRESENTATION OF FORCES. VECTORS

Suppose we are to slide a box along the floor by pulling it with a string or pushing it with a stick, as in Fig. 1-2. That is, we are to slide it by exerting a force on it. The point of view which we now adopt is that the motion of the box is caused not by the *objects* which push or pull on it, but by the *forces* which these exert. For concreteness, assume the magnitude of the push or pull to be 10 lb.

It is clear that simply to write "10 lb" on the diagram would not completely describe the force, since it would not indicate the direction in which the force was acting. One might write "10 lb, 30° above horizontal to the right," or "10 lb, 45° below horizontal to the right," but all the above information may be conveyed more briefly if we adopt the convention of representing a force by an arrow. The length of the arrow, to some chosen scale, indicates the size or *magnitude* of the force, and the direction in which the arrow points indicates the *direction* of the force. Thus Fig. 1–3 is the force diagram corresponding to Fig. 1–2. (There are other forces acting on the box, but these are not shown in the figure.)

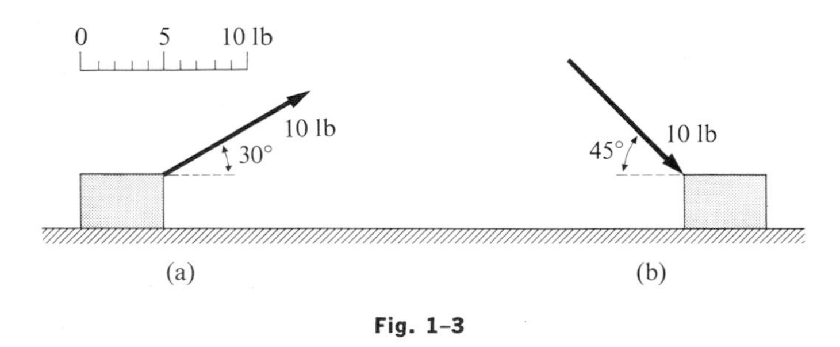

Fig. 1–3

Force is not the only physical quantity which requires the specification of a direction in space as well as a magnitude. For example, the velocity of an aircraft is not completely specified by stating that it is 300 miles per hour; we need to know the direction also. The concept of volume, on the other hand, has no direction associated with it.

Quantities like volume, which involve a magnitude only, are called *scalars*. Those like force and velocity, which involve both a magnitude and a direction in space, are called *vector quantities*. Any vector quantity can be represented by an arrow, and this arrow is called a vector (or if a more specific statement is needed, a force vector or a velocity vector).

Some vector quantities, of which force is one, are not *completely* specified by their magnitude and direction alone. Thus the effect of a force depends also on its *line of action* and its *point of application*. (The line of action is a line of indefinite length, of which the force vector is a segment.) For example, if one is pushing horizontally against a door, the effectiveness of a force of given magnitude and direction depends on the distance of its line of action from the hinges. If a body is deformable, as all bodies are to a greater or lesser extent, the deformation depends on the point of application of the force. However, since many actual objects are deformed only very slightly by the forces acting on them, we shall assume for the present that all objects considered are perfectly rigid. The point of application of a given force acting on a rigid body may be transferred to any other point on the line of action without altering the effect of the force. Thus *a force applied to a rigid body may be regarded as acting anywhere along its line of action.*

A vector quantity is represented by a letter in boldface type. The same letter in ordinary type represents the magnitude of the quantity. Thus the magnitude of a force ***F*** is represented by *F*.

1–6 VECTOR ADDITION. RESULTANT OF A SET OF FORCES

The sciences of arithmetic and algebra deal with pure numbers. Similarly, in the science of *vector analysis*, another branch of pure mathematics, a vector is considered simply as an arrow or a "directed line segment," without any physical significance. However, just as the laws of arithmetic and algebra are found to describe certain operations that can be carried out with some physical quantities, so the laws of vector algebra are found to represent some (but not all) aspects of the behavior of other physical quantities.

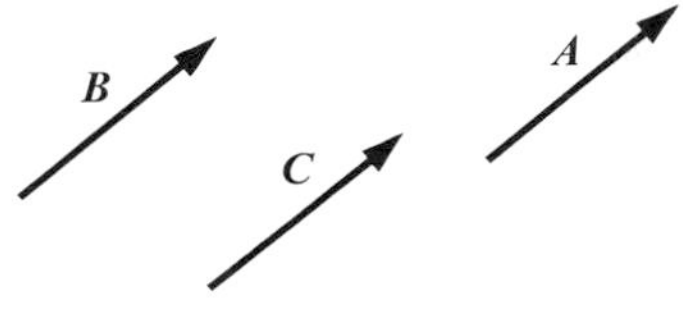

1–4 The vectors, ***A***, ***B***, and ***C*** are mathematically equal.

For example, two (mathematical) vectors are considered equal, by definition, if they have the same magnitude and direction. Thus in Fig. 1–4 the vectors ***A***, ***B***, and ***C*** are all equal. It follows that in mathematics a given vector may be moved around at will, provided its length and direction are not changed. However, if the vectors in Fig. 1–4 represent forces acting on a body, the forces are not physically equivalent, since they have different points of application and different lines of action.

The *vector sum* of two (mathematical) vectors is defined as follows. Let ***A*** and ***B*** in Fig. 1–5(a) be two given

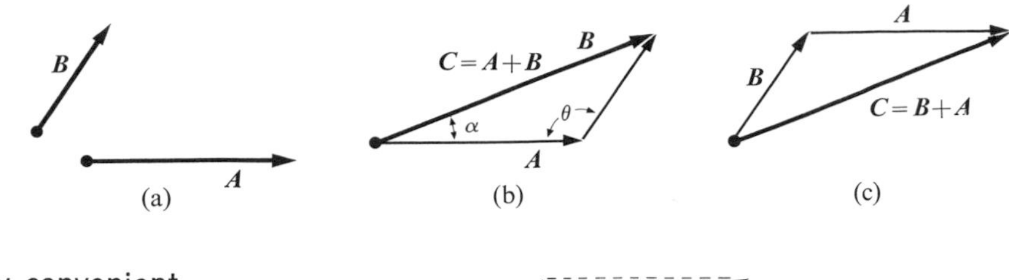

1-5 Vector $\boldsymbol{C}$ is the vector sum of vectors $\boldsymbol{A}$ and $\boldsymbol{B}$. $\boldsymbol{C} = \boldsymbol{A} + \boldsymbol{B} = \boldsymbol{B} + \boldsymbol{A}$.

vectors. Draw the vectors as in (b) at any convenient point, with the initial point of $\boldsymbol{B}$ at the endpoint of $\boldsymbol{A}$. The vector sum $\boldsymbol{C}$ is then defined as the vector from the initial point of $\boldsymbol{A}$ to the endpoint of $\boldsymbol{B}$. The symbol for vector addition is the same as for algebraic addition, and we write

$$\boldsymbol{C} = \boldsymbol{A} + \boldsymbol{B}.$$

Alternatively, the given vectors can be drawn as in Fig. 1-5(c), with the initial point of $\boldsymbol{A}$ at the endpoint of $\boldsymbol{B}$. The vector $\boldsymbol{C}$ has the same magnitude and direction as in (b) and hence the two vector sums are mathematically equal. The order in which the vectors are added is therefore immaterial, and vector addition obeys the same commutative law as algebraic addition:

$$\boldsymbol{A} + \boldsymbol{B} = \boldsymbol{B} + \boldsymbol{A}.$$

The magnitude and direction of the vector sum $\boldsymbol{C}$ can be found from measurements on a carefully drawn diagram. They can also be computed by the methods of trigonometry. Thus if θ represents the angle between vectors $\boldsymbol{A}$ and $\boldsymbol{B}$, as in Fig. 1-5(b), the magnitude of $\boldsymbol{C}$ is given by

$$C^2 = A^2 + B^2 - 2AB \cos \theta.$$

The angle α between $\boldsymbol{C}$ and $\boldsymbol{A}$ can be found from the relation

$$\frac{\sin \alpha}{B} = \frac{\sin \theta}{C}.$$

Another useful method of finding the sum of two vectors is shown in Fig. 1-6, where vectors $\boldsymbol{A}$ and $\boldsymbol{B}$ are both drawn from a common point. The vector sum $\boldsymbol{C}$ is the concurrent diagonal of a parallelogram of which the given vectors form two sides.

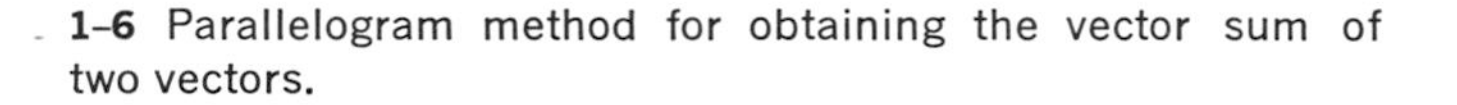

1-6 Parallelogram method for obtaining the vector sum of two vectors.

Figure 1-7 illustrates a special case in which two vectors are parallel, as in (a), or antiparallel, as in (b). If they are parallel, the magnitude of the vector sum $\boldsymbol{C}$ equals the sum of the magnitudes of $\boldsymbol{A}$ and $\boldsymbol{B}$. If they are antiparallel, the magnitude of the vector sum equals the difference of the magnitudes of $\boldsymbol{A}$ and $\boldsymbol{B}$. (The vectors in Fig. 1-7 have been displaced slightly sidewise from their line of action to show them more clearly. Actually, all vectors lie along the same geometrical line.)

When more than two vectors are to be added, we may first find the vector sum of any two, add this vectorially to the third, and so on. This process is illustrated in Fig. 1-8, which shows in part (a) four vectors $\boldsymbol{A}$, $\boldsymbol{B}$, $\boldsymbol{C}$, and $\boldsymbol{D}$. In Fig. 1-8(b), vectors $\boldsymbol{A}$ and $\boldsymbol{B}$ are first added by the triangle method, giving a vector sum $\boldsymbol{E}$; vectors $\boldsymbol{E}$ and $\boldsymbol{C}$ are then added by the same process to obtain the vector sum $\boldsymbol{F}$; finally, $\boldsymbol{F}$ and $\boldsymbol{D}$ are added to obtain the vector sum

$$\boldsymbol{G} = \boldsymbol{A} + \boldsymbol{B} + \boldsymbol{C} + \boldsymbol{D}.$$

Evidently the vectors $\boldsymbol{E}$ and $\boldsymbol{F}$ need not have been drawn; we need only draw the given vectors in succession, with the tail of each at the head of the one preceding it, and complete the polygon by a vector $\boldsymbol{G}$ from the tail of the first to the head of the last vector. The order in which

1-7 Vector sum of (a) two parallel vectors, (b) two antiparallel vectors.

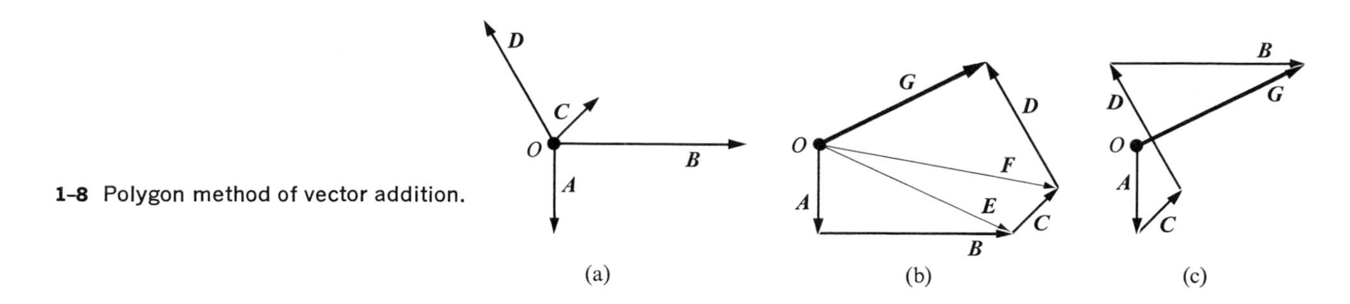

1-8 Polygon method of vector addition.

the vectors are drawn makes no difference, as shown in Fig. 1-8(c).

Now consider the following physical problem. Two forces, represented by the vectors $\boldsymbol{F}_1$ and $\boldsymbol{F}_2$ in Fig. 1-9, are simultaneously applied at the same point A of a body. Is it possible to produce the same effect by applying a single force at A, and if so, what should be its magnitude and direction? The question can be answered only by experiment, and experimental results show that a single force, represented in magnitude, direction, and line of action by the vector sum $\boldsymbol{R}$ of the original forces, is in all respects equivalent to them. This single force is called the *resultant* of the original forces. Hence the mathematical process of *vector addition* of two force vectors corresponds to the physical operation of finding the *resultant* of two forces, simultaneously applied at a given point.

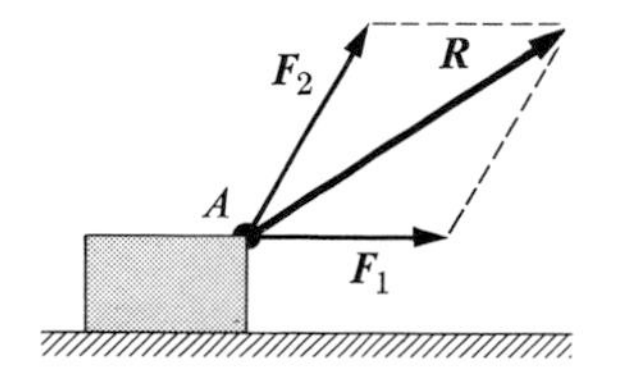

1-9 A force represented by the vector $\boldsymbol{R}$, equal to the vector sum of $\boldsymbol{F}_1$ and $\boldsymbol{F}_2$, produces the same effect as the forces $\boldsymbol{F}_1$ and $\boldsymbol{F}_2$ acting simultaneously.

Consider next the more general case in which two forces are applied at different points of a rigid body, as in Fig. 1-10, where the points of application of forces $\boldsymbol{F}_1$ and $\boldsymbol{F}_2$ are at A and B. We shall consider only the special case in which the forces lie in the same plane, or are *coplanar.* Since a force applied to a rigid body can be displaced along its line of action, let the forces be transferred to point C, where their action lines intersect. (The point of intersection may lie outside the body on which the forces act; see Fig. 2-9.) The resultant $\boldsymbol{R}$ is then obtained as in Fig. 1-9, and this force, applied at any point on its line of action, such as point D, is physically equivalent to forces $\boldsymbol{F}_1$ and $\boldsymbol{F}_2$ acting simultaneously.

The vector sum of $\boldsymbol{F}_1$ and $\boldsymbol{F}_2$ can also be obtained by the construction of Fig. 1-5, where the vectors are drawn tail-to-head at any convenient point. The vector sum then has the same magnitude and direction as the resultant $\boldsymbol{R}$, but not necessarily the same line of action. This again illustrates that although a mathematical vector may be displaced in any way (retaining its original magnitude and direction) a force acting on a rigid body can be displaced only along its action line.

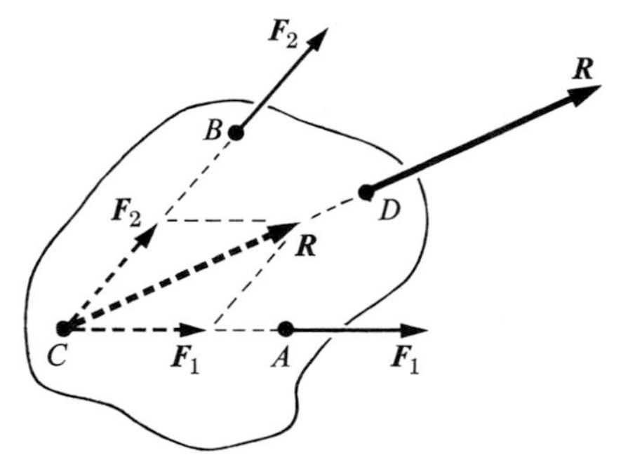

1-10 Method of finding the resultant $\boldsymbol{R}$ of two coplanar forces $\boldsymbol{F}_1$ and $\boldsymbol{F}_2$ having different points of application.

The resultant of more than two coplanar forces can be found by a repetition of the process in Fig. 1-10. We first find the resultant of any two forces, then combine this with a third, and so on. Since vectors obey the commutative law of addition, the order in which they are added makes no difference.

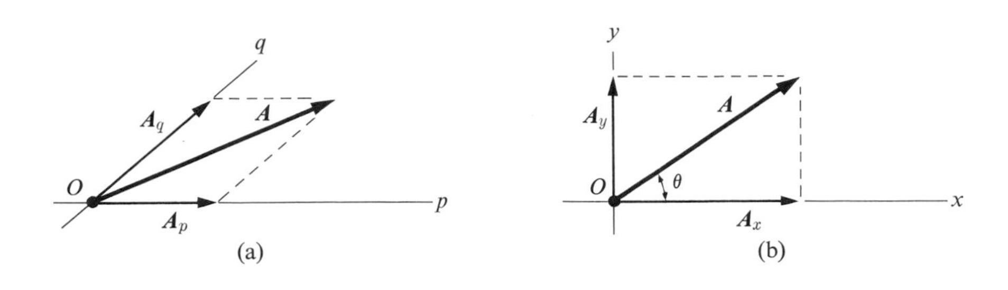

1-11 (a) Vectors $\boldsymbol{A}_p$ and $\boldsymbol{A}_q$ are the components of $\boldsymbol{A}$ in the directions Op and Oq. (b) Vectors $\boldsymbol{A}_x$ and $\boldsymbol{A}_y$ are the rectangular components of $\boldsymbol{A}$ in the directions of the x- and y- axes.

1-7 COMPONENTS OF A VECTOR

Any two vectors whose vector sum equals a given vector are called the *components* of that vector. In Fig. 1-6, for example, vectors $\boldsymbol{A}$ and $\boldsymbol{B}$ are components of the vector $\boldsymbol{C}$. Evidently, a given vector has an infinite number of pairs of possible components. If the *directions* of the components are specified, however, the problem of finding the components, or of *resolving* the vector into components, has a unique solution. Thus suppose we are given the vector $\boldsymbol{A}$ in Fig. 1-11(a) and we wish to resolve it into components in the directions of the lines Op and Oq. From the tip of vector $\boldsymbol{A}$ draw the dotted construction lines parallel to Op and Oq, forming a parallelogram. Vectors $\boldsymbol{A}_p$ and $\boldsymbol{A}_q$, from O to the points of intersection of the construction lines with Op and Oq, are then the desired components, since they are in the specified directions and the given vector is their vector sum.

The special case in which the specified directions are at right angles to each other is of particular importance. In Fig. 1-11(b), lines Ox and Oy are the axes of a rectangular coordinate system. The parallelogram obtained by drawing the dotted construction lines from the tip of vector $\boldsymbol{A}$ then becomes a rectangle, and the components $\boldsymbol{A}_x$ and $\boldsymbol{A}_y$ are called the *rectangular components* of $\boldsymbol{A}$.

The magnitudes of the rectangular components of a vector are easily computed. If θ is the angle which the vector $\boldsymbol{A}$ makes with the x-axis, then

$$A_x = A\cos\theta, \qquad A_y = A\sin\theta,$$

where A, A_x, and A_y are the *magnitudes* of the corresponding vectors.

The application of the concepts above to a physical problem is illustrated in Fig. 1-12, where a force $\boldsymbol{F}$ is exerted on a body at point O. The rectangular components of $\boldsymbol{F}$ in the directions Ox and Oy are $\boldsymbol{F}_x$ and $\boldsymbol{F}_y$, and it is found that simultaneous application of the forces $\boldsymbol{F}_x$ and $\boldsymbol{F}_y$, as in Fig. 1-12(b), is equivalent in all respects to the effect of the original force. *Any force can be replaced by its rectangular components.*

As a numerical example, let

$$F = 10\text{ lb}, \qquad \theta = 30°.$$

Then

$$F_x = F\cos\theta = 10\text{ lb} \times 0.866 = 8.66\text{ lb},$$
$$F_y = F\sin\theta = 10\text{ lb} \times 0.500 = 5.00\text{ lb},$$

and the effect of the original 10-lb force is equivalent to

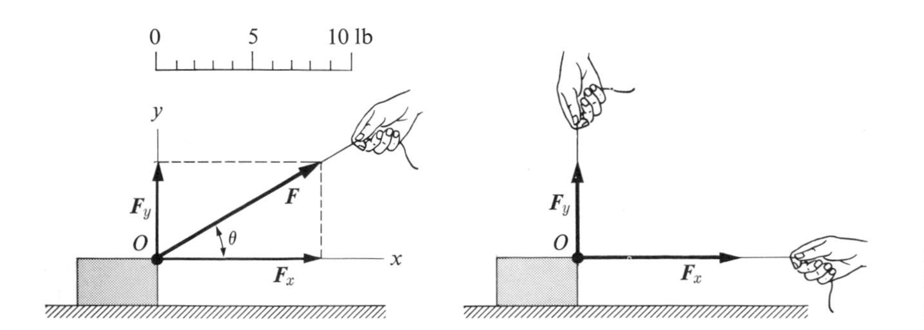

1-12 The inclined force $\boldsymbol{F}$ may be replaced by its rectangular components $\boldsymbol{F}_x$ and $\boldsymbol{F}_y$. $F_x = F\cos\theta$, $F_y = F\sin\theta$.

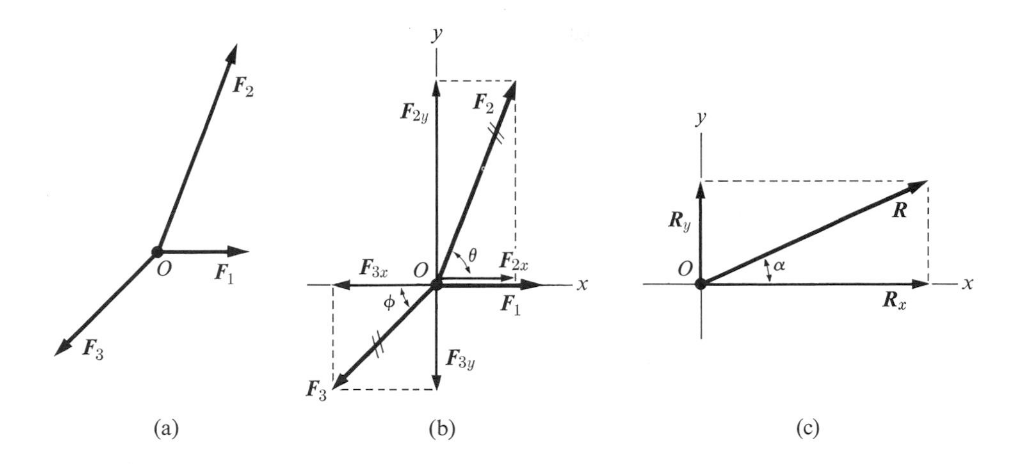

1-13 Vector $\boldsymbol{R}$, the resultant of $\boldsymbol{F}_1$, $\boldsymbol{F}_2$, and $\boldsymbol{F}_3$, is obtained by the method of rectangular resolution. The magnitudes of the rectangular components of $\boldsymbol{R}$ are $R_x = \sum F_x$, $R_y = \sum F_y$.

the simultaneous application of a horizontal force of 8.66 lb and a lifting force of 5.00 lb.

1-8 RESULTANT BY RECTANGULAR RESOLUTION

Although the polygon method is a satisfactory graphical one for finding the resultant of a number of forces, it is awkward for computation because one must, in general, solve a number of oblique triangles. Therefore the usual analytical method of finding the resultant is first to resolve all forces into rectangular components along any convenient pair of axes and then combine these into a single resultant. This makes it possible to work with *right* triangles only.

Figure 1-13(a) shows three concurrent forces $\boldsymbol{F}_1$, $\boldsymbol{F}_2$, and $\boldsymbol{F}_3$, whose resultant we wish to find. Let a pair of rectangular axes be constructed in any arbitrary direction. Simplification results if one axis coincides with one of the forces, which is always possible. In Fig. 1-13(b), the x-axis coincides with $\boldsymbol{F}_1$. Let us first resolve each of the given forces into x- and y-components. According to the usual conventions of analytic geometry, x-components toward the right are considered positive and those toward the left, negative. Upward y-components are positive and downward y-components are negative

Force $\boldsymbol{F}_1$ lies along the x-axis and need not be resolved. The components of $\boldsymbol{F}_2$ are $F_{2x} = F_2 \cos\theta$, $F_{2y} = F_2 \sin\theta$. Both of these are positive, and $\boldsymbol{F}_{2x}$ has been slightly displaced upward to show it more clearly. The components of $\boldsymbol{F}_3$ are $F_{3x} = F_3 \cos\phi$, $F_{3y} = F_3 \sin\phi$. Both of these are negative.

We now imagine $\boldsymbol{F}_2$ and $\boldsymbol{F}_3$ to be removed and replaced by their rectangular components. To indicate this, the vectors $\boldsymbol{F}_2$ and $\boldsymbol{F}_3$ are crossed out lightly. All of the x-components can now be combined into a single force $\boldsymbol{R}_x$ whose magnitude equals the algebraic sum of the magnitudes of the x-components, or $\sum F_x$, and all of the y-components can be combined into a single force $\boldsymbol{R}_y$ of magnitude $\sum F_y$:

$$R_x = \sum F_x, \qquad R_y = \sum F_y.$$

Finally, these can be combined as in part (c) of the figure to form the resultant $\boldsymbol{R}$ whose magnitude, since $\boldsymbol{R}_x$ and $\boldsymbol{R}_y$ are perpendicular to each other, is

$$R = \sqrt{R_x^2 + R_y^2}.$$

The angle α between $\boldsymbol{R}$ and the x-axis can now be found from any one of its trigonometric functions. For example, $\tan\alpha = R_y/R_x$.

Example. In Fig. 1-13, let $F_1 = 120$ lb, $F_2 = 200$ lb, $F_3 = 150$ lb, $\theta = 60°$, $\phi = 45°$. The computations can be arranged systematically as in the table on page 10.

Force	Angle	x-component	y-component
$F_1 = 120$ lb	0	+120 lb	0
$F_2 = 200$ lb	60°	+100 lb	+173 lb
$F_3 = 150$ lb	45°	−106 lb	−106 lb
		$\sum F_x = +114$ lb	$\sum F_y = +67$ lb

$$R = \sqrt{(114\text{ lb})^2 + (67\text{ lb})^2} = 132\text{ lb},$$

$$\alpha = \tan^{-1}\frac{67\text{ lb}}{114\text{ lb}} = \tan^{-1} 0.588 = 30.4°.$$

1-9 VECTOR DIFFERENCE

It is sometimes necessary to subtract one vector from another. The process of subtracting one *algebraic* quantity from another is equivalent to adding the negative of the quantity to be subtracted. That is,

$$a - b = a + (-b).$$

Similarly, the process of subtracting one *vector* quantity from another is equivalent to adding (vectorially) the negative of the vector to be subtracted, where the negative of a given vector is defined as a vector of the same length but in the opposite direction. That is, if $\boldsymbol{A}$ and $\boldsymbol{B}$ are two vectors,

$$\boldsymbol{A} - \boldsymbol{B} = \boldsymbol{A} + (-\boldsymbol{B}).$$

Vector subtraction is illustrated in Fig. 1-14. The given vectors are shown in part (a). In part (b), the vector sum of $\boldsymbol{A}$ and $-\boldsymbol{B}$, or the vector difference $\boldsymbol{A} - \boldsymbol{B}$, is found by the parallelogram method. Part (c) shows a second method; vectors $\boldsymbol{A}$ and $\boldsymbol{B}$ are drawn from a common origin, and the vector difference $\boldsymbol{A} - \boldsymbol{B}$ is the vector from the tip of $\boldsymbol{B}$ to the tip of $\boldsymbol{A}$. The vector difference $\boldsymbol{A} - \boldsymbol{B}$ is therefore seen to be the vector that must be added to $\boldsymbol{B}$ in order to give $\boldsymbol{A}$, since

$$\boldsymbol{B} + (\boldsymbol{A} - \boldsymbol{B}) = \boldsymbol{A}.$$

Vector differences may also be found by the method of rectangular resolution. Both vectors are resolved into x- and y-components. The difference between the x-components is the x-component of the desired vector difference, and the difference between the y-components is the y-component of the vector difference.

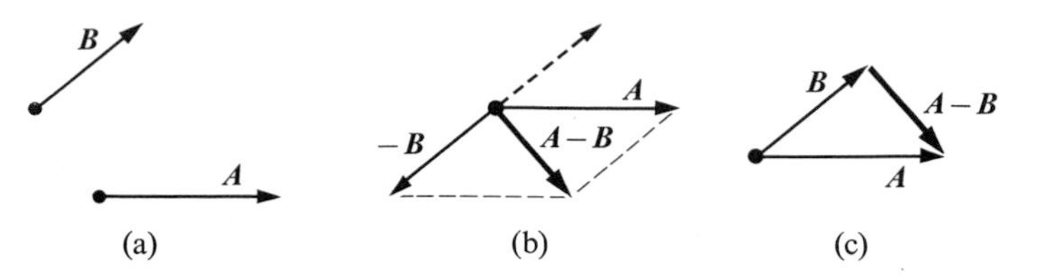

1-14 The vector difference $\boldsymbol{A} - \boldsymbol{B}$ is found in part (b) by the parallelogram method and in part (c) by the triangle method.

Vector subtraction is not often used when dealing with forces, but we shall use it frequently in connection with velocities and accelerations.

1-10 A WORD ABOUT PROBLEMS

To obtain a benefit commensurate with the time expended, problem-solving should be considered as much more than merely substituting numbers for the symbols in a formula or fitting together the pieces of a jigsaw puzzle. To merely thumb through the book until you find a formula that seems to fit, or a worked-out example that resembles the problem, is a waste of time and effort. One purpose which problems can serve is to enable you to find out for yourself whether or not you understand the assigned material, after having listened to a lecture and studied the text. Do your studying *before* you tackle the problems, instead of beginning with the problems and not referring to your book or lecture notes until you find yourself stymied. No problem assignment of reasonable length can hope to cover *every* important point, and you will miss a great deal if you read only enough to enable you to "do" your problems.

Every problem involves one or more general physical laws or definitions. After reading the statement of a problem, ask yourself what these laws or definitions are and be sure that you know them. This means that you should be able to state them to yourself clearly and explicitly, and not be satisfied with the comforting feeling that you "understand" Newton's second law although you can't quite put it in so many words.

Nearly every quantity of physical interest is expressed in terms of a unit of that quantity (some quantities are pure numbers). No answer is complete unless the proper units are given.

Unless otherwise stated, the numerical data given in the problems are to be assumed correct to three significant figures (for example, 2 ft implies 2.00 ft). Numerical answers should therefore be rounded off to three significant figures. That is, if long division leads to the result 6.574938 . . . ft s^{-1}, give the answer as 6.57 ft s^{-1}. This is automatically taken care of if you use a slide rule.

Answers are provided to about half the problems, and are rounded off to two or three significant figures.

One of the best ways of making sure you understand the principles covered by a particular problem is to work it backwards. That is, if the problem gives x and asks you to compute y, then make up another problem in which y is given and x is to be found. Another helpful procedure is to ask yourself how the result would be altered if the given conditions had been somewhat different. Suppose the friction force had been 10 lb instead of 5 lb? Suppose the slope of the plane had been 60° instead of 30°?

Helpful advice on the techniques of studying and of solving problems will be found in *How to Study Physics* by Chapman and *How to Study, How to Solve* by Dadourian (published by Addison-Wesley).

Problems

1-1 Find graphically the magnitude and direction of the resultant of the three forces in Fig. 1-15. Use the polygon method.

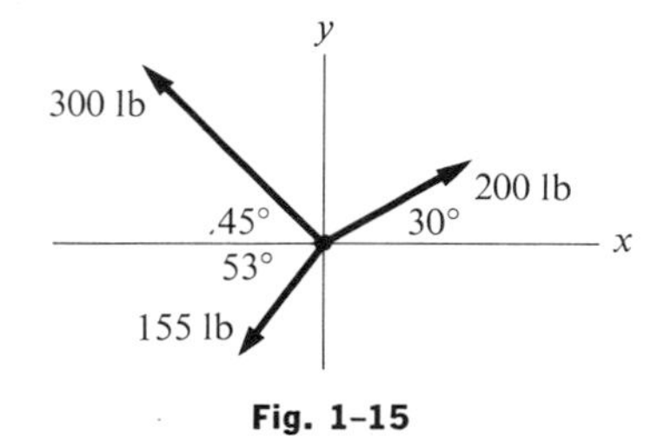

Fig. 1-15

1-2 Two men and a boy want to push a crate in the direction marked x in Fig. 1-16. The two men push with forces F_1 and F_2 whose magnitudes and directions are indicated in the figure. Find the magnitude and direction of the smallest force which the boy should exert.

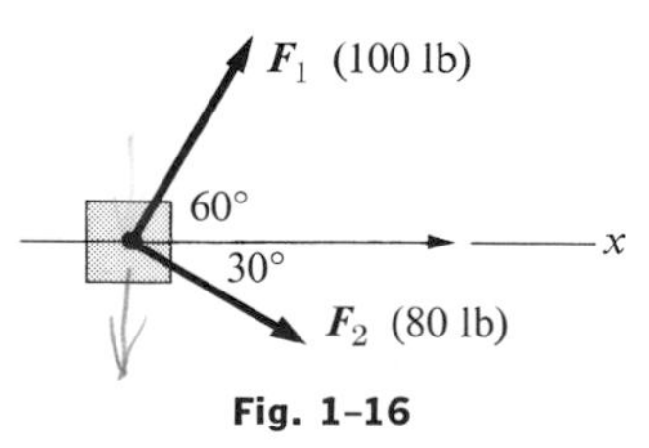

Fig. 1-16

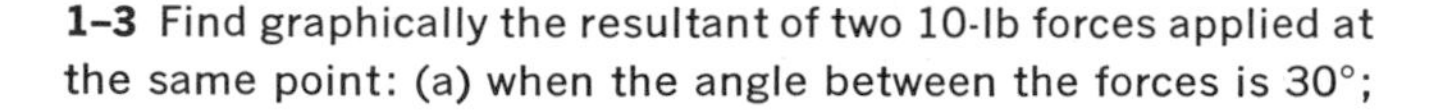
1-3 Find graphically the resultant of two 10-lb forces applied at the same point: (a) when the angle between the forces is 30°; (b) when the angle between them is 130°. Use any convenient scale.

1-4 Two men pull horizontally on ropes attached to a post, the angle between the ropes being 45°. If man A exerts a force of 150 lb and man B a force of 100 lb, find the magnitude of the resultant force and the angle it makes with A's pull. Solve: (a) graphically, by the parallelogram method, and (b) graphically, by the triangle method. Let 1 in. = 50 lb in (a) and (b).

1-5 Find graphically the vector sum $\boldsymbol{A} + \boldsymbol{B}$ and the vector difference $\boldsymbol{A} - \boldsymbol{B}$ in Fig. 1-17.

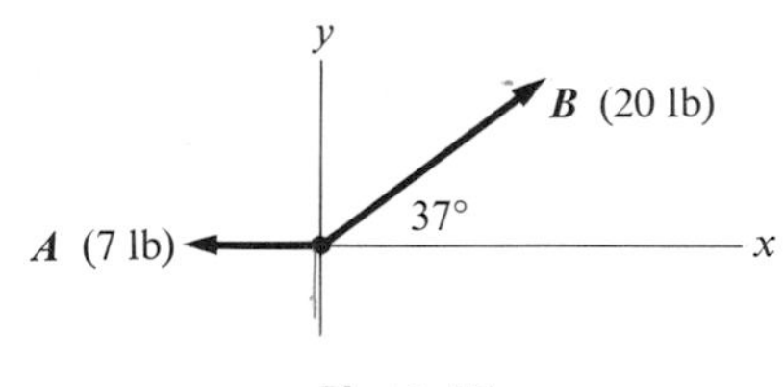

Fig. 1-17

1-6 Vector $\boldsymbol{A}$ is 2 in. long and is 60° above the x-axis in the first quadrant. Vector $\boldsymbol{B}$ is 2 in. long and is 60° below the x-axis in the fourth quadrant. Find graphically (a) the vector sum $\boldsymbol{A} + \boldsymbol{B}$, and (b) the vector differences $\boldsymbol{A} - \boldsymbol{B}$ and $\boldsymbol{B} - \boldsymbol{A}$.

1-7 (a) Find graphically the horizontal and vertical components of a 40-lb force the direction of which is 50° above the horizontal to the right. Let $\frac{1}{16}$ in. = 1 lb. (b) Check your results by calculating the components.

1–8 A box is pushed along the floor as in Fig. 1–2 by a force of 40 lb making an angle of 30° with the horizontal. Using a scale of 1 in. = 10 lb, find the horizontal and vertical components of the force by the graphical method. Check your results by calculating the components.

1–9 A block is dragged up an inclined plane of slope angle 20° by a force $\boldsymbol{F}$ making an angle of 30° with the plane. (a) How large a force $\boldsymbol{F}$ is necessary in order that the component F_x parallel to the plane shall be 16 lb? (b) How large will the component F_y then be? Solve graphically, letting 1 in. = 8 lb.

1–10 The three forces shown in Fig. 1–15 act on a body located at the origin. (a) Find the x- and y-components of each of the three forces. (b) Use the method of rectangular resolution to find the resultant of the forces. (c) Find the magnitude and direction of a fourth force which must be added to make the resultant force zero. Indicate the fourth force by a diagram.

1–11 Use the method of rectangular resolution to find the resultant of the following set of forces and the angle it makes with the horizontal: 200 lb, along the x-axis toward the right; 300 lb, 60° above the x-axis to the right; 100 lb, 45° above the x-axis to the left; 200 lb, vertically down.

1–12 A vector $\boldsymbol{A}$ of length 10 units makes an angle of 30° with a vector $\boldsymbol{B}$ of length 6 units. Find the magnitude of the vector difference $\boldsymbol{A} - \boldsymbol{B}$ and the angle it makes with vector $\boldsymbol{A}$: (a) by the parallelogram method; (b) by the triangle method; (c) by the method of rectangular resolution.

1–13 Two forces, $\boldsymbol{F}_1$ and $\boldsymbol{F}_2$, act at a point. The magnitude of $\boldsymbol{F}_1$ is 8 lb and its direction is 60° above the x-axis in the first quadrant. The magnitude of $\boldsymbol{F}_2$ is 5 lb and its direction is 53° below the x-axis in the fourth quadrant. (a) What are the horizontal and vertical components of the resultant force? (b) What is the magnitude of the resultant? (c) What is the magnitude of the vector difference $\boldsymbol{F}_1 - \boldsymbol{F}_2$?

1–14 Two forces, $\boldsymbol{F}_1$ and $\boldsymbol{F}_2$, act upon a body in such a manner that the resultant force $\boldsymbol{R}$ has a magnitude equal to that of $\boldsymbol{F}_1$ and makes an angle of 90° with $\boldsymbol{F}_1$. Let $F_1 = R = 10$ lb. Find the magnitude of the second force, and its direction (relative to $\boldsymbol{F}_1$).

1–15 The resultant of four forces is 1000 lb in the direction 30° west of north. Three of the forces are 400 lb, 60° north of east; 300 lb, south; and 400 lb, 53° west of south. Find the rectangular components of the fourth force.

1–16 The velocity of an airplane relative to the surface of the earth, $\boldsymbol{v}_{PE}$, equals the vector sum of its velocity relative to the air, $\boldsymbol{v}_{PA}$, and the velocity of the air relative to the earth, $\boldsymbol{v}_{AE}$. Thus

$$\boldsymbol{v}_{PE} = \boldsymbol{v}_{PA} + \boldsymbol{v}_{AE}.$$

(a) Find graphically the magnitude and direction of the velocity $\boldsymbol{v}_{PE}$ if the velocity $\boldsymbol{v}_{PA}$ is 100 mi hr^{-1} due north, and if the wind velocity $\boldsymbol{v}_{AE}$ is 40 mi hr^{-1} from east to west. Let 1 cm = 20 mi hr^{-1}. (b) Find the direction of the velocity $\boldsymbol{v}_{PA}$ if its magnitude is 100 mi hr^{-1}, if the direction of $\boldsymbol{v}_{PE}$ is due north, and if $\boldsymbol{v}_{AE}$ is 40 mi hr^{-1} from east to west. What is then the magnitude of $\boldsymbol{v}_{PE}$?

2 Equilibrium of a Particle

2-1 INTRODUCTION

The science of mechanics is based on three natural laws which were clearly stated for the first time by Sir Isaac Newton (1643–1727) and were published in 1686 in his *Philosophiae Naturalis Principia Mathematica* ("The Mathematical Principles of Natural Science"). It should not be inferred, however, that the science of mechanics began with Newton. Many men had preceded him in this field, the most outstanding being Galileo Galilei (1564–1642), who in his studies of accelerated motion had laid much of the groundwork for Newton's three laws.

In this chapter we shall make use of only two of Newton's laws, the first and the third. Newton's second law will be discussed in Chapter 5.

2-2 EQUILIBRIUM. NEWTON'S FIRST LAW

One effect of a force is to alter the dimensions or shape of a body on which the force acts; another is to alter the state of motion of the body.

The motion of a body can be considered as made up of its motion as a whole, or its *translational* motion, together with any *rotational* motion the body may have. In the most general case, a single force acting on a body produces a change in both its translational and rotational motion. However, when several forces act on a body simultaneously, their effects can compensate one another, with the result that there is no change in either the translational or rotational motion. When this is the case, the body is said to be in *equilibrium*. This means that (1) the body as a whole either remains at rest or moves in a straight line with constant speed, and (2) that the body is either not rotating at all or is rotating at a constant rate.

Let us consider some (idealized) experiments from which the laws of equilibrium can be deduced. Figure 2-1 represents a flat, rigid object of arbitrary shape on a level surface having negligible friction. If a single force F_1 acts on the body, as in Fig. 2-1(a), and if the body is originally at rest, it at once starts to move and to rotate clockwise. If originally in motion, the effect of the force is to change the translational motion of the body in magnitude or direction (or both) and to increase or decrease its rate of rotation. In either case, the body does not remain in equilibrium.

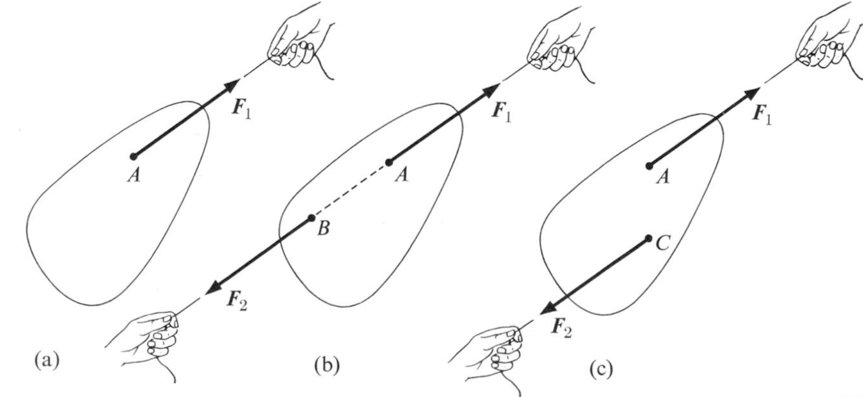

2-1 A rigid body acted on by two forces is in equilibrium if the forces are equal in magnitude, opposite in direction, and have the same line of action, as in part (b).

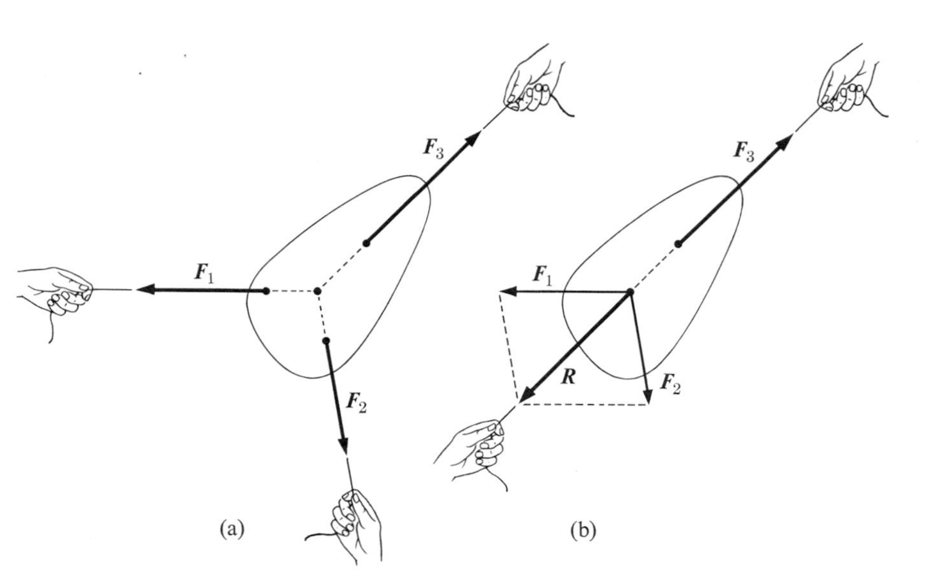

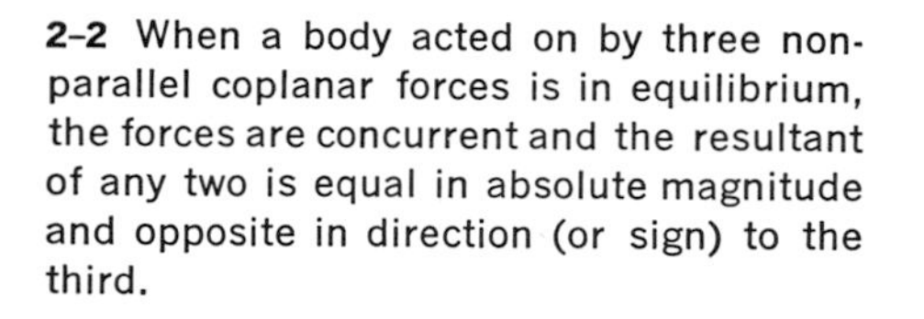
2-2 When a body acted on by three nonparallel coplanar forces is in equilibrium, the forces are concurrent and the resultant of any two is equal in absolute magnitude and opposite in direction (or sign) to the third.

Equilibrium can be maintained, however, by the application of a second force $\boldsymbol{F}_2$ as in Fig. 2-1(b), provided that $\boldsymbol{F}_2$ is *equal in magnitude* to $\boldsymbol{F}_1$, is *opposite in direction* to $\boldsymbol{F}_1$, and has the *same line of action* as $\boldsymbol{F}_1$. The resultant of $\boldsymbol{F}_1$ and $\boldsymbol{F}_2$ is then zero. If the lines of action of the two forces are not the same, as in Fig. 2-1(c), the body will be in translational but not in rotational equilibrium. (The forces then form a *couple*; see Section 3-5.)

Strictly speaking, we should say that the forces $\boldsymbol{F}_1$ and $\boldsymbol{F}_2$ are equal in *absolute* magnitude and opposite in sign, or that one equals the negative of the other. That is,

$$\boldsymbol{F}_2 = -\boldsymbol{F}_1.$$

Then if $\boldsymbol{R}$ represents the resultant of $\boldsymbol{F}_1$ and $\boldsymbol{F}_2$,

$$\boldsymbol{R} = \boldsymbol{F}_1 + \boldsymbol{F}_2 = \boldsymbol{F}_1 - \boldsymbol{F}_1 = 0.$$

For brevity, we shall often speak of two forces as simply being "equal and opposite," meaning that their absolute magnitudes are equal and that one is the negative of the other.

In Fig. 2-2(a), a body is acted on by three nonparallel coplanar forces, $\boldsymbol{F}_1$, $\boldsymbol{F}_2$, and $\boldsymbol{F}_3$. Any force applied to a rigid body may be regarded as acting anywhere along its line of action. Therefore, let any two of the force vectors, say $\boldsymbol{F}_1$ and $\boldsymbol{F}_2$, be transferred to the point of intersection of their lines of action and their resultant $\boldsymbol{R}$ obtained as in Fig. 2-2(b). The forces are now reduced to two, $\boldsymbol{R}$ and $\boldsymbol{F}_3$, and for equilibrium these must (1) be equal in magnitude, (2) be opposite in direction, and (3) have the same line of action. It follows from the first two conditions that the resultant of the three forces is zero. The third condition can be fulfilled only if the line of action of $\boldsymbol{F}_3$ passes through the point of intersection of the lines of action of $\boldsymbol{F}_1$ and $\boldsymbol{F}_2$. In other words, the three forces must be *concurrent.*

The construction in Fig. 2-2 provides a satisfactory graphical method for the solution of problems in equilibrium. For an analytical solution, it is usually simpler to deal with the rectangular components of the forces. We have shown that the magnitudes of the rectangular components of the resultant $\boldsymbol{R}$ of any set of coplanar forces are

$$R_x = \sum F_x, \qquad R_y = \sum F_y.$$

When a body is in equilibrium, the resultant of all of the forces acting on it is zero. Both rectangular components are then zero and hence, for a body in equilibrium,

$$\boldsymbol{R} = 0, \qquad \text{or} \qquad \sum F_x = 0, \qquad \sum F_y = 0.$$

These equations are called the *first condition of equilibrium.*

The *second condition of equilibrium* is an equation to be developed in Chapter 3 that is a mathematical statement of the following facts:

a) When a rigid body is in equilibrium under the action of only two forces, they must have the same line of action.
b) When a rigid body is in equilibrium under the action of three forces, they must be concurrent.

The first condition of equilibrium ensures that a body shall be in translational equilibrium; the second, that it be in rotational equilibrium. The statement that a body is in complete equilibrium when both conditions are satisfied is the essence of *Newton's first law of motion*. Newton did not state his first law in exactly these words. His original statement (translated from the Latin in which the *Principia* was written) reads:

"*Every body continues in its state of rest, or of uniform motion in a straight line, unless it is compelled to change that state by forces impressed on it.*"

Although rotational motion was not explicitly mentioned by Newton, it is clear from his work that he fully understood the conditions that the forces must satisfy when the rotation is zero or is constant.

2-3 DISCUSSION OF NEWTON'S FIRST LAW OF MOTION

Newton's first law of motion is not as self-evident as it may seem. In the first place, this law asserts that in the absence of any impressed force a body either remains at rest or moves uniformly in a straight line. It follows that *once a body has been set in motion it is no longer necessary to exert a force on it to maintain it in motion.* This assertion appears to be contradicted by everyday experience. Suppose we exert a force with the hand to move a book along a level table top. After the book has left our hand, and we are no longer exerting a force on it, it does *not* continue to move indefinitely, but slows down and eventually comes to rest. If we wish to keep it moving uniformly we must continue to exert some forward force on it. But this force is required only because a frictional force is exerted on the sliding book by the table top, in a direction opposite to the motion of the book. The smoother the surfaces of book and table, the smaller the frictional force and the smaller the force we must exert to keep the book moving. The first law asserts that if the frictional force could be eliminated completely, no forward force at all would be required to keep the book moving, once it had been set in motion. More than that, however, the law implies that if the *resultant* force on the book is zero, as it is when the frictional force is balanced by an equal forward force, the book also continues to move uniformly. In other words, *zero resultant force is equivalent to no force at all.*

In the second place, the first law defines by implication what is known as an *inertial reference system*. To understand what is meant by this term we must recognize that the motion of a given body can be specified only with respect to, or relative to, some other body. Its motion relative to one body may be very different from that relative to another. Thus a passenger in an aircraft which is making its take-off run may continue at rest relative to the aircraft but be moving faster and faster relative to the earth.

A *reference system* means a set of coordinate axes attached to (or moving with) some specified body or bodies. Suppose we consider a reference system attached to the aircraft referred to above. Everyone knows that during the take-off run, while the aircraft is going faster and faster, a passenger feels the back of his seat pushing him forward, although he remains at rest relative to a reference system attached to the aircraft. Newton's first law does not therefore correctly describe the situation; a forward force *does* act on the passenger, but nevertheless (relative to the aircraft) he remains at rest.

Suppose, on the other hand, that the passenger is standing in the aisle on roller skates. He then starts to move *backward*, relative to the aircraft, when the take-off run begins, even though no backward force acts on him. Newton's first law again does not correctly describe the facts.

We can now define an inertial reference system as one relative to which a body *does* remain at rest or move uniformly in a straight line when no force (or no resultant force) acts on it. That is, *an inertial reference system is one in which Newton's first law correctly describes the motion of a body not acted on by any force.*

An aircraft making its take-off run is evidently *not* an inertial system. For many purposes, a reference system attached to the earth can be considered an inertial system, although we shall show in a later chapter that because of the earth's rotation and its other motions this is not exactly true. Newton himself believed that it was possible to conceive of a reference system in a state of "absolute rest," anchored in some way to empty space. When Newton referred to a state of rest or uniform motion he was using such a system as a reference system. The principles of relativity have led us to believe that the concepts of "absolute rest" and "absolute motion" have no physical meaning. The best solution seems to be to select a reference system at rest relative to the so-called *fixed stars*, stars that are so far away that their motions relative to one another cannot be detected. Newton's first law appears to describe correctly the motions of bodies relative to the fixed stars, and they may therefore be considered an inertial system.

Thirdly, Newton's first law contains a qualitative definition of the concept of *force*, or at any rate of one aspect of the force concept, as "that which changes the state of motion of a body." (This does not mean that forces do not produce other effects also, such as changing the length of a coil spring.) When a body at rest relative to the earth is observed to start moving, or when a moving body speeds up, slows down, or changes its direction, we can conclude that a force is acting on it. This effect of a force can be used to define the ratio of two forces and to define a unit force, and we shall show in Chapter 5 how this is done.

2-4 STABLE, UNSTABLE, AND NEUTRAL EQUILIBRIUM

When a body in equilibrium is displaced slightly, the magnitudes, directions, and lines of action of the forces acting on it may all change. If the forces in the displaced position are such as to return the body to its original position, the equilibrium is *stable*. If the forces act to increase the displacement still further, the equilibrium is *unstable*. If the body is still in equilibrium in the displaced position, the equilibrium is *neutral*. Whether a given equilibrium state is stable, unstable, or neutral can be determined only by considering other states slightly displaced from the first.

A right circular cone on a level surface affords an example of the three types of equilibrium. When the

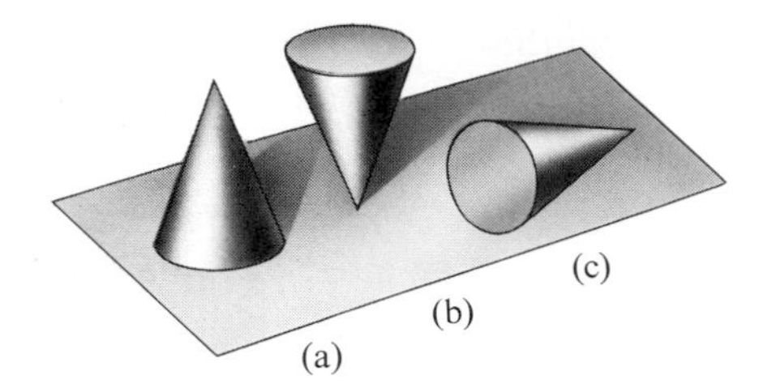

2-3 (a) Stable, (b) unstable, and (c) neutral equilibrium.

cone rests on its base, as in Fig. 2-3(a), the equilibrium is stable. When balanced on its apex, as in part (b), the equilibrium is unstable. When resting on its side, as in part (c), the equilibrium is neutral.

2-5 NEWTON'S THIRD LAW OF MOTION

Any given force is but one aspect of a mutual interaction between *two* bodies. It is found that *whenever one body exerts a force on another, the second always exerts on the first a force which is equal in absolute magnitude, opposite in direction (or sign), and has the same line of action.** A single, isolated force is therefore an impossibility.

The two forces involved in every interaction between two bodies are often called an "action" and a "reaction," but this does not imply any difference in their nature, or that one force is the "cause" and the other its "effect." *Either* force may be considered the "action," and the other the "reaction" to it.

This property of forces was stated by Newton in his *third law of motion*. In his own words, "*To every action there is always opposed an equal reaction: or, the mutual actions of two bodies upon each other are always equal, and directed to contrary parts.*"

As an example, suppose that a man pulls on one end of a rope attached to a block, as in Fig. 2-4. The weight of the block and the force exerted on it by the surface are not shown. The block may or may not be in equilibrium. The resulting action-reaction pairs of forces are indicated in the figure. (Actually, the lines of action of all the forces lie along the rope. The force vectors have been offset from this line to show them more

* This statement must be modified when forces of electromagnetic origin are considered. It is correct, however, for any of the forces we shall encounter in mechanics.

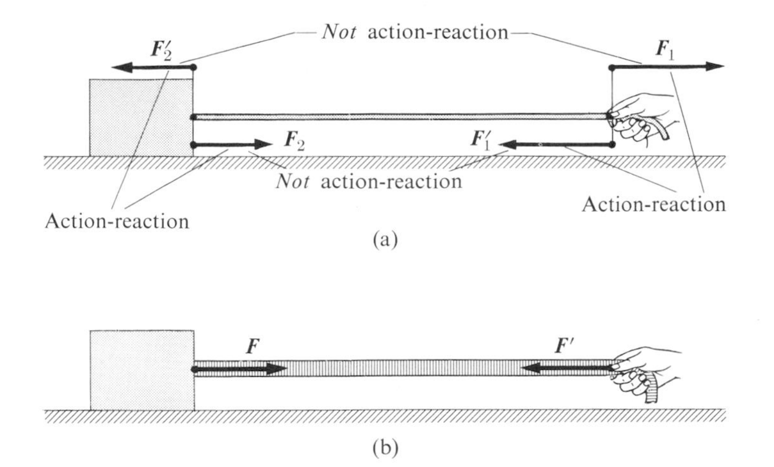

2-4 (a) Forces F_1 and F_1' form one action–reaction pair, and forces F_2 and F_2' another. F_1 is always equal to F_1', F_2 is always equal to F_2'. F_1 and F_2' are equal only if the rope is in equilibrium, and force F_2' is not the reaction to F_1. (Actually all forces lie along the rope.) (b) If the rope is in equilibrium, it can be considered to transmit a force from the man to the block, and vice versa.

clearly.) Vector $\boldsymbol{F}_1$ represents the force exerted on the rope by the man. Its reaction is the equal and opposite force $\boldsymbol{F}_1'$ exerted on the man by the rope. Vector $\boldsymbol{F}_2$ represents the force exerted on the block by the rope. The reaction to it is the equal and opposite force $\boldsymbol{F}_2'$, exerted on the rope by the block:

$$\boldsymbol{F}_1' = -\boldsymbol{F}_1, \qquad \boldsymbol{F}_2' = -\boldsymbol{F}_2.$$

It is very important to realize that the forces $\boldsymbol{F}_1$ and $\boldsymbol{F}_2'$, although they are opposite in direction and have the same line of action, do *not* constitute an action-reaction pair. For one thing, both of these forces act on the *same* body (the rope) while an action and its reaction necessarily act on *different* bodies. Furthermore, the forces $\boldsymbol{F}_1$ and $\boldsymbol{F}_2'$ are not *necessarily* equal in magnitude. If the block and rope are moving to the right with increasing speed, the rope is not in equilibrium and $\boldsymbol{F}_1$ is greater than $\boldsymbol{F}_2'$. Only in the special case when the rope remains at rest or moves with constant speed are the forces $\boldsymbol{F}_1$ and $\boldsymbol{F}_2'$ equal in magnitude, but this is an example of Newton's *first* law, not his *third*. Even when the speed of the rope is changing, however, the action-reaction forces $\boldsymbol{F}_1$ and $\boldsymbol{F}_1'$ are equal in magnitude to *each other*, and the action-reaction forces $\boldsymbol{F}_2$ and $\boldsymbol{F}_2'$ are equal in magnitude to *each other*, although then $\boldsymbol{F}_1$ is not equal to $\boldsymbol{F}_2'$.

In the special case when the rope is in equilibrium, and when no forces act on it except those at its ends, $\boldsymbol{F}_2'$ equals $\boldsymbol{F}_1$ by Newton's *first* law. Since $\boldsymbol{F}_2$ *always* equals $\boldsymbol{F}_2'$, by Newton's *third* law, then in this special case $\boldsymbol{F}_2$ also equals $\boldsymbol{F}_1$ and the force exerted on the block by the rope is equal to the force exerted on the rope by the man. The rope can therefore be considered to "transmit" to the block, without change, the force exerted on it by the man. This point of view is often useful, but it is important to remember that it applies only under the restricted conditions as stated.

If we adopt this point of view, the rope itself need not be considered and we have the simpler force diagram of Fig. 2–4(b), where the man is considered to exert a force $\boldsymbol{F}$ directly on the block. The reaction is the force $\boldsymbol{F}'$ exerted by the block on the man. The only effect of the rope is to transmit these forces from one body to the other.

A body, like the rope in Fig. 2–4, which is subjected to pulls at its ends is said to be in *tension*. The tension at any point equals the force exerted at that point. Thus in Fig. 2–4(a) the tension at the right-hand end of the rope equals the magnitude of $\boldsymbol{F}_1$ (or of $\boldsymbol{F}_1'$) and the tension at the left-hand end equals the magnitude of $\boldsymbol{F}_2$ (or of $\boldsymbol{F}_2'$). If the rope is in equilibrium and if no forces act except at its ends, as in Fig. 2–4(b), the tension is the same at both ends. If, for example, in Fig. 2–4(b) the magnitudes of $\boldsymbol{F}$ and $\boldsymbol{F}'$ are each 50 lb, the tension in the rope is 50 lb (*not* 100 lb).

2-6 EQUILIBRIUM OF A PARTICLE

The materials found in nature and the processes which occur are rarely simple. As a first step in dealing with a problem of nature it is necessary to *idealize* the material and to make *simplifying assumptions* concerning the processes. Suppose, for example, a baseball is thrown into the air and it is desired to calculate where it lands and with what velocity. The first step in treating this problem consists in idealizing the baseball by ignoring the details of its surface and all departures from sphericity during its motion. In other words, we replace the baseball by an ideal object, i.e., a rigid smooth sphere. The next step is to ignore the rotation of the baseball and the forces brought into play by the air dragged around by the rotating ball. A further step is to regard as negligible the buoyancy and the resistance of the air. We are then left with a problem quite different from the original one; in fact we may be accused of having robbed the problem of almost all of its reality. This is true, but what remains approximates the original problem at low velocities, and has the virtue that it is amenable to simple mathematical treatment, whereas the original problem at high velocities would require the most advanced methods.

In general, the forces acting on a rigid body do not all pass through one point (they are *nonconcurrent*) and, as a result, the rigid body undergoes rotational as well as translational motion. There are, however, many situations of great interest where the rotation of a body is of only minor consequence and is not pertinent in the solution of the problem. The planetary motion, for example, of the earth about the sun taking place under the action of the gravitational force between the two bodies may be studied alone without regard to the earth's rotation. *A body whose rotation is ignored as irrelevant is called a particle. A particle may be so small that it is an approximation to a point, or it may be of any size, provided that the action lines of all the forces acting on it intersect in one point.*

In the remainder of this chapter, we shall limit ourselves to examples and problems involving the equilibrium of particles only. It is surprising how many situations of interest and importance in engineering, in the life and earth sciences, and in everyday life involve the equilibrium of particles. In many of these, it is important to know how to calculate one or two of the forces acting on a particle when the others are given. In order to do this, it is best to adhere scrupulously to the following rules:

1. Make a simple line sketch of the apparatus or structure, showing dimensions and angles.
2. Choose some object (like a knot in a rope, for example) as the particle in equilibrium. Draw a separate diagram of this object and show by arrows (use a colored pencil) *all* of the forces exerted *on* it by other bodies. This is called the *force diagram* or *free-body* diagram. When a system is composed of several particles, it may be necessary to construct a separate free-body diagram for each. Do *not* show, in the free-body diagram of a chosen particle, any of the forces exerted *by* it. These forces (which are the reactions to the forces acting *on* the chosen particle) all act *on other bodies* and appear in the free-body diagrams of those bodies.
3. Construct a set of rectangular axes and resolve any inclined forces into rectangular components. Cross out lightly those forces which have been resolved.
4. Set the algebraic sum of all x-forces (or force components) equal to zero, and the algebraic sum of all y-forces (or components) equal to zero. This provides two independent equations which can be solved simultaneously for two unknown quantities (which may be forces, angles, distances, etc.).

A force which will be encountered in many problems is the *weight* of a body, that is, the force of gravitational attraction exerted on the body by the earth. We shall show in the next chapter that the line of action of this force always passes through a point called the *center of gravity* of the body.

The force of gravitational attraction exerted on a body by the earth is but one aspect of a mutual interaction between the earth and the body. That is, the earth attracts the body and at the same time the body attracts the earth. The force exerted on the earth by the body is opposite in direction and equal in magnitude to the force exerted on the body by the earth. Thus if a body weighs 10 lb (that is, if the earth pulls down on it with a force of 10 lb), the body pulls up on the earth with an equal force of 10 lb. The equal and opposite forces exerted on body and earth are another example of an action-reaction pair.

Example 1. To begin with a simple example, consider the body in Fig. 2–5(a), hanging at rest from the ceiling by a vertical cord. Part (b) of the figure is the free-body diagram for the body. The forces on it are its weight $\boldsymbol{w}_1$ and the upward force $\boldsymbol{T}_1$ exerted on it by the cord. If we take the x-axis horizontal and the y-axis

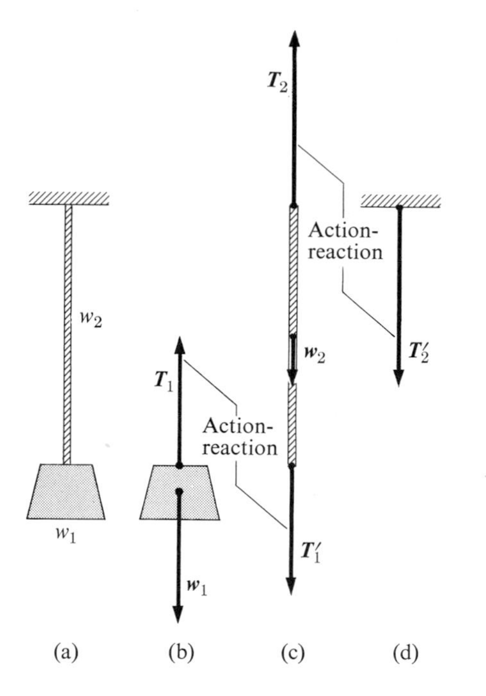

2-5 (a) Block hanging at rest from vertical cord. (b) The block is isolated and all forces acting on it are shown. (c) Forces on the cord. (d) Downward force on the ceiling. Lines connect action–reaction pairs.

vertical, there are no x-components of force, and the y-components are the forces $\boldsymbol{w}_1$ and $\boldsymbol{T}_1$. Then from the condition that $\sum F_y = 0$, we have

$$\sum F_y = T_1 - w_1 = 0,$$
$$T_1 = w_1. \qquad \text{(First law)}$$

In order that both forces have the same line of action, the center of gravity of the body must lie vertically below the point of attachment of the cord.

Let us emphasize again that the forces $\boldsymbol{w}_1$ and $\boldsymbol{T}_1$ are *not* an action-reaction pair, although they are equal in magnitude, opposite in direction, and have the same line of action. The weight $\boldsymbol{w}_1$ is a force of attraction exerted on the body by the earth. Its reaction is an equal and opposite force of attraction exerted on the earth by the body. This reaction is one of the set of forces acting *on the earth*, and therefore it does not appear in the free-body diagram of the suspended block.

The reaction to the force $\boldsymbol{T}_1$ is an equal downward force, $\boldsymbol{T}_2'$, exerted *on the cord* by the suspended body:

$$T_1 = T_1'. \qquad \text{(Third law)}$$

The force $\boldsymbol{T}_1'$ is shown in part (c), which is the free-body diagram of the cord. The other forces on the cord are its own weight $\boldsymbol{w}_2$ and the upward force $\boldsymbol{T}_2$ exerted on its upper end by the ceiling. Since the cord is also in equilibrium,

$$\sum F_y = T_2 - w_2 - T_1' = 0,$$
$$T_2 = w_2 + T_1'. \qquad \text{(First law)}$$

The reaction to $\boldsymbol{T}_2$ is the downward force $\boldsymbol{T}_2'$ in part (d), exerted on the ceiling by the cord:

$$T_2 = T_2'. \qquad \text{(Third law)}$$

As a numerical example, let the body weigh 20 lb and the cord weigh 1 lb. Then

$$T_1 = w_1 = 20 \text{ lb},$$
$$T_1' = T_1 = 20 \text{ lb},$$
$$T_2 = w_2 + T_1'$$
$$= 1 \text{ lb} + 20 \text{ lb} = 21 \text{ lb},$$
$$T_2' = T_2 = 21 \text{ lb}.$$

If the weight of the cord were so small as to be negligible, then in effect no forces would act on it except at its ends. The forces $\boldsymbol{T}_2$ and $\boldsymbol{T}_2'$ would then each equal 20 lb and, as explained earlier, the cord could be considered to transmit a 20-lb force from one end to the other without change. We could then consider the upward pull of the cord on the block as an "action" and the downward pull on the ceiling as its "reaction." The tension in the cord would then be 20 lb.

Example 2. In Fig. 2-6(a), a block of weight w hangs from a cord which is knotted at O to two other cords fastened to the ceiling. We wish to find the tensions in these three cords. The weights of the cords are negligible.

In order to use the conditions of equilibrium to compute an unknown force, we must consider some body which is in equilibrium and on which the desired force acts. The hanging block is one such body and, as shown in the preceding example, the tension in the vertical cord supporting the block is equal to the weight of the block. The inclined cords do not exert forces on the block, but they do act on the knot at O. Hence we consider the *knot* as a particle in equilibrium whose own weight is negligible.

The free-body diagrams for the block and the knot are shown in Fig. 2-6(b), where $\boldsymbol{T}_1$, $\boldsymbol{T}_2$, and $\boldsymbol{T}_3$ represent the forces exerted *on the knot* by the three cords and $\boldsymbol{T}_1'$, $\boldsymbol{T}_2'$, and $\boldsymbol{T}_3'$ are the reactions to these forces.

Consider first the hanging block. Since it is in equilibrium,

$$T_1' = w. \qquad \text{(First law)}$$

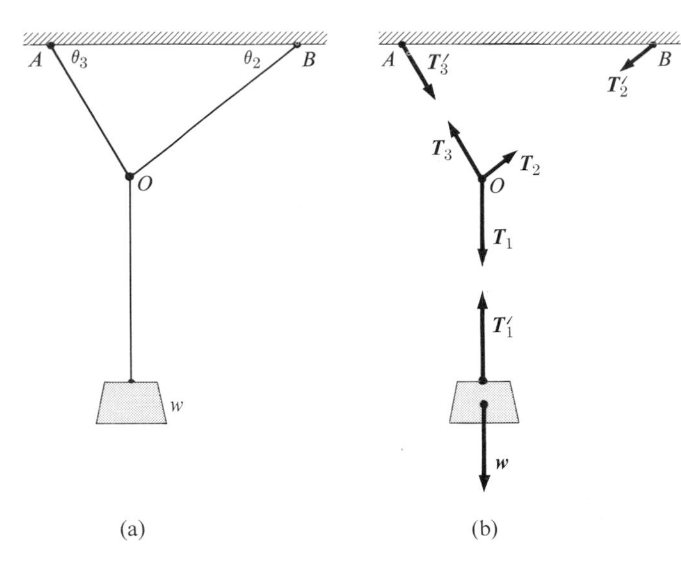

2-6 (a) Block hanging in equilibrium. (b) Forces acting on the block, on the knot, and on the ceiling.

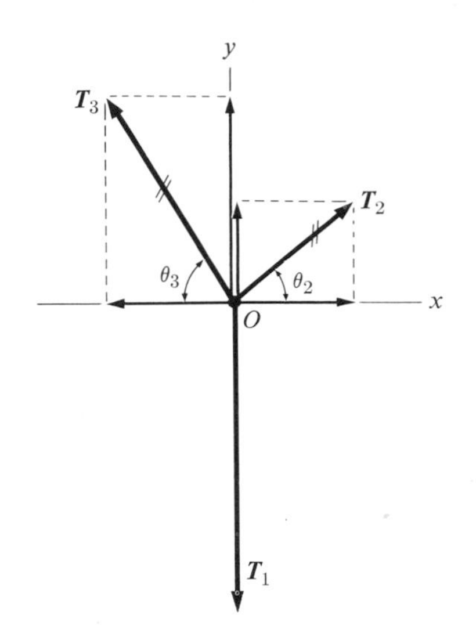

2-7 Forces on the knot in Fig. 2-6, resolved in *x*- and *y*-components.

Since $\boldsymbol{T}_1$ and $\boldsymbol{T}_1'$ form an action-reaction pair,

$$T_1' = T_1. \qquad \text{(Third law)}$$

Hence

$$T_1 = w.$$

To find the forces $\boldsymbol{T}_2$ and $\boldsymbol{T}_3$, we resolve these forces (see Fig. 2-7) into rectangular components. Then, from Newton's *first* law,

$$\sum F_x = T_2 \cos \theta_2 - T_3 \cos \theta_3 = 0,$$
$$\sum F_y = T_2 \sin \theta_2 + T_3 \sin \theta_3 - T_1 = 0.$$

As a numerical example, let

$$w = 50 \text{ lb}, \qquad \theta_2 = 30°, \qquad \theta_3 = 60°.$$

Then $T_1 = 50$ lb, and from the two preceding equations,

$$T_2 = 25 \text{ lb}, \qquad T_3 = 43.3 \text{ lb}.$$

Finally, we know from Newton's *third* law that the inclined cords exert on the ceiling the forces $\boldsymbol{T}_2'$ and $\boldsymbol{T}_3'$, equal and opposite to $\boldsymbol{T}_2$ and $\boldsymbol{T}_3$, respectively.

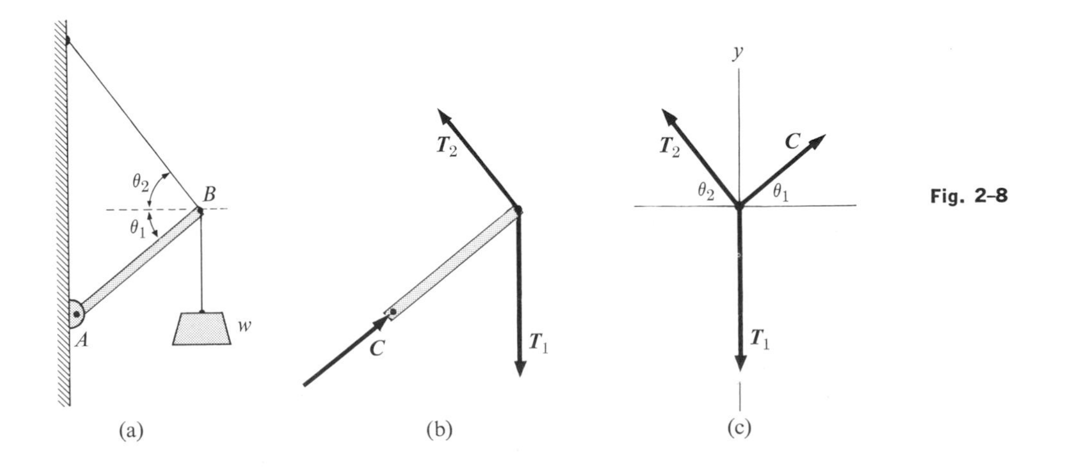

Fig. 2-8

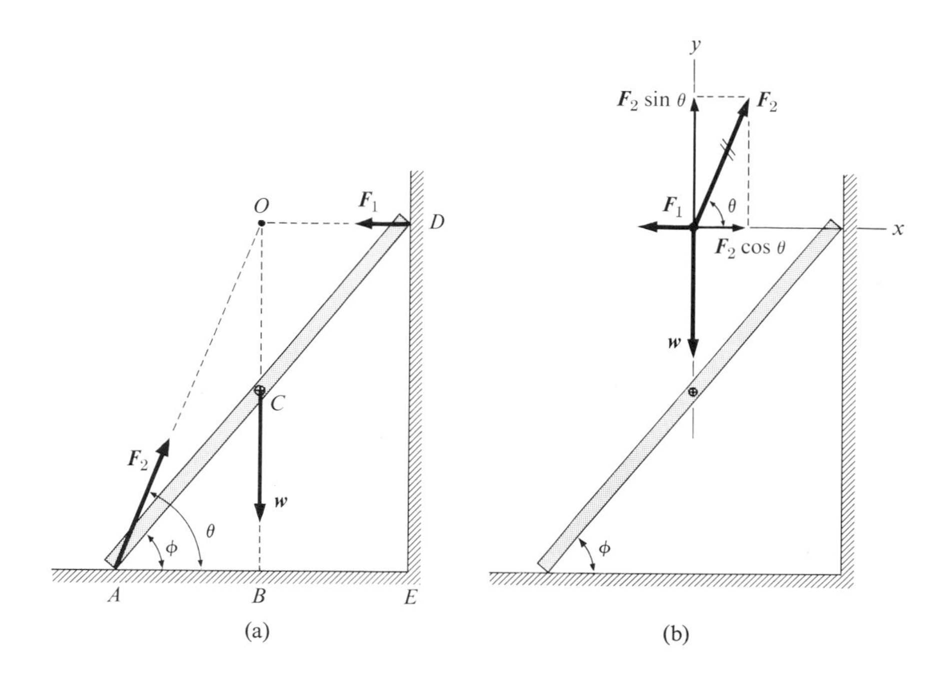

2-9 Forces on a ladder leaning against a vertical frictionless wall.

Example 3. Figure 2-8(a) shows a strut AB, pivoted at end A, attached to a wall by a cable, and carrying a load w at end B. The weights of the strut and of the cable are negligible.

Figure 2-8(b) shows the forces acting on the strut: $\mathbf{T}_1$ is the force exerted by the vertical cable, $\mathbf{T}_2$ is the force exerted by the inclined cable, and $\mathbf{C}$ is the force exerted by the pivot. Force $\mathbf{T}_1$ is known both in magnitude and in direction, force $\mathbf{T}_2$ is known in direction only, and neither the magnitude nor the direction of $\mathbf{C}$ is known. However, the forces $\mathbf{T}_1$ and $\mathbf{T}_2$ must intersect at the outer end of the strut, and since the strut is in equilibrium under three forces, the line of action of force $\mathbf{C}$ must also pass through the outer end of the strut. In other words, the direction of force $\mathbf{C}$ is along the line of the strut.

Hence the resultant of $\mathbf{T}_1$ and $\mathbf{T}_2$ is also along this line and the strut, in effect, is acted on by forces at its ends, directed toward each other along the line of the strut. The effect of these forces is to compress the strut, and it is said to be in *compression.*

If the forces acting on a strut are not all applied at its ends, the direction of the resultant force at the ends is not along the line of the strut. This is illustrated in the next example.

In Fig. 2-8(c), the force $\mathbf{C}$ has been transferred along its line of action to the point of intersection B of the three forces. The force diagram for a particle at B is exactly like that of Fig. 2-7, and the problem is solved in the same way. The conditions of equilibrium provide two independent equations among the five quantities T_1, T_2, C, θ_1, and θ_2. Hence if any three are known, the others may be calculated.

Example 4. In Fig. 2-9 a ladder, which is in equilibrium, leans against a vertical frictionless wall. The forces on the ladder are (1) its weight $\mathbf{w}$, (2) the force $\mathbf{F}_1$ exerted on the ladder by the vertical wall and which is perpendicular to the wall if there is no friction, and (3) the force $\mathbf{F}_2$ exerted by the ground on the base of the ladder. The force $\mathbf{w}$ is known in magnitude and in direction, the force $\mathbf{F}_1$ is known in direction only, and the force $\mathbf{F}_2$ is unknown in both magnitude and direction. As in the preceding example, the ladder is in equilibrium under three forces, *which must be concurrent.* Since the lines of action of $\mathbf{F}_1$ and $\mathbf{w}$ are known, their point of intersection (point O) can be located. The line of action of $\mathbf{F}_2$ must then pass through this point also. Note that neither the direction of $\mathbf{F}_1$ nor that of $\mathbf{F}_2$ lies along the line of the ladder. In part (b) the forces have been transferred to the point of intersection of their lines of action, and applying the equations for the equilibrium of a particle at this point, we get

$$\sum F_x = F_2 \cos\theta - F_1 = 0, \tag{2-1}$$

$$\sum F_y = F_2 \sin\theta - w = 0. \tag{2-2}$$

As a numerical example, suppose the ladder weighs 80 lb, is 20 ft long, has its center of gravity at its center, and makes an angle $\phi = 53°$ with the ground. We wish to find the angle θ and

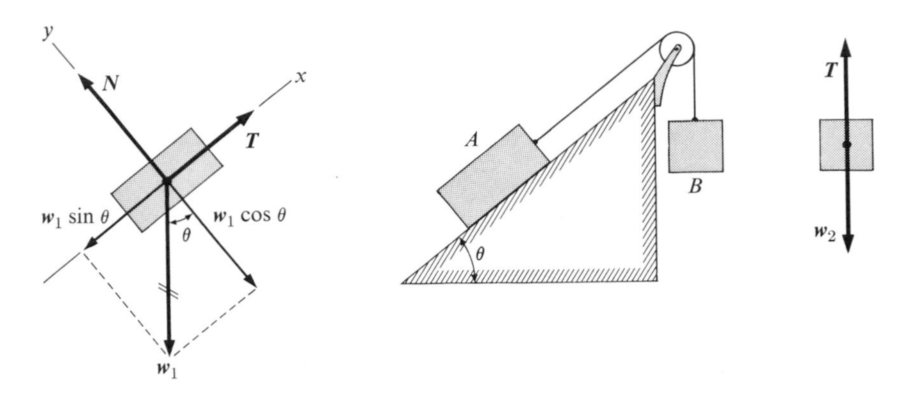

2-10 Forces on a block in equilibrium on a frictionless inclined plane.

the forces $\boldsymbol{F}_1$ and $\boldsymbol{F}_2$. To calculate θ, we first find the lengths of AB and BO. From the right triangle ABC, we have

$$\overline{AB} = \overline{AC} \cos\phi = 10 \text{ ft} \times 0.60 = 6.0 \text{ ft},$$

and from the right triangle AED,

$$\overline{DE} = \overline{AD} \sin\phi = 20 \text{ ft} \times 0.80 = 16 \text{ ft}.$$

Then, from the right triangle AOB, since $\overline{OB} = \overline{DE}$, we have

$$\tan\theta = \frac{\overline{OB}}{\overline{AB}} = \frac{16 \text{ ft}}{6.0 \text{ ft}} = 2.67.$$

Then

$$\theta = 69.5°, \quad \sin\theta = 0.937, \quad \cos\theta = 0.350.$$

$$F_2 = \frac{w}{\sin\theta} = \frac{80 \text{ lb}}{0.937} = 85.5 \text{ lb},$$

and from Eq. (2–1),

$$F_1 = F_2 \cos\theta = 85.5 \text{ lb} \times 0.350 = 30 \text{ lb}.$$

The ladder presses against the wall and the ground with forces which are equal and opposite to $\boldsymbol{F}_1$ and $\boldsymbol{F}_2$, respectively.

In the next chapter we shall explain another method of solving this problem, using the concept of the *moment* of a force.

Example 5. In Fig. 2–10, block A of weight w_1 rests on a frictionless inclined plane of slope angle θ. The center of gravity of the block is at its center. A flexible cord is attached to the center of the right face of the block, passes over a frictionless pulley, and is attached to a second block B of weight w_2. The weight of the cord and friction in the pulley are negligible. If w_1 and θ are given, find the weight w_2 for which the system is in equilibrium, that is, for which it remains at rest or moves in either direction at constant speed.

The free-body diagrams for the two blocks are shown at the left and right. The forces on block B are its weight $\boldsymbol{w}_2$ and the force $\boldsymbol{T}$ exerted on it by the cord. Since it is in equilibrium,

$$T = w_2. \quad \text{(First law)} \tag{2–3}$$

Block A is acted on by its weight $\boldsymbol{w}_1$, the force $\boldsymbol{T}$ exerted on it by the cord, and the force $\boldsymbol{N}$ exerted on it by the plane. We can use the same symbol $\boldsymbol{T}$ for the force exerted on each block by the cord because, as explained in Section 2–5, these forces are equivalent to an action-reaction pair and have the same magnitude. The force $\boldsymbol{N}$, if there is no friction, is perpendicular or *normal* to the surface of the plane. Since the lines of action of $\boldsymbol{w}_1$ and $\boldsymbol{T}$ intersect at the center of gravity of the block, the line of action of $\boldsymbol{N}$ passes through this point also. It is simplest to choose x- and y-axes parallel and perpendicular to the surface of the plane, because then only the weight $\boldsymbol{w}_1$ needs to be resolved into components. The conditions of equilibrium give

$$\left.\begin{aligned} \sum F_x &= T - w_1 \sin\theta = 0, \quad (2\text{–}4) \\ \sum F_y &= N - w_1 \cos\theta = 0. \quad (2\text{–}5) \end{aligned}\right\} \text{(First law)}$$

Thus if $\boldsymbol{w}_1 = 100$ lb and $\theta = 30°$, we have from Eqs. (2–3) and (2–4),

$$w_2 = T = w_1 \sin\theta = 100 \text{ lb} \times 0.500 = 50 \text{ lb},$$

and from Eq. (2–5),

$$N = w_1 \cos\theta = 100 \text{ lb} \times 0.866 = 86.6 \text{ lb}.$$

Note carefully that *in the absence of friction* the same weight w_2 of 50 lb is required whether the system remains at rest or moves with constant speed in *either* direction. This is not the case when friction is present.

2–7 FRICTION

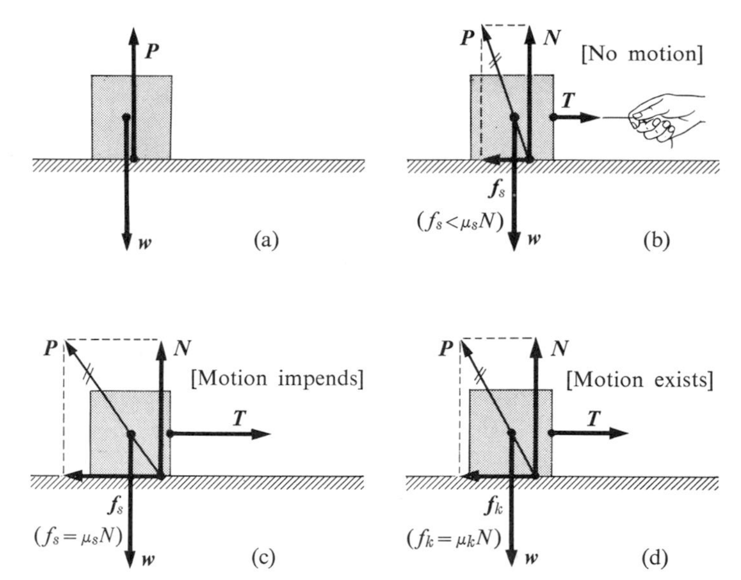

2–11 The magnitude of the friction force f is less than or equal to $\mu_s N$ when there is no relative motion, and is equal to $\mu_k N$ when motion exists.

Whenever the surface of one body slides over that of another, *each* body exerts a frictional force on the other, parallel to the surfaces. The force *on* each body is opposite to the direction of its motion relative to the other. Thus when a block slides from left to right along the surface of a table, a frictional force to the left acts on the block and an equal force toward the right acts on the table. Frictional forces may also act when there is no relative motion. A horizontal force on a heavy packing case resting on the floor may not be enough to set the case in motion, because of an equal and opposite frictional force exerted on the case by the floor.

The origin of these frictional forces is not fully understood, and the study of them is an important field of research. When one unlubricated metal slides over another, there appears to be an actual momentary welding of the metals together at the "high spots" where they make contact. The observed friction force is the force required to break these tiny welds. The mechanism of the friction force between two blocks of wood, or between two bricks, must be quite different.

In Fig. 2–11(a) a block is at rest on a horizontal surface, in equilibrium under the action of its weight $\mathbf{w}$ and the upward force $\mathbf{P}$ exerted on it by the surface. The lines of action of $\mathbf{w}$ and $\mathbf{P}$ have been displaced slightly to show these forces more clearly.

Suppose now that a cord is attached to the block as in Fig. 2–11(b) and the tension $\mathbf{T}$ in the cord is gradually increased. Provided the tension is not too great, the block remains at rest. The force $\mathbf{P}$ exerted on the block by the surface is inclined toward the left as shown, since the three forces $\mathbf{P}$, $\mathbf{w}$, and $\mathbf{T}$ must be concurrent. The component of $\mathbf{P}$ parallel to the surface is called the *force of static friction*, $\mathbf{f}_s$. The other component, $\mathbf{N}$, is the *normal* force exerted on the block by the surface. From the conditions of equilibrium, the force of static friction $\mathbf{f}_s$ equals the force $\mathbf{T}$, and the normal force $\mathbf{N}$ equals the weight $\mathbf{w}$.

As the force $\mathbf{T}$ is increased further, a limiting value is reached at which the block breaks away from the surface and starts to move. In other words, there is a certain maximum value which the force of static friction $\mathbf{f}_s$ can have. Figure 2–11(c) is the force diagram when $\mathbf{T}$ is just below its limiting value and motion impends. If the force $\mathbf{T}$ exceeds this limiting value, the block is no longer in equilibrium.

For a given pair of surfaces, the magnitude of the maximum value of $\mathbf{f}_s$ is nearly proportional to that of the normal force $\mathbf{N}$. The actual force of static friction can therefore have any magnitude between zero (when there is no applied force parallel to the surface) and a maximum value proportional to N or equal to $\mu_s N$. The factor μ_s is called the *coefficient of static friction*. Thus,

$$f_s \leqq \mu_s N. \tag{2-6}$$

The equality sign holds only when the applied force $\mathbf{T}$, parallel to the surface, has such a value that motion is about to start (Fig. 2–11c). When $\mathbf{T}$ is less than this value (Fig. 2–11b), the inequality sign holds and the magnitude of the friction force must be computed from the conditions of equilibrium.

As soon as sliding begins, it is found that the friction force decreases. This new friction force, for a given pair of surfaces, is also nearly proportional to the magnitude of the normal force. The proportionality factor, μ_k, is called the *coefficient of sliding friction* or *kinetic* friction.

Thus, when the block is in motion, the force of sliding or kinetic friction is given by

$$f_k = \mu_k N. \qquad (2\text{-}7)$$

This is illustrated in Fig. 2-11(d).

The coefficients of static and sliding friction depend primarily on the nature of the surfaces in contact, being relatively large if the surfaces are rough and small if they are smooth. The coefficient of sliding friction varies somewhat with the relative velocity, but for simplicity we shall assume it to be independent of velocity. It is also nearly independent of the contact area. However, since two real surfaces actually touch each other only at a relatively small number of high spots, the true contact area is very different from the overall area. Equations (2-6) and (2-7) are useful empirical relations, but do not represent fundamental physical laws like Newton's laws. Typical numerical values are given in Table 2-1.

TABLE 2-1 COEFFICIENTS OF FRICTION

Materials	Static, μ_s	Kinetic, μ_k
Steel on steel	0.74	0.57
Aluminum on steel	0.61	0.47
Copper on steel	0.53	0.36
Brass on steel	0.51	0.44
Zinc on cast iron	0.85	0.21
Copper on cast iron	1.05	0.29
Glass on glass	0.94	0.4
Copper on glass	0.68	0.53
Teflon on Teflon	0.04	0.04
Teflon on steel	0.04	0.04

In Table 2-1, the coefficients of friction only of solids are listed. Liquids and gases show frictional effects also but the simple equation $f = \mu N$ does not hold. It will be shown in Chapter 14 that there exists a property of liquids and gases called *viscosity* which determines the friction force between two surfaces sliding over each other with a layer of liquid or gas between them. Gases have the lowest viscosities of all materials at normal temperatures, and therefore, to reduce friction to a value close to zero, it is convenient to have an object slide on a layer of gas.

Two methods for doing this are shown in Fig. 2-12. In Fig. 2-12(a) is a development by John Stull of Alfred University of a frictionless track conceived originally by H. V. Neher and R. B. Leighton at the California Institute of Technology. Elastic bumpers are provided for the inverted V-shaped riders. When one rider only is placed on the air track and is given a push, it will hit the stationary bumper and then proceed to move back and forth many times before coming to rest. The friction is incredibly small. The frictionless air table shown in Fig. 2-12(b) was first developed by Harold A. Daw of New Mexico State University. The pucks are made of plastic and may be unscrewed to admit extra disks when different masses are required. The table is operated from a large volume source of low-pressure air. Air is distributed around the four edges of the table, then moves under the top surface through a distribution layer, and is finally discharged through over 1600 tiny air jets about an inch apart. Two-dimensional collisions, both elastic and inelastic, may be demonstrated on this table.

Example 1. In Fig. 2-11, suppose that the block weighs 20 lb, that the tension $\boldsymbol{T}$ can be increased to 8 lb before the block starts to slide, and that a force of 4 lb will keep the block moving at constant speed once it has been set in motion. Find the coefficients of static and kinetic friction.

From Fig. 2-11(c) and the data above, we have

$$\left.\begin{aligned}\textstyle\sum F_y &= N - w = N - 20\text{ lb} = 0,\\ \textstyle\sum F_x &= T - f_s = 8\text{ lb} - f_s = 0,\end{aligned}\right\}\quad\text{(First law)}$$

$$f_s = \mu_s N \quad \text{(motion impends).}$$

Hence we have

$$\mu_s = \frac{f_s}{N} = \frac{8\text{ lb}}{20\text{ lb}} = 0.40.$$

From Fig. 2-11(d), we have

$$\left.\begin{aligned}\textstyle\sum F_y &= N - w = N - 20\text{ lb} = 0,\\ \textstyle\sum F_x &= T - f_k = 4\text{ lb} - f_k = 0,\end{aligned}\right\}\quad\text{(First law)}$$

$$f_k = \mu_k N \quad \text{(motion exists).}$$

Hence

$$\mu_k = \frac{f_k}{N} = \frac{4\text{ lb}}{20\text{ lb}} = 0.20.$$

Example 2. What is the friction force if the block in Fig. 2-11(b) is at rest on the surface and a horizontal force of 5 lb is exerted on it?

We have

$$\textstyle\sum F_x = T - f_s = 5\text{ lb} - f_s = 0, \qquad \text{(First law)}$$

$$f_s = 5\text{ lb}.$$

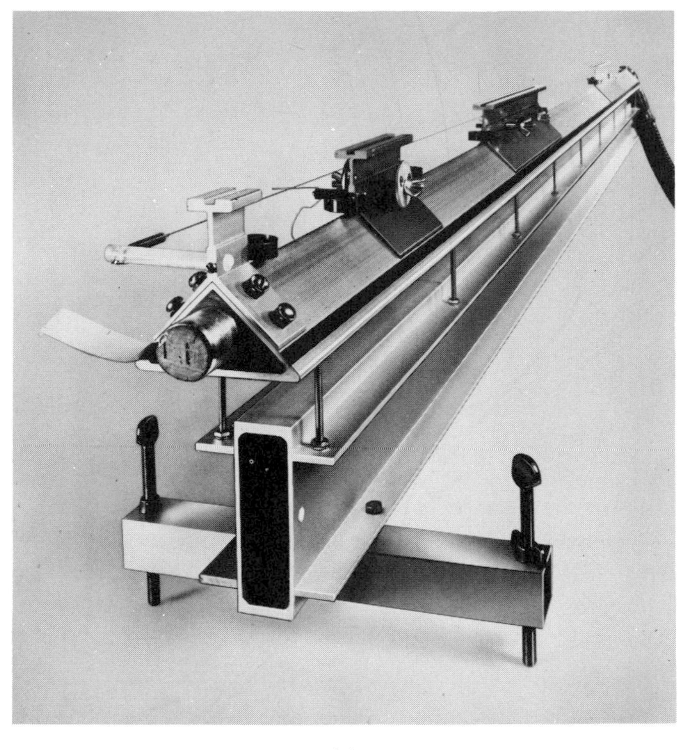

(a)

(b)

2–12 (a) The Ealing-Stull linear air track. Inverted Y-shaped sliders ride on a layer of air streaming through many fine holes in the inverted V-shaped surface. (b) The Ealing-Daw two-dimensional air table. Plastic pucks slide on a cushion of air issuing from more than a thousand minute holes in the tabletop. [Courtesy of the Ealing Corporation.]

Note that in this case $f_s < \mu_s N$.

Example 3. What force $\boldsymbol{T}$, at an angle of 30° above the horizontal, is required to drag a block weighing 20 lb to the right at constant speed, as in Fig. 2–13, if the coefficient of kinetic friction between block and surface is 0.20?

The forces on the block are shown in the diagram. From the first condition of equilibrium,

$$\sum F_x = T \cos 30° - 0.2N = 0,$$
$$\sum F_y = T \sin 30° + N - 20 \text{ lb} = 0.$$

Simultaneous solution gives

$$T = 4.15 \text{ lb},$$
$$N = 17.9 \text{ lb}.$$

Note that in this example the normal force $\boldsymbol{N}$ is not equal to the weight of the block, but is less than the weight by the vertical component of the force $\boldsymbol{T}$.

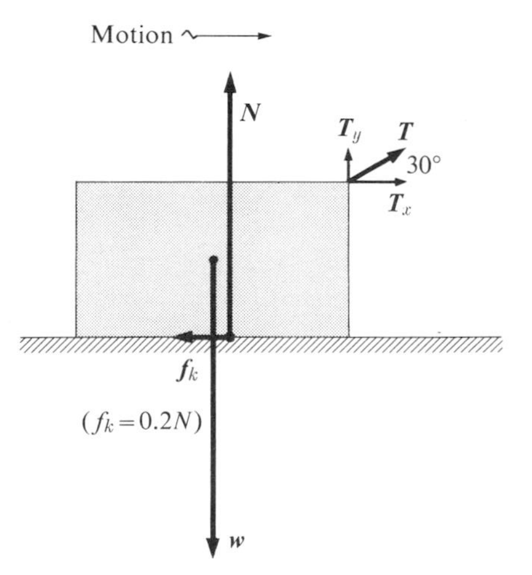

2–13 Forces on a block being dragged to the right on a level surface at constant speed.

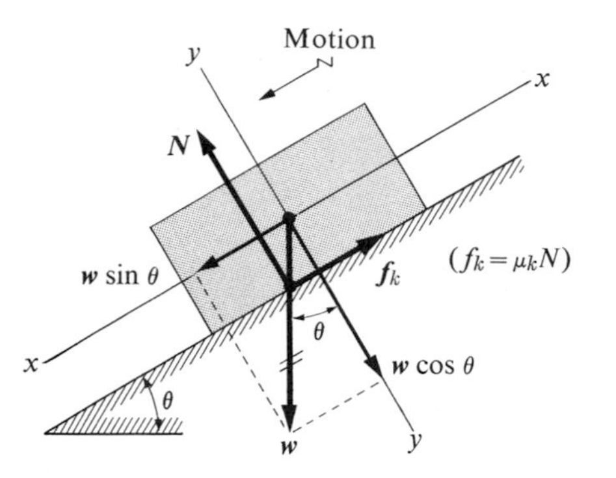

2–14 Forces on a block sliding down an inclined plane (with friction) at constant speed.

Example 4. In Fig. 2–14, a block has been placed on an inclined plane and the slope angle θ of the plane has been adjusted until the block slides down the plane at constant speed, once it has been set in motion. Find the angle θ.

The forces on the block are its weight $\boldsymbol{w}$ and the normal and frictional components of the force exerted by the plane. Take axes perpendicular and parallel to the surface of the plane. Then

$$\left.\begin{aligned}\sum F_x &= \mu_k N - w \sin\theta = 0,\\ \sum F_y &= N - w\cos\theta = 0.\end{aligned}\right\} \quad \text{(First law)}$$

Hence

$$\mu_k N = w \sin\theta, \qquad N = w\cos\theta.$$

Dividing the former by the latter, we get

$$\mu_k = \tan\theta.$$

It follows that a block, regardless of its weight, slides down an inclined plane with constant speed if the tangent of the slope angle of the plane equals the coefficient of kinetic friction. Measurement of this angle then provides a simple experimental method for determining the coefficient of kinetic friction.

Problems

Section 2–6 should be read carefully before beginning the solution of problems in this chapter. If difficulties arise, they will not result from any complicated mathematics or from a failure to remember the "formula" (after all, the only "formulas" are

$$\sum F_x = 0, \qquad \sum F_y = 0)$$

but rather from (1) a failure to *select some one body to talk about*, and (2) a failure to *recognize precisely what forces are exerted* **on** *the selected body*. Once the forces acting on the body have been clearly and correctly shown in a force diagram, the physics of the problem is finished; the rest of the computations might be (and in practice often are) turned over to a machine.

2–1 Imagine that you are holding a book weighing 4 lb at rest on the palm of your hand. Complete the following sentences.

(a) A downward force of magnitude 4 lb is exerted on the book by________.

(b) An upward force of magnitude ________ is exerted on ________ by the hand.

(c) Is the upward force (b) the reaction to the downward force (a)?

(d) The reaction to force (a) is a force of magnitude ________, exerted on ________ by ________. Its direction is ________.

(e) The reaction to force (b) is a force of magnitude ________, exerted on ________ by ________. Its direction is ________.

(f) That the forces (a) and (b) are equal and opposite is an example of Newton's ________ law.

(g) That forces (b) and (e) are equal and opposite is an example of Newton's ________ law.

Suppose now that you exert an upward force of magnitude 5 lb on the book.

(h) Does the book remain in equilibrium?

(i) Is the force exerted on the book by the hand equal and opposite to the force exerted on the book by the earth?

(j) Is the force exerted on the book by the earth equal and opposite to the force exerted on the earth by the book?

(k) Is the force exerted on the book by the hand equal and opposite to the force exerted on the hand by the book?

Finally, suppose that you snatch your hand away while the book is moving upward.

(l) How many forces then act on the book?

(m) Is the book in equilibrium?

(n) What balances the downward force exerted on the book by the earth?

2-2 A block is given a push along a table top, and slides off the edge of the table. (a) What force or forces are exerted on it while it is falling from the table to the floor? (b) What is the reaction to each force, that is, on what body and by what body is the reaction exerted? Neglect air resistance.

2-3 Two 10-lb weights are suspended at opposite ends of a rope which passes over a light frictionless pulley. The pulley is attached to a chain which goes to the ceiling. (a) What is the tension in the rope? (b) What is the tension in the chain?

2-4 In Fig. 2-6, let the weight of the hanging block be 50 lb. Find the tensions T_2 and T_3, (a) if $\theta_2 = \theta_3 = 60°$, (b) if $\theta_2 = \theta_3 = 10°$, (c) if $\theta_2 = 60°$, $\theta_3 = 0$, and (d) if $AB = 10$ ft, $AO = 6$ ft, $OB = 8$ ft.

2-5 Find the tension in each cord in Fig. 2-15 if the weight of the suspended body is 200 lb.

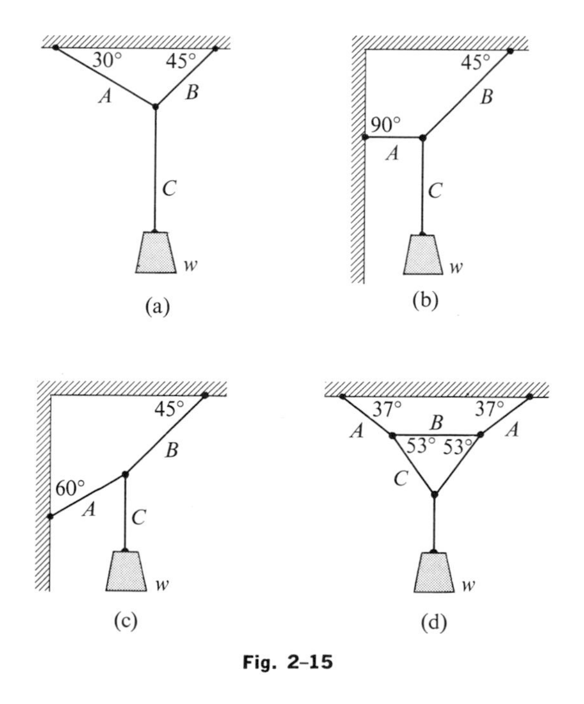

Fig. 2-15

2-6 Find the tension T in the cable, and the magnitude and direction of the force C exerted on the strut by the pivot in the arrangements in Fig. 2-16. Let the weight of the suspended object in each case be 1000 lb. Neglect the weight of the strut.

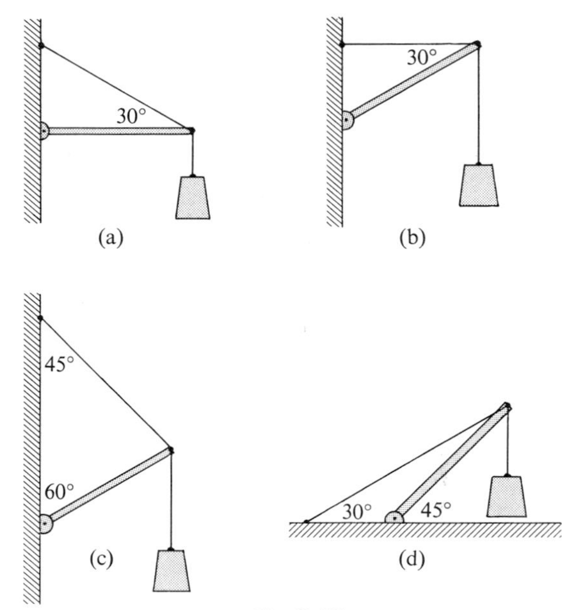

Fig. 2-16

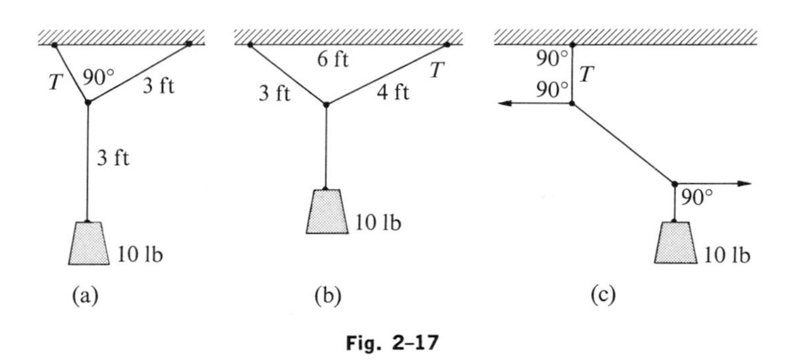

Fig. 2-17

2-7 (a) In which of the arrangements in Fig. 2-17 can the tension T be computed, if the only quantities known are those explicitly given? (b) For each case in which insufficient information is given, state one additional quantity, a knowledge of which would permit solution.

2-8 A horizontal boom 8 ft long is hinged to a vertical wall at one end, and a 500-lb body hangs from its outer end. The boom is supported by a guy wire from its outer end to a point on the wall directly above the boom. (a) If the tension in this wire is not to exceed 1000 lb, what is the minimum height above the boom at which it may be fastened to the wall? (b) By how many pounds would the tension be increased if the wire were fastened 1 ft below this point, the boom remaining horizontal? Neglect the weight of the boom.

2–9 One end of a rope 50 ft long is attached to an automobile. The other end is fastened to a tree. A man exerts a force of 100 lb at the midpoint of the rope, pulling it 2 ft to the side. What is the force exerted on the automobile?

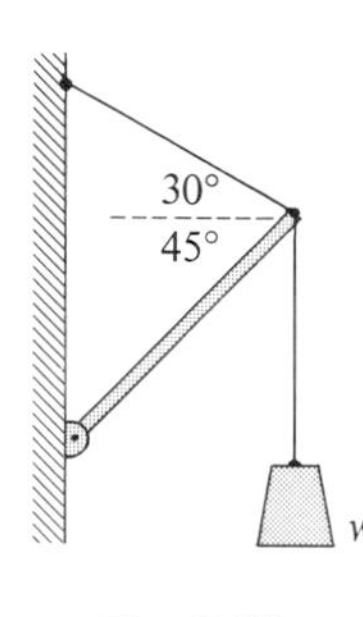

Fig. 2–18

2–10 Find the largest weight w which can be supported by the structure in Fig. 2–18 if the maximum tension the upper rope can withstand is 1000 lb and the maximum compression the strut can withstand is 2000 lb. The vertical rope is strong enough to carry any load required. Neglect the weight of the strut.

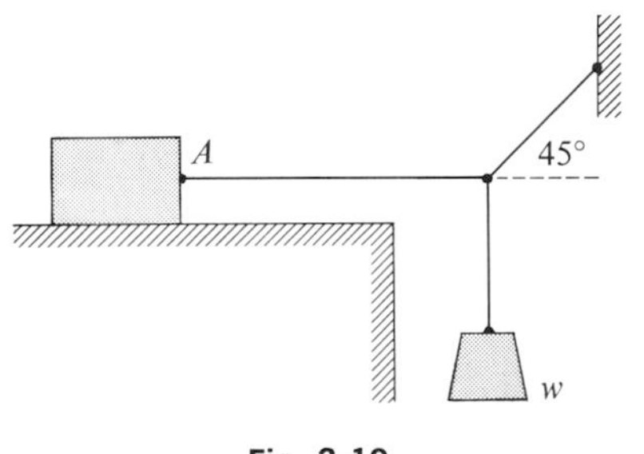

Fig. 2–19

2–11 (a) Block A in Fig. 2–19 weighs 100 lb. The coefficient of static friction between the block and the surface on which it rests is 0.30. The weight w is 20 lb and the system is in equilibrium. Find the friction force exerted on block A. (b) Find the maximum weight w for which the system will remain in equilibrium.

2–12 A block hangs from a cord 10 ft long. A second cord is tied to the midpoint of the first, and a horizontal pull equal to half the weight of the block is exerted on it, the second cord being always kept horizontal. (a) How far will the block be pulled to one side? (b) How far will it be lifted?

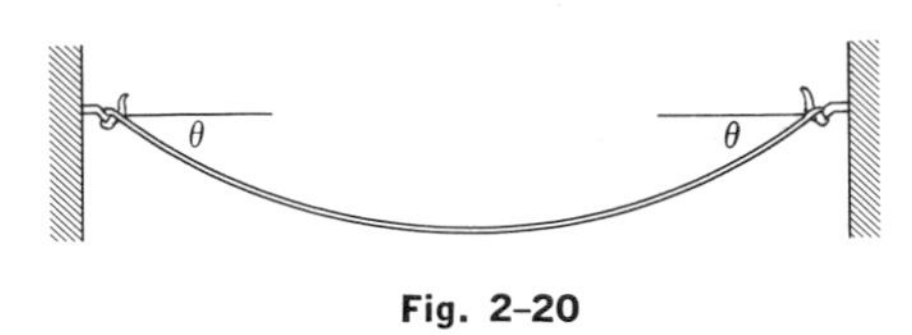

Fig. 2–20

2–13 A flexible chain of weight w hangs between two hooks at the same height, as shown in Fig. 2–20. At each end the chain makes an angle θ with the horizontal. (a) What is the magnitude and direction of the force $\boldsymbol{F}$ exerted by the chain on the hook at the left? (b) What is the tension $\boldsymbol{T}$ in the chain at its lowest point?

2–14 A 30-lb block is pulled at constant speed up a frictionless inclined plane by a weight of 10 lb hanging from a cord attached to the block and passing over a frictionless pulley at the top of the plane. (See Fig. 2–10.) Find (a) the slope angle of the plane, (b) the tension in the cord, and (c) the normal force exerted on the block by the plane.

2–15 (a) A block rests upon a rough horizontal surface. A horizontal force $\boldsymbol{T}$ is applied to the block and is slowly increased from zero. Draw a graph with T along the x-axis and the friction force f along the y-axis, starting at $T = 0$ and showing the region of no motion, the point where motion impends, and the region where motion exists. (b) A block of weight w rests on a rough horizontal plank. The slope angle of the plank θ is gradually increased until the block starts to slip. Draw two graphs, both with θ along the x-axis. In one graph show the ratio of the normal force to the weight N/w as a function of θ. In the second graph, show the ratio of the friction force to the weight f/w. Indicate the region of no motion, the point where motion impends, and the region where motion exists.

2–16 A block weighing 20 lb rests on a horizontal surface. The coefficient of static friction between block and surface is 0.40 and the coefficient of sliding friction is 0.20. (a) How large is the friction force exerted on the block? (b) How great will the friction force be if a horizontal force of 5 lb is exerted on the block? (c) What is the minimum force which will start the block in motion? (d) What is the minimum force which will keep the block in motion once it has been started? (e) If the horizontal force is 10 lb, how great is the friction force?

2–17 A block is pulled to the right at constant velocity by a 10-lb force acting 30° above the horizontal. The coefficient of sliding friction between the block and the surface is 0.5. What is the weight of the block?

2–18 A block weighing 14 lb is placed on an inclined plane and connected to a 10-lb block by a cord passing over a small frictionless pulley, as in Fig. 2–10. The coefficient of sliding friction between the block and the plane is 1/7. For what two values of

θ will the system move with constant velocity? [*Hint:* $\cos\theta = \sqrt{1 - \sin^2\theta}$.]

2–19 A block weighing 100 lb is placed on an inclined plane of slope angle 30° and is connected to a second hanging block of weight w by a cord passing over a small frictionless pulley, as in Fig. 2–10. The coefficient of static friction is 0.40 and the coefficient of sliding friction is 0.30. (a) Find the weight w for which the 100-lb block moves up the plane at constant speed. (b) Find the weight w for which it moves down the plane at constant speed. (c) For what range of values of w will the block remain at rest?

2–20 What force P at an angle ϕ above the horizontal is needed to drag a box of weight w at constant speed along a level floor if we are given that the coefficient of sliding friction between box and floor is μ?

2–21 A safe weighing 600 lb is to be lowered at constant speed down skids 8 ft long, from a truck 4 ft high. (a) If the coefficient of sliding friction between safe and skids is 0.30, will the safe need to be pulled down or held back? (b) How great a force parallel to the skids is needed?

2–22 (a) If a force of 86 lb parallel to the surface of a 20° inclined plane will push a 120-lb block up the plane at constant speed, what force parallel to the plane will push it down at constant speed? (b) What is the coefficient of sliding friction?

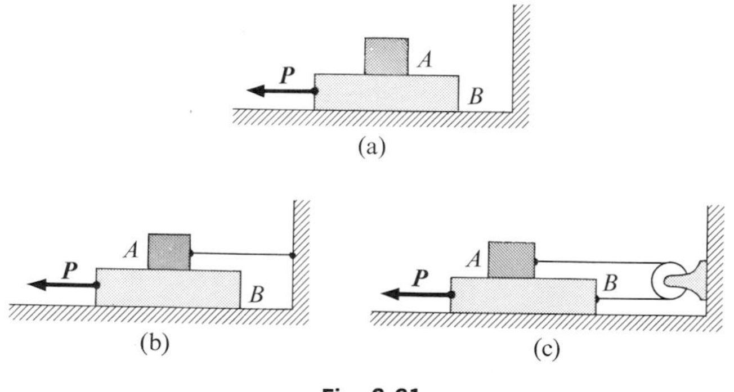

Fig. 2–21

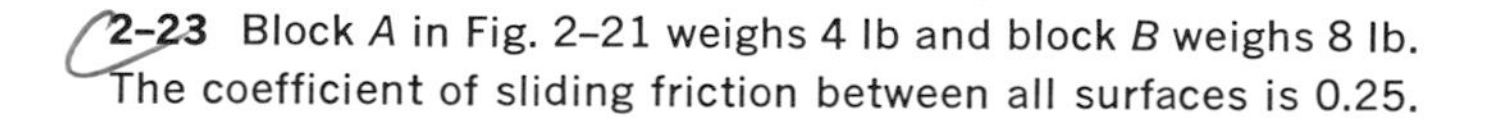

2–23 Block A in Fig. 2–21 weighs 4 lb and block B weighs 8 lb. The coefficient of sliding friction between all surfaces is 0.25. Find the force $\boldsymbol{P}$ necessary to drag block B to the left at constant speed (a) if A rests on B and moves with it, (b) if A is held at rest, and (c) if A and B are connected by a light flexible cord passing around a fixed frictionless pulley.

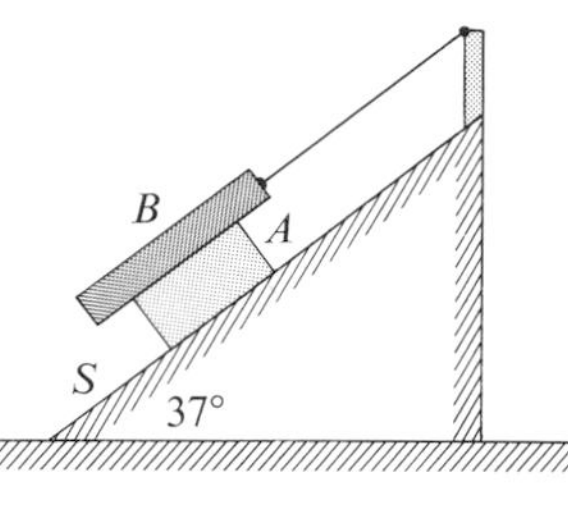

Fig. 2–22

2–24 Block A, of weight w, slides down an inclined plane S of slope angle 37° at constant velocity while the plank B, also of weight w, rests on top of A. The plank is attached by a cord to the top of the plane (Fig. 2–22). (a) Draw a diagram of all the forces acting on block A. (b) If the coefficient of kinetic friction is the same between the surfaces A and B and between S and A, determine its value.

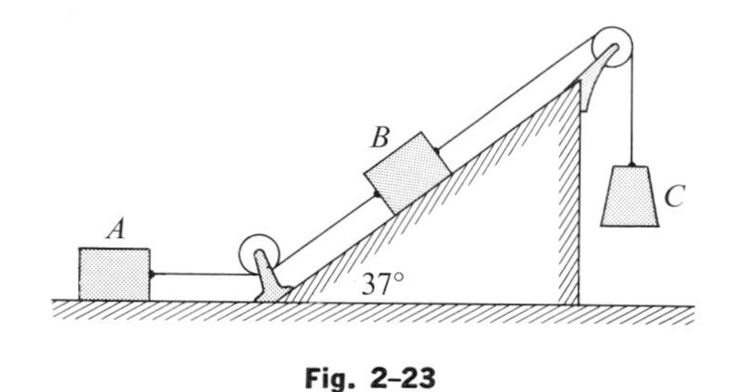

Fig. 2–23

2–25 Two blocks, A and B, are placed as in Fig. 2–23 and connected by ropes to block C. Both A and B weigh 20 lb and the coefficient of sliding friction between each block and the surface is 0.5. Block C descends with constant velocity. (a) Draw two separate force diagrams showing the forces acting on A and B. (b) Find the tension in the rope connecting blocks A and B. (c) What is the weight of block C?

3 Equilibrium. Moment of a Force

3-1 MOMENT OF A FORCE

The effect produced on a body by a force of given magnitude and direction depends on the position of the *line of action* of the force. Thus in Fig. 3-1 the force $\boldsymbol{F}_1$ would produce a counterclockwise rotation (together with a translation toward the right), while $\boldsymbol{F}_2$ would produce a clockwise rotation.

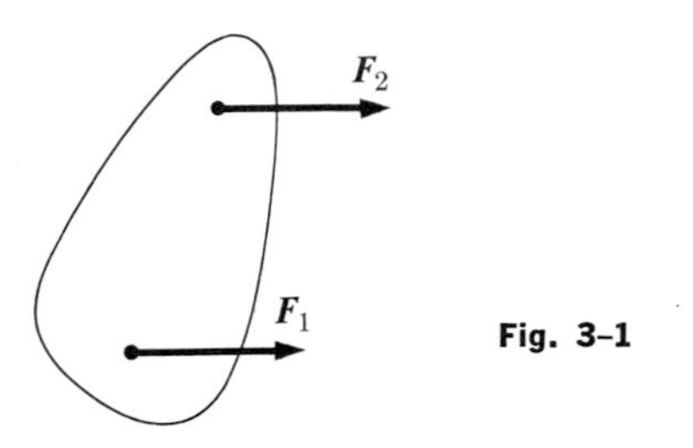

Fig. 3-1

The line of action of a force can be specified by giving the perpendicular distance from some reference point to the line of action. In many instances, we shall be studying the motion of a body which is free to rotate about some axis, and which is acted on by a number of coplanar forces all lying in a plane perpendicular to the axis. It is then most convenient to select as the reference point the point at which the axis intersects the plane of the forces. The perpendicular distance from this point to the line of action of a force is called the *force arm* or the *moment arm* of the force about the axis. The product of the magnitude of a force and its force arm is called the *moment* of the force about the axis, or the *torque*.

Thus Fig. 3-2 is a top view of a flat object, pivoted about an axis perpendicular to the plane of the diagram and passing through point O. The body is acted on by the forces $\boldsymbol{F}_1$ and $\boldsymbol{F}_2$, lying in the plane of the diagram. The moment arm of $\boldsymbol{F}_1$ is the perpendicular distance OA, of length ℓ_1, and the moment arm of $\boldsymbol{F}_2$ is the perpendicular distance OB of length ℓ_2.

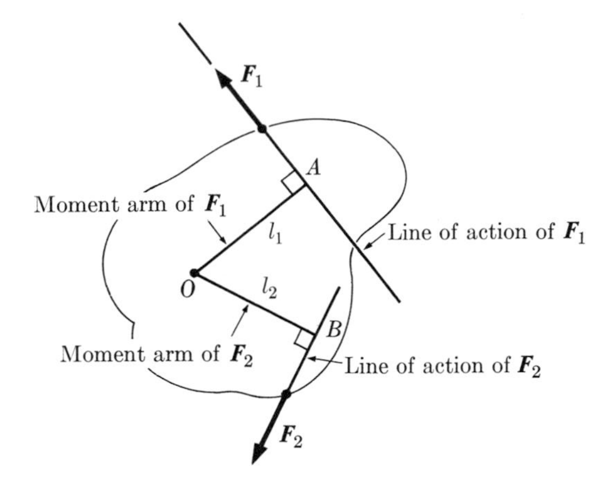

3-2 The moment of a force about an axis is the product of the force and its moment arm.

The effect of the force $\boldsymbol{F}_1$ is to produce counterclockwise rotation about the axis, while that of $\boldsymbol{F}_2$ is to produce clockwise rotation. To distinguish between these directions of rotation, we shall adopt the convention that counterclockwise moments are positive, and that clockwise moments are negative. Hence the moment Γ_1 (Greek "gamma") of the force $\boldsymbol{F}_1$ about the axis through O is

$$\Gamma_1 = +F_1\ell_1,$$

and the moment Γ_2 of $\boldsymbol{F}_2$ is

$$\Gamma_2 = -F_2\ell_2.$$

If forces are expressed in pounds and lengths in feet, torques are expressed in pound feet.

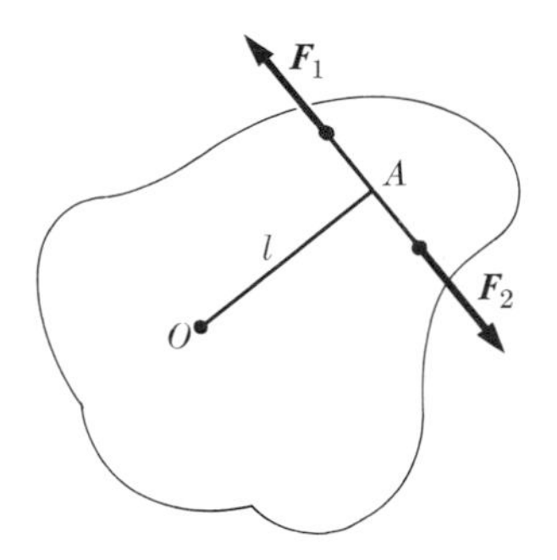

3-3 When two forces are in equilibrium, their resultant moment about any axis is zero.

3-2 THE SECOND CONDITION OF EQUILIBRIUM

We saw in Section 2-2 that when a body is acted on by any number of coplanar forces, the forces can always be reduced to two, as in Fig. 2-2. If the body is in equilibrium, these forces must (a) be equal in magnitude and opposite in direction, and (b) must have the same line of action.

Requirement (a) is satisfied by the *first condition of equilibrium*,

$$\sum F_x = 0, \qquad \sum F_y = 0.$$

Requirement (b), which is the second condition of equilibrium, can be simply expressed in terms of the moments of the forces. Figure 3-3 again shows a flat object acted on by two forces $\boldsymbol{F}_1$ and $\boldsymbol{F}_2$. If the object is in equilibrium, the magnitudes of $\boldsymbol{F}_1$ and $\boldsymbol{F}_2$ are equal and both forces have the same line of action. Hence they have the same moment arm OA, of length ℓ, about an axis perpendicular to the plane of the body and passing through any arbitrary point O. Their moments about the axis are therefore equal in magnitude and opposite in sign, and the algebraic sum of their moments is zero. The necessary and sufficient condition, therefore, that two equal and opposite forces have the same line of action is that the algebraic sum of their moments, about any axis, shall be zero. The *second condition of equilibrium* may therefore be expressed analytically as

$$\sum \Gamma = 0 \text{ (about any arbitrary axis).}$$

It is not necessary to first reduce a set of coplanar forces to two forces in order to calculate the sum of their moments. One need only calculate the moment of each force separately, and then add these moments algebraically. Thus if a body is in equilibrium under the action of any number of coplanar forces, the algebraic sum of the torques about any arbitrary axis is zero.

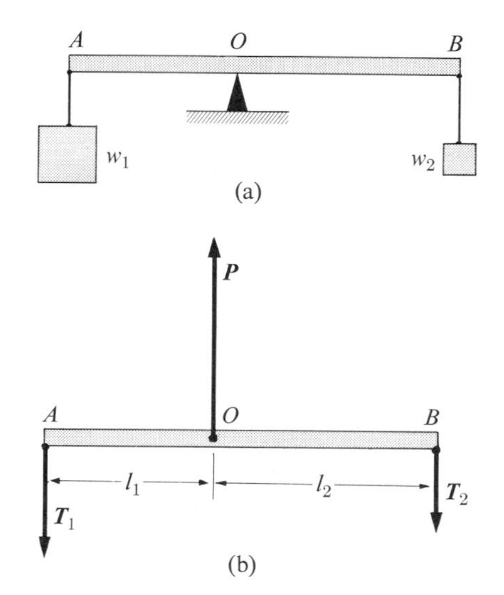

3-4 A rod in equilibrium under three parallel forces.

Example 1. A rigid rod whose own weight is negligible (Fig. 3-4) is pivoted at point O and carries a body of weight w_1 at end A. Find the weight w_2 of a second body which must be attached at end B if the rod is to be in equilibrium, and find the force exerted on the rod by the pivot at O.

Figure 3-4(b) is the free-body diagram of the rod. The forces T_1 and T_2 are equal respectively to w_1 and w_2. The conditions of equilibrium, taking moments about an axis through O, perpendicular to the diagram, give

$$\sum F_y = P - T_1 - T_2 = 0, \qquad \text{(First condition)}$$
$$\sum \Gamma_O = T_1\ell_1 - T_2\ell_2 = 0. \qquad \text{(Second condition)}$$

Let $\ell_1 = 3$ ft, $\ell_2 = 4$ ft, $w_1 = 4$ lb. Then, from the equations above,

$$P = 7 \text{ lb}, \qquad T_2 = w_2 = 3 \text{ lb}.$$

To illustrate that the resultant moment about *any* axis is zero, let us compute moments about an axis through point A:

$$\sum \Gamma_A = P\ell_1 - T_2(\ell_1 + \ell_2) = 7 \text{ lb} \times 3 \text{ ft} - 3 \text{ lb} \times 7 \text{ ft} = 0.$$

The point about which moments are computed need not lie on the rod. To verify this, let the reader calculate the resultant moment about a point 1 ft to the left of A and 1 ft above it.

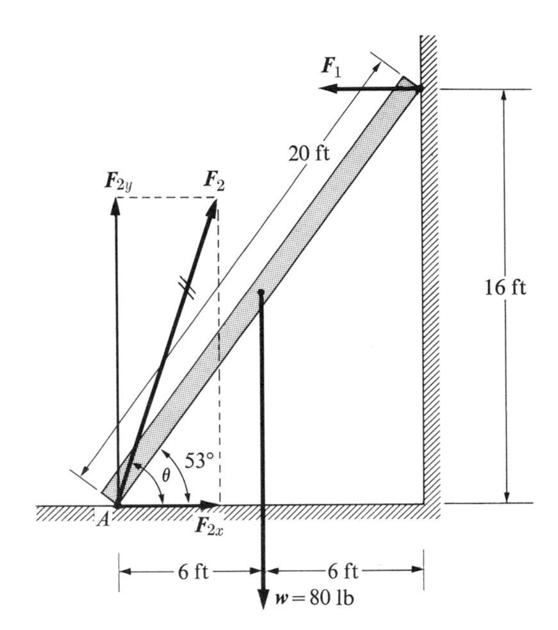

3-5 Forces on a ladder in equilibrium, leaning against a frictionless wall.

Example 2. In Fig. 3-5 a ladder 20 ft long, of weight 80 lb, with its center of gravity at its center, in equilibrium, leans against a vertical frictionless wall and makes an angle of 53° with the horizontal. We wish to find the magnitudes and directions of the forces $\boldsymbol{F}_1$ and $\boldsymbol{F}_2$.

If the wall is frictionless, $\boldsymbol{F}_1$ is horizontal. The direction of $\boldsymbol{F}_2$ is unknown (except in special cases, its direction does *not* lie along the ladder). Instead of considering its magnitude and direction as unknowns, it is simpler to resolve the force $\boldsymbol{F}_2$ into x- and y-components and solve for these. The magnitude and direction of $\boldsymbol{F}_2$ may then be computed. The first condition of equilibrium therefore provides the equations

$$\left.\begin{aligned} \sum F_x &= F_2 \cos\theta - F_1 = 0, \\ \sum F_y &= F_2 \sin\theta - 80 \text{ lb} = 0. \end{aligned}\right\} \qquad \text{(First condition)}$$

In writing the second condition, moments may be computed about an axis through any point. The resulting equation is simplest if one selects a point through which two or more forces pass, since these forces then do not appear in the equation. Let us therefore take moments about an axis through point A.

$$\sum \Gamma_A = F_1 \times 16 \text{ ft} - 80 \text{ lb} \times 6 \text{ ft} = 0. \qquad \text{(Second condition)}$$

From the second equation, $F_2 \sin\theta = 80$ lb, and from the third,

$$F_1 = \frac{480 \text{ lb ft}}{16 \text{ ft}} = 30 \text{ lb}.$$

Then from the first equation,

$$F_2 \cos\theta = 30 \text{ lb}.$$

Hence

$$F_2 = \sqrt{(80 \text{ lb})^2 + (30 \text{ lb})^2} = 85.5 \text{ lb},$$

$$\theta = \tan^{-1} \frac{80 \text{ lb}}{30 \text{ lb}} = 69.5°.$$

[Note that this problem has already been solved by a different method in Example 4 in Section 2-6.]

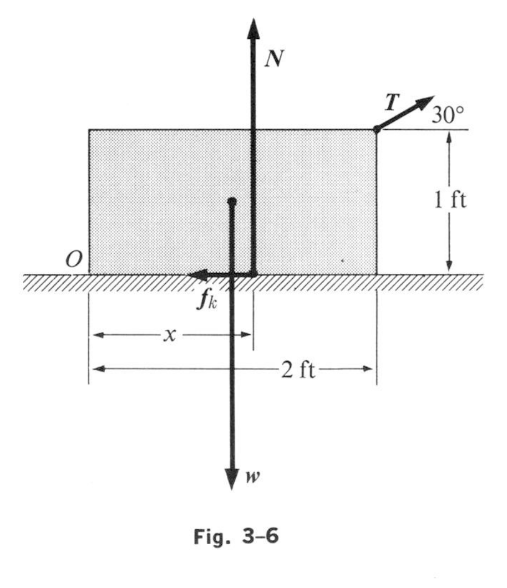

Fig. 3-6

Example 3. Figure 3-6 illustrates a problem that has already been solved, in part, in Example 3 at the end of Section 2-7. (a) What force T, at an angle of 30° above the horizontal, is required to drag a block of weight $w = 20$ lb to the right at constant speed along a level surface if the coefficient of sliding friction between block and surface is 0.20? (b) Determine the line of action of the normal force $\boldsymbol{N}$ exerted on the block by the surface. The block is 1 ft high, 2 ft long, and its center of gravity is at its center.

From the first condition of equilibrium,

$$\left.\begin{aligned}\sum F_x &= T\cos 30^\circ - f_k = 0,\\ \sum F_y &= T\sin 30^\circ + N - 20\text{ lb} = 0.\end{aligned}\right\}\quad\text{(First condition)}$$

Let x represent the distance from point O to the line of action of $\boldsymbol{N}$, and take moments about an axis through O. Then, from the second condition of equilibrium,

$$\begin{aligned}\sum\Gamma_O = T\sin 30^\circ \times 2\text{ ft} - T\cos 30^\circ \times 1\text{ ft}\\ + N \times x - 20\text{ lb} \times 1\text{ ft} = 0.\quad\text{(Second condition)}\end{aligned}$$

From the first two equations, we get

$$T = 4.15\text{ lb},\qquad N = 17.9\text{ lb},$$

and from the third equation,

$$x = 1.08\text{ ft}.$$

The line of action of $\boldsymbol{N}$ therefore lies 0.08 ft to the right of the center of gravity.

3–3 RESULTANT OF PARALLEL FORCES

The direction of the resultant of a set of parallel forces is the same as that of the forces, and its magnitude equals the sum of their magnitudes. The line of action of the resultant can be found from the requirement that the moment of the resultant, about any axis, shall equal the sum of the moments of the given forces.

Consider the parallel forces $\boldsymbol{F}_1$ and $\boldsymbol{F}_2$ in Fig. 3–7. Point O is any arbitrary point and the x-axis has been taken at right angles to the directions of the forces. The forces have no x-components, so the magnitude of the resultant is

$$R = \sum F_y = F_1 + F_2.$$

If x_1 and x_2 are the perpendicular distances from O to the lines of action of the forces, their resultant moment about an axis through O is

$$\sum\Gamma_O = x_1F_1 + x_2F_2.$$

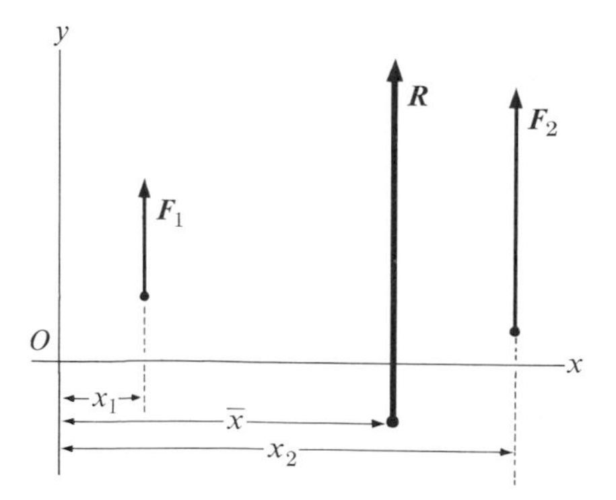

3–7 Vector $\boldsymbol{R}$ represents the resultant of the parallel forces $\boldsymbol{F}_1$ and $\boldsymbol{F}_2$, in magnitude, direction, and line of action.

Let $\bar{x}$ represent the distance from O to the line of action of the resultant. The moment of the resultant about O is then

$$R\bar{x} = (F_1 + F_2)\bar{x},$$

and since this equals the resultant moment, we have

$$(F_1 + F_2)\bar{x} = F_1x_1 + F_2x_2.$$

Therefore

$$\bar{x} = \frac{F_1x_1 + F_2x_2}{F_1 + F_2},$$

and the magnitude, direction, and line of action of the resultant are determined.

The resultant of any number of parallel forces is found in the same way. The magnitude of the resultant is

$$R = \sum F,$$

and, if the forces are parallel to the y-axis, the x-coordinate of its line of action is

$$x = \frac{\sum Fx}{\sum F} = \frac{\sum Fx}{R}.$$

Example. When a body is in equilibrium under the action of three forces, the resultant of any two is equal and opposite to the third and has the same line of action. Show that these conditions are satisfied by the three parallel forces in Fig. 3–4(b).

It was shown in Example 1 at the end of Section 3–2 that if $\ell_1 = 3$ ft, $\ell_2 = 4$ ft, and $T_1 = 4$ lb, then $T_2 = 3$ lb and $P = 7$ lb.

Let us first find the resultant of T_1 and T_2. Take the x-axis along the rod with origin at point A. The magnitude of the resultant is

$$R = \sum F = -4\text{ lb} - 3\text{ lb} = -7\text{ lb}.$$

The coordinate of its line of action is

$$\bar{x} = \frac{\sum Fx}{\sum F} = \frac{4\text{ lb} \times 0 - 3\text{ lb} \times 7\text{ ft}}{-7\text{ lb}} = 3\text{ ft}.$$

Hence the resultant of T_1 and T_2 is equal and opposite to P and has the same line of action.

The resultant of P and T_2 has a magnitude

$$R = \sum F = 7\text{ lb} - 3\text{ lb} = 4\text{ lb}.$$

The coordinate of its line of action is

$$\bar{x} = \frac{\sum Fx}{\sum F} = \frac{7\text{ lb} \times 3\text{ ft} - 3\text{ lb} \times 7\text{ ft}}{4\text{ lb}} = 0,$$

so the resultant of P and I_2 is equal and opposite to T_1 and has the same line of action.

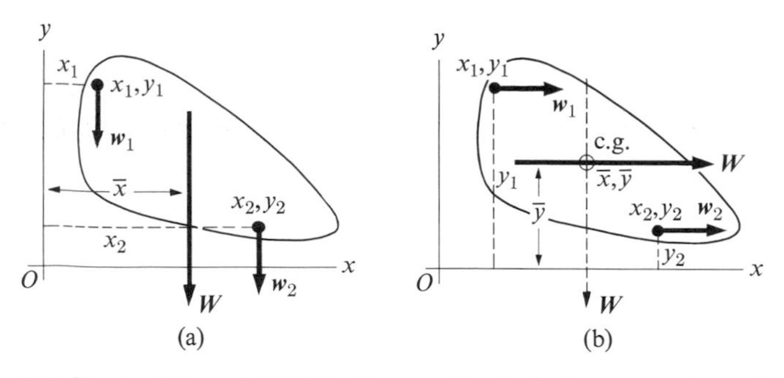

3-8 The body's weight W is the resultant of a large number of parallel forces. The line of action of $\boldsymbol{W}$ always passes through the center of gravity.

3-4 CENTER OF GRAVITY

Every particle of matter in a body is attracted by the earth, and the single force which we call the *weight* of the body is the resultant of all these forces of attraction. The direction of the force on each particle is toward the center of the earth, but the distance to the earth's center is so great that for all practical purposes the forces can be considered parallel to one another. Hence the weight of a body is the resultant of a large number of parallel forces.

Figure 3-8(a) shows a flat object of arbitrary shape in the xy-plane, the y-axis being vertical. Let the body be subdivided into a large number of small particles of weights $\boldsymbol{w}_1$, $\boldsymbol{w}_2$, etc., and let the coordinates of these particles be x_1 and y_1, x_2 and y_2, etc. The total weight W of the object is

$$W = w_1 + w_2 + \cdots = \sum w. \tag{3-1}$$

The x-coordinate of the line of action of $\boldsymbol{W}$ is

$$\bar{x} = \frac{w_1x_1 + w_2x_2 + \cdots}{w_1 + w_2 + \cdots} = \frac{\sum wx}{\sum w} = \frac{\sum wx}{W}. \tag{3-2}$$

Now let the object and the reference axes be rotated 90° clockwise or, which amounts to the same thing, let us consider the gravitational forces to be rotated 90° counterclockwise as in Fig. 3-8(b). The total weight W is unaltered and the y-coordinate of its line of action is

$$\bar{y} = \frac{w_1y_1 + w_2y_2 + \cdots}{w_1 + w_2 + \cdots} = \frac{\sum wy}{\sum w} = \frac{\sum wy}{W}. \tag{3-3}$$

The point of intersection of the lines of action of $\boldsymbol{W}$ in the two parts of Fig. 3-8 has the coordinates $\bar{x}$ and $\bar{y}$ and is called the *center of gravity* of the object. By considering some arbitrary orientation of the object, one can show that the line of action of $\boldsymbol{W}$ *always* passes through the center of gravity.

If the centers of gravity of each of a number of bodies have been determined, the coordinates of the center of gravity of the combination can be computed from Eqs. (3-1) and (3-2), letting w_1, w_2, etc., be the weights of the bodies and x_1 and y_1, and y_2, etc., be the coordinates of the center of gravity of each.

Symmetry considerations are often useful in finding the position of the center of gravity. Thus the center of gravity of a homogeneous sphere, cube, circular disk, or rectangular plate is at its center. That of a cylinder or right circular cone is on the axis of symmetry, and so on.

Example. Locate the center of gravity of the machine part in Fig. 3-9, consisting of a disk 2 in. in diameter and 1 in. long, and a rod 1 in. in diameter and 6 in. long, constructed of a homogeneous material.

By symmetry, the center of gravity lies on the axis and the center of gravity of each part is midway between its ends. The volume of the disk is π in^3 and that of the rod is $3\pi/2$ in^3. Since the weights of the two parts are proportional to their volumes,

$$\frac{w\text{ (disk)}}{w\text{ (rod)}} = \frac{w_1}{w_2} = \frac{\pi}{3\pi/2} = \frac{2}{3}.$$

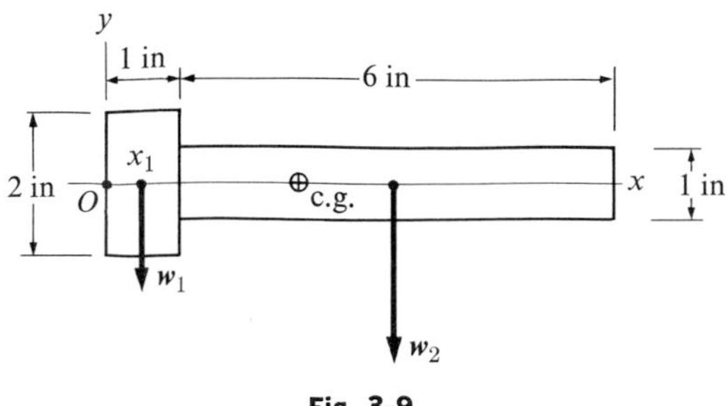

Fig. 3–9

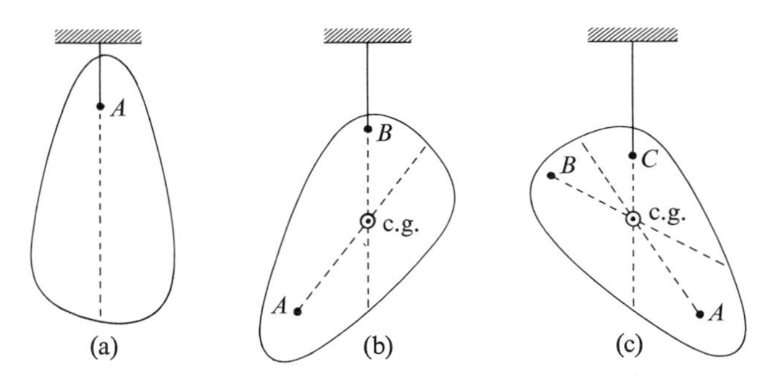

3–10 Locating the center of gravity of a flat object.

Take the origin O at the left face of the disk, on the axis. Then

$$x_1 = 0.5 \text{ in.}, \qquad x_2 = 4.0 \text{ in.},$$

and

$$\bar{x} = \frac{w_1 \times 0.5 \text{ in.} + \frac{3}{2}w_1 \times 4.0 \text{ in.}}{w_1 + \frac{3}{2}w_1} = 2.6 \text{ in.}$$

The center of gravity is on the axis, 2.6 in. to the right of O.

The center of gravity of a flat object can be located experimentally as shown in Fig. 3–10. In part (a) the body is suspended from some arbitrary point A. When allowed to come to equilibrium, the center of gravity must lie on a vertical line through A. When the object is suspended from a second point B, as in part (b), the center of gravity lies on a vertical line through B and hence lies at the point of intersection of this line and the first. If the object is now suspended from a third point C, as in part (c), a vertical line through C will be found to pass through the point of intersection of the first two lines.

The center of gravity of a body has another important property. A force $\boldsymbol{F}$ whose line of action lies at one side or the other of the center of gravity, as in Fig. 3–11(a), will change both the translational and rotational motion of the body on which it acts. However, if the line of action passes through the center of gravity, as in part (b), only the translational motion is affected and the body remains in rotational equilibrium. Thus when a body is tossed in the air with a whirling motion, it continues to rotate at a constant rate, since the line of action of its weight passes through the center of gravity.

It may also be pointed out that when one object rests on or slides over another, the normal and frictional forces are sets of parallel forces distributed over the area in contact. The single vectors which have been used to represent these forces are therefore actually the resultants of sets of parallel forces.

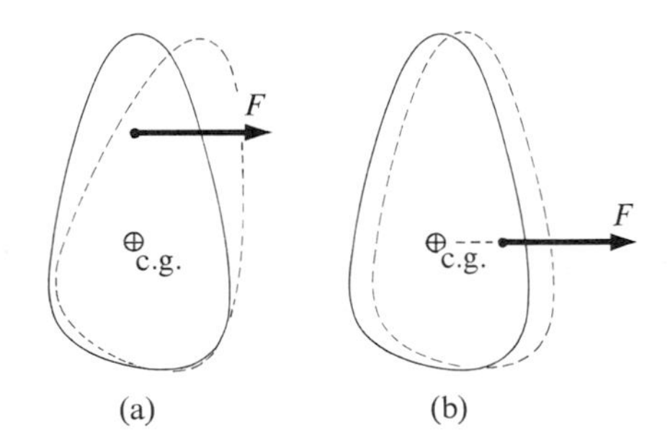

3–11 A body is in rotational but not translational equilibrium when acted on by a force whose line of action passes through the center of gravity, as in (b).

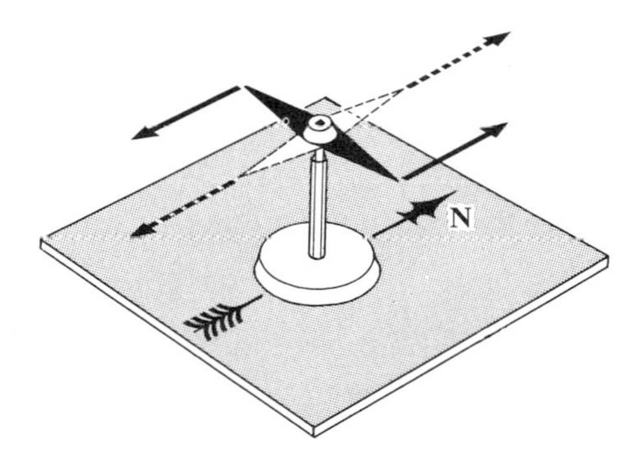

3–12 Forces on the poles of a compass needle.

3–5 COUPLES

It often happens that the forces on a body reduce to two forces of equal magnitude and opposite direction, having lines of action which are parallel but which do not coincide. Such a pair of forces is called a *couple*. A common example is afforded by the forces on a compass needle in the earth's magnetic field as shown in Fig. 3–12.

The north and south poles of the needle are acted on by equal forces, one toward the north and the other toward the south. Except when the needle points in the N-S direction, the two forces do not have the same line of action.

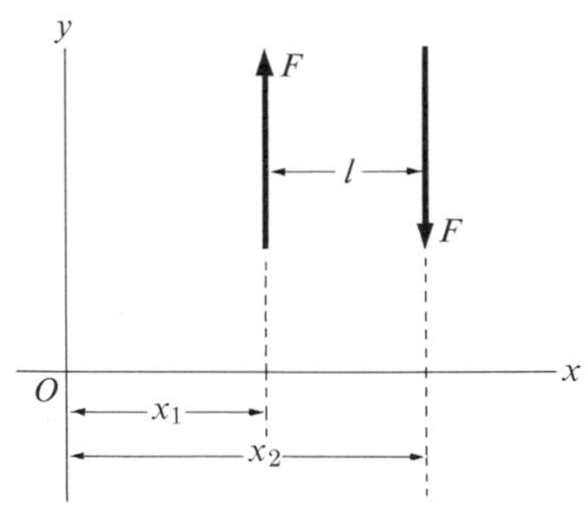

3–13 Two equal and opposite forces having different lines of action are called a couple. The moment of the couple is the same about all points, and is equal to ℓF.

Figure 3–13 shows a couple consisting of two forces, each of magnitude F, separated by a perpendicular distance l. The resultant R of the forces is

$$R = F - F = 0.$$

The fact that the resultant force is zero means that a couple has no effect in producing translation as a whole of the body on which it acts. The only effect of a couple is to produce rotation.

The resultant torque of the couple in Fig. 3–13, about an arbitrary point O, is

$$\begin{aligned}\sum\Gamma_O &= x_1F - x_2F \\ &= x_1F - (x_1 + \ell)F \\ &= -\ell F.\end{aligned}$$

Since the distances x_1 and x_2 do not appear in the result, we conclude that the torque of the couple is the same about *all* points in the plane of the forces forming the couple and is equal to the product of the magnitude of either force and the perpendicular distance between their lines of action.

A body acted on by a couple can be kept in equilibrium only by another couple of the same moment and in the opposite direction. As an example, the ladder in Fig. 3–5 can be considered as acted on by two couples, one formed by the forces $F_2 \sin\theta$ and w, the other by the forces $F_2 \cos\theta$ and F_1. The moment of the first is

$$\Gamma_1 = 6 \text{ ft} \times 80 \text{ lb} = 480 \text{ lb ft}.$$

The moment of the second is

$$\Gamma_2 = 16 \text{ ft} \times 30 \text{ lb} = 480 \text{ lb ft}.$$

The first moment is clockwise and the second is counterclockwise.

Problems

Section 2–6 and the note at the beginning of the problems in Chapter 2 are equally applicable to these problems. The only difference is that now we have a third "formula," $\sum\Gamma$ (about any axis) $= 0$.

3–1 A force F in the xy-plane has components of magnitudes F_x and F_y, respectively. The coordinates of its point of application are x, y. The magnitude of the moment of the force about the origin is $\Gamma_O = xF_y - yF_x$. Show that this equation is correct if the point x, y lies in any one of the four quadrants, and also that it is correct regardless of the direction of the force $\boldsymbol{F}$.

3–2 You are given (a) a meter stick through which a number of holes have been bored so that its center of gravity is not at its center, (b) a knife edge on which the meter stick can be pivoted, (c) a body whose weight is known to be w, and (d) a spool of thread. Using this equipment only, explain with the aid of a diagram how you would determine the weight of the meter stick.

3–3 A uniform plank 30 ft long, weighing 80 lb, rests symmetrically on two supports 16 ft apart, as shown in Fig. 3–14. A boy weighing 128 lb starts at point A and walks toward the right. (a) Construct in the same diagram two graphs showing

Fig. 3–14

the upward forces F_A and F_B exerted on the plank at points A and B, as functions of the coordinate x of the boy. Let 1 in. = 50 lb vertically, and 1 in. = 5 ft horizontally. (b) Find from your diagram how far beyond point B the boy can walk before the plank tips. (c) How far from the right end of the plank should support B be placed in order that the boy can just walk to the end of the plank without causing it to tip?

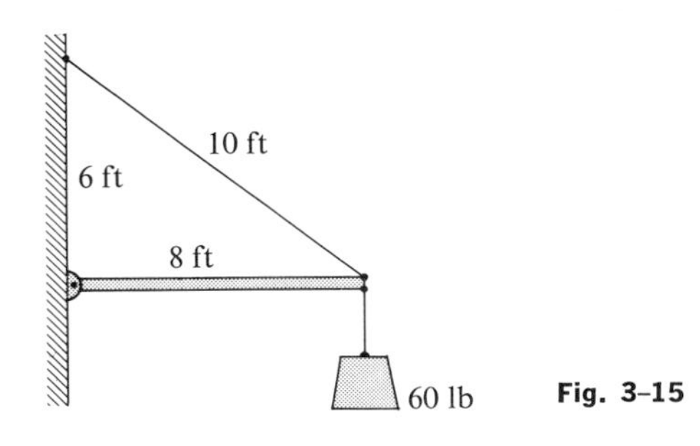

Fig. 3–15

3–4 The strut in Fig. 3–15 weighs 40 lb and its center of gravity is at its center. Find (a) the tension in the cable and (b) the horizontal and vertical components of the force exerted on the strut at the wall.

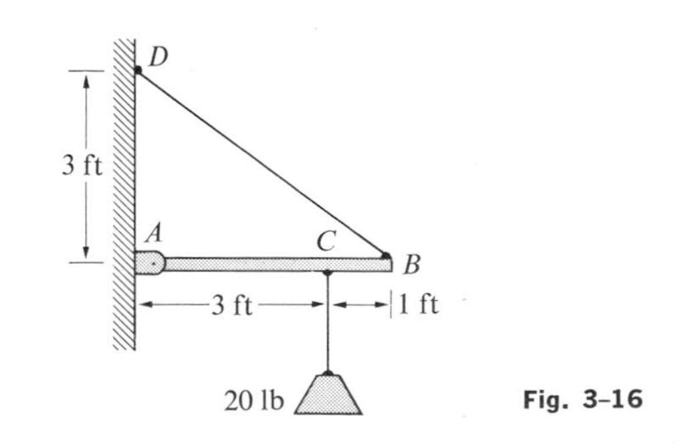

Fig. 3–16

3–5 Find the tension in the cable BD in Fig. 3–16, and the horizontal and vertical components of the force exerted on the strut AB at pin A, (a) by using the first and second conditions of equilibrium ($\sum F_x = 0$, $\sum F_y = 0$, $\sum \Gamma = 0$), taking moments about an axis through point A perpendicular to the plane of the diagram, and (b) by using the second condition of equilibrium only, taking moments first about an axis through A, then about an axis through B, and finally about an axis through D. The weight of the strut can be neglected. (c) Represent the computed forces by vectors in a diagram drawn to scale, and show that the lines of action of the forces exerted on the strut at points A, B, and C intersect at a common point.

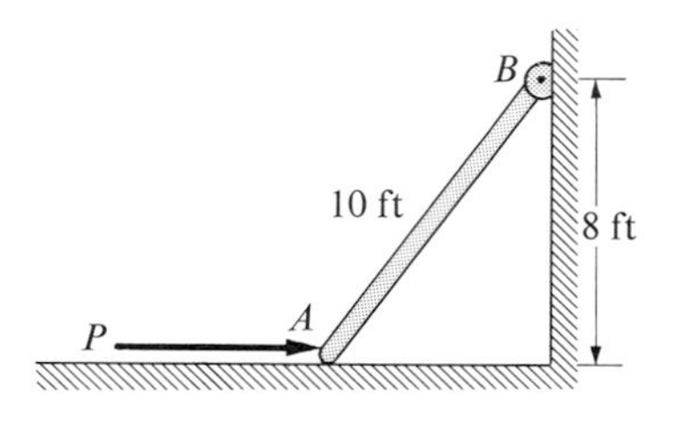

Fig. 3–17

3–6 End A of the bar AB in Fig. 3–17 rests on a frictionless horizontal surface, while end B is hinged. A horizontal force $\boldsymbol{P}$ of 12 lb is exerted on end A. Neglect the weight of the bar. What are the horizontal and vertical components of the force exerted by the bar on the hinge at B?

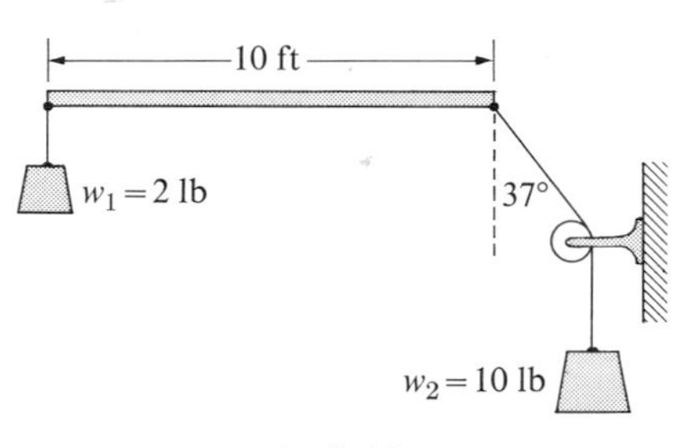

Fig. 3–18

3–7 A single force is to be applied to the bar in Fig. 3–18 to maintain it in equilibrium in the position shown. The weight of the bar can be neglected. (a) What are the x- and y-components of the required force? (b) What is the tangent of the angle which the force must make with the bar? (c) What is the magnitude of the required force? (d) Where should the force be applied?

3–8 A circular disk 1 ft in diameter, pivoted about a horizontal axis through its center, has a cord wrapped around its rim. The cord passes over a frictionless pulley P and is attached to a body of weight 48 lb. A uniform rod 4 ft long is fastened to the disk with one end at the center of the disk. The apparatus is in equilibrium, with the rod horizontal, as shown in Fig. 3–19. (a) What is the weight of the rod? (b) What is the new equilibrium

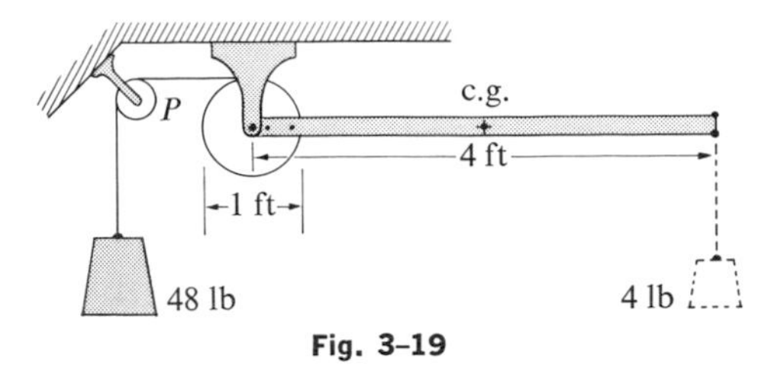

Fig. 3–19

direction of the rod when a second body weighing 4 lb is suspended from the outer end of the rod, as shown by the dotted line?

3–9 A roller whose diameter is 20 in. weighs 72 lb. What horizontal force is necessary to pull the roller over a brick 2 in. high when (a) the force is applied at the center, (b) at the top?

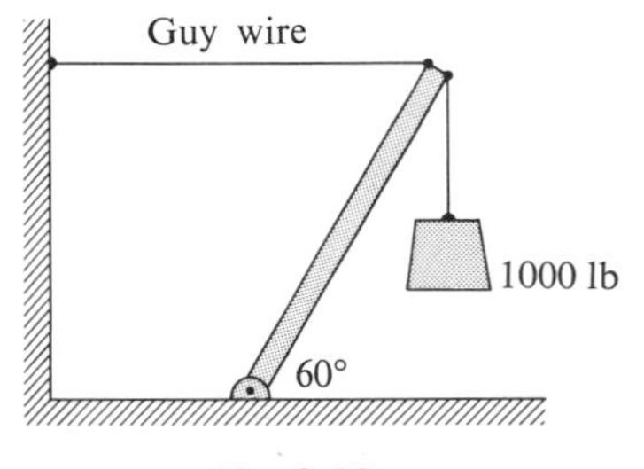

Fig. 3–20

3–10 The boom in Fig. 3–20 is uniform and weighs 500 lb. (a) Find the tension in the guy wire, and the horizontal and vertical components of the force exerted on the boom at its lower end. (b) Does the line of action of this force lie along the boom?

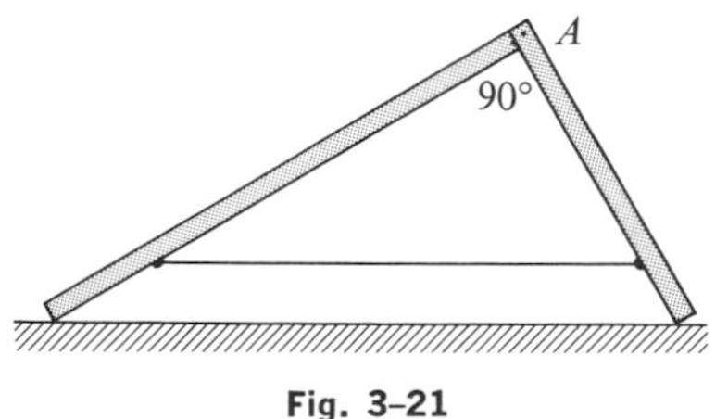

Fig. 3–21

3–11 Two ladders, 20 ft and 15 ft long, respectively, are hinged at point A and tied together by a horizontal rope 3 ft above the floor, as in Fig. 3–21. The ladders weigh 80 lb and 60 lb, respectively, and the center of gravity of each is at its center. If the floor is frictionless, find (a) the upward force at the bottom of each ladder, (b) the tension in the rope, and (c) the force which one ladder exerts on the other at point A. (d) If a load of 200 lb is now suspended from point A, find the tension in the rope.

3–12 A uniform ladder 20 ft long rests against a vertical frictionless wall with its lower end 12 ft from the wall. The ladder weighs 80 lb. The coefficient of static friction between the foot of the ladder and the ground is 0.40. A man weighing 160 lb climbs slowly up the ladder. (a) What is the maximum frictional force which the ground can exert on the ladder at its lower end? (b) What is the actual frictional force when the man has climbed 10 ft along the ladder? (c) How far along the ladder can the man climb before the ladder starts to slip?

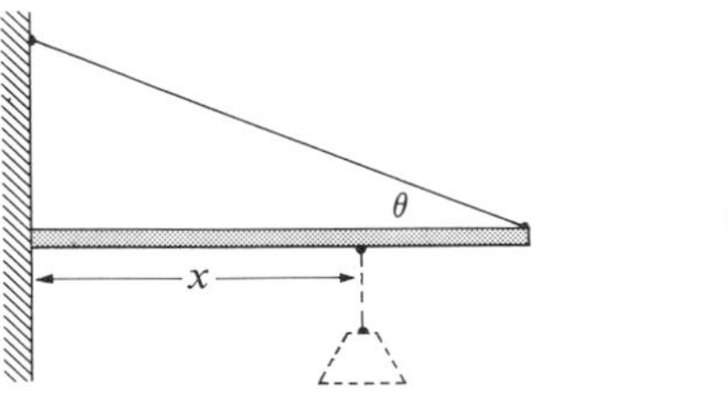

Fig. 3–22

3–13 One end of a meter stick is placed against a vertical wall, as in Fig. 3–22. The other end is held by a light cord making an angle θ with the stick. The coefficient of static friction between the end of the meter stick and the wall is 0.30. (a) What is the maximum value the angle θ can have if the stick is to remain in equilibrium? (b) Let the angle θ be 10°. A body of the same weight as the meter stick is suspended from the stick as shown by dotted lines, at a distance x from the wall. What is the minimum value of x for which the stick will remain in equilibrium? (c) When $\theta = 10°$, how large must the coefficient of static friction be so that the body can be attached at the left end of the stick without causing it to slip?

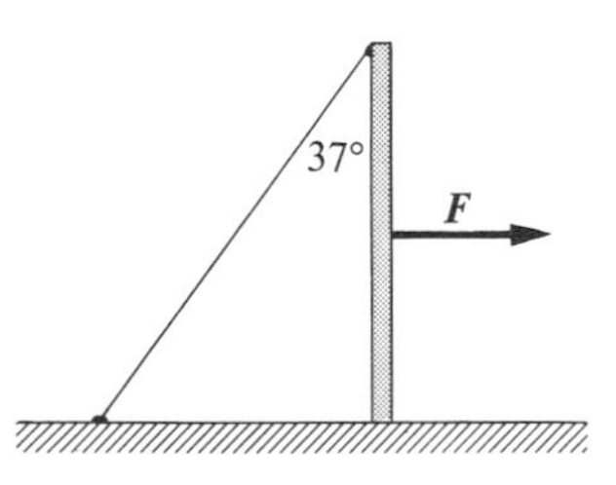

Fig. 3–23

3–14 One end of a post weighing 100 lb rests on a rough horizontal surface with $\mu_s = 0.3$. The upper end is held by a rope fastened to the surface and making an angle of 37° with the post, as in Fig. 3–23. A horizontal force $\boldsymbol{F}$ is exerted on the post as shown. (a) If the force $\boldsymbol{F}$ is applied at the midpoint of the post, what is the largest value it can have without causing the post to slip? (b) How large can the force be, without causing the post to slip, if its height above the surface equals $\frac{9}{10}$ of the length of the post?

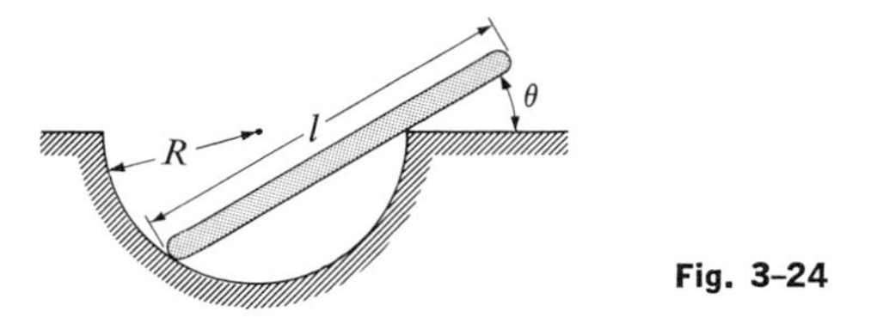

Fig. 3-24

3-15 A uniform smooth rod of length ℓ and weight w rests at equilibrium in a smooth semispherical bowl of radius R, as shown in Fig. 3-24, where $R < \ell/2 < 2R$. If θ is the angle of equilibrium and P is the force exerted by the edge of the bowl on the rod, prove that (a) $P = (\ell/4R)w$, (b) $\cos 2\theta/\cos\theta = \ell/4R$.

3-16 A door 7 ft high and 3 ft wide is hung from hinges 6 ft apart and 6 inches from the top and bottom of the door. The door weighs 60 lb, its center of gravity is at its center, and each hinge carries half the weight of the door. Find the horizontal component of the force exerted on the door at each hinge.

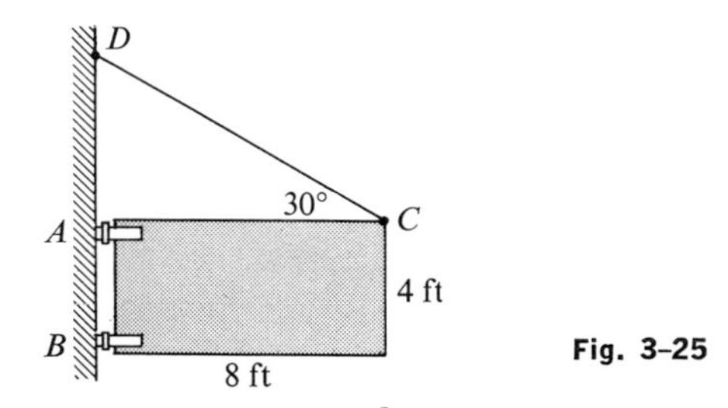

Fig. 3-25

3-17 A gate 8 ft long and 4 ft high weighs 80 lb. Its center of gravity is at its center, and it is hinged at A and B. To relieve the strain on the top hinge a wire CD is connected as shown in Fig. 3-25. The tension in CD is increased until the horizontal force at hinge A is zero. (a) What is the tension in the wire CD? (b) What is the magnitude of the horizontal component of force at hinge B? (c) What is the combined vertical force exerted by hinges A and B?

3-18 A uniform rectangular block, 2 ft high and 1 ft wide, rests on a plank AB, as in Fig. 3-26. The coefficient of static friction between block and plank is 0.40. (a) In a diagram drawn to scale, show the line of action of the resultant normal force exerted on the block by the plank when the angle $\theta = 15°$. (b) If end B of the plank is slowly raised, will the block start to slide down the plank before it tips over? Find the angle θ at which it

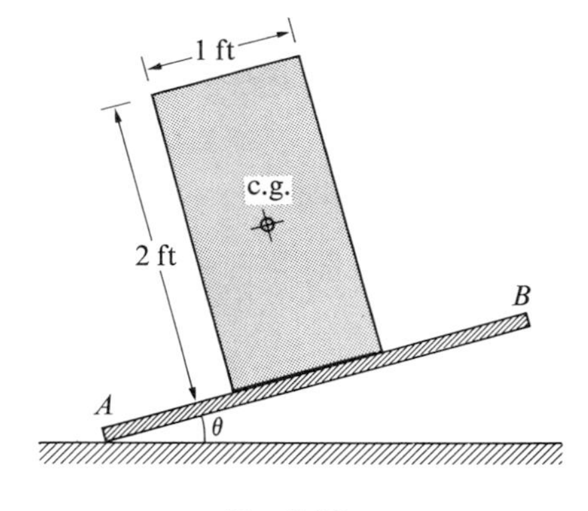

Fig. 3-26

starts to slide, or at which it tips. (c) What would be the answer to part (b) if the coefficient of static friction were 0.60? If it were 0.50?

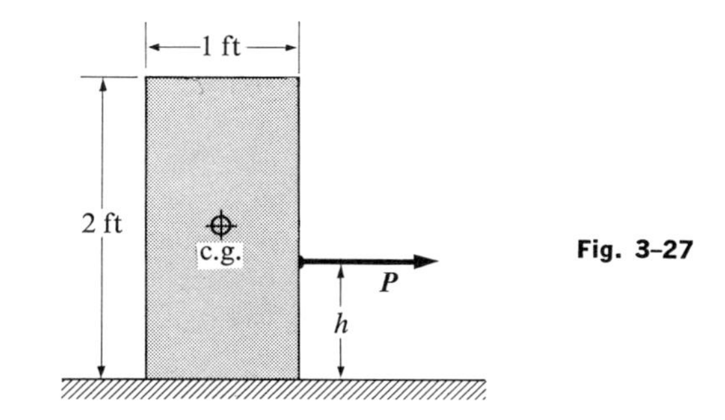

Fig. 3-27

3-19 A rectangular block 1 ft wide and 2 ft high is dragged to the right along a level surface at constant speed by a horizontal force $\boldsymbol{P}$, as shown in Fig. 3-27. The coefficient of sliding friction is 0.40, the block weighs 50 lb, and its center of gravity is at its center. (a) Find the magnitude of the force $\boldsymbol{P}$. (b) Find the position of the line of action of the normal force $\boldsymbol{N}$ exerted on the block by the surface, if the height $h = 6$ inches. (c) Find the value of h at which the block just starts to tip.

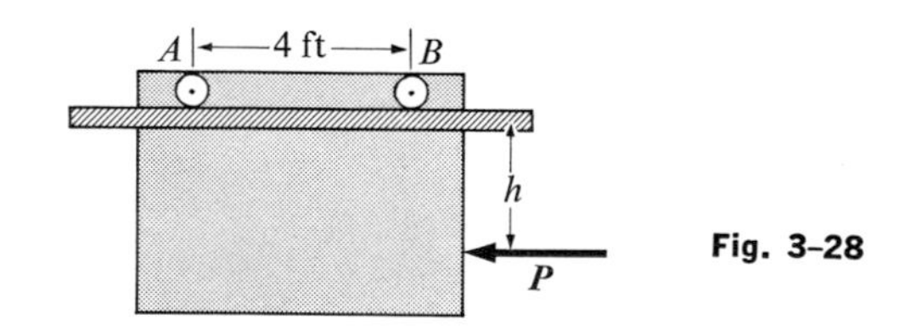

Fig. 3-28

3-20 A garage door is mounted on an overhead rail as in Fig. 3-28. The wheels at A and B have rusted so that they do not roll, but slide along the track. The coefficient of sliding friction is 0.5. The distance between the wheels is 4 ft, and each is 1 ft in from the vertical sides of the door. The door is symmetrical and weighs 160 lb. It is pushed to the left at constant velocity by a horizontal force $\boldsymbol{P}$. (a) If the distance h is 3 ft, what is the vertical component of the force exerted on each wheel by the track? (b) Find the maximum value h can have without causing one wheel to leave the track.

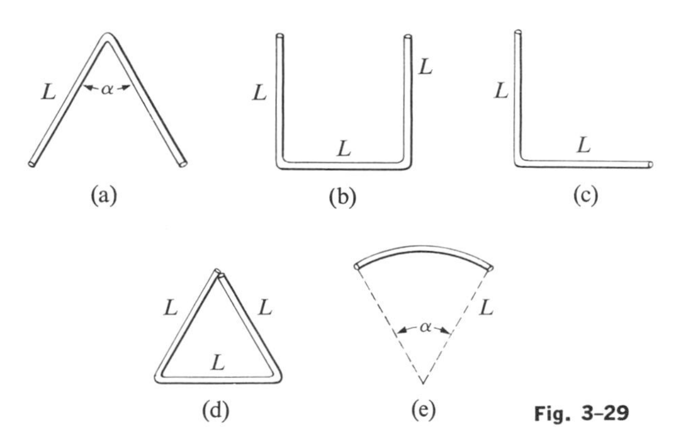

Fig. 3-29

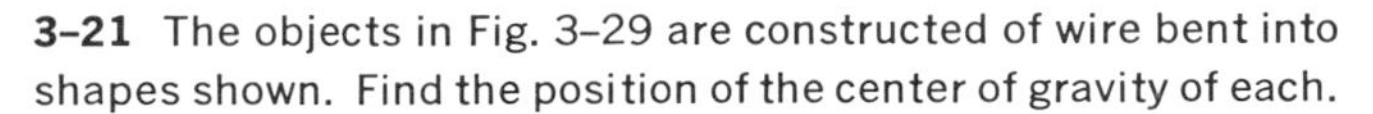

3-21 The objects in Fig. 3-29 are constructed of wire bent into shapes shown. Find the position of the center of gravity of each.

3-22 Figure 3-30 is a top view of a triangular flat plate ABC resting on a horizontal frictionless surface. The length of side AB is 20 cm and that of CB is 15 cm. The plate is acted on by a couple consisting of the horizontal forces $\boldsymbol{F}_1$ and $\boldsymbol{F}_1'$, each of magnitude 150 units. (a) In part (a) of the figure, the plate is kept in equilibrium by a second couple consisting of the forces $\boldsymbol{F}_2$ and $\boldsymbol{F}_2'$. Find the magnitude of each if the length $ab = 10$ cm. (b) Show that the plate is still in equilibrium (that is, the first and second conditions of equilibrium are satisfied) if the couple formed by $\boldsymbol{F}_2$ and $\boldsymbol{F}_2'$ is displaced to the right, as in part (b). (c) Can equilibrium be maintained by the couple formed by the forces $\boldsymbol{F}_3$ and $\boldsymbol{F}_3'$ in part (c) of the diagram? If so, find the magnitude of each. (d) Show in a diagram how the plate could be kept in equilibrium by a couple consisting of two forces perpendicular to side AC and applied at corners A and C. Find the magnitude of the required forces.

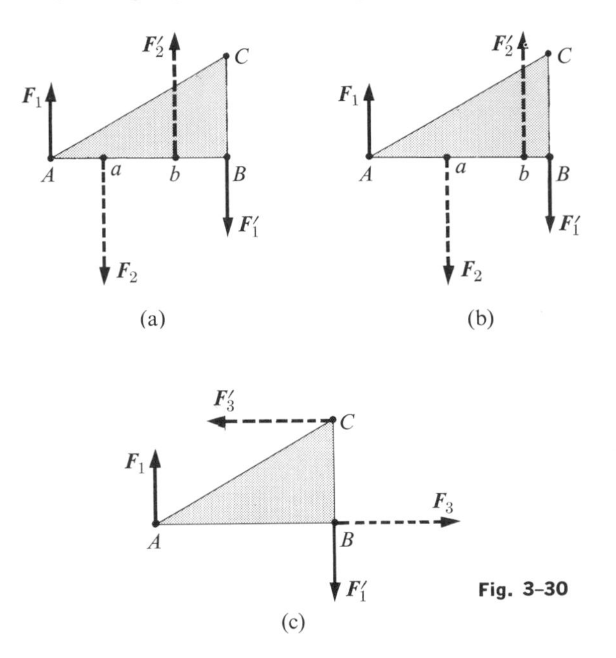

Fig. 3-30

3-23 Where is the center of gravity of the system made up of the earth and the moon?

4 Rectilinear Motion

4–1 MOTION

At the beginning of Chapter 1 it was stated that mechanics deals with the relations of force, matter, and motion. The preceding chapters have been concerned with forces, and we are now ready to discuss the mathematical methods of describing motion. This branch of mechanics is called *kinematics*.

Motion may be defined as a continuous change of position. In most actual motions, different points in a body move along different paths. The complete motion is known if we know how each point in the body moves, so to begin with we consider only a moving point, or a very small body called a *particle*.

The position of a particle is conveniently specified by its projections onto the three axes of a rectangular coordinate system. As the particle moves along any path in space, its projections move in straight lines along the three axes. The actual motion can be reconstructed from the motions of these three projections, so we shall begin by discussing the motion of a single particle along a straight line, or *rectilinear motion*.

4–2 AVERAGE VELOCITY

Consider a particle moving along the x-axis, as in Fig. 4–1(a). The curve in Fig. 4–1(b) is a graph of its coordinate x plotted as a function of time t. At a time t_1 the particle is at point P in Fig. 4–1(a), where its coordinate is x_1, and at a later time t_2 it is at point Q, whose coordinate is x_2. The corresponding points on the coordinate-time graph in part (b) are lettered p and q.

The *displacement* of a particle as it moves from one point of its path to another is defined as the vector $\Delta\mathbf{x}$ drawn from the first point to the second. Thus in Fig. 4–1(a) the vector PQ, of magnitude $x_2 - x_1 = \Delta x$, is the displacement. The *average velocity* of the particle is defined as the ratio of the displacement to the time interval $t_2 - t_1 = \Delta t$. We shall represent average velocity by the symbol $\bar{\mathbf{v}}$ (the bar signifying an average value).

$$\bar{\mathbf{v}} = \frac{\Delta \mathbf{x}}{\Delta t}.$$

Average velocity is a vector, since the ratio of a vector to a scalar is itself a vector. Its direction is the

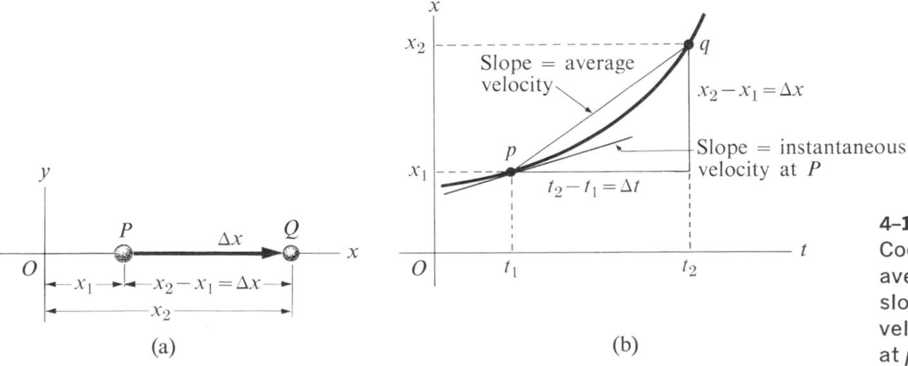

4–1 (a) Particle moving on the x-axis. (b) Coordinate-time graph of the motion. The average velocity between t_1 and t_2 equals the slope of the chord pq. The instantaneous velocity at p equals the slope of the tangent at p.

same as that of the displacement vector. The magnitude of the average velocity is therefore

$$\bar{v} = \frac{x_2 - x_1}{t_2 - t_1} = \frac{\Delta x}{\Delta t}. \tag{4-1}$$

In Fig. 4-1(b), the average velocity is represented by the slope of the chord *pq* (due allowance being made for the scales to which x and t are plotted), since the slope is the ratio of the "rise," $x_2 - x_1$ or Δx, to the "run," $t_2 - t_1$ or Δt.

Equation (4-1) can be cleared of fractions and written

$$x_2 - x_1 = \bar{v}(t_2 - t_1). \tag{4-2}$$

Since our time-measuring device can be started at any instant, we can let $t_1 = 0$ and let t_2 be any arbitrary time t. Then if x_0 is the coordinate when $t = 0$ (x_0 is called the *initial position*), and x is the coordinate at time t, Eq. (4-2) becomes

$$x - x_0 = \bar{v}t. \tag{4-3}$$

If the particle is at the origin when $t = 0$, then $x_0 = 0$ and Eq. (4-3) simplifies further to

$$x = \bar{v}t. \tag{4-4}$$

4-3 INSTANTANEOUS VELOCITY

The velocity of a particle at some one instant of time, or at some one point of its path, is called its *instantaneous velocity*. This concept requires careful definition.

Suppose we wish to find the instantaneous velocity of the particle in Fig. 4-1 at the point P. The average velocity between points P and Q is associated with the entire displacement $\Delta \mathbf{x}$, and with the entire time interval Δt. Imagine the second point Q to be taken closer and closer to the first point P, and let the average velocity be computed over these shorter and shorter displacements and time intervals. The instantaneous velocity at the first point can then be defined as the *limiting value* of the average velocity when the second point is taken closer and closer to the first. Although the displacement then becomes extremely small, the time interval by which it must be divided becomes small also and the quotient therefore is not necessarily a small quantity.

In the notation of calculus, the limiting value of $\Delta x/\Delta t$, as Δt approaches zero, is written dx/dt and is called the *derivative* of x with respect to t.

Then if $\mathbf{v}$ represents the instantaneous velocity, its magnitude is

$$\boxed{v = \lim_{\Delta t \to 0} \frac{\Delta x}{\Delta t} = \frac{dx}{dt}.} \tag{4-5}$$

Instantaneous velocity is also a vector, whose direction is the limiting direction of the displacement vector $\Delta \mathbf{x}$. Since Δt is necessarily positive, it follows that v has the same algebraic sign as Δx. Hence a positive velocity indicates motion toward the right along the x-axis, if we use the usual convention of signs.

As point Q approaches point P in Fig. 4-1(a), point q approaches point p in Fig. 4-1(b). In the limit, the slope of the chord *pq* equals the slope of the *tangent* to the curve at point p, due allowance being made for the scales to which x and t are plotted. *The instantaneous velocity at any point of a coordinate-time graph therefore equals the slope of the tangent to the graph at that point.* If the tangent slopes upward to the right, its slope is positive, the velocity is positive, and the motion is toward the right. If the tangent slopes downward to the right, the velocity is negative. At a point where the tangent is horizontal, its slope is zero and the velocity is zero.

If we express distance in feet and time in seconds, velocity is expressed in *feet per second* (ft s^{-1}). Other common units of velocity are meters per second (m s^{-1}), centimeters per second (cm s^{-1}), miles per hour (mi hr^{-1}), and knots (1 knot = 1 nautical mile per hour).

Example. Suppose the motion of the particle in Fig. 4-1 is described by the equation $x = a + bt^2$, where $a = 20$ cm and $b = 4$ cm s^{-2}. (a) Find the displacement of the particle in the time interval between $t_1 = 2$ s and $t_2 = 5$ s.

At time $t_1 = 2$ s,

$$x_1 = 20\ \text{cm} + 4\ \text{cm s}^{-2} \times (2\ \text{s})^2 = 36\ \text{cm}.$$

At time $t_2 = 5$ s,

$$x_2 = 20\ \text{cm} + 4\ \text{cm s}^{-2} \times (5\ \text{s})^2 = 120\ \text{cm}.$$

The displacement is therefore

$$x_2 - x_1 = 120\ \text{cm} - 36\ \text{cm} = 84\ \text{cm}.$$

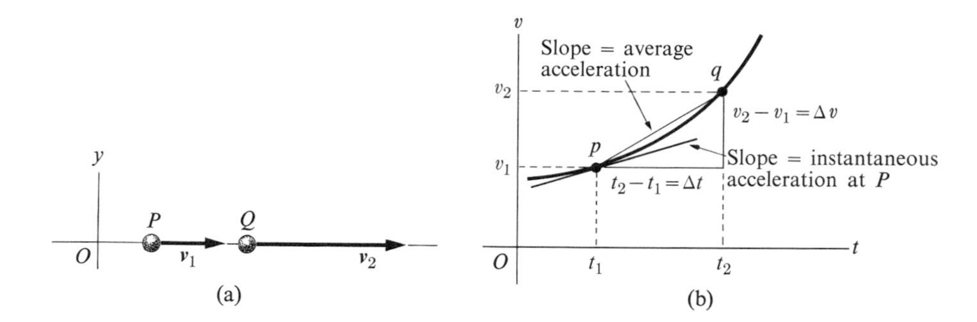

4–2 (a) Particle moving on the x-axis. (b) Velocity-time graph of the motion. The average acceleration between t_1 and t_2 equals the slope of the chord pq. The instantaneous acceleration at p equals the slope of the tangent at p.

(b) Find the average velocity in this time interval.

$$\bar{v} = \frac{x_2 - x_1}{t_2 - t_1} = \frac{84 \text{ cm}}{3 \text{ s}} = 28 \text{ cm s}^{-1}.$$

This corresponds to the slope of the chord pq in Fig. 4–1(b).

(c) Find the instantaneous velocity at time $t_1 = 2$ s. The instantaneous velocity is given by

$$v = \frac{dx}{dt} = \frac{d}{dt}(a + bt^2) = 2bt.$$

Hence at time $t_1 = 2$ s,

$$v = 2 \times 4 \text{ cm s}^{-2} \times 2 \text{ s} = 16 \text{ cm s}^{-1}.$$

This corresponds to the slope of the tangent at point p in Fig. 4–1(b).

The term *speed* has two different meanings. It is sometimes used to mean the *magnitude* of the instantaneous velocity. In this sense, two automobiles traveling at 50 mi hr^{-1}, one north and the other south, are both said to have a speed of 50 mi hr^{-1}. In another sense, the *average speed* of a body means the total length of path covered, divided by the elapsed time. Thus if an automobile travels a total distance of 90 mi in 3 hr, its average speed is said to be 30 mi hr^{-1} even if the trip starts and ends at the same point. The average *velocity*, in the latter case, would be zero, since the displacement is zero.

4–4 AVERAGE AND INSTANTANEOUS ACCELERATION

Except in certain special cases, the velocity of a moving body changes continuously as the motion proceeds. When this is the case the body is said to move with *accelerated motion*, or to have an *acceleration*.

Figure 4–2(a) shows a particle moving along the x-axis. The vector $\mathbf{v}_1$ represents its instantaneous velocity at point P, and the vector $\mathbf{v}_2$ represents its instantaneous velocity at point Q. Figure 4–2(b) is a graph of the instantaneous velocity v plotted as a function of time, points p and q corresponding to P and Q in part (a).

The *average acceleration* of the particle as it moves from P to Q is defined as the ratio of the change in velocity to the elapsed time.

$$\bar{\mathbf{a}} = \frac{\mathbf{v}_2 - \mathbf{v}_1}{t_2 - t_1} = \frac{\Delta \mathbf{v}}{\Delta t}, \tag{4–6}$$

where t_1 and t_2 are the times corresponding to the velocities $\mathbf{v}_1$ and $\mathbf{v}_2$. Since $\mathbf{v}_1$ and $\mathbf{v}_2$ are vectors, the quantity $\mathbf{v}_2 - \mathbf{v}_1$ is a *vector difference* and must be found by the methods explained in Section 1–9. However, since in rectilinear motion both vectors lie in the same straight line, the magnitude of the vector difference in this special case equals the difference between the magnitudes of the vectors. The more general case, in which $\mathbf{v}_1$ and $\mathbf{v}_2$ are not in the same direction, will be considered in Chapter 6.

In Fig. 4–2(b), the magnitude of the average acceleration is represented by the slope of the chord pq.

The *instantaneous acceleration* of a body, that is, its acceleration at some one instant of time or at some one point of its path, is defined in the same way as instantaneous velocity. Let the second point Q in Fig. 4–2(a) be taken closer and closer to the first point P, and let the average acceleration be computed over shorter and shorter intervals of time. The instantaneous acceleration at the first point is defined as the limiting value of the average acceleration when the second point is taken closer and closer to the first:

$$a = \lim_{\Delta t \to 0} \frac{\Delta v}{\Delta t} = \frac{dv}{dt}. \tag{4-7}$$

The direction of the instantaneous acceleration is the limiting direction of the vector change in velocity, $\Delta \mathbf{v}$.

Instantaneous acceleration plays an important part in the laws of mechanics. Average acceleration is less frequently used. Hence from now on when the term "acceleration" is used we shall understand it to mean instantaneous acceleration.

The definition of acceleration just given applies to motion along any path, straight or curved. When a particle moves in a curved path, the *direction* of its velocity changes and this change in direction also gives rise to an acceleration, as will be explained in Chapter 6.

As point Q approaches point P in Fig. 4-2(a), point q approaches point p in Fig. 4-2(b) and the slope of the chord pq approaches the slope of the tangent to the velocity-time graph at point p. *The instantaneous acceleration at any point of the graph therefore equals the slope of the tangent to the graph at that point.*

The acceleration $a = dv/dt$ can be expressed in various ways. Since $v = dx/dt$, it follows that

$$a = \frac{dv}{dt} = \frac{d}{dt}\left(\frac{dx}{dt}\right) = \frac{d^2x}{dt^2}.$$

The acceleration is therefore the *second* derivative of the coordinate with respect to time.

We can also use the chain rule and write

$$a = \frac{dv}{dt} = \frac{dv}{dx}\frac{dx}{dt} = v\frac{dv}{dx}, \tag{4-8}$$

which expresses the acceleration in terms of the *space* rate of change of velocity, dv/dx.

If we express velocity in feet per second and time in seconds, acceleration is expressed in feet per second, per second (ft s^{-1} s^{-1}). This is usually written as ft s^{-2}, and is read "feet per second squared." Other common units of acceleration are meters per second squared (m s^{-2}) and centimeters per second squared (cm s^{-2}).

When the absolute value of the magnitude of the velocity of a body is decreasing (in other words, when the body is slowing down), the body is said to be *decelerated* or to have a *deceleration*.

Example. Suppose the velocity of the particle in Fig. 4-2 is given by the equation

$$v = m + nt^2,$$

where $m = 10$ cm s^{-1} and $n = 2$ cm s^{-3}. (a) Find the change in velocity of the particle in the time interval between $t_1 = 2$ s and $t_2 = 5$ s.

At time $t_1 = 2$ s,

$$v_1 = 10 \text{ cm s}^{-1} + 2 \text{ cm s}^{-3} \times (2 \text{ s})^2$$
$$= 18 \text{ cm s}^{-1}.$$

At time $t_2 = 5$ s,

$$v_2 = 10 \text{ cm s}^{-1} + 2 \text{ cm s}^{-3} \times (5 \text{ s})^2$$
$$= 60 \text{ cm s}^{-1}.$$

The change in velocity is therefore

$$v_2 - v_1 = 60 \text{ cm s}^{-1} - 18 \text{ cm s}^{-1}$$
$$= 42 \text{ cm s}^{-1}.$$

(b) Find the average acceleration in this time interval.

$$\bar{a} = \frac{v_2 - v_1}{t_2 - t_1} = \frac{42 \text{ cm s}^{-1}}{3 \text{ s}}$$
$$= 14 \text{ cm s}^{-2}.$$

This corresponds to the slope of the chord pq in Fig. 4-2(b).

(c) Find the instantaneous acceleration at time $t_1 = 2$ s.

The instantaneous acceleration is given by

$$a = \frac{dv}{dt} = \frac{d}{dt}(m + nt^2)$$
$$= 2nt.$$

Hence when $t = 2$ s,

$$a = 2 \times 2 \text{ cm s}^{-3} \times 2 \text{ s}$$
$$= 8 \text{ cm s}^{-2}.$$

This corresponds to the slope of the tangent at point p in Fig. 4-2(b).

4-5 RECTILINEAR MOTION WITH CONSTANT ACCELERATION

The simplest kind of accelerated motion is rectilinear motion in which the acceleration is constant, that is, in which the velocity changes at the same rate throughout the motion. The velocity-time graph is then a straight line as in Fig. 4-3, the velocity increasing by equal amounts in equal intervals of time. The slope of a chord between any two points on the line is the same as the slope of a tangent at any point, and the average and instantaneous accelerations are equal. Hence in Eq. (4-6) the average acceleration $\bar{a}$ can be replaced by the constant acceleration a, and we have

$$a = \frac{v_2 - v_1}{t_2 - t_1}.$$

Now let $t_1 = 0$ and let t_2 be any arbitrary time t. Let v_0 represent the velocity when $t = 0$ (v_0 is called the *initial* velocity), and let v be the velocity at time t. Then the preceding equation becomes

$$a = \frac{v - v_0}{t - 0},$$

or

$$\boxed{v = v_0 + at.} \qquad (4\text{-}9)$$

This equation can be interpreted as follows. The acceleration a is the constant rate of change of velocity, or the change per unit time. The term at is the product of the change in velocity per unit time, a, and the duration of the time interval, t. Therefore it equals the total change in velocity. The velocity v at the time t then equals the velocity v_0 at the time $t = 0$, plus the change in velocity at. Graphically, the ordinate v at time t, in Fig. 4-3, can be considered as the sum of two segments: one of length v_0 equal to the initial velocity, the other of length at equal to the change in velocity in time t.

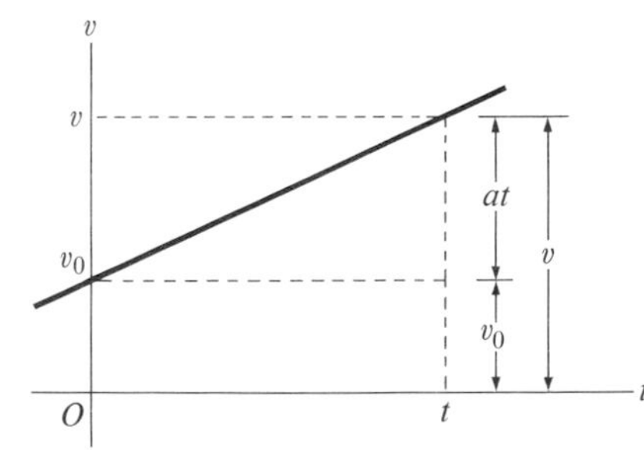

4-3 Velocity-time graph for rectilinear motion with constant acceleration.

To find the displacement of a particle moving with constant acceleration, we make use of the fact that when the acceleration is constant and the velocity-time graph is a straight line, as in Fig. 4-3, the average velocity in any time interval equals one-half the sum of the velocities at the beginning and the end of the interval. Hence the average velocity between zero and t is

$$\bar{v} = \frac{v_0 + v}{2}. \qquad (4\text{-}10)$$

This is *not* true, in general, when the acceleration is not constant and the velocity-time graph is curved as in Fig. 4-2.

We have shown that for a particle which is at the origin when $t = 0$, the coordinate x at any time t is

$$x = \bar{v}t, \qquad (4\text{-}11)$$

where $\bar{v}$ is the average velocity. Hence from the two preceding equations,

$$\boxed{x = \frac{v_0 + v}{2} \times t.} \qquad (4\text{-}12)$$

Two more very useful equations can be obtained from Eqs. (4-9) and (4-12), first by eliminating v and then by eliminating t. When we substitute in Eq. (4-12) the expression for v in Eq. (4-9), we get

$$x = \frac{v_0 + v_0 + at}{2} \times t,$$

or

$$\boxed{x = v_0 t + \tfrac{1}{2}at^2.} \qquad (4\text{-}13)$$

When Eq. (4-9) is solved for t and the result substituted in Eq. (4-12), we have

$$x = \frac{v_0 + v}{2} \times \frac{v - v_0}{a} = \frac{v^2 - v_0^2}{2a}$$

or, finally,

$$\boxed{v^2 = v_0^2 + 2ax.} \qquad (4\text{-}14)$$

Equations (4–9), (4–12), (4–13), and (4–14) are the *equations of motion with constant acceleration*, for the special case where the particle is at the origin when $t = 0$.

The curve in Fig. 4–4 is the coordinate-time graph for motion with constant acceleration. That is, it is a graph of Eq. (4–13). The curve is a *parabola*. The slope of the tangent at $t = 0$ equals the initial velocity v_0, and the slope of the tangent at time t equals the velocity v at that time. It is evident that the slope continually increases, and measurements would show that the *rate* of increase is constant, that is, that the acceleration is constant.

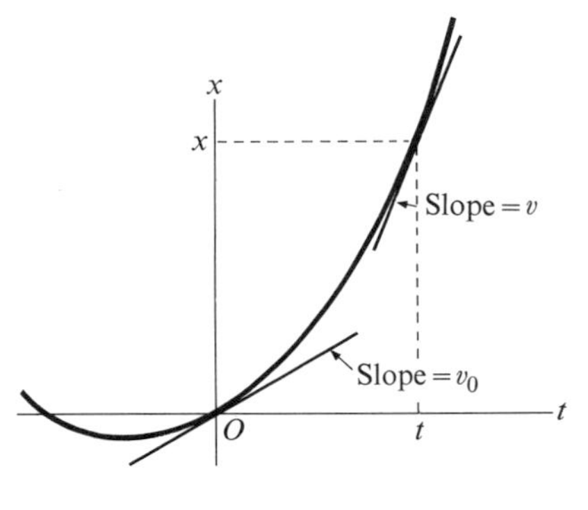

Fig. 4–4

A special case of motion with constant acceleration is that in which the acceleration is zero. The *velocity* is then constant and the equations of motion become simply

$$v = \text{constant},$$
$$x = vt.$$

4-6 VELOCITY AND COORDINATE BY INTEGRATION

If the coordinate x of a particle moving on the x-axis is given as a function of time, the velocity can be found by differentiation, from the definition $v = dx/dt$. A second differentiation gives the acceleration, since $a = dv/dt$. We now consider the converse process: given the acceleration, to find the velocity and the coordinate. This can be done by the methods of *integral* calculus. We shall discuss first the indefinite and then the definite integral.

Suppose we are given the acceleration $a(t)$ as a function of time. Then, since

$$\frac{dv}{dt} = a(t),$$

we have

$$dv = a(t)\,dt, \qquad \int dv = \int a(t)\,dt,$$

and

$$v = \int a(t)\,dt + C_1, \tag{4–15}$$

where C_1 is an integration constant whose value can be determined if the velocity is known at any time. It is customary to express C_1 in terms of the velocity v_0 when $t = 0$.

When the integral above has been evaluated, we have the velocity $v(t)$ as a function of time. Then, since

$$\frac{dx}{dt} = v(t),$$

we have

$$dx = v(t)\,dt, \qquad \int dx = \int v(t)\,dt,$$
$$x = \int v(t)\,dt + C_2, \tag{4–16}$$

where C_2 is a second integration constant whose value can be determined if the coordinate is known at any time. It is customary to express C_2 in terms of the coordinate x_0 when $t = 0$.

If the acceleration is given as a function of x, we can use Eq. (4–8):

$$v\frac{dv}{dx} = a(x), \qquad \int v\,dv = \int a(x)\,dx,$$
$$\frac{v^2}{2} = \int a(x)\,dx + C_3. \tag{4–17}$$

Example. Derive the equations of motion with constant acceleration, using the indefinite integral.

If a is constant, then from Eq. (4–15),

$$v = at + C_1.$$

But $v = v_0$ when $t = 0$, so

$$v_0 = 0 + C_1$$

and

$$v = v_0 + at,$$

which is Eq. (4–9).

From Eq. (4–16), when a is constant,

$$x = \int (v_0 + at)\, dt = v_0 t + \tfrac{1}{2}at^2 + C_2.$$

If $x = 0$ when $t = 0$, then $C_2 = 0$ and

$$x = v_0 t + \tfrac{1}{2}at^2,$$

which is Eq. (4–13).

From Eq. (4–17), when a is constant,

$$\frac{v^2}{2} = ax + C_3.$$

If $v = v_0$ when $x = 0$, then $C_3 = v_0^2/2$ and

$$v^2 = v_0^2 + 2ax,$$

which is Eq. (4–14).

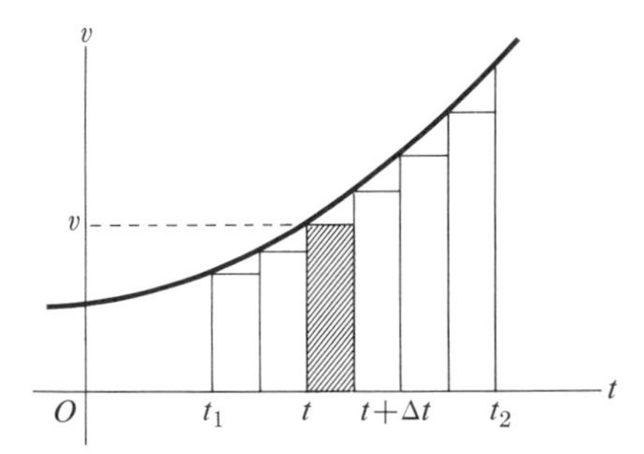

4–5 The area under a velocity-time graph equals the displacement.

Consider next the definite integral. Let the area under the velocity-time graph in Fig. 4–5, between the vertical lines at t_1 and t_2, be subdivided into narrow rectangular strips of width Δt. The ordinate of the graph at any time t equals the instantaneous velocity v at that time. If the velocity remained constant at this value, the displacement Δx in the time interval between t and $t + \Delta t$ would equal $v\,\Delta t$. But this is the *area* of the shaded strip, since its height is v and its width is Δt. The *sum* of the areas of all such rectangles, between t_1 and t_2, is approximately equal to the *total* displacement $x_2 - x_1$ in this time interval:

$$x_2 - x_1 \approx \sum v\,\Delta t.$$

The smaller the time intervals Δt, the more closely does $v\,\Delta t$ approach the actual displacement. In the limit, as Δt approaches zero, the sum of the areas becomes exactly equal to the total area under the curve, and to the total displacement $x_2 - x_1$. The limit of the sum of the areas is the definite integral from t_1 to t_2, so

$$x_2 - x_1 = \int_{t_1}^{t_2} v\, dt. \tag{4–18}$$

The displacement in any time interval is therefore equal to the area between a velocity-time graph and the time axis, bounded by vertical lines at the beginning and end of the interval.

In the same way, the area under an acceleration-time graph can be subdivided into vertical strips of height a and width Δt. If the acceleration remained constant, the change in velocity Δv in time Δt would equal $a\,\Delta t$, the area of a rectangular strip. The total change in velocity, $v_2 - v_1$, in the time interval from t_1 to t_2, is approximately equal to the sum of all such areas.

$$v_2 - v_1 \approx \sum a\,\Delta t.$$

In the limit, as Δt approaches zero,

$$v_2 - v_1 = \int_{t_1}^{t_2} a\, dt. \tag{4–19}$$

The change in velocity in any time interval is therefore equal to the area between an acceleration-time graph and the time axis, bounded by vertical lines at the beginning and end of the interval.

Example. Use the definite integral to find the velocity and coordinate, at any time t, of a body moving on the x-axis with constant acceleration. The initial velocity is v_0 and the initial coordinate is zero.

As the limits of integration, we take $t_1 = 0$ and $t_2 = t$. Then, from Eq. (4–19),

$$v - v_0 = \int_0^t a\, dt = at,$$

which is Eq. (4–9).

From Eq. (4–18),

$$x - 0 = \int_0^t (v_0 + at)\, dt = v_0 t + \tfrac{1}{2}at^2,$$

which is Eq. (4–13).

It is not always necessary to use the methods of integral calculus to find the area under a graph. Figure 4–6 is the velocity-time graph for motion with constant acceleration. The area under the graph, between $t = 0$ and $t = t$, can be subdivided

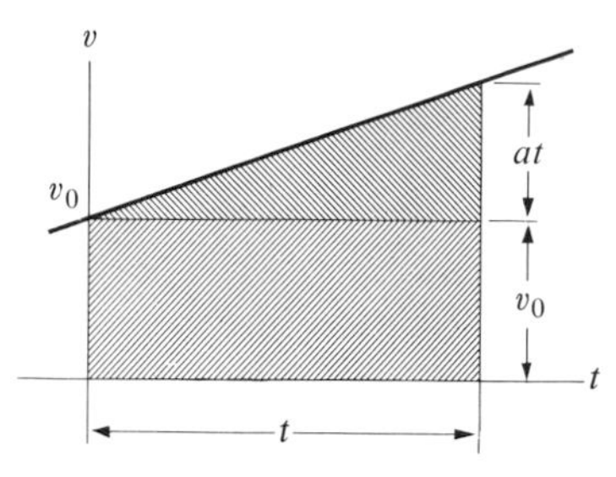

4-6 The area under a velocity-time graph equals the displacement $x - x_0$.

into a rectangle and a triangle. The area of the rectangle is $v_0 t$ and that of the triangle is $\frac{1}{2} \times t \times at = \frac{1}{2}at^2$. Since the displacement equals the total area,

$$x - x_0 = v_0 t + \tfrac{1}{2}at^2,$$

which reduces to Eq. (4-13) if $x_0 = 0$.

4-7 FREELY FALLING BODIES

The most common example of motion with (nearly) constant acceleration is that of a body falling toward the earth. In the absence of air resistance it is found that all bodies, regardless of their size or weight, fall with the same acceleration at the same point on the earth's surface, and if the distance covered is not too great, the acceleration remains constant throughout the fall. The effect of air resistance and the decrease in acceleration with altitude will be neglected. This idealized motion is spoken of as "free fall," although the term includes rising as well as falling.

The acceleration of a freely falling body is called the acceleration due to gravity, or the acceleration of gravity, and is denoted by the letter g. At or near the earth's surface its magnitude is approximately 32 ft s^{-2}, 9.8 m s^{-2}, or 980 cm s^{-2}. More precise values, and small variations with latitude and elevation, will be considered later.

Note. The quantity "g" is sometimes referred to simply as "gravity," or as "the force of gravity," both of which are incorrect. "Gravity" is a phenomenon, and the "force of gravity" means the force with which the earth attracts a body, otherwise known as the weight of the body. The letter "g" represents the *acceleration* caused by the force resulting from the phenomenon of gravity.

Example 1. A body is released from rest and falls freely. Compute its position and velocity after 1, 2, 3, and 4 s. Take the origin O at the elevation of the starting point, the y-axis vertical, and the upward direction as positive.

The initial coordinate y_0 and the initial velocity v_0 are both zero. The acceleration is downward, in the negative y-direction, so $a = -g = -32$ ft s^{-2}.

From Eqs. (4-13) and (4-9),

$$y = v_0 t + \tfrac{1}{2}at^2 = 0 - \tfrac{1}{2}gt^2 = -16 \text{ ft s}^{-2} \times t^2,$$
$$v = v_0 + at = 0 - gt = -32 \text{ ft s}^{-2} \times t.$$

When $t = 1$ s, $y = -16$ ft s$^{-2} \times 1$ s$^2 = -16$ ft, and $v = -32$ ft s$^{-2} \times 1$ s $= -32$ ft s^{-1}. The body is therefore 16 ft below the origin (y is negative) and has a downward velocity (v is negative) of magnitude 32 ft s^{-1}.

The position and velocity at 2, 3, and 4 s are found in the same way. The results are illustrated in Fig. 4-7.

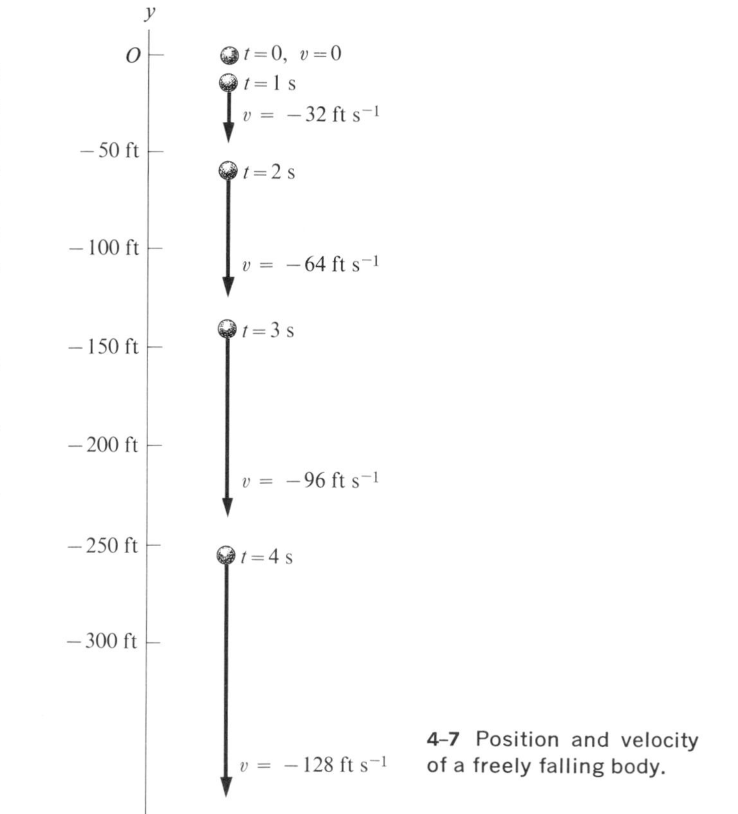

4-7 Position and velocity of a freely falling body.

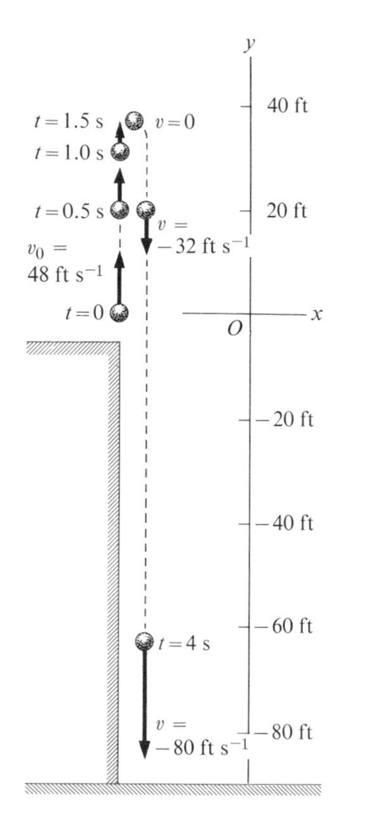

4–8 Position and velocity of a body thrown vertically upward.

Example 2. A ball is thrown (nearly) vertically upward from the cornice of a tall building, leaving the thrower's hand with a speed of 48 ft s^{-1} and just missing the cornice on the way down. (See Fig. 4–8. The dotted line does not represent the actual path of the body.) Find (a) the position and velocity of the ball, 1 s and 4 s after leaving the thrower's hand; (b) the velocity when the ball is 20 ft above its starting point; (c) the maximum height reached and the time at which it is reached. Take the origin at the elevation at which the ball leaves the thrower's hand, the y-axis vertical and positive upward.

The initial position y_0 is zero. The initial velocity v_0 is $+48$ ft s^{-1}, and the acceleration is -32 ft s^{-2}.

The velocity at any time is

$$v = v_0 + at = 48 \text{ ft s}^{-1} - (32 \text{ ft s}^{-2} \times t). \qquad (4\text{–}20)$$

The coordinate at any time is

$$y = v_0 t + \tfrac{1}{2}at^2 = (48 \text{ ft s}^{-1} \times t) - (16 \text{ ft s}^{-2} \times t^2).$$

The velocity at any coordinate is

$$v^2 = v_0^2 + 2ay = (48 \text{ ft s}^{-1})^2 - (64 \text{ ft s}^{-2} \times y). \qquad (4\text{–}21)$$

(a) When $t = 1$ s,

$$y = +32 \text{ ft},$$
$$v = +16 \text{ ft s}^{-1}.$$

The ball is 32 ft above the origin (y is positive) and it has an upward velocity (v is positive) of 16 ft s^{-1}.

When $t = 4$ s,

$$y = -64 \text{ ft}, \qquad v = -80 \text{ ft s}^{-1}.$$

The ball has passed its highest point and is 64 ft *below* the origin (y is negative). It has a *downward* velocity (v is negative) of magnitude 80 ft s^{-1}. Note that it is not necessary to find the highest point reached, or the time at which it was reached. The equations of motion give the position and velocity at *any* time, whether the ball is on the way up or the way down.

(b) When the ball is 20 ft above the origin,

$$y = +20 \text{ ft}$$

and

$$v^2 = 1024 \text{ ft}^2 \text{ s}^{-2}, \qquad v = \pm 32 \text{ ft s}^{-1}.$$

The ball passes this point twice, once on the way up and again on the way down. The velocity on the way up is $+32$ ft s^{-1}, and on the way down it is -32 ft s^{-1}.

(c) At the highest point, $v = 0$. Hence

$$y = +36 \text{ ft}.$$

The time can now be found either from Eq. (4–20), setting $v = 0$, or from Eq. (4–21), setting $y = 36$ ft. From either equation, we get

$$t = 1.5 \text{ s}.$$

Figure 4–9 is a "multiflash" photograph of a freely falling golf ball. This photograph was taken with the aid of the ultra-high-speed stroboscopic light source developed by Dr. Harold E. Edgerton of the Massachusetts Institute of Technology. By means of this source a series of intense flashes of light can be produced. The interval between successive flashes is controllable at will, and the duration of each flash is so short (a few millionths of a second) that there is no blur in the image of even a rapidly moving body. The camera shutter is left open during the entire motion, and as each flash occurs, the

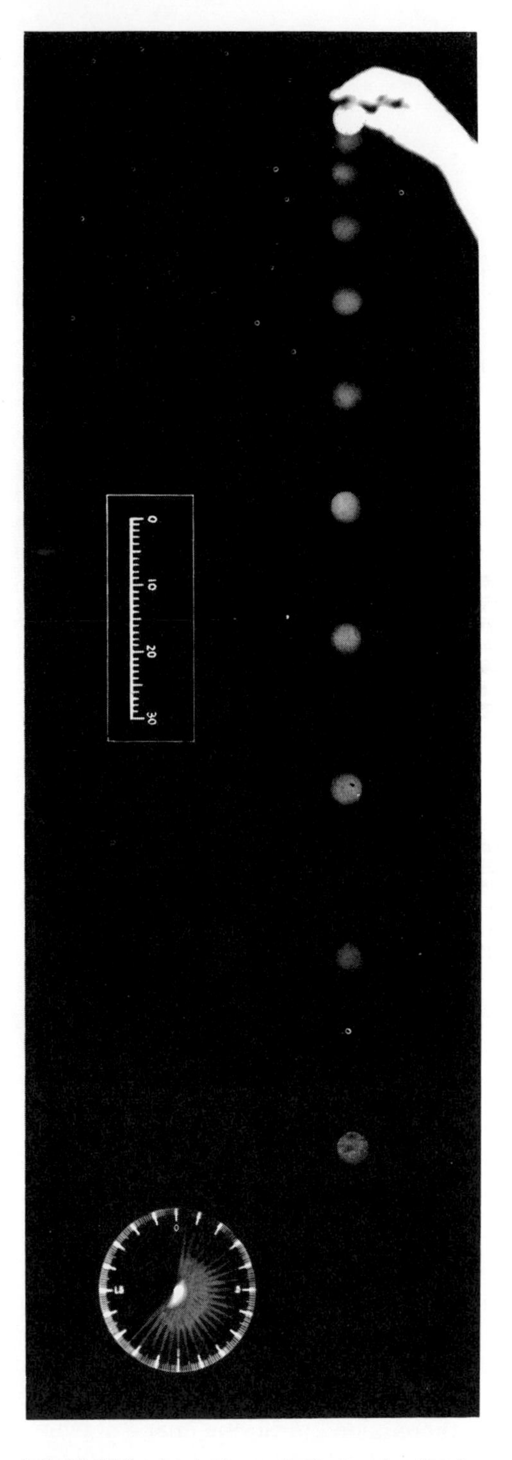

4-9 Multiflash photograph (retouched) of freely falling golf ball.

position of the ball at that instant is recorded on the photographic film.

The equally spaced light flashes subdivide the motion into equal time intervals Δt. Since the time intervals are all equal, the velocity of the ball between any two flashes is directly proportional to the separation of its corresponding images in the photograph. If the velocity were constant, the images would be equally spaced. The increasing separation of the images during the fall shows that the velocity is continually increasing or the motion is accelerated. By comparing two successive displacements of the ball, the *change* in velocity in the corresponding time interval can be found. Careful measurements, preferably on an enlarged print, show that this change in velocity is the same in each time interval. In other words, the motion is one of *constant* acceleration.

4-8 RECTILINEAR MOTION WITH VARIABLE ACCELERATION

Motion with constant acceleration approximates the motion of some falling bodies, and of cars and of airplanes at the start of their journey. There are many important types of motion, however, in which the acceleration is variable, and it is worth while to develop the technique of dealing with such cases. Consider, for example, the motion of a body in the positive x-direction with an acceleration whose direction is opposite that of the velocity and whose magnitude is proportional to the speed. For such a motion

$$a = -kv,$$

where k is a constant. If the initial speed is v_0, how do the speed and the distance vary with the time?

Since $a = dv/dt$, we have

$$\frac{dv}{dt} = -kv, \qquad \text{or} \qquad \frac{dv}{v} = -k\,dt,$$

and since $v = v_0$ when $t = 0$,

$$\int_{v_0}^{v} \frac{dv}{v} = -k \int_0^t dt.$$

Integration yields the result

$$\ln \frac{v}{v_0} = -kt,$$

which may be written as

$$v = v_0 e^{-kt}, \qquad (4\text{-}22)$$

showing that the velocity decays exponentially with the time. Exponential curves are found frequently in many diverse fields of physics. The decrease in activity of a radioactive substance, the discharge of a capacitor, and the dying out of mechanical, acoustical, and electrical oscillations are all described by exponential equations. Equation (4-22) shows that an infinite time would be required to reduce the speed to zero.

To find the distance x as a function of the time, we replace v by dx/dt. Thus

$$\frac{dx}{dt} = v_0 e^{-kt}.$$

Suppose that $x = 0$ when $t = 0$. Then

$$\int_0^x dx = v_0 \int_0^t e^{-kt} dt,$$

and

$$x = -\frac{v_0}{k}\left[e^{-kt}\right]_0^t,$$

$$x = \frac{v_0}{k}(1 - e^{-kt}). \qquad (4\text{-}23)$$

It follows from this equation that although an infinite time is necessary for the body to come to rest, in this infinite time the body will have gone only a finite distance v_0/k.

Other examples of variable acceleration will be found in the problems at the end of this chapter and also in Chapter 5.

4-9 VELOCITY COMPONENTS. RELATIVE VELOCITY

Velocity is a vector quantity involving both magnitude and direction. A velocity may therefore be resolved into components, or a number of velocity components combined into a resultant. As an example of the former process, suppose that a ship is steaming 30° E of N at 20 mi hr^{-1} in still water. Its velocity may be represented by the arrow in Fig. 4-10, and one finds by the usual method that its velocity component toward the east is 10 mi hr^{-1}, while toward the north it is 17.3 mi hr^{-1}.

The velocity of a body, like its position, can be specified only relative to some other body. The second body may be in motion relative to a third, and so on.

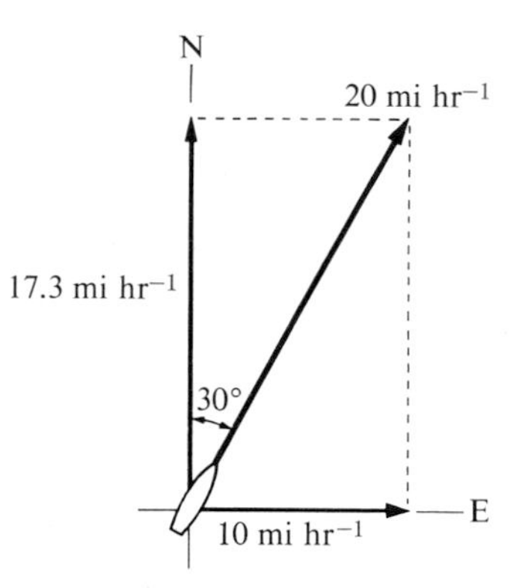

4-10 Resolution of a velocity vector into components.

Thus when we speak of "the velocity of an automobile," we usually mean its velocity relative to the earth. But the earth is in motion relative to the sun, the sun is in motion relative to some other star, and so on.

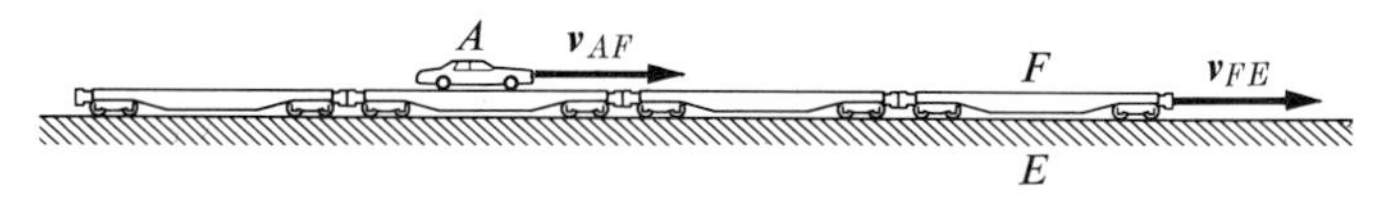

4-11 Vector V_{FE} is the velocity of the flatcars F relative to the earth E, and V_{AF} is the velocity of automobile A relative to the flatcars.

Suppose a long train of flatcars is moving to the right along a straight level track as in Fig. 4-11, and that a daring automobile driver is driving to the right along the flatcars. The vector $\mathbf{v}_{FE}$ in Fig. 4-11 represents the velocity of the flatcars F relative to the earth E, and the vector $\mathbf{v}_{AF}$ the velocity of the automobile A relative to the flatcars F. The velocity of the automobile relative to the earth, v_{AE}, is evidently equal to the sum of the relative velocities v_{AF} and v_{FE}:

$$v_{AE} = v_{AF} + v_{FE}. \qquad (4\text{-}24)$$

Thus if the flatcars are traveling relative to the earth at 30 mi hr^{-1} ($= v_{FE}$) and the automobile is traveling relative to the flatcars at 40 mi hr^{-1} ($= v_{AF}$), the velocity of the automobile relative to the earth (v_{AE}) is 70 mi hr^{-1}.

Imagine now that the flatcars are wide enough so that the automobile can drive on them in any direction. The velocity of the automobile relative to the earth is then the *vector* sum of its velocity relative to the flatcars and the velocity of the flatcars relative to the earth. The preceding equation is therefore a special case of the more general *vector* equation,

$$\mathbf{v}_{AE} = \mathbf{v}_{AF} + \mathbf{v}_{FE}. \tag{4-25}$$

Thus if the automobile were driving transversely across the flatcars at 40 mi hr^{-1}, its velocity relative to the earth would be 50 mi hr^{-1} at an angle of 53° with the railway track.

In the special case in which $\mathbf{v}_{AF}$ and $\mathbf{v}_{FE}$ are along the same line, as in Fig. 4-11, the velocity v_{AE} is the *algebraic* sum of v_{AF} and v_{FE}. Thus if the automobile were traveling to the left with a velocity of 40 mi hr^{-1} relative to the flatcars, $v_{AF} = -40$ mi hr^{-1} and the velocity of the automobile relative to the earth would be -10 mi hr^{-1}. That is, it would be traveling to the left, relative to the earth.

Equation (4-25) can be extended to include any number of relative velocities. For example, if a bug B crawls along the floor of the automobile with a velocity relative to the automobile of $\mathbf{v}_{BA}$, his velocity relative to the earth is the vector sum of his velocity relative to the automobile and that of the velocity of the automobile relative to the earth:

$$\mathbf{v}_{BE} = \mathbf{v}_{BA} + \mathbf{v}_{AE}.$$

When this is combined with Eq. (4-25), we get

$$\mathbf{v}_{BE} = \mathbf{v}_{BA} + \mathbf{v}_{AF} + \mathbf{v}_{FE}. \tag{4-26}$$

This equation illustrates the general rule for combining relative velocities. (1) Write each velocity with a double subscript in the *proper order*, meaning "velocity of (first subscript) relative to (second subscript)." (2) When adding relative velocities, the first letter of any subscript is to be the same as the last letter of the preceding subscript. (3) The first letter of the subscript of the first velocity in the sum, and the second letter of the subscript of the last velocity, are the subscripts, in that order, of the relative velocity represented by the sum. This somewhat lengthy statement should be clear when it is compared with Eq. (4-26).

Any of the relative velocities in an equation like Eq. (4-26) can be transferred from one side of the equation to the other, with sign reversed. Thus Eq. (4-25) can be written

$$\mathbf{v}_{AF} = \mathbf{v}_{AE} - \mathbf{v}_{FE}.$$

The velocity of the automobile relative to the flatcar equals the *vector difference* between the velocities of automobile and flatcar, both relative to the earth.

One more point should be noted. The velocity of body A relative to body B, $\mathbf{v}_{AB}$, is the negative of the velocity of B relative to A, $\mathbf{v}_{BA}$:

$$\mathbf{v}_{AB} = -\mathbf{v}_{BA}.$$

That is, $\mathbf{v}_{AB}$ is equal in magnitude and opposite in direction to $\mathbf{v}_{BA}$. If the automobile in Fig. 4-11 is traveling to the right at 40 mi hr^{-1} relative to the flatcars, the flatcars are traveling to the left at 40 mi hr^{-1}, relative to the automobile.

Example 1. An automobile driver A traveling relative to the earth at 65 mi hr^{-1} on a straight level road is ahead of motorcycle officer B traveling in the same direction at 80 mi hr^{-1}. What is the velocity of B relative to A?

We have given

$$v_{AE} = 65 \text{ mi hr}^{-1}, \qquad v_{BE} = 80 \text{ mi hr}^{-1},$$

and we wish to find v_{BA}.

From the rule for combining velocities (along the same line)

$$v_{BA} = v_{BE} + v_{EA}.$$

But

$$v_{EA} = -v_{AE},$$

so

$$\begin{aligned} v_{BA} &= v_{BE} - v_{AE} \\ &= 80 \text{ mi hr}^{-1} - 65 \text{ mi hr}^{-1} = 15 \text{ mi hr}^{-1}, \end{aligned}$$

and the officer is overtaking the driver at 15 mi hr^{-1}.

Example 2. How would the relative velocity be altered if B were ahead of A?

Not at all. The relative *positions* of the bodies do not matter. The velocity of B relative to A is still $+15$ mi hr^{-1}, but he is now pulling ahead of A at this rate.

Example 3. The compass of an aircraft indicates that it is headed due north, and its airspeed indicator shows that it is moving through the air at 120 mi hr^{-1}. If there is a wind of 50 mi hr^{-1} from west to east, what is the velocity of the aircraft relative to the earth?

Let subscript A refer to the aircraft, and subscript F to the moving air (which now corresponds to the flatcar in Fig. 4-11). Subscript E refers to the earth. We have given

$$\mathbf{v}_{AF} = 120 \text{ mi hr}^{-1}, \text{ due north}$$

$$\mathbf{v}_{FE} = 50 \text{ mi hr}^{-1}, \text{ due east},$$

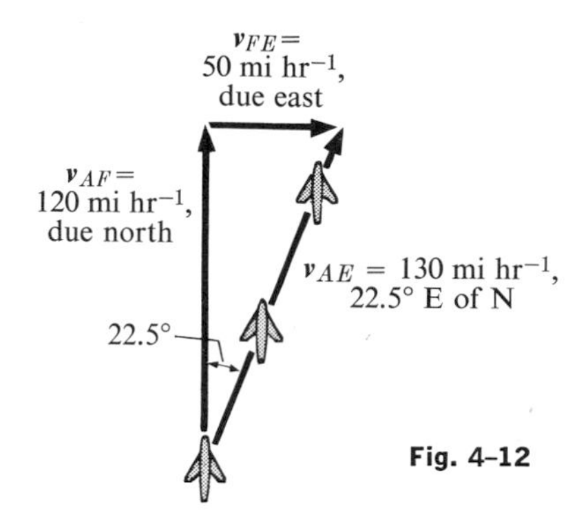

Fig. 4-12

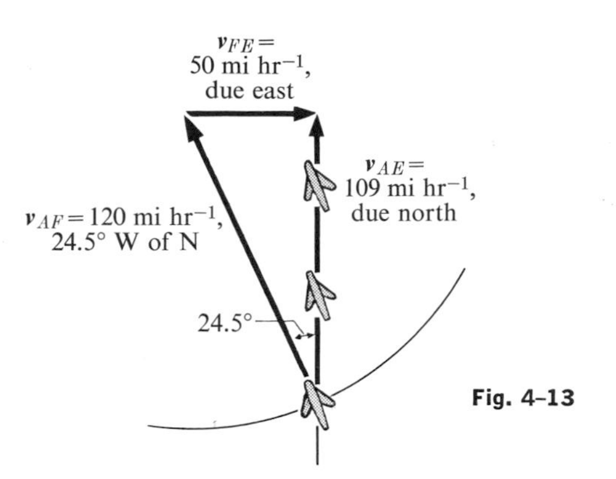

Fig. 4-13

and we wish to find the magnitude and direction of $\mathbf{v}_{AE}$:

$$\mathbf{v}_{AE} = \mathbf{v}_{AF} + \mathbf{v}_{FE}.$$

The three relative velocities are shown in Fig. 4-12. It follows from this diagram that

$$\mathbf{v}_{AE} = 130 \text{ mi hr}^{-1}, 22.5° \text{ E of N}.$$

Example 4. In what direction should the pilot head in order to travel due north? What will then be his velocity relative to the earth? The magnitude of his airspeed, and the wind velocity, are the same as in the preceding example.

We now have given:

$$\mathbf{v}_{AF} = 120 \text{ mi hr}^{-1}, \text{ direction unknown},$$
$$\mathbf{v}_{FE} = 50 \text{ mi hr}^{-1}, \text{ due east},$$

and we wish to find $\mathbf{v}_{AE}$, whose magnitude is unknown but whose direction is due north. (Note that both this and the preceding example require us to determine two unknown quantities. In the former example, these were the *magnitude and direction* of $\mathbf{v}_{AE}$. In this example, the unknowns are the *direction* of $\mathbf{v}_{AF}$ and the *magnitude* of $\mathbf{v}_{AE}$.)

The three relative velocities must still satisfy the *vector equation*

$$\mathbf{v}_{AE} = \mathbf{v}_{AF} + \mathbf{v}_{FE}.$$

The problem can be solved graphically as follows. First construct the vector $\mathbf{v}_{FE}$ (see Fig. 4-13), known in magnitude and direction. At the head of this vector draw a construction line of indefinite length due north, in the known direction of $\mathbf{v}_{AE}$. With the tail of $\mathbf{v}_{FE}$ as a center, construct a circular arc of radius equal to the known magnitude of $\mathbf{v}_{AF}$. Vectors $\mathbf{v}_{AF}$ and $\mathbf{v}_{AE}$ may then be drawn from the point of intersection of this arc and the construction line to the ends of the vector $\mathbf{v}_{FE}$. We find on solving the right triangle that the magnitude of $\mathbf{v}_{AE}$ is 109 mi hr^{-1} and the direction of $\mathbf{v}_{AF}$ is 24.5° W of N. That is, the pilot should head 24.5° W of N, and his ground speed will be 109 mi hr^{-1}.

Problems

4-1 A ball is released from rest and rolls down an inclined plane, requiring 4 s to cover a distance of 100 cm. (a) What was its acceleration in cm s^{-2}? (b) How many centimeters would it have fallen vertically in the same time?

4-2 The "reaction time" of the average automobile driver is about 0.7 s. (The reaction time is the interval between the perception of a signal to stop and the application of the brakes.) If an automobile can decelerate at 16 ft s^{-2}, compute the total distance covered in coming to a stop after a signal is observed: (a) from an initial velocity of 30 mi hr^{-1}, (b) from an initial velocity of 60 mi hr^{-1}.

4-3 At the instant the traffic lights turn green, an automobile that has been waiting at an intersection starts ahead with a constant acceleration of 6 ft s^{-2}. At the same instant a truck, traveling with a constant velocity of 30 ft s^{-1}, overtakes and passes the automobile. (a) How far beyond its starting point will the automobile overtake the truck? (b) How fast will it be traveling?

4-4 The engineer of a passenger train traveling at 100 ft s^{-1} sights a freight train whose caboose is 600 ft ahead on the same track. The freight train is traveling in the same direction as the passenger train with a velocity of 30 ft s^{-1}. The engineer of the passenger train immediately applies the brakes, causing a constant deceleration of 4 ft s^{-2}, while the freight train continues with constant speed. (a) Will there be a collision? (b) If so, where will it take place?

4-5 A sled starts from rest at the top of a hill and slides down with a constant acceleration. The sled is 140 ft from the top of the hill 2 s after passing a point which is 92 ft from the top. Four seconds after passing the 92-ft point it is 198 ft from the top, and 6 s after passing the point it is 266 ft from the top. (a) What is the average velocity of the sled during each of the 2-s intervals after passing the 92-ft point? (b) What is the acceleration of the sled? (c) What was the velocity of the sled when it passed the 92-ft point? (d) How long did it take to go from the top to the 92-ft point? (e) How far did the sled go during the first second after passing the 92-ft point? (f) How long does it take the sled to go from the 92-ft point to the midpoint between the 92-ft and the 140-ft mark? (g) What is the velocity of the sled as it passes the midpoint in part (f)?

4-6 A subway train starts from rest at a station and accelerates at a rate of 4 ft s^{-2} for 10 s. It then runs at constant speed for 30 s, and decelerates at 8 ft s^{-2} until it stops at the next station. Find the *total* distance covered.

4-7 A body starts from rest, moves in a straight line with constant acceleration, and covers a distance of 64 ft in 4 s. (a) What was the final velocity? (b) How long a time was required to cover half the total distance? (c) What was the distance covered in one-half the total time? (d) What was the velocity when half the total distance had been covered? (e) What was the velocity after one-half the total time?

4-8 The speed of an automobile going north is reduced from 45 to 30 mi hr^{-1} in a distance of 264 ft. Find (a) the magnitude and direction of the acceleration, assuming it to be constant, (b) the elapsed time, (c) the distance in which the car can be brought to rest from 30 mi hr^{-1}, assuming the acceleration of part (a).

4-9 An automobile and a truck start from rest at the same instant, with the automobile initially at some distance behind the truck. The truck has a constant acceleration of 4 ft s^{-2} and the automobile an acceleration of 6 ft s^{-2}. The automobile overtakes the truck after the truck has moved 150 ft. (a) How long does it take the auto to overtake the truck? (b) How far was the auto behind the truck initially? (c) What is the velocity of each when they are abreast?

4-10 (a) With what velocity must a ball be thrown vertically upward in order to rise to a height of 50 ft? (b) How long will it be in the air?

4-11 A ball is thrown vertically downward from the top of a building, leaving the thrower's hand with a velocity of 30 ft s^{-1}. (a) What will be its velocity after falling for 2 s? (b) How far will it fall in 2 s? (c) What will be its velocity after falling 30 ft? (d) If it moved a distance of 3 ft while in the thrower's hand, find its acceleration while in his hand. (e) If the ball was released at a point 120 ft above the ground, in how many seconds will it strike the ground? (f) What will the velocity of the ball be when it strikes the ground?

4-12 A balloon, rising vertically with a velocity of 16 ft s^{-1}, releases a sandbag at an instant when the balloon is 64 ft above the ground. (a) Compute the position and velocity of the sandbag at the following times after its release: $\frac{1}{4}$ s, $\frac{1}{2}$ s, 1 s, 2 s. (b) How many seconds after its release will the bag strike the ground? (c) With what velocity will it strike?

4-13 A stone is dropped from the top of a tall cliff, and 1 s later a second stone is thrown vertically downward with a velocity of 60 ft s^{-1}. How far below the top of the cliff will the second stone overtake the first?

4-14 A ball dropped from the cornice of a building takes 0.25 s to pass a window 9 ft high. How far is the top of the window below the cornice?

4-15 A ball is thrown nearly vertically upward from a point near the cornice of a tall building. It just misses the cornice on the way down, and passes a point 160 ft below its starting point 5 s after it leaves the thrower's hand. (a) What was the initial velocity of the ball? (b) How high did it rise above its starting point? (c) What were the magnitude and direction of its velocity at the highest point? (d) What were the magnitude and direction of its acceleration at the highest point? (e) What was the magnitude of its velocity as it passed a point 64 ft below the starting point?

4-16 A juggler performs in a room whose ceiling is 9 ft above the level of his hands. He throws a ball vertically upward so that it just reaches the ceiling. (a) With what initial velocity does he throw the ball? (b) What time is required for the ball to reach the ceiling?

He throws a second ball upward with the same initial velocity, at the instant that the first ball is at the ceiling. (c) How long after the second ball is thrown do the two balls pass each other? (d) When the balls pass each other, how far are they above the juggler's hands?

4-17 An object is thrown vertically upward. It has a speed of 32 ft s^{-1} when it has reached one-half its maximum height. (a) How high does it rise? (b) What is its velocity and acceleration 1 s after it is thrown? (c) 3 s after? (d) What is the average velocity during the first half second?

4-18 A student determined to test the law of gravity for himself walks off a skyscraper 900 ft high, stopwatch in hand, and starts his free fall (zero initial velocity). Five seconds later, Superman arrives at the scene and dives off the roof to save the student. (a) What must Superman's initial velocity be in order that he catch the student just before the ground is reached? (b) What must be the height of the skyscraper so that even Superman can't save him? (Assume that Superman's acceleration is that of any freely falling body.)

4-19 A ball is thrown vertically upward from the ground and a student gazing out of the window sees it moving upward past him at 16 ft s^{-1}. The window is 32 ft above the ground. (a) How high does the ball go above the ground? (b) How long does it take to go from a height of 32 ft to its highest point? (c) Find its velocity and acceleration $\frac{1}{2}$ s after it left the ground, and 2 s after it left the ground.

4-20 A ball is thrown vertically upward from the ground with a velocity of 80 ft s^{-1}. (a) How long will it take to rise to its highest point? (b) How high does the ball rise? (c) How long after projection will the ball have a velocity of 16 ft s^{-1} upward? (d) Of 16 ft s^{-1} downward? (e) When is the displacement of the ball zero? (f) When is the magnitude of the ball's velocity equal to half its velocity of projection? (g) When is the magnitude of the ball's displacement equal to half the greatest height to which it rises? (h) What are the magnitude and direction of the acceleration while the ball is moving upward? (i) While moving downward? (j) When at the highest point?

4-21 A ball rolling on an inclined plane moves with a constant acceleration. One ball is released from rest at the top of an inclined plane 18 m long and reaches the bottom 3 s later. At the same instant that the first ball is released, a second ball is projected upward along the plane from its bottom with a certain initial velocity. The second ball is to travel part way up the plane, stop, and return to the bottom so that it arrives simultaneously with the first ball. (a) Find the acceleration. (b) What must be the initial velocity of the second ball? (c) How far up the plane will it travel?

4-22 The rocket-driven sled Sonic Wind No. 2, used for investigating the physiological effects of large accelerations and decelerations, runs on a straight, level track 3500 ft long. Starting from rest, a speed of 1000 mi hr^{-1} can be reached in 1.8 s. (a) Compute the acceleration, assuming it to be constant. (b) What is the ratio of this acceleration to that of a freely-falling body, g? (c) What is the distance covered? (d) A magazine article states that at the end of a certain run the speed of the sled was decreased from 632 mi hr^{-1} to zero in 1.4 s, and that as the sled decelerated its passenger was subjected to more than 40 times the pull of gravity (that is, the deceleration was greater than $40g$). Are these figures consistent?

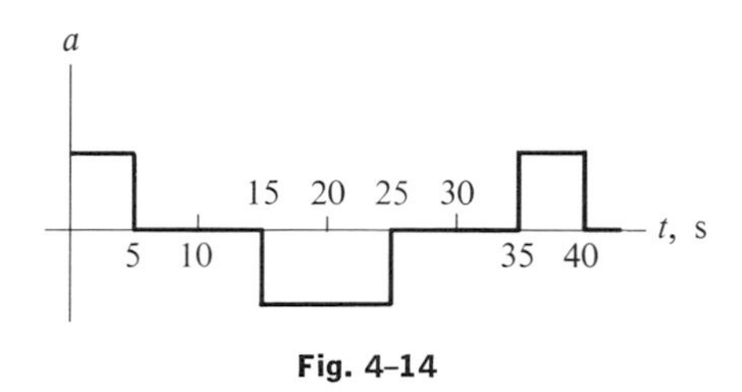

Fig. 4-14

4-23 Figure 4-14 is a graph of the acceleration of a body moving on the x-axis. Sketch the graphs of its velocity and coordinate as functions of time, if $x = v = 0$ when $t = 0$.

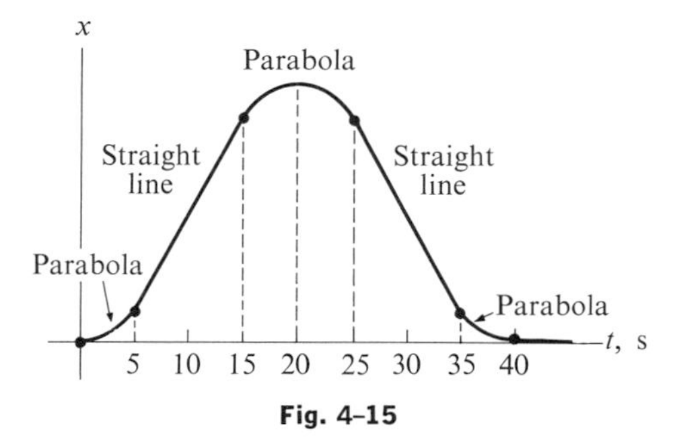

Fig. 4-15

4-24 Figure 4-15 is a graph of the coordinate of a body moving on the x-axis. Sketch the graphs of its velocity and acceleration as functions of the time.

4-25 The first stage of a rocket to launch an earth satellite will, if fired vertically upward, attain a speed of 4000 mi hr^{-1} at a height of 36 mi above the earth's surface, at which point its fuel supply will be exhausted. (a) Assuming constant acceleration, find the time to reach a height of 36 mi. (b) How much higher would the rocket rise if it continued to "coast" vertically upward?

4-26 Suppose the acceleration of gravity were only 3.2 ft s^{-2}, instead of 32 ft s^{-2}. (a) Estimate the height to which you could jump vertically from a standing start. (b) How high could you throw a baseball? (c) Estimate the maximum height of a window from which you would care to jump to a concrete sidewalk below. (Each story of an average building is about 10 ft high.) (d) With what speed, in miles per hour, would you strike the sidewalk? (e) How many seconds would be required?

4–27 After the engine of a moving motorboat is cut off, the boat has an acceleration in the opposite direction to its velocity and directly proportional to the square of its velocity. That is, $dv/dt = -kv^2$, where k is a constant. (a) Show that the magnitude v of the velocity at a time t after the engine is cut off is given by

$$\frac{1}{v} = \frac{1}{v_0} + kt.$$

(b) Show that the distance x traveled in a time t is

$$x = \frac{1}{k} \ln (v_0kt + 1).$$

(c) Show that the velocity after traveling a distance x is

$$v = v_0e^{-kx}.$$

As a numerical example, suppose the engine is cut off when the velocity $v_0 = 20$ ft s^{-1}, and that the velocity decreases to 10 ft s^{-1} in a time of 15 s. (d) Find the numerical value of the constant k, and the unit in which it is expressed. (e) Find the acceleration at the instant the engine is cut off. (f) Construct graphs of x, v, and a for a time of 20 s. Let 1 in. = 5 s horizontally, and 1 in. = 100 ft, 5 ft s^{-1}, and 0.5 ft s^{-2} vertically.

4–28 The equation of motion of a body suspended from a spring and oscillating vertically is $y = A \sin \omega t$, where A and ω are constants. (a) Find the velocity of the body as a function of time. (b) Find its acceleration as a function of time. (c) Find its velocity as a function of its coordinate. (d) Find its acceleration as a function of its coordinate. (e) What is the maximum distance of the body from the origin? (f) What is its maximum velocity? (g) What is its maximum acceleration? (h) Sketch graphs of y, v, and a as functions of time.

4–29 The acceleration of a body suspended from a spring and oscillating vertically is $a = -Ky$, where K is a constant and y is the coordinate measured from the equilibrium position. Suppose that a body moving in this way is given an initial velocity v_0 at the coordinate y_0. Find the expression for the velocity v of the body as a function of its coordinate y. [*Hint:* Use the expression $a = v\, dv/dy$.]

4–30 The motion of a body falling from rest in a resisting medium is described by the equation

$$\frac{dv}{dt} = A - Bv,$$

where A and B are constants. In terms of A and B, find (a) the initial acceleration, and (b) the velocity at which the acceleration becomes zero (the terminal velocity). (c) Show that the velocity at any time t is given by

$$v = \frac{A}{B}(1 - e^{-Bt}).$$

4–31 A passenger on a ship traveling due east with a speed of 18 knots observes that the stream of smoke from the ship's funnels makes an angle of 20° with the ship's wake. The wind is blowing from south to north. Assume that the smoke acquires a velocity (with respect to the earth) equal to the velocity of the wind, as soon as it leaves the funnels. Find the velocity of the wind.

4–32 An airplane pilot wishes to fly due north. A wind of 60 mi hr^{-1} is blowing toward the west. If the flying speed of the plane (its speed in still air) is 180 mi hr^{-1}, in what direction should the pilot head? What is the speed of the plane over the ground? Illustrate with a vector diagram.

4–33 An airplane pilot sets a compass course due west and maintains an air speed of 120 mi hr^{-1}. After flying for $\frac{1}{2}$ hr, he finds himself over a town which is 75 mi west and 20 mi south of his starting point. (a) Find the wind velocity, in magnitude and direction. (b) If the wind velocity were 60 mi hr^{-1} due south, in what direction should the pilot set his course in order to travel due west? Take the same air speed of 120 mi hr^{-1}.

4–34 When a train has a speed of 10 mi hr^{-1} eastward, raindrops which are falling vertically with respect to the earth make traces which are inclined 30° to the vertical on the windows of the train. (a) What is the horizontal component of a drop's velocity with respect to the earth? with respect to the train? (b) What is the velocity of the raindrop with respect to the earth? with respect to the train?

4–35 A river flows due north with a velocity of 3 mi hr^{-1}. A man rows a boat across the river, his velocity relative to the water being 4 mi hr^{-1} due east. (a) What is his velocity relative to the earth? (b) If the river is 1 mi wide, how far north of his starting point will he reach the opposite bank? (c) How long a time is required to cross the river?

4–36 (a) In what direction should the rowboat in Problem 4–35 be headed in order to reach a point on the opposite bank directly east from the start? (b) What will be the velocity of the boat relative to the earth? (c) How long a time is required to cross the river?

4–37 A motorboat is observed to travel 10 mi hr^{-1} relative to the earth in the direction 37° north of east. If the velocity of the boat due to the wind is 2 mi hr^{-1} eastward and that due to the current is 4 mi hr^{-1} southward, what is the magnitude and direction of the velocity of the boat due to its own power?

5 Newton's Second Law. Gravitation

5-1 INTRODUCTION

In the preceding chapters we have discussed separately the concepts of force and acceleration. We have made use, in problems in equilibrium, of Newton's first law, which states that when the resultant force on a body is zero, the acceleration of the body is also zero. The next logical step is to ask how a body behaves when the resultant force on it is *not* zero. The answer to this question is contained in Newton's second law, which states that when the resultant force is not zero the body moves with accelerated motion, and that the acceleration, with a given force, depends on a property of the body known as its *mass*.

This part of mechanics, which includes the study both of motion and of the forces that bring about the motion, is called *dynamics*. In its broadest sense, dynamics includes nearly the whole of mechanics. Statics treats of special cases in which the acceleration is zero, and kinematics deals with motion only.

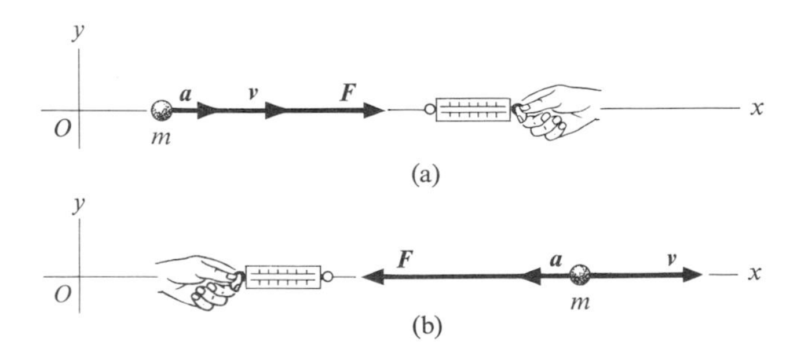

5-1 The acceleration $\boldsymbol{a} = d\boldsymbol{v}/dt$ is proportional to the force $\boldsymbol{F}$ and is in the same direction as the force.

It will be assumed in this chapter that all velocities are small compared with the velocity of light, so that relativistic considerations do not arise. We shall also assume that unless otherwise stated all velocities and accelerations are measured relative to an inertial reference system (Section 2-3), and shall consider rectilinear motion only. Motion in a curved path will be discussed in the next chapter.

5-2 NEWTON'S SECOND LAW. MASS

Figure 5-1(a) is a top view of a small body (a particle) on a level frictionless surface, moving to the right along the x-axis of an inertial reference system. A horizontal force $\boldsymbol{F}$, measured by a spring balance calibrated as described in Section 1-4, is exerted on the body. We find that the velocity of the body increases as long as the force acts. In other words, the body has an acceleration $\boldsymbol{a} = d\boldsymbol{v}/dt$, toward the right. If the magnitude of the force $\boldsymbol{F}$ is kept constant, the velocity increases at a constant rate. If the force is altered, the rate of change of velocity alters in the same proportion. Doubling the force doubles the rate of change of velocity, halving the force halves the rate of change, etc. If the force is reduced to zero, the rate of change of velocity is zero and the body continues to move with constant velocity.

Before the time of Galileo and Newton, it was generally believed that a force was necessary just to keep a body moving, even on a level frictionless surface or in "outer space." The great contribution of Newton to this part of mechanics was his realization that *no* force is necessary to keep a body moving, once it has been set in motion, and that the effect of a force is not to *maintain* the velocity of a body, but to *change* its velocity. The *rate of change* of velocity, for a given body, is directly proportional to the force exerted on it.

In Fig. 5–1(b), the velocity of the body is also toward the right, but the force is toward the left. Under these conditions the body moves more slowly (if the force continues to act, it will ultimately reverse its direction of motion). The acceleration is now toward the *left*, in the same direction as the force $\boldsymbol{F}$. Hence we conclude not only that the *magnitude* of the acceleration is proportional to that of the force, but that the *direction* of the acceleration is the *same* as that of the force, regardless of the direction of the velocity.

To say that the rate of change of the velocity of a body is directly proportional to the force exerted on it is to say that the ratio of the force to the rate of change of velocity is a constant, regardless of the magnitude of the force. This constant ratio of force to rate of change of velocity is called the *mass* m of the body. Thus

$$m = \frac{F}{dv/dt} = \frac{F}{a},$$

or

$$\boldsymbol{F} = m\frac{d\boldsymbol{v}}{dt} = m\boldsymbol{a}. \qquad (5\text{–}1)$$

By writing this as a *vector* equation, we automatically include the experimental fact that the direction of the acceleration $\boldsymbol{a}$ is the same as that of the force $\boldsymbol{F}$.

The mass of a body can be thought of as the *force per unit of acceleration*. For example, if the acceleration of a certain body is found to be 5 ft s^{-2} when the force is 20 lb, the mass of the body is

$$m = \frac{20\ \text{lb}}{5\ \text{ft s}^{-2}} = 4\ \text{lb ft}^{-1}\ \text{s}^2,$$

and a force of 4 lb must be exerted on the body for every ft s^{-2} of acceleration.

In *rectilinear* motion, the force $\boldsymbol{F}$ acting on a body, and its velocity $\boldsymbol{v}$, always have the same action line, as in Fig. 5–1. If the direction of the force is *not* the same as that of the velocity, the body is deflected sidewise and moves in a curved path. We shall see in the next chapter, however, that Eq. (5–1) applies in this case also, except that the change in velocity $d\boldsymbol{v}$, or the acceleration $\boldsymbol{a}$, includes the change in *direction* as well as the change in *magnitude* of the velocity. In every case, then, *the vector force equals the product of the mass and the vector acceleration.*

When two vectors are equal, their rectangular components are equal also. Hence the vector equation (5–1) is equivalent (for forces and accelerations in the *xy*-plane) to the pair of scalar equations

$$F_x = m\frac{dv_x}{dt} = ma_x, \qquad F_y = m\frac{dv_y}{dt} = ma_y. \qquad (5\text{–}2)$$

This means that each component of the force can be considered to produce its own component of acceleration. It follows that if any number of forces act on a body simultaneously, as is usually the case in problems of practical interest, the forces may be resolved into *x*- and *y*-components, the algebraic sums $\sum F_x$ and $\sum F_y$ may be computed, and the components of acceleration are given by

$$\sum F_x = m\frac{dv_x}{dt} = ma_x, \qquad \sum F_y = m\frac{dv_y}{dt} = ma_y.$$

This pair of equations is equivalent to the single vector equation

$$\boxed{\sum \boldsymbol{F} = m\frac{d\boldsymbol{v}}{dt} = m\boldsymbol{a},} \qquad (5\text{–}3)$$

where we write the left-hand side explicitly as $\sum \boldsymbol{F}$ to emphasize that the acceleration is determined by the *resultant* of *all* of the *external* forces acting *on* the body.

Equation (5–3) is the mathematical statement of Newton's *second law of motion*. If we think of the equation as solved for $\boldsymbol{a}$ or $d\boldsymbol{v}/dt$, the law can be stated: *The rate of change of the velocity of a particle, or its acceleration, is equal to the resultant of all external forces exerted on the particle divided by the mass of the particle, and is in the same direction as the resultant force.* The acceleration is to be measured relative to an inertial system.

It is necessary to state the law for a *particle*, because when a resultant force acts on an extended body, the body may be set in rotation and not all particles in it have the same acceleration. We shall discuss this more fully later on, but it may be said at this point that the acceleration of the center of gravity of a body is the same as that of a particle whose mass equals that of the body.

5-3 SYSTEMS OF UNITS

Nothing was said in the preceding discussion regarding the *units* in which force, mass, and acceleration are to be expressed. It is evident, however, from the equation $\boldsymbol{F} = m\boldsymbol{a}$, that these must be such that *unit force imparts unit acceleration to unit mass.*

In the *meter-kilogram-second* (mks) system, the mass of the standard kilogram is the unit of mass, and the unit of acceleration is 1 m s^{-2}. The unit force in this system is then *that force which gives a standard kilogram an acceleration of* 1 m s^{-2}. This force is called *one newton* (1 N). One newton is approximately equal to one-quarter of a pound-force (more precisely, 1 N = 0.22481 lb). Thus in the mks system,

$$F\ (\text{N}) = m\ (\text{kg}) \times a\ (\text{m s}^{-2}).$$

The mass unit in the *centimeter-gram-second* (cgs) system is one gram, equal to 1/1000 kg, and the unit of acceleration is 1 cm s^{-2}. The unit force in this system is then *that force which gives a body whose mass is one gram, an acceleration of* 1 cm s^{-2}. This force is called *one dyne* (1 dyn). Since 1 kg = 10^3 g and 1 m s^{-2} = 10^2 cm s^{-2}, it follows that 1 N = 10^5 dyn. In the cgs system,

$$F\ (\text{dyn}) = m\ (\text{g}) \times a\ (\text{cm s}^{-2}).$$

In setting up the mks and cgs systems, we first selected units of mass and acceleration, and defined the unit of force in terms of these. In the *British engineering system*, we first select a unit of *force* (1 lb) and a unit of acceleration (1 ft s^{-2}) and then define the unit of mass as *the mass of a body whose acceleration is* 1 ft s^{-2} *when the resultant force on the body is* 1 lb. This unit of mass is called *one slug*. (The origin of the name is obscure; it may have arisen from the concept of mass as inertia or *sluggishness*.) Then in the engineering system,

$$F\ (\text{lb}) = m\ (\text{slugs}) \times a\ (\text{ft s}^{-2}).$$

The units of force, mass, and acceleration in the three systems are summarized in Table 5-1.

Example 1. The acceleration of a certain body is found to be 5 ft s^{-2} when the resultant force on the body is 20 lb. The mass of the body is

$$m = \frac{F}{a} = \frac{20\ \text{lb}}{5\ \text{ft s}^{-2}} = 4\ \text{lb ft}^{-1}\ \text{s}^2 = 4\ \text{slugs}.$$

TABLE 5-1

System of units	Force	Mass	Acceleration
mks	newton (N)	kilogram (kg)	m s^{-2}
cgs	dyne (dyn)	gram (g)	cm s^{-2}
engineering	pound (lb)	slug	ft s^{-2}

Example 2. A constant horizontal force of 2 N is applied to a body of mass 4 kg, resting on a level frictionless surface. The acceleration of the body is

$$a = \frac{F}{m} = \frac{2\ \text{N}}{4\ \text{kg}} = 0.5\ \text{m s}^{-2}.$$

Since the force is constant, the acceleration is constant also. Hence if the initial position and velocity of the body are known, the velocity and position at any later time can be found from the equations of motion with constant acceleration.

Example 3. A body of mass 200 g is given an initial velocity of 40 cm s^{-1} toward the right along a level laboratory table top. The body is observed to slide a distance of 100 cm along the table before coming to rest. What was the magnitude and direction of the friction force f acting on it?

In the absence of further information, let us assume that the friction force is constant. The acceleration is then constant also and from the equations of motion with constant acceleration, we have

$$v^2 = v_0^2 + 2ax, \qquad 0 = (40\ \text{cm s}^{-1})^2 + (2a \times 100\ \text{cm}),$$
$$a = -8\ \text{cm s}^{-2}.$$

The negative sign means that the acceleration is toward the *left* (although the velocity is toward the right). The friction force on the body is

$$f = ma = 200\ \text{g} \times (-8\ \text{cm s}^{-2}) = -1600\ \text{dyn},$$

and is toward the left also. (A force of equal magnitude, but directed toward the right, is exerted on the table by the sliding body.)

5-4 NEWTON'S LAW OF UNIVERSAL GRAVITATION

Throughout our study of mechanics we have been continually encountering the force of gravitational attraction between a body and the earth. We now wish to study this phenomenon of gravitation in more detail.

The law of universal gravitation was discovered by Newton, and was published by him in 1686. There seems to be some evidence that Newton was led to deduce the law from speculations concerning the fall of an apple toward the earth, but his first published calculations to justify its correctness had to do with the motion of the moon around the earth.

Newton's law of gravitation may be stated: *Every particle of matter in the universe attracts every other particle with a force which is directly proportional to the product of the masses of the particles and inversely proportional to the square of the distance between them.* Thus

$$\boxed{F_g = G\frac{mm'}{r^2},} \tag{5-4}$$

where F_g is the gravitational force on either particle, m and m' are their masses, r is the distance between them, and G is a universal constant called the *gravitational constant*, whose numerical value depends on the units in which force, mass, and length are expressed.

The gravitational forces acting on the particles form an action-reaction pair. Although the masses of the particles may be different, forces of *equal* magnitude act on each, and the action line of both forces lies along the line joining the particles.

Newton's law of gravitation refers to the force between two *particles*. How can it be applied to the force between a small body and the *earth*, or between the earth and the moon, since the particles that make up these bodies are at different distances from one another and their forces of attraction are in different directions? Newton delayed publication of his law for 11 years after he became convinced of its validity, because he could not prove mathematically that the force of attraction exerted on or by a homogeneous sphere is the same as if the mass of the sphere were concentrated at its center. (To prove this, he had to invent the methods of calculus.) The proof is not difficult but is too long to give here, and we shall simply state as a fact that *the gravitational force exerted on or by a homogeneous sphere is the same as if the entire mass of the sphere were concentrated in a point at its center.* Thus if the earth were a homogeneous sphere, the force exerted by it on a small body of mass m, at a distance r from its center, would be

$$F_g = G\frac{mm_E}{r^2},$$

where m_E is the mass of the earth. A force of the same magnitude would be exerted *on* the earth by the body.

The magnitude of the gravitational constant G can be found experimentally by measuring the force of gravitational attraction between two bodies of known masses m and m', at a known separation. For bodies of moderate size the force is extremely small, but it can be measured with an instrument which was invented by the Rev. John Michell, although it was first used for this purpose by Sir Henry Cavendish in 1798. The same type of instrument was also used by Coulomb for studying forces of electrical and magnetic attraction and repulsion.

The Cavendish balance consists of a light, rigid T-shaped member (Fig. 5-2) supported by a fine vertical fiber such as a quartz thread or a thin metallic ribbon. Two small spheres of mass m are mounted at the ends of the horizontal portion of the T, and a small mirror M, fastened to the vertical portion, reflects a beam of light onto a scale. To use the balance, two large spheres of mass m' are brought up to the positions shown. The forces of gravitational attraction between the large and small spheres result in a *couple* which twists the system through a small angle, thereby moving the reflected light beam along the scale.

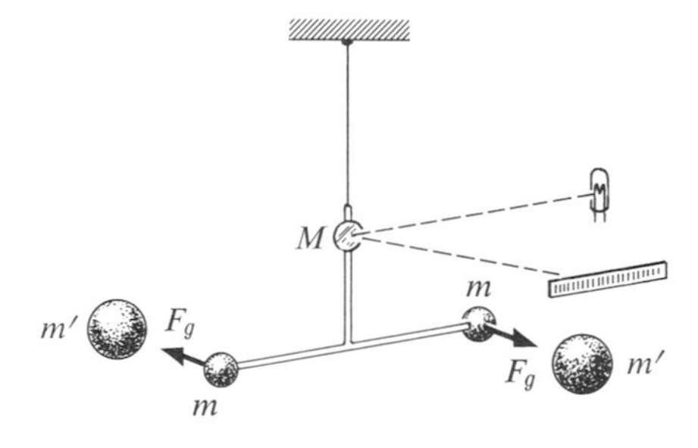

5-2 Principle of the Cavendish balance.

By using an extremely fine fiber, the deflection of the mirror may be made sufficiently large so that the gravitational force can be measured quite accurately. The gravitational constant, measured in this way, is found to be

$$\begin{aligned} G &= 6.670 \times 10^{-11}\ \text{N m}^2\ \text{kg}^{-2} \\ &= 6.670 \times 10^{-8}\ \text{dyn cm}^2\ \text{g}^{-2}. \end{aligned}$$

Example 1. The mass m of one of the small spheres of a Cavendish balance is 1 g, the mass m' of one of the large spheres is

500 g, and the center-to-center distance between the spheres is 5 cm. The gravitational force on each sphere is

$$F_g = 6.67 \times 10^{-8} \text{ dyn cm}^2 \text{ g}^{-2} \frac{1 \text{ g} \times 500 \text{ g}}{(5 \text{ cm})^2}$$
$$= 1.33 \times 10^{-6} \text{ dyn},$$

or about one-millionth of a dyne!

Example 2. Suppose the spheres in Example 1 are placed 5 cm from each other at a point in space far removed from all other bodies. What is the acceleration of each, relative to an inertial system?

The acceleration a of the smaller sphere is

$$a = \frac{F_g}{m} = \frac{1.33 \times 10^{-6} \text{ dyn}}{1 \text{ g}} = 1.33 \times 10^{-6} \text{ cm s}^{-2}.$$

The acceleration a' of the larger sphere is

$$a' = \frac{F_g}{m'} = \frac{1.33 \times 10^{-6} \text{ dyn}}{500 \text{ g}} = 2.67 \times 10^{-9} \text{ cm s}^{-2}.$$

In this case, the accelerations are *not* constant because the gravitational force increases as the spheres approach each other.

5-5 MASS AND WEIGHT

The *weight* of a body can now be defined more generally than in the preceding chapters as *the resultant gravitational force exerted on the body by all other bodies in the universe.* At or near the surface of the earth, the force of the earth's attraction is so much greater than that of any other body that for practical purposes all other gravitational forces can be neglected and the weight can be considered as arising solely from the gravitational attraction of the earth. Similarly, at the surface of the moon, or of another planet, the weight of a body results almost entirely from the gravitational attraction of the moon or the planet. Thus if the earth were a homogeneous sphere of radius R, the weight w of a small body at or near its surface would be

$$w = F_g = \frac{G\, mm_E}{R^2}. \tag{5-5}$$

There is no general agreement among physicists as to the precise definition of "weight." Some prefer to use this term for a quantity we shall define later and call the "apparent weight" or "relative weight." In the absence of a generally accepted definition we shall continue to use the term as defined above.

Because of inhomogeneities in the composition of the earth, and because the earth is not a perfect sphere but a spheroid flattened at the poles, the weight of a given body varies slightly from point to point of the earth's surface. Also, the weight of a given body decreases inversely with the square of its distance from the earth's center, and at a radial distance of two earth radii, for example, it has decreased to one-quarter of its value at the earth's surface.

The *apparent* weight of a body at the surface of the earth differs slightly in magnitude and direction from the earth's force of gravitational attraction because of the rotation of the earth about its axis. For the present, we shall ignore the small difference between the apparent weight of a body and the force of the earth's gravitational attraction, and shall assume that the earth is an inertial reference system. Then when a body is allowed to fall freely, the force accelerating it is its weight $\boldsymbol{w}$ and the acceleration produced by this force is the acceleration due to gravity, $\boldsymbol{g}$. The general relation

$$\boldsymbol{F} = m\boldsymbol{a}$$

therefore becomes, for the special case of a freely falling body,

$$\boldsymbol{w} = m\boldsymbol{g}. \tag{5-6}$$

Since

$$w = mg = G\frac{mm_E}{R^2},$$

it follows that

$$g = \frac{Gm_E}{R^2}, \tag{5-7}$$

showing that the acceleration due to gravity is the same for *all* bodies (since m cancelled out) and is very nearly constant (since G and m_E are constants and R varies only slightly from point to point on the earth).

The weight of a body is a force, and must be expressed in terms of the unit of force in the particular system of units one is using. Thus in the mks system the unit of weight is 1 N, in the cgs system it is 1 dyn, and in the engineering system it is 1 lb. Equation (5-6) indicates the relation between the mass and weight of a body in any consistent set of units.

For example, the weight of a standard kilogram, at a point where $g = 9.80\ \text{m s}^{-2}$, is

$$w = mg = 1\ \text{kg} \times 9.80\ \text{m s}^{-2} = 9.80\ \text{N}.$$

At a second point where $g = 9.78\ \text{m s}^{-2}$, the weight is

$$w = 9.78\ \text{N}.$$

Thus, unlike the mass of a body, which is a constant, the weight varies from one point to another.

The weight of a body whose mass is 1 g, at a point where $g = 980\ \text{cm s}^{-2}$, is

$$w = mg = 1\ \text{g} \times 980\ \text{cm s}^{-2} = 980\ \text{dyn}.$$

The weight of a body whose mass is 1 slug, at a point where $g = 32.0\ \text{ft s}^{-2}$, is

$$w = mg = 1\ \text{slug} \times 32.0\ \text{ft s}^{-2} = 32.0\ \text{lb},$$

and the mass of a man who weighs 160 lb at this point is

$$m = \frac{w}{g} = \frac{160\ \text{lb}}{32.0\ \text{ft s}^{-2}} = 5\ \text{slugs}.$$

If we insert for the weight w in the equation $w = mg$, the gravitational force F_g as given by Newton's law of gravitation, we obtain, after cancelling the mass m,

$$m_E = \frac{R^2 g}{G},$$

where R is the earth's radius. All of the quantities on the right are known, so the mass of the earth, m_E, can be calculated. Taking $R = 6370\ \text{km} = 6.37 \times 10^6\ \text{m}$, and $g = 9.80\ \text{m s}^{-2}$, we find

$$m_E = 5.98 \times 10^{24}\ \text{kg} = 5.98 \times 10^{27}\ \text{g}.$$

The volume of the earth is

$$V = \tfrac{4}{3}\pi R^3 = 1.09 \times 10^{21}\ \text{m}^3 = 1.09 \times 10^{27}\ \text{cm}^3.$$

The mass of a body divided by its volume is known as its average *density*. (The density of water is $1\ \text{g cm}^{-3} = 1000\ \text{kg m}^{-3}$.) The average density of the earth is therefore

$$\frac{m_E}{V} = 5.5\ \text{g cm}^{-3} = 5500\ \text{kg m}^{-3}.$$

This is considerably larger than the average density of the material near the earth's surface (the density of rock is about $3\ \text{g cm}^{-3} = 3000\ \text{kg m}^{-3}$) so the interior of the earth must be of much higher density.

As with many other physical quantities, the mass of a body can be measured in several different ways. One is to use the relation by which the quantity is defined, which in this case is the ratio of the force on the body to its acceleration. A measured force is applied to the body, its acceleration is measured, and the unknown mass is obtained by dividing the force by the acceleration. This method is used exclusively to measure masses of atomic particles.

The second method consists of finding by trial some other body whose mass (a) is equal to that of the given body, and (b) is already known. Consider first a method of determining when two masses are equal. It will be recalled that at the same point on the earth's surface all bodies fall freely with the same acceleration g. Since the weight w of a body equals the product of its mass m and the acceleration g, it follows that if, at the same point, the weights of two bodies are equal, their masses are equal also. The *equal-arm balance* is an instrument by means of which one can determine very precisely when the weights of two bodies are equal, and hence when their masses are equal.

It will be seen from the preceding discussion that the property of matter called *mass* makes itself evident in two very different ways. The force of gravitational attraction between two particles is said to be proportional to the product of their masses, and in this sense mass can be considered as *that property of matter by virtue of which every particle exerts a force of attraction on every other particle.* We may call this property *gravitational mass.* On the other hand, Newton's second law is concerned with an entirely different property of matter, namely, the fact that a force (not necessarily gravitational) must be exerted on a particle in order to accelerate it, i.e., to change its velocity, either in magnitude or direction. This property can be called *inertial mass.* It is not at all obvious that the gravitational mass of a particle should be the same as its inertial mass, but experiment shows that the two are in fact the same or, better, that one is directly proportional to the other. That is, if we have to push twice as hard on body A as we do on body B to produce a given acceleration, then the force of gravitational attraction between body A and some third body C is twice as great as the gravitational attraction between body B and body C, the distance between them being the same. Since the two kinds of mass are proportional, it is customary to consider them equal. Thus when an equal-arm balance is used to compare masses, it is actually *gravitational* mass that is being measured. By convention, we assign the same numerical value to the inertial mass. Therefore the property represented by m in Newton's second law can be operationally defined as the result obtained by the prescribed methods of using an equal-arm balance.

5-6 APPLICATIONS OF NEWTON'S SECOND LAW

We now give a number of applications of Newton's second law to specific problems. In all these examples, and in the problems at the end of the chapter, it will be assumed that the acceleration due to gravity is 9.80 m s^{-2} or 32.0 ft s^{-2}, unless otherwise specified.

Example 1. A block whose mass is 10 kg rests on a horizontal surface. What constant horizontal force $\boldsymbol{T}$ is required to give it a velocity of 4 m s^{-1} in 2 s, starting from rest, if the friction force between the block and the surface is constant and is equal to 5 N? Assume that all forces act at the center of the block. (See Fig. 5-3.)

The mass of the block is given. Its y-acceleration is zero. Its x-acceleration can be found from the data on the velocity

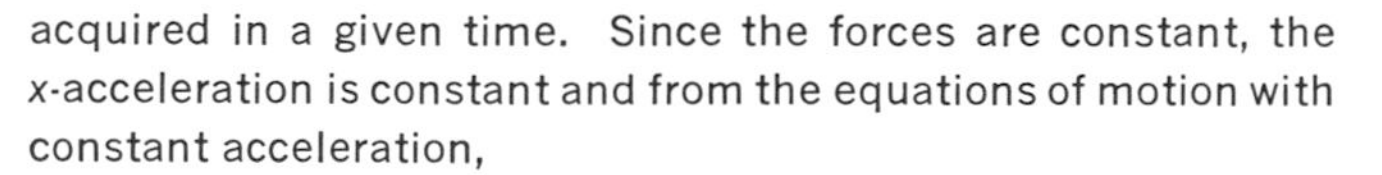

Fig. 5-3

acquired in a given time. Since the forces are constant, the x-acceleration is constant and from the equations of motion with constant acceleration,

$$a_x = \frac{v - v_0}{t} = \frac{4 \text{ m s}^{-1} - 0}{2 \text{ s}} = 2 \text{ m s}^{-2}.$$

The resultant of the x-forces is

$$\sum F_x = T - f,$$

and that of the y-forces is

$$\sum F_y = N - w.$$

Hence from Newton's second law in component form,

$$T - f = ma_x, \qquad N - w = ma_y = 0.$$

From the second equation, we find that

$$N = w = mg = 10 \text{ kg} \times 9.80 \text{ m s}^{-2} = 98.0 \text{ N}.$$

and from the first,

$$T = f + ma_x = 5 \text{ N} + (10 \text{ kg} \times 2 \text{ m s}^{-2}) = 25 \text{ N}.$$

Example 2. An elevator and its load weigh a total of 1600 lb. Find the tension $\boldsymbol{T}$ in the supporting cable when the elevator, originally moving downward at 20 ft s^{-1}, is brought to rest with constant acceleration in a distance of 50 ft. (See Fig. 5-4.)

The mass of the elevator is

$$m = \frac{w}{g} = \frac{1600 \text{ lb}}{32 \text{ ft s}^{-2}} = 50 \text{ slugs}.$$

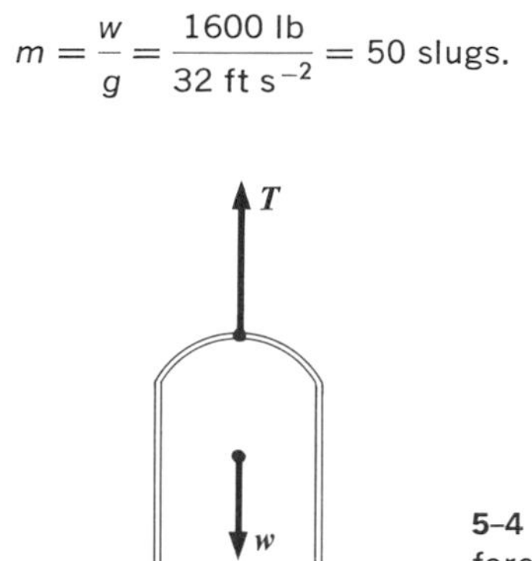

5-4 The resultant force is $\boldsymbol{T} - \boldsymbol{w}$.

From the equations of motion with constant acceleration,

$$v^2 = v_0^2 + 2ay, \qquad a = \frac{v^2 - v_0^2}{2y}.$$

The initial velocity v_0 is -20 ft s^{-1}; the velocity v is zero. If we take the origin at the point where the deceleration begins, then $y = -50$ ft. Hence

$$a = \frac{0 - (-20 \text{ ft s}^{-1})^2}{-2 \times 50 \text{ ft}} = 4 \text{ ft s}^{-2}.$$

The acceleration is therefore positive (upward). From the free-body diagram (Fig. 5-4) the resultant force is

$$\sum F = T - w = T - 1600 \text{ lb}.$$

Since

$$\sum F = ma, \qquad T - 1600 \text{ lb} = 50 \text{ slugs} \times 4 \text{ ft s}^{-2} = 200 \text{ lb},$$

$$T = 1800 \text{ lb}.$$

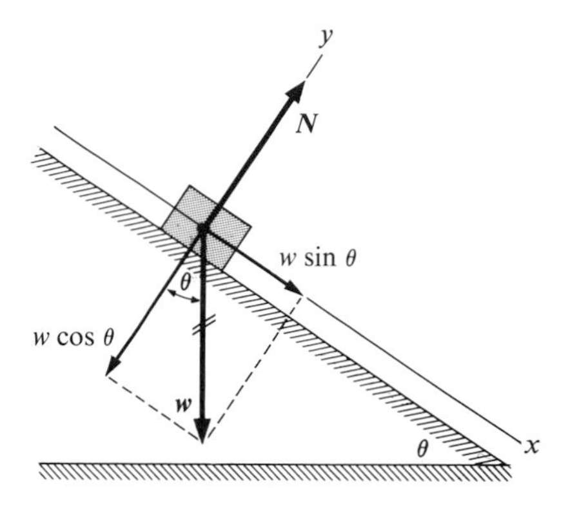

5-5 A block on a frictionless inclined plane.

Example 3. What is the acceleration of a block on a frictionless plane inclined at an angle θ with the horizontal?

The only forces acting on the block are its weight $\boldsymbol{w}$ and the normal force $\boldsymbol{N}$ exerted by the plane (Fig. 5-5).

Take axes parallel and perpendicular to the surface of the plane and resolve the weight into x- and y-components. Then

$$\sum F_y = N - w \cos \theta, \qquad \sum F_x = w \sin \theta.$$

But we know that $a_y = 0$, so from the equation $\sum F_y = ma_y$ we find that $N = w \cos \theta$. From the equation $\sum F_x = ma_x$, we have

$$w \sin \theta = ma_x,$$

and since $w = mg$,

$$a_x = g \sin \theta.$$

The mass does not appear in the final result, which means that any block, regardless of its mass, will slide on a frictionless inclined plane with an acceleration down the plane of $g \sin \theta$. (Note that the *velocity* is not necessarily down the plane.)

Example 4. Refer to Fig. 2-4 (Chapter 2). Let the mass of the block be 4 kg and that of the rope be 0.5 kg. If the force $\boldsymbol{F}_1$ is 9 N, what are the forces $\boldsymbol{F}_1'$, $\boldsymbol{F}_2$, and $\boldsymbol{F}_2'$? The surface on which the block moves is level and frictionless.

We know from Newton's third law that $F_1 = F_1'$ and that $F_2 = F_2'$. Hence $F_1' = 9$ N. The force F_2 could be computed by applying Newton's second law to the block, if its acceleration were known, or the force F_2' could be computed by applying this law to the rope if its acceleration were known. The acceleration is not given, but it can be found by considering the block and rope together as a single system. The vertical forces on this system need not be considered. Since there is no friction, the resultant *external* force acting *on* the system is the force F_1. (The forces F_2 and F_2' are *internal* forces when we consider block and rope as a single system, and the force F_1' does not act on the system, but *on the man*.) Then, from Newton's second law,

$$\sum F = ma,$$
$$9 \text{ N} = (4 \text{ kg} + 0.5 \text{ kg}) \times a,$$
$$a = 2 \text{ m s}^{-2}.$$

We can now apply Newton's second law to the block.

$$\sum F = ma,$$
$$F_2 = 4 \text{ kg} \times 2 \text{ m s}^{-2} = 8 \text{ N}.$$

If we consider the rope alone, the resultant force on it is

$$\sum F = F_1 - F_2' = 9 \text{ N} - F_2',$$

and from the second law,

$$9 \text{ N} - F_2' = 0.5 \text{ kg} \times 2 \text{ m s}^{-2} = 1 \text{ N},$$
$$F_2' = 8 \text{ N}.$$

In agreement with Newton's third law, which was tacitly used when the forces F_2 and F_2' were omitted in considering the system as a whole, we find that F_2 and F_2' are equal in magnitude. Note, however, that the forces F_1 and F_2 are *not* equal and opposite (the rope is not in equilibrium) and that these forces are *not* an action-reaction pair.

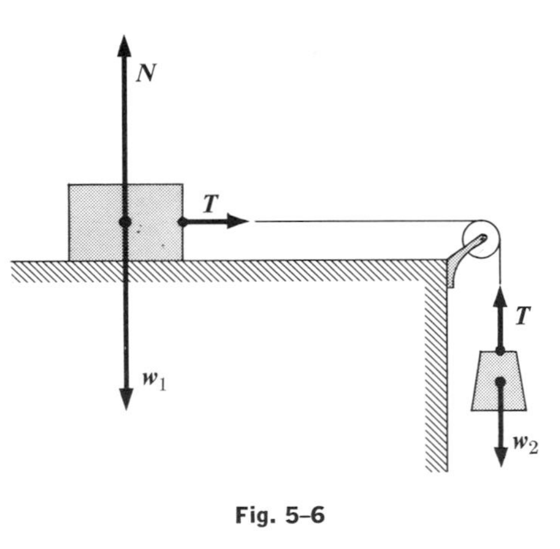

Fig. 5-6

Example 5. In Fig. 5-6, a block of weight w_1 (mass $= m_1$) moves on a level frictionless surface, connected by a light flexible cord passing over a small frictionless pulley to a second hanging block of weight w_2 (mass $= m_2$). What is the acceleration of the system, and what is the tension in the cord connecting the two blocks?

The diagram shows the forces acting on each block. The forces exerted on the blocks by the cord can be considered an action-reaction pair, so we have used the same symbol $\boldsymbol{T}$ for each. For the block on the surface,

$$\sum F_x = T = m_1 a,$$
$$\sum F_y = N - w_1 = 0.$$

Since the cord connecting the two blocks is inextensible, the accelerations are the same. Applying Newton's second law to the hanging block, we obtain

$$\sum F_y = w_2 - T = m_2 a.$$

Addition of the first and third equations gives

$$w_2 = (m_1 + m_2)a,$$

or

$$a = \frac{w_2}{m_1 + m_2},$$

which says that the acceleration of the *entire system* equals the *resultant external force* ($\mathbf{w}_2$) divided by the *total mass* ($m_1 + m_2$). Since $w_2 = m_2 g$,

$$a = g\frac{m_2}{m_1 + m_2}.$$

Eliminating a from the first and third equations, we get

$$T = w_2 \frac{m_1}{m_1 + m_2},$$

so that T is *only a fraction* of w_2. Although the earth pulls on the *hanging* block with a force $\mathbf{w}_2$, the force exerted on the *sliding* block is only a fraction of $\mathbf{w}_2$. It is not the earth that pulls on the sliding block, but the cord, whose tension *must be less than* $\mathbf{w}_2$ if the hanging block is to accelerate downward.

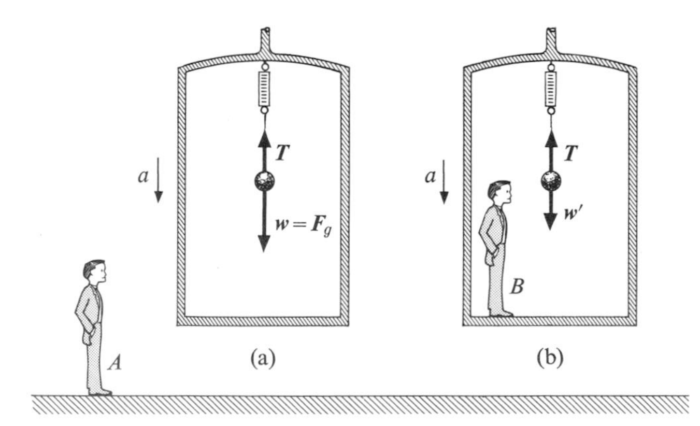

5-7 (a) To observer A, the body has a downward acceleration a, and he writes $w - T = ma$. (b) To observer B, the acceleration of the body is zero. He writes $w' = T$.

Example 6. A body of mass m is suspended from a spring balance attached to the roof of an elevator, as shown in Fig. 5-7. What is the reading of the balance if the elevator has an acceleration $\mathbf{a}$, relative to the earth? Consider the earth's surface to be an inertial reference system.

The forces on the body are its weight $\mathbf{w}$ (the gravitational force $\mathbf{F}_g$ exerted on it by the earth) and the upward force $\mathbf{T}$ exerted on it by the balance. The body is at rest relative to the elevator, and hence has an acceleration $\mathbf{a}$ relative to the earth. (We take the downward direction as positive.) The resultant force on the body is $w - T$, so from Newton's second law,

$$w - T = ma, \qquad T = w - ma.$$

By Newton's third law, the body pulls down on the balance with a force equal and opposite to $\mathbf{T}$, or equal to $w - ma$, so the balance reading equals $w - ma$.

If the same body were suspended in equilibrium from a balance attached to the earth, the balance reading would equal the weight w. To an observer riding in the elevator, the body *appears* to be in equilibrium and hence *appears* to be acted on by a downward force $\mathbf{w}'$ equal in magnitude to the balance reading, as in Fig. 5-7(b). This *apparent* force $\mathbf{w}'$ can be called the *apparent weight* of the body. The gravitational force $\mathbf{w}$ will be called the *true weight*. Then

$$w' = w - ma. \tag{5-8}$$

If the elevator is at rest, or moving vertically (either up or down) with constant velocity, $a = 0$ and the apparent weight equals the true weight. If the acceleration is downward, as in Fig. 5-7, so that a is positive, the apparent weight is less than the true weight; the body appears "lighter." If the acceleration is *upward*, a is negative, the apparent weight is greater than the true weight, and the body appears "heavier." If the elevator falls freely, $a = g$, and since the true weight w also equals mg, the apparent weight is zero and the body *appears* "weightless." It is in this sense that an astronaut orbiting the earth in a space capsule is said to be "weightless."

Example 7. Figure 5-8(a) represents a simple *accelerometer*. A small body is fastened at one end of a light rod pivoted freely at the point P. When the system has an acceleration a toward the right, the rod makes an angle θ with the vertical. (In a practical instrument, some form of damping must be provided to avoid violent swinging of the rod when the acceleration changes. The rod might hang in a tank of mineral oil.)

As shown in the free-body diagram, Fig. 5-8(b), two forces are exerted on the body; its weight $\mathbf{w}$ and the tension $\mathbf{T}$ in the rod. (We neglect the weight of the rod.) The resultant horizontal force is

$$\sum F_x = T \sin\theta,$$

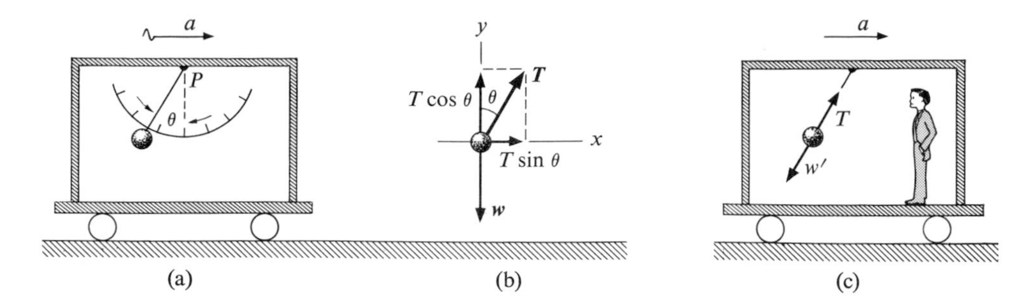

5-8 (a) A simple accelerometer. (b) The forces on the body are $\boldsymbol{w}$ and $\boldsymbol{T}$. (c) The apparent weight $\boldsymbol{w}'$ is equal and opposite to $\boldsymbol{T}$.

and the resultant vertical force is

$$\sum F_y = T\cos\theta - w.$$

The x-acceleration is the acceleration a of the system, and the y-acceleration is zero. Hence

$$T\sin\theta = ma, \qquad T\cos\theta = w.$$

When the first equation is divided by the second, and w replaced by mg, we get

$$a = g\tan\theta,$$

and the acceleration a is proportional to the tangent of the angle θ.

Let us consider this problem from the standpoint of an observer riding with the accelerometer, as in Fig. 5-8(c). To him, the body *appears* in equilibrium and hence appears to be acted on by a force $\boldsymbol{w}'$, equal and opposite to $\boldsymbol{T}$, and which he calls the *apparent weight* of the body. To obtain the expression for $\boldsymbol{w}'$, let us write Newton's second law in general vector form. The resultant force on the body is the vector sum of $\boldsymbol{w}$ and $\boldsymbol{T}$, and hence

$$\boldsymbol{w} + \boldsymbol{T} = m\boldsymbol{a}, \qquad \boldsymbol{T} = -\boldsymbol{w} + m\boldsymbol{a}.$$

The apparent weight $\boldsymbol{w}'$ is equal and opposite to $\boldsymbol{T}$, so $\boldsymbol{w}' = -\boldsymbol{T}$ and

$$\boldsymbol{w}' = \boldsymbol{w} - m\boldsymbol{a}. \tag{5-9}$$

This is the general form of Eq. (5-8) for the apparent weight $\boldsymbol{w}'$ of a body, relative to a system which has an acceleration $\boldsymbol{a}$ relative to an inertial system. In this example, *the apparent weight differs in both magnitude and direction from the true weight.*

Up to this point, applications of Newton's second law have been limited to cases where the resultant force acting on a body was constant, thereby imparting to the body a constant acceleration. Such cases are very important, but make only slight demands on one's mathematical knowledge. When the resultant force is variable, however, the acceleration is not constant and the simple equations of motion with constant acceleration do not apply. We conclude this section with two examples of motion under the action of a variable force.

Example 8. Assume the earth to be a nonrotating, homogeneous sphere. Discuss the motion of a freely falling body (or a body projected vertically upward), taking into account the variation of the gravitational force on the body with its distance from the earth's center. Neglect air resistance.

The gravitational force on the body at a distance r from the earth's center is Gmm_E/r^2, and from Newton's second law its acceleration is

$$g = \frac{w}{m} = -\frac{Gm_E}{r^2},$$

where the positive direction is upward (or better, radially outward).

It was shown in Eq. (4-8) that the acceleration can be expressed as

$$g = v\frac{dv}{dr}.$$

Then

$$v\frac{dv}{dr} = -\frac{Gm_E}{r^2}, \qquad \int_{v_1}^{v_2} v\,dv = -Gm_E\int_{r_1}^{r_2}\frac{dr}{r^2},$$

where v_1 and v_2 are the velocities at the radial distances r_1 and r_2. It follows that

$$v_2^2 - v_1^2 = 2Gm_E\left(\frac{1}{r_2} - \frac{1}{r_1}\right). \tag{5-10}$$

As an illustration, let us find the initial velocity v_1 required to project a body vertically upward so that it rises to a height above the earth's surface equal to the earth's radius R. Then

$v_2 = 0$, $r_1 = R$, $r_2 = 2R$, and

$$v_1^2 = \frac{Gm_E}{R}. \tag{5-11}$$

Let g_0 represent the acceleration of gravity at the earth's surface, where $r = R$. Then

$$Gm_E = g_0 R^2,$$

and Eq. (5-11) can be written

$$v_1^2 = g_0 R. \tag{5-12}$$

How does this compare with the velocity that would be required if the acceleration had the *constant* value g_0?

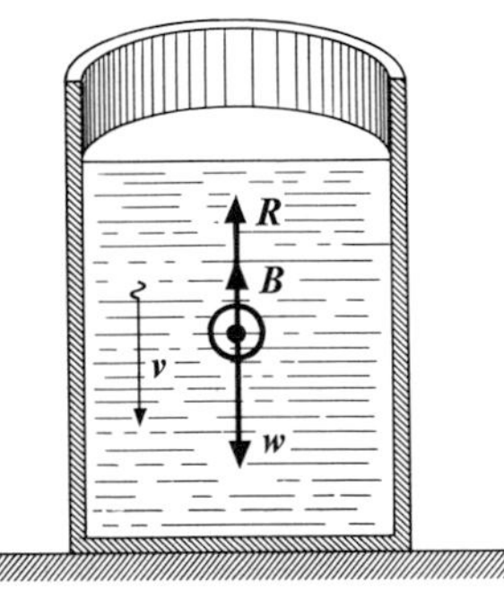

5-9 Forces acting on a small sphere falling through a viscous fluid.

Example 9. A resisting force that varies directly with the speed is found frequently in nature. Any small spherical body of radius r, like a raindrop, an oil droplet, or a steel sphere, moving with a small velocity $\mathbf{v}$ through a viscous fluid (liquid or gas) is subjected to a force $\mathbf{R}$, where

$$\mathbf{R} = -6\pi\eta r\mathbf{v},$$

and η is the viscosity. This relation is known as *Stokes' law*. Letting

$$k = 6\pi\eta r,$$

we may write Stokes' law simply as

$$\mathbf{R} = -k\mathbf{v}.$$

A small sphere falling through a viscous fluid is subjected to three vertical forces, as shown in Fig. 5-9: the weight $\mathbf{w}$, the buoyant force $\mathbf{B}$, and the resisting force $\mathbf{R}$.

Let us suppose that the sphere starts from rest and that the positive y-direction is downward. Then

$$\sum F_y = w - B - kv = ma.$$

At first, when $v = 0$, the resisting force is zero and the initial acceleration a_0 is positive:

$$a_0 = \frac{w - B}{m}. \tag{5-13}$$

The sphere speeds up and, after a while, when v becomes large enough, the resisting force equals $w - B$ and there is no resultant force acting on the sphere. At this moment the acceleration is zero and the speed undergoes no further increase. The maximum or *terminal speed* v_T may therefore be calculated by setting $a = 0$; thus

$$w - B - kv_T = 0$$

or

$$v_T = \frac{w - B}{k}. \tag{5-14}$$

To find the relation between the speed and the time during the interval before the terminal speed is reached, we go back to Newton's second law,

$$m\frac{dv}{dt} = w - B - kv.$$

After rearranging terms and replacing $(w - B)/k$ by v_T, we get

$$\frac{dv}{v - v_T} = -\frac{k}{m}\,dt.$$

Since $v = 0$ when $t = 0$,

$$\int_0^v \frac{dv}{v - v_T} = -\frac{k}{m}\int_0^t dt,$$

whence

$$\ln\frac{v_T - v}{v_T} = -\frac{k}{m}t,$$

or

$$1 - \frac{v}{v_T} = e^{-(k/m)t},$$

and finally

$$v = v_T(1 - e^{-(k/m)t}). \tag{5-15}$$

An important concept related to an exponentially varying quantity is the *relaxation time*, t_R, whose meaning can be seen from Fig. 5-10. Suppose the acceleration remained constant at the initial value a_0, as indicated by the dotted line. The relaxation time can be defined as the time that *would* be required to reach the terminal velocity with this constant acceleration, and evidently

$$t_R = \frac{v_T}{a_0} = \frac{(w - B)/k}{(w - B)/m} = \frac{m}{k}.$$

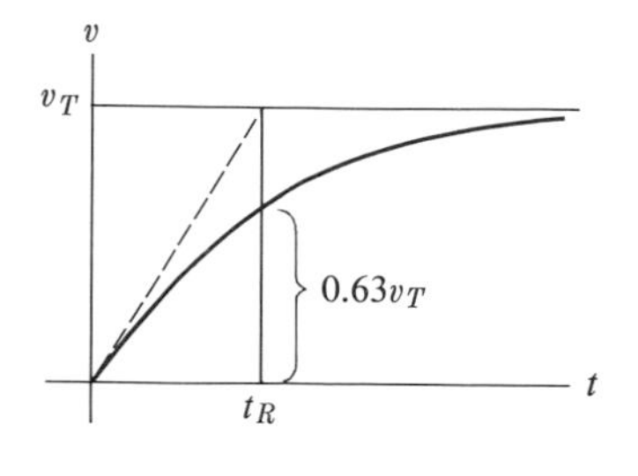

5-10 Velocity-time graph for a body falling in a viscous medium.

Equation (5-15) can now be written more simply as

$$v = v_T(1 - e^{-t/t_R}). \tag{5-16}$$

At the instant when t equals the relaxation time, $t/t_R = 1$ and

$$v = v_T(1 - e^{-1}) = 0.63v_T.$$

Thus about 63% of the final velocity is actually acquired in a time equal to the relaxation time.

Problems

For problem work, use the approximate values of $g = 32 \text{ ft s}^{-2} = 9.80 \text{ m s}^{-2} = 980 \text{ cm s}^{-2}$. A force diagram should be constructed for each problem.

5-1 (a) What is the mass of a body which weighs 1 N at a point where $g = 9.80 \text{ m s}^{-2}$? (b) What is the mass of a body which weighs 1 dyn at a point where $g = 980 \text{ cm s}^{-2}$? (c) What is the mass of a standard pound body?

5-2 (a) At what distance from the earth's center would a standard kilogram weigh 1 N? (b) At what distance would a 1-g body weigh 1 dyn? (c) At what distance would a 1-slug body weigh 1 lb?

5-3 If action and reaction are always equal in magnitude and opposite in direction, why don't they always cancel each other and leave no net force for accelerating a body?

5-4 The mass of a certain object is 10 g. (a) What would its mass be if taken to the planet Mars? (b) Is the expression $F = ma$ valid on Mars? (c) Newton's second law is sometimes written in the form $F = wa/g$ instead of $F = ma$. Would this expression be valid on Mars? (d) If a Martian scientist hangs a standard pound body on a spring balance calibrated correctly on the earth, would the spring balance read 1 lb? Explain.

5-5 A constant horizontal force of 10 lb acts on a body on a smooth horizontal plane. The body starts from rest and is observed to move 250 ft in 5 s. (a) What is the mass of the body? (b) If the force ceases to act at the end of 5 s, how far will the body move in the next 5 s?

5-6 A .22 rifle bullet, traveling at 36,000 cm s^{-1}, strikes a block of soft wood, which it penetrates to a depth of 10 cm. The mass of the bullet is 1.8 g. Assume a constant retarding force. (a) How long a time was required for the bullet to stop? (b) What was the decelerating force, in dynes? in pounds?

5-7 An electron (mass $= 9 \times 10^{-28}$ g) leaves the cathode of a radio tube with zero initial velocity and travels in a straight line to the anode, which is 1 cm away. It reaches the anode with a velocity of $6 \times 10^8 \text{ cm s}^{-1}$. If the accelerating force was constant, compute (a) the accelerating force, in dynes, (b) the time to reach the anode, (c) the acceleration. The gravitational force on the electron may be neglected.

5-8 Give arguments either for or against the statement that "the only reason an apple falls downward to meet the earth instead of the earth falling upward to meet the apple is that the earth, being so much more massive, exerts the greater pull."

5-9 In an experiment using the Cavendish balance to measure the gravitational constant G, it is found that a sphere of mass 800 g attracts another sphere of mass 4 g with a force of 13×10^{-6} dyn when the distance between the centers of the spheres is 4 cm. The acceleration of gravity at the earth's surface is 980 cm s^{-2}, and the radius of the earth is 6400 km. Compute the mass of the earth from these data.

5-10 Two spheres, each of mass 6400 g, are fixed at points A and B (Fig. 5-11). Find the magnitude and direction of the initial acceleration of a sphere of mass 10 g if released from rest at point P and acted on only by forces of gravitational attraction of the spheres at A and B.

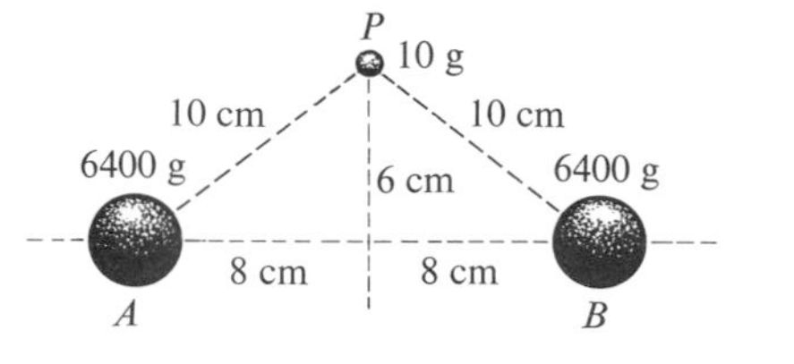

Fig. 5-11

5-11 The mass of the moon is one eighty-first, and its radius one-fourth, that of the earth. What is the acceleration due to gravity on the surface of the moon?

5-12 In round numbers, the distance from the earth to the moon is 250,000 mi, the distance from the earth to the sun is 93 million miles, the mass of the earth is 6×10^{27} g, and the mass of the sun is 2×10^{33} g. Approximately, what is the ratio of the gravitational pull of the sun on the moon to that of the earth on the moon?

5-13 A 5-kg block is supported by a cord and pulled upward with an acceleration of 2 m s^{-2}. (a) What is the tension in the cord? (b) After the block has been set in motion, the tension in the cord is reduced to 49 N. What sort of motion will the block perform? (c) If the cord is now slackened completely, the block is observed to move up 2 m farther before coming to rest. With what velocity was it traveling?

5-14 A block weighing 10 lb is held up by a string which can be moved up or down. What conclusions can you draw regarding magnitude and direction of the acceleration and velocity of the upper end of the string when the tension in the string is (a) 5 lb, (b) 10 lb, (c) 15 lb?

5-15 A body hangs from a spring balance supported from the roof of an elevator. (a) If the elevator has an upward acceleration of 4 ft s^{-2} and the balance reads 45 lb, what is the true weight of the body? (b) Under what circumstances will the balance read 35 lb? (c) What will the balance read if the elevator cable breaks?

5-16 A transport plane is to take off from a level landing field with two gliders in tow, one behind the other. Each glider weighs 2400 lb, and the friction force or drag on each may be assumed constant and equal to 400 lb. The tension in the towrope between the transport plane and the first glider is not to exceed 2000 lb. (a) If a velocity of 100 ft s^{-1} is required for the take-off, how long a runway is needed? (b) What is the tension in the towrope between the two gliders while the planes are accelerating for the take-off?

5-17 If the coefficient of friction between tires and road is 0.5, what is the shortest distance in which an automobile can be stopped when traveling at 60 mi hr^{-1}?

5-18 An 80-lb packing case is on the floor of a truck. The coefficient of static friction between the case and the truck floor is 0.30, and the coefficient of sliding friction is 0.20. Find the magnitude and direction of the friction force acting on the case (a) when the truck is accelerating at 6 ft s^{-2}, (b) when it is decelerating at 10 ft s^{-2}.

5-19 A balloon is descending with a constant acceleration a, less than the acceleration due to gravity g. The weight of the balloon, with its basket and contents, is w. What weight, W, of ballast should be released so that the balloon will begin to be accelerated upward with constant acceleration a? Neglect air resistance.

5-20 A 64-lb block is pushed up a 37° inclined plane by a horizontal force of 100 lb. The coefficient of sliding friction is 0.25. Find (a) the acceleration, (b) the velocity of the block after it has moved a distance of 20 ft along the plane, and (c) the normal force exerted by the plane. Assume that all forces act at the center of the block.

5-21 A block rests on an inclined plane which makes an angle θ with the horizontal. The coefficient of sliding friction is 0.50, and the coefficient of static friction is 0.75. (a) As the angle θ is increased, find the minimum angle at which the block starts to slip. (b) At this angle, find the acceleration once the block has begun to move. (c) How long a time is required for the block to slip 20 ft along the inclined plane?

5-22 (a) What constant horizontal force is required to drag a 16-lb block along a horizontal surface with an acceleration of 4 ft s^{-2} if the coefficient of sliding friction between block and surface is 0.5? (b) What weight, hanging from a cord attached to the 16-lb block and passing over a small frictionless pulley, will produce this acceleration?

5-23 A block weighing 8 lb resting on a horizontal surface is connected by a cord passing over a light frictionless pulley to a hanging block weighing 8 lb. The coefficient of friction between the block and the horizontal surface is 0.5. Find (a) the tension in the cord, and (b) the acceleration of each block.

5-24 A block having a mass of 2 kg is projected up a long 30° incline with an initial velocity of 22 m s^{-1}. The coefficient of friction between the block and the plane is 0.3. (a) Find the friction force acting on the block as it moves up the plane. (b) How long does the block move up the plane? (c) How far does the block move up the plane? (d) How long does it take the block to slide from its position in part (c) to its starting point? (e) With what velocity does it arrive at this point? (f) If the mass of the block had been 5 kg instead of 2 kg, would the answers in the preceding parts be changed?

5-25 A 30-lb block on a level frictionless surface is attached by a cord passing over a small frictionless pulley to a hanging block originally at rest 4 ft above the floor. The hanging block strikes the floor in 2 s. Find (a) the weight of the hanging block, (b) the tension in the string while both blocks were in motion.

5-26 Two blocks, each having mass 20 kg, rest on frictionless surfaces as shown in Fig. 5-12. Assuming the pulleys to be light and frictionless, compute (a) the time required for block A to move 1 m down the plane, starting from rest, (b) the tension in the cord connecting the blocks.

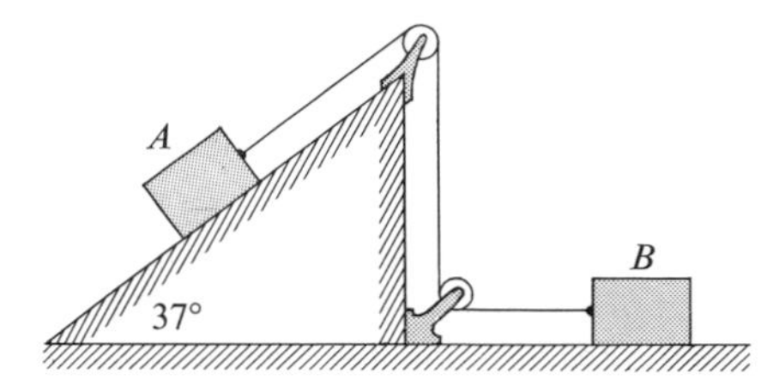

Fig. 5-12

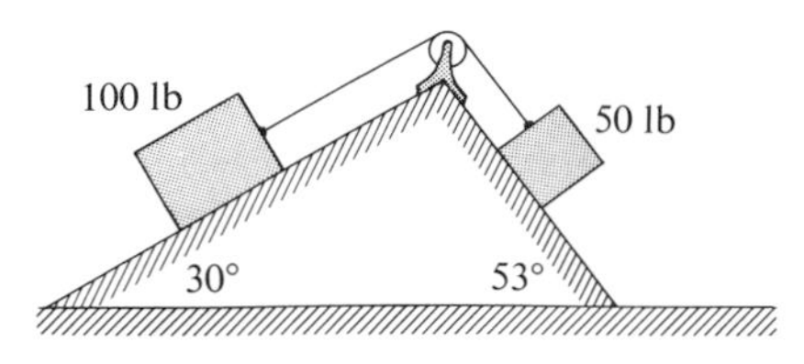

Fig. 5-15

5-27 A block of mass 200 g rests on the top of a block of mass 800 g. The combination is dragged along a level surface at constant velocity by a hanging block of mass 200 g as in Fig. 5-13(a). (a) The first 200-g block is removed from the 800-g block and attached to the hanging block, as in Fig. 5-13(b). What is now the acceleration of the system? (b) What is the tension in the cord attached to the 800-g block in part (b)?

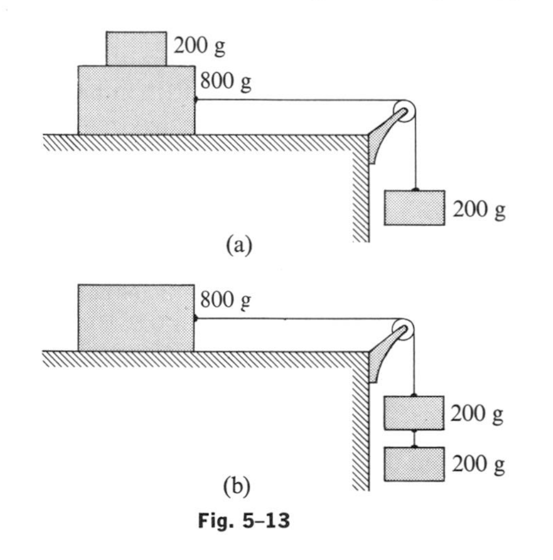

Fig. 5-13

5-28 Block *A* in Fig. 5-14 weighs 3 lb and block *B* weighs 30 lb. The coefficient of friction between *B* and the horizontal surface is 0.1. (a) What is the weight of block *C* if the acceleration of *B* is 6 ft s^{-2} toward the right? (b) What is the tension in each cord when *B* has the acceleration stated above?

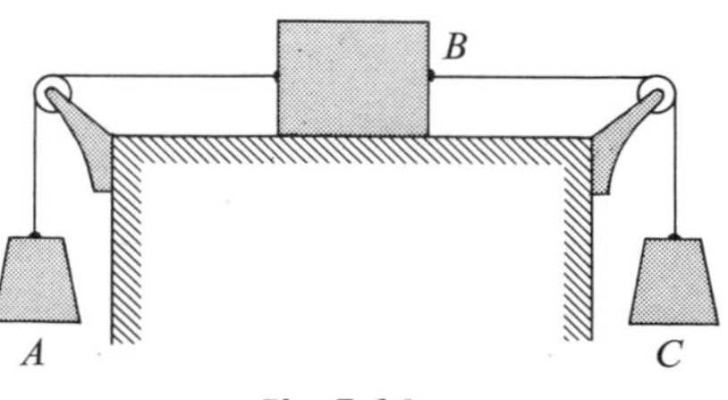

Fig. 5-14

5-29 Two blocks connected by a cord passing over a small frictionless pulley rest on frictionless planes as shown in Fig. 5-15. (a) Which way will the system move? (b) What is the acceleration of the blocks? (c) What is the tension in the cord?

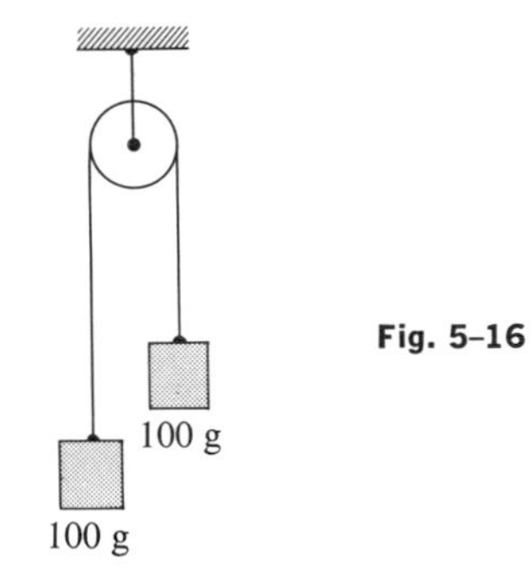

Fig. 5-16

5-30 Two 100-g blocks hang at the ends of a light flexible cord passing over a small frictionless pulley as in Fig. 5-16. A 40-g block is placed on the block on the right, and removed after 2 s. (a) How far will each block move in the first second after the 40-g block is removed? (b) What was the tension in the cord before the 40-g block was removed? After it was removed? (c) What was the tension in the cord supporting the pulley before the 40-g block was removed? Neglect the weight of the pulley.

5-31 Two 10-lb blocks hang at the ends of a cord as in Fig. 5-16. What weight must be added to one of the blocks to cause it to move down a distance of 4 ft in 2 s?

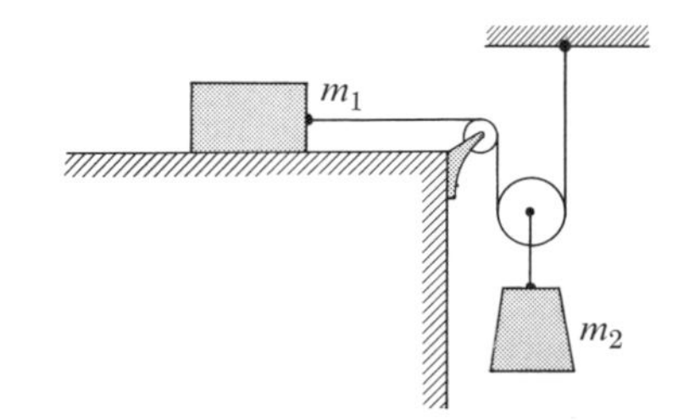

Fig. 5-17

5-32 In terms of m_1, m_2, and g, find the accelerations of both blocks in Fig. 5-17. Neglect all friction and the masses of the pulleys.

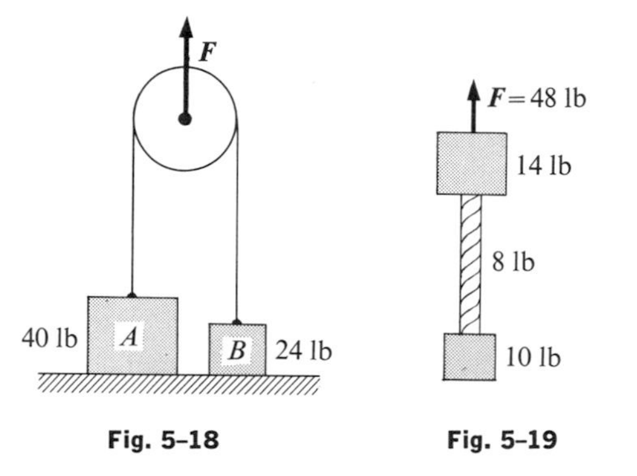

Fig. 5-18 Fig. 5-19

5-33 The bodies A and B in Fig. 5-18 weigh 40 lb and 24 lb, respectively. They are initially at rest on the floor and are connected by a weightless string passing over a weightless and frictionless pulley. An upward force $\boldsymbol{F}$ is applied to the pulley. Find the accelerations a_1 of body A and a_2 of body B when F is (a) 24 lb, (b) 40 lb, (c) 72 lb, (d) 90 lb, (e) 120 lb.

5-34 The two blocks in Fig. 5-19 are connected by a heavy uniform rope which weighs 8 lb. An upward force of 48 lb is applied as shown. (a) What is the acceleration of the system? (b) What is the tension at the top of the 8-lb rope? (c) What is the tension at the midpoint of the rope?

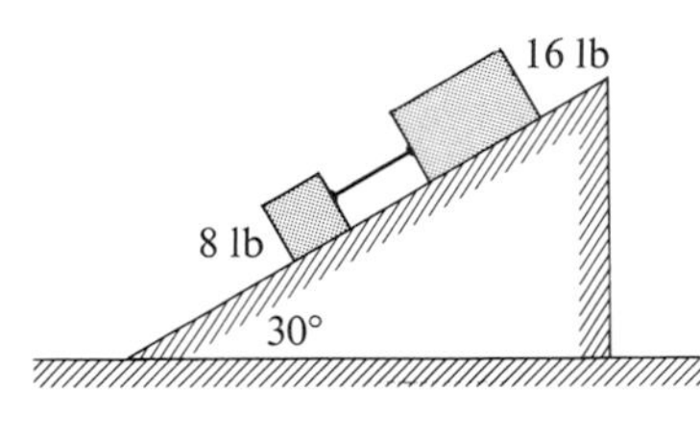

Fig. 5-20

5-35 Two blocks, weighing 8 and 16 lb, respectively, are connected by a string and slide down a 30° inclined plane, as in Fig. 5-20. The coefficient of sliding friction between the 8-lb block and the plane is 0.25, and between the 16-lb block and the plane it is 0.50. (a) Calculate the acceleration of each block. (b) Calculate the tension in the string.

5-36 Two bodies weighing 10 lb and 6 lb, respectively, hang 4 ft above the floor from the ends of a cord 12 ft long passing over a frictionless pulley. Both bodies start from rest. Find the maximum height reached by the 6-lb body.

5-37 A man who weighs 160 lb stands on a platform which weighs 80 lb. He pulls a rope which is fastened to the platform and runs over a pulley on the ceiling. With what force does he have to pull in order to give himself and the platform an upward acceleration of 2 ft s^{-2}?

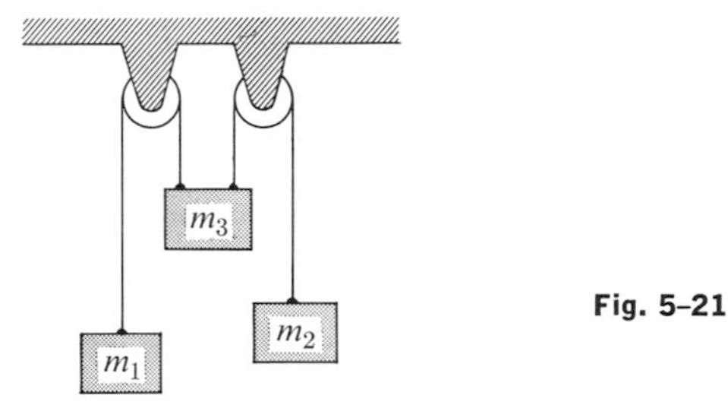

Fig. 5-21

5-38 In Fig. 5-21 there is depicted a double Atwood's machine. Calculate the acceleration of the system and the tensions in the strings supporting the bodies of mass m_1 and m_2. Neglect friction and the masses of the pulleys.

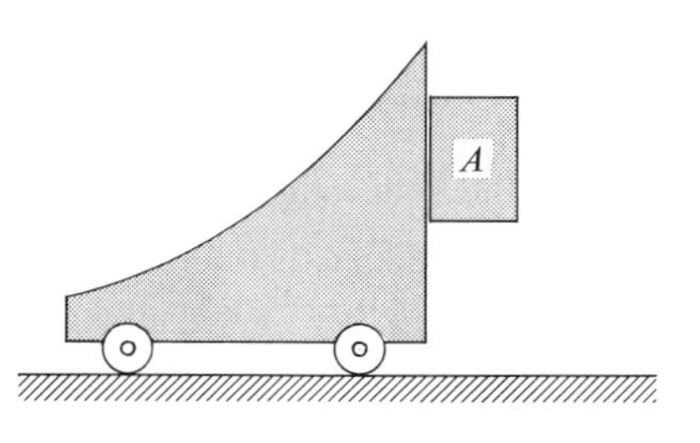

Fig. 5-22

5-39 What acceleration must the cart in Fig. 5-22 have in order that the block A will not fall? The coefficient of friction between the block and the cart is μ. How would the behavior of the block be described by an observer on the cart?

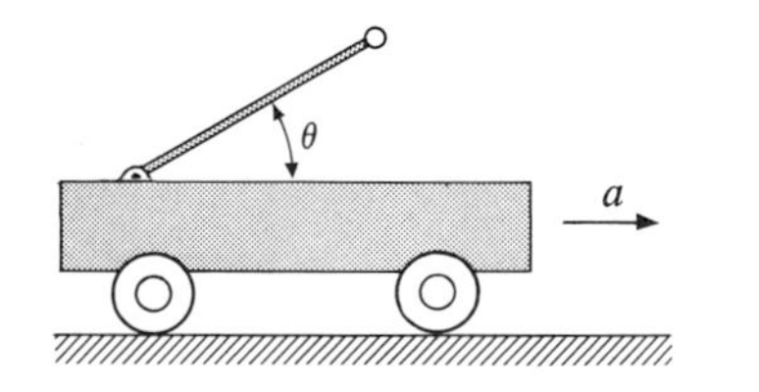

Fig. 5-23

5-40 The left end of the weightless rod shown in Fig. 5-23 is hinged to a cart. A heavy particle is attached to the right end. If the cart has an acceleration a to the right, find the angle θ.

How would this situation be described by an observer on the cart?

5-41 Which way will the accelerometer in Fig. 5-8 deflect under the following conditions? (a) The cart is moving toward the right and traveling faster. (b) The cart is moving toward the right and traveling slower. (c) The cart is moving toward the left and traveling faster. (d) The cart is moving toward the left and traveling slower. (e) The cart is at rest on a sloping surface. (f) The cart is given an upward velocity on a frictionless inclined plane. It first moves up, then stops, and then moves down. What is the deflection in each stage of the motion?

5-42 (a) In terms of the acceleration of gravity at the surface of the earth, g_0, and the earth's radius R, find the velocity with which a body must be projected vertically upward, in the absence of air resistance, to rise to an infinite distance above the earth's surface. This is called the *escape velocity*. (b) In terms of the same quantities, find the velocity with which a body will strike the earth's surface if it falls from rest toward the earth from an infinitely distant point. (c) Compute both of these velocities in miles per hour. (d) Explain why the velocities are not infinitely large.

5-43 The mass of the motorboat in Problem 4-37 is 100 slugs. Find the force decelerating the boat when its speed is (a) 20 ft per sec, and (b) 10 ft per sec. (c) If the boat is being towed at 10 ft per sec, what is the tension in the towline?

5-44 A body of mass $m = 5$ kg falls from rest in a viscous medium. The body is acted on by a net constant downward force of 20 N, and by a viscous retarding force proportional to its speed and equal to $5v$, where v is the speed in meters per second. (a) Find the initial acceleration, a_0. (b) Find the acceleration when the speed is 3 m s^{-1}. (c) Find the speed when the acceleration equals $0.1a_0$. (d) Find the terminal velocity, v_T. (e) Find the relaxation time, t_R. (f) Find the coordinate, velocity, and acceleration 2 s after the start of the motion. (g) Find the time required to reach a speed $0.9v_T$. (h) Construct a graph of v versus t, for a time of 3 s.

5-45 A body falls from rest through a medium which exerts a resisting force that varies directly with the square of the velocity ($R = -kv^2$). (a) Draw a diagram showing the direction of motion and indicate with the aid of vectors all of the forces acting on the body. (b) Apply Newton's second law and infer from the resulting equation the general properties of the motion. (c) Show that the body acquires a terminal velocity and calculate it. (d) Find the relaxation time. (e) Derive the equation for the velocity at any time.

5-46 A particle of mass m, originally at rest, is subjected to a force whose direction is constant but whose magnitude varies with the time according to the relation

$$F = F_0\left[1 - \left(\frac{t - T}{T}\right)^2\right],$$

where F_0 and T are constants. The force acts only for the time interval $2T$. (a) Make a rough graph of F versus t. (b) Prove that the speed v of the particle after a time $2T$ has elapsed is equal to $4F_0T/3m$. (c) Choose numbers for v, T, and m that might be appropriate to a batted baseball, and calculate the force F_0. Judge whether the answer is sensible.

6 Motion in a Plane

6-1 MOTION IN A PLANE

Thus far we have discussed only motion along a straight line, or *rectilinear motion*. In this chapter we shall consider *plane motion*, that is, motion in a curved path which lies in a fixed plane. Examples of such motion are the flight of a thrown or batted baseball, a projectile shot from a gun, a ball whirled at the end of a cord, the motion of the moon or of a satellite around the earth, and the motion of the planets around the sun.

If the motion is referred to a set of rectangular coordinate axes x and y, the equation of the path expresses y as a function of x, $y = f(x)$. Very often one is interested in the position of the moving body as a function of time. If s is the distance along the path from some fixed point to the position of the body, its position at any time is given by an equation of the form $s = f(t)$. It is usually simpler, however, to deal with the x- and y-coordinates separately and to describe the motion by the two equations

$$x = f_1(t), \qquad y = f_2(t). \tag{6-1}$$

These can be considered as *parametric equations* of the path, expressing the coordinates x and y in terms of the parameter t.

Problems in plane motion can be divided into two classes. In one, the motion of the particle is known and we wish to determine its velocity and acceleration and the resultant force acting on it. An example is that of a ball attached to a cord and whirled in a circle at constant speed. What is the tension in the cord? In the other, the force acting on a particle is known at every point of space and we wish to find the equation of motion of the particle. Examples are the orbit of a planet around the sun, or the path followed by a rocket.

6-2 AVERAGE AND INSTANTANEOUS VELOCITY

Consider a particle moving along the curved path in Fig. 6-1(a). Points P and Q represent two positions of the particle. Its displacement as it moves from P to Q is the vector $\Delta \boldsymbol{s}$. Just as in the case of rectilinear motion,

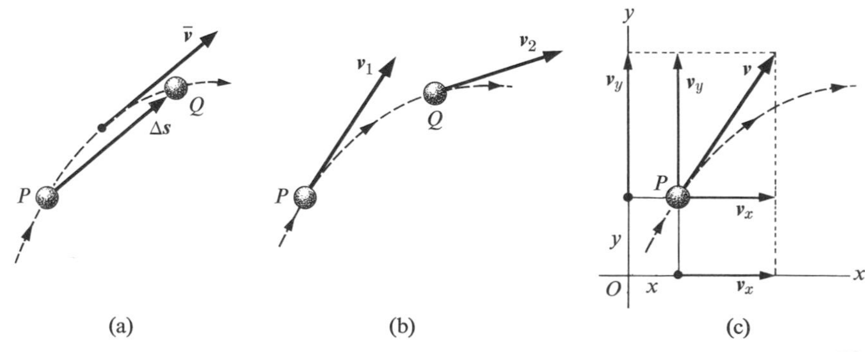

6-1 (a) The vector $\boldsymbol{v} = \Delta s/\Delta t$ represents the average velocity between P and Q. (b) Vectors $\boldsymbol{v}_1$ and $\boldsymbol{v}_2$ represent the instantaneous velocities at P and Q. (c) The velocities $\boldsymbol{v}_x$ and $\boldsymbol{v}_y$ of the projections of P are the rectangular components of $\boldsymbol{v}$.

the average velocity $\bar{\boldsymbol{v}}$ of the particle is defined as the vector displacement $\Delta \boldsymbol{s}$ divided by the elapsed time Δt:

$$\text{Average velocity } \bar{\boldsymbol{v}} = \Delta \boldsymbol{s}/\Delta t. \tag{6-2}$$

The average velocity would be the same for any path that would take the particle from P to Q in the time interval Δt.

Average velocity is a vector quantity, in the same direction as the vector $\Delta \boldsymbol{s}$. Because it is associated with the entire displacement $\Delta \boldsymbol{s}$, the vector $\bar{\boldsymbol{v}}$ has been constructed in Fig. 6-1(a) at a point midway between P and Q.

The *instantaneous* velocity $\boldsymbol{v}$ at the point P is defined in magnitude and direction as the limit approached by the average velocity when point Q is taken closer and closer to point P:

$$\text{Instantaneous velocity } \boldsymbol{v} = \lim_{\Delta t \to 0} \frac{\Delta \boldsymbol{s}}{\Delta t} = \frac{d\boldsymbol{s}}{dt}. \tag{6-3}$$

As point Q approaches point P, the direction of the vector $\Delta \boldsymbol{s}$ approaches that of the tangent to the path at P, so that the instantaneous velocity vector at any point is tangent to the path at that point. The instantaneous velocities at points P and Q are shown in Fig. 6-1(b).

In Fig. 6-1(c), the motion of a particle is referred to a rectangular coordinate system. As the particle moves along its path, its projections onto the x- and y-axes move along these axes in rectilinear motion, with velocities of magnitudes

$$v_x = \frac{dx}{dt}, \qquad v_y = \frac{dy}{dt}.$$

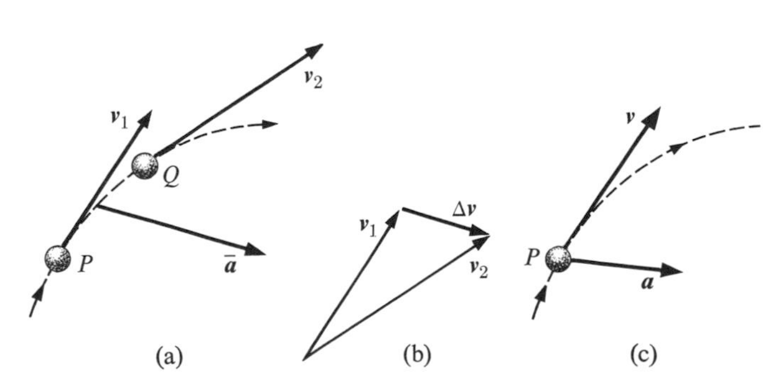

6-2 (a) The vector $\boldsymbol{a} = \Delta \boldsymbol{v}/\Delta t$ represents the average acceleration between P and Q. (b) Construction for obtaining $\Delta \boldsymbol{v} = \boldsymbol{v}_2 - \boldsymbol{v}_1$. (c) Instantaneous velocity $\boldsymbol{v}$ and instantaneous acceleration $\boldsymbol{a}$ at point P. Vector $\boldsymbol{v}$ is tangent to path; vector $\boldsymbol{a}$ points toward concave side of path.

The velocities of the projections, however, are also the rectangular components of the velocity $\boldsymbol{v}$ of the particle. Thus if the x- and y-coordinates of the particle are known as functions of time, the components v_x and v_y can be found by differentiation and these can then be added vectorially as in Fig. 6-1(c) to obtain the magnitude and direction of the velocity $\boldsymbol{v}$. The direction of the path is then determined also, since $\boldsymbol{v}$ is tangent to the path.

Conversely, if the velocities v_x and v_y are known as functions of time, the coordinates x and y can be found by integration.

6-3 AVERAGE AND INSTANTANEOUS ACCELERATION

In Fig. 6-2(a), the vectors $\boldsymbol{v}_1$ and $\boldsymbol{v}_2$ represent the instantaneous velocities, at points P and Q, of a particle moving in a curved path. The velocity $\boldsymbol{v}_2$ necessarily differs in *direction* from the velocity $\boldsymbol{v}_1$. The diagram has been constructed for a case in which it differs in *magnitude* also, although in special cases the magnitude of the velocity may remain constant.

The *average acceleration* $\bar{\boldsymbol{a}}$ of the particle as it moves from P to Q is defined, just as in the case of rectilinear motion, as the *vector change in velocity*, $\Delta \boldsymbol{v}$, divided by the time interval Δt:

$$\text{Average acceleration } \bar{\boldsymbol{a}} = \frac{\Delta \boldsymbol{v}}{\Delta t}. \tag{6-4}$$

Average acceleration is a vector quantity, in the same direction as the vector $\Delta \boldsymbol{v}$.

The vector change in velocity, $\Delta \boldsymbol{v}$, means the vector difference $\boldsymbol{v}_2 - \boldsymbol{v}_1$:

$$\Delta \boldsymbol{v} = \boldsymbol{v}_2 - \boldsymbol{v}_1,$$

or

$$\boldsymbol{v}_2 = \boldsymbol{v}_1 + \Delta \boldsymbol{v}.$$

As explained in Section 1-9, the vector difference $\Delta \boldsymbol{v}$ can be found by drawing the vectors $\boldsymbol{v}_1$ and $\boldsymbol{v}_2$ from a common point, as in Fig. 6-2(b), and constructing the vector from the tip of $\boldsymbol{v}_1$ to the tip of $\boldsymbol{v}_2$. Then $\boldsymbol{v}_2$ is the vector sum of $\boldsymbol{v}_1$ and $\Delta \boldsymbol{v}$.

The average acceleration vector, $\bar{\boldsymbol{a}} = \Delta \boldsymbol{v}/\Delta t$, is shown in Fig. 6-2(a) at a point midway between P and Q.

The *instantaneous acceleration* $\boldsymbol{a}$ at point P is defined in magnitude and direction as the limit approached by

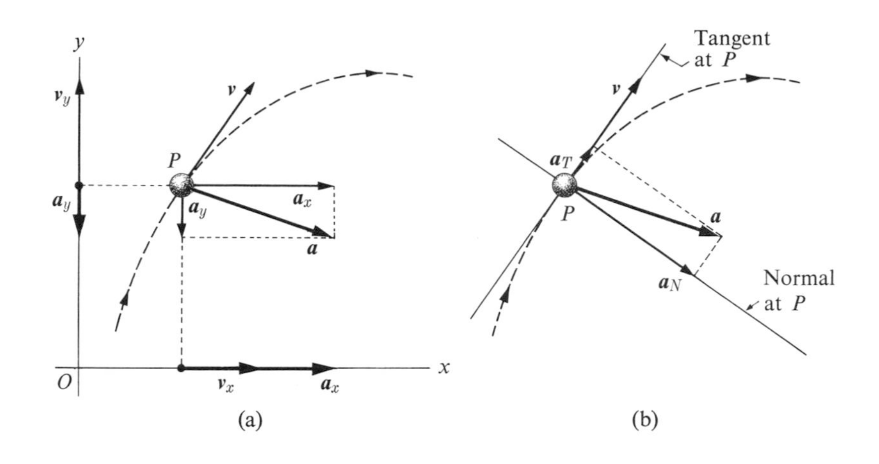

6-3 In (a) the acceleration $\boldsymbol{a}$ is resolved into rectangular components $\boldsymbol{a}_x$ and $\boldsymbol{a}_y$. In (b) it is resolved into a normal component $\boldsymbol{a}_N$ and a tangential component $\boldsymbol{a}_T$.

the average acceleration when point Q approaches point P and $\Delta \boldsymbol{v}$ and Δt both approach zero:

$$\text{Instantaneous acceleration } \boldsymbol{a} = \lim_{\Delta t \to 0} \frac{\Delta \boldsymbol{v}}{\Delta t} = \frac{d\boldsymbol{v}}{dt}. \quad (6\text{-}5)$$

The instantaneous acceleration vector at point P is shown in Fig. 6-2(c). Note that it does *not* have the same direction as the velocity vector. Reference to the construction of Fig. 6-2(b) will show that the acceleration vector must always lie on the *concave* side of the curved path.

6-4 COMPONENTS OF ACCELERATION

Figure 6-3(a) again shows the motion of a particle referred to a rectangular coordinate system. The accelerations of the projections of the particle onto the x- and y-axes are

$$a_x = \frac{dv_x}{dt} = \frac{d^2x}{dt^2}, \qquad a_y = \frac{dv_y}{dt} = \frac{d^2y}{dt^2}.$$

These accelerations, however, are also the rectangular components of the acceleration $\boldsymbol{a}$ of the particle. Thus if the x- and y-coordinates of the particle are known as functions of time, the components a_x and a_y can be found by differentiation and these can then be combined vectorially as in Fig. 6-3(a) to obtain the magnitude and direction of the acceleration $\boldsymbol{a}$. When this is known, the force on the particle can be found, in magnitude and direction, from Newton's second law,

$$\boldsymbol{F} = m\frac{d\boldsymbol{v}}{dt} = m\boldsymbol{a}.$$

Conversely, if the force $\boldsymbol{F}$ is known at every point, the acceleration $\boldsymbol{a}$ and its components $\boldsymbol{a}_x$ and $\boldsymbol{a}_y$ can be found from Newton's second law. One integration then gives the velocity components v_x and v_y, and a second integration gives the coordinates x and y.

The acceleration of a particle moving in a curved path can also be resolved into rectangular components $\boldsymbol{a}_N$ and $\boldsymbol{a}_T$, in directions *normal* and *tangential* to the path, as shown in Fig. 6-3(b). Unlike the rectangular components referred to a set of fixed axes, the normal and tangential components do not have fixed directions in space. They do, however, have the following physical significance, namely, the tangential component $\boldsymbol{a}_T$ arises from a change in the *magnitude* of the velocity vector $\boldsymbol{v}$, while the normal component $\boldsymbol{a}_N$ arises from a change in the *direction* of the velocity.

This is illustrated in Fig. 6-4, which corresponds to Fig. 6-2(b). The vector from O to A represents the velocity $\boldsymbol{v}_1$ of a particle at point P in Fig. 6-2(a), and the vector from O to B represents the velocity $\boldsymbol{v}_2$ at point Q. The change in velocity, $\Delta \boldsymbol{v}$, is given by the vector from A to B.

Note that the vectors $\Delta \boldsymbol{v}$ in Figs. 6-4 and 6-2(b) have the same direction and (apart from the difference in scale of the two diagrams) the same length.

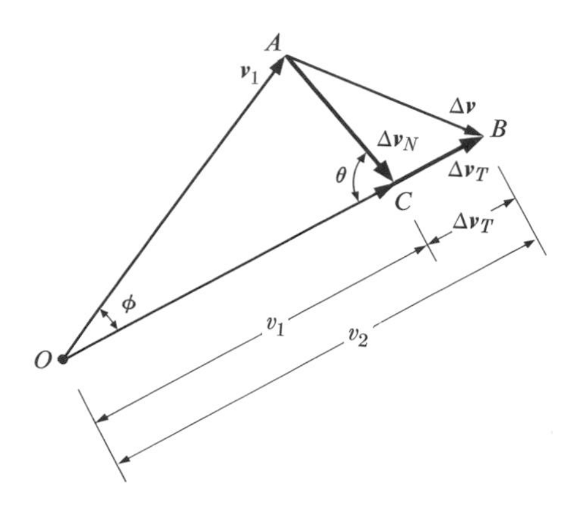

6–4 The vector $\Delta \boldsymbol{v}$ is resolved into normal and tangential components $\Delta \boldsymbol{v}_N$ and $\Delta \boldsymbol{v}_T$. The normal component is the change in velocity resulting from the change in direction of $\boldsymbol{v}$; the tangential component is the change resulting from a change in the magnitude of $\boldsymbol{v}$.

The vector from O to C in Fig. 6–4 has the same length as $\boldsymbol{v}_1$. The vector $\Delta \boldsymbol{v}$ can be resolved into components represented by the vectors from A to C, and from C to B. The length of the vector from C to B equals the difference in *length* between the vectors $\boldsymbol{v}_2$ and $\boldsymbol{v}_1$. That is, this vector represents the change in *magnitude* of the velocity, and when divided by Δt gives the component of average acceleration resulting from this change in magnitude.

If the magnitude of the velocity did *not* change between points P and Q, the velocity $\boldsymbol{v}_2$ at point Q would be represented by the vector from O to C in Fig. 6–4. In this case, however, *there would still be a change in the vector velocity*, represented by the vector from A to C. This change would result from the change in *direction* of the velocity vector, and when divided by Δt would give the component of average acceleration resulting from this change in direction. That is, *motion in a curved path with constant speed is accelerated motion*, because velocity is a vector quantity which can change in magnitude, in direction, or, as in Fig. 6–4, in both.

Now suppose that point Q in Fig. 6–2(a) approaches point P. The vector from O to B in Fig. 6–4 then swings upward toward the vector $\boldsymbol{v}_1$. The angle ϕ becomes smaller and smaller and the angle θ approaches 90°. The vector from A to C becomes more and more nearly perpendicular to $\boldsymbol{v}_1$ and the vector from C to B becomes more and more nearly parallel to $\boldsymbol{v}_1$. In the limit, the vector from A to C becomes normal to $\boldsymbol{v}_1$ (and hence normal to the path) and the vector from C to B becomes parallel to $\boldsymbol{v}_1$ (and hence tangent to the path). Thus although the vectors labeled $\Delta \boldsymbol{v}_N$ and $\Delta \boldsymbol{v}_T$ in Fig. 6–4 are not normal and parallel to $\boldsymbol{v}_1$ in this figure, they become so in the limit as point Q approaches point P. The limiting value of $\Delta \boldsymbol{v}_N/\Delta t$ equals the normal component of acceleration $\boldsymbol{a}_N$, and the limiting value of $\Delta \boldsymbol{v}_T/\Delta t$ equals the tangential component of acceleration $\boldsymbol{a}_T$.

Is the acceleration resulting from a change in the *direction* of a velocity as "real" as that arising from a change in its *magnitude*? From a purely kinematical viewpoint the answer is of course "Yes," since by definition acceleration equals the vector rate of change of velocity. A more satisfying answer is that a force must be exerted on a body to change its direction of motion, as well as to increase or decrease its speed. In the absence of an external force, a body continues to move not only with constant speed but also *in a straight line*. When a body moves in a *curved* path, a transverse force must be exerted on it to deviate it sidewise, and the ratio of transverse force to normal acceleration is found to equal the ratio of longitudinal force to longitudinal acceleration. That is, if $\boldsymbol{F}_N$ and $\boldsymbol{F}_T$ are the normal and tangential components of the force $\boldsymbol{F}$ on a body moving in a curved path, Newton's second law (with the same value of m) applies to both of these components:

$$F_N = ma_N, \qquad F_T = ma_T.$$

In fact, the usual experimental method of measuring the mass of an individual ion is to project it into a magnetic field. The magnetic field exerts a transverse force on the ion and the mass of the ion is obtained by dividing this force by the measured transverse acceleration.

It follows from the discussion above that if the force on a particle is always normal to the path, then $F_T = 0$, $a_T = 0$, and the particle has no tangential component of acceleration. The *magnitude* of the velocity then remains constant and the only effect of the force is to change the direction of motion, that is, to deviate the particle sidewise.

If the force has no normal component, then $F_N = 0$, $a_N = 0$, and there is no change in the *direction* of the velocity; that is, the particle moves in a straight line.

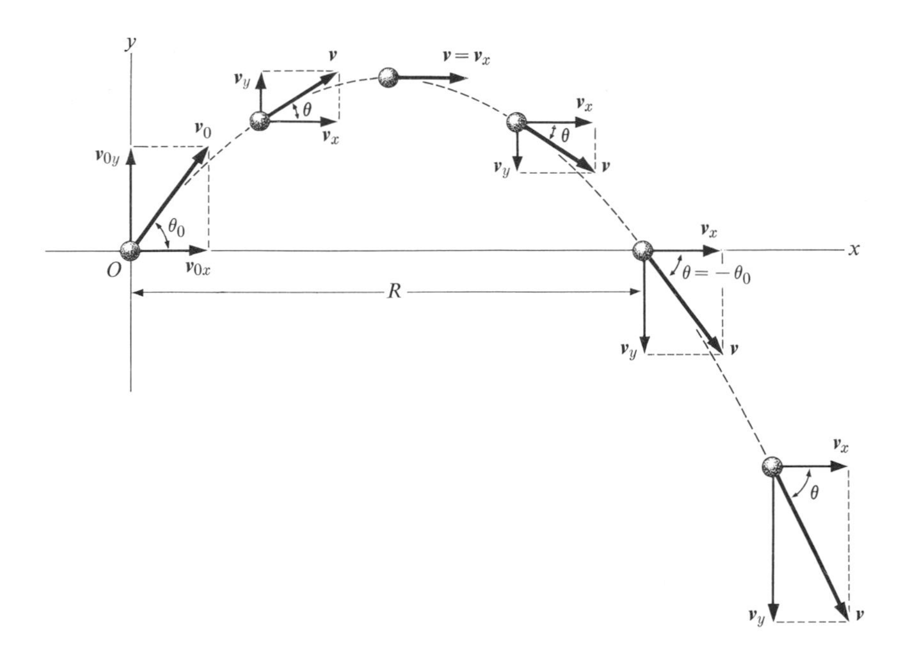

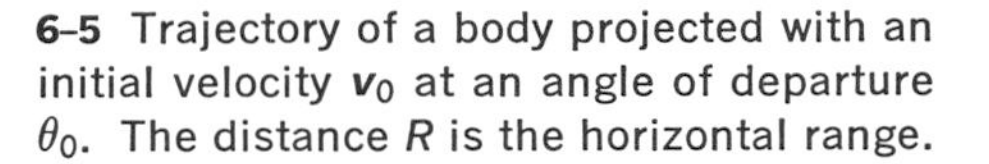

6–5 Trajectory of a body projected with an initial velocity $\boldsymbol{v}_0$ at an angle of departure θ_0. The distance R is the horizontal range.

6–5 MOTION OF A PROJECTILE

Any object that is given an initial velocity and which subsequently follows a path determined by the gravitational force acting on it and by the frictional resistance of the atmosphere is called a *projectile*. Thus the term applies to a missile shot from a gun, a rocket after its fuel is exhausted, a bomb released from an airplane, or a thrown or batted baseball. The motion of a freely falling body discussed in Chapter 4 is a special case of projectile motion. The path followed by a projectile is called its *trajectory*.

The gravitational force on a projectile is directed toward the center of the earth and is inversely proportional to the square of the distance from the earth's center. Here we shall consider only trajectories which are of sufficiently short range so that the gravitational force can be considered constant in magnitude and direction. The motion will be referred to axes fixed with respect to the earth. Since this is not an inertial system, it is not strictly correct to use Newton's second law to relate the force on the projectile to its acceleration. However, for trajectories of short range, the error is very small. Finally, all effects of air resistance will be ignored, so that our results apply only to motion in a vacuum on a flat, nonrotating earth.

Since the only force on a projectile in this idealized case is its weight, considered constant in magnitude and direction, the motion is best referred to a set of rectangular coordinate axes. We shall take the x-axis horizontal, the y-axis vertical, and the origin at the point where the projectile starts its free flight, for example at the muzzle of a gun or the point where it leaves the thrower's hand. The x-component of the force on the projectile is then zero and the y-component is the weight of the projectile, $-mg$. Then from Newton's second law,

$$a_x = \frac{F_x}{m} = 0, \qquad a_y = \frac{F_y}{m} = \frac{-mg}{m} = -g.$$

That is, the horizontal component of acceleration is zero and the vertical component is downward and equal to that of a freely falling body. The forward component of velocity does not "support" the projectile in flight. Since zero acceleration means constant velocity, the motion can be described as a combination of *horizontal motion with constant velocity* and *vertical motion with constant acceleration*.

Consider next the velocity of the projectile. In Fig. 6–5, x- and y-axes have been constructed with the origin at the point where the projectile begins its free flight. We shall set $t = 0$ at this point. The velocity at the origin is represented by the vector $\mathbf{v}_0$, called the *initial velocity* or the *muzzle velocity* if the projectile is shot from a gun. The angle θ_0 is the *angle of departure*. The initial velocity has been resolved into a horizontal component $\mathbf{v}_{0x}$, of magnitude $v_0 \cos\theta_0$, and a vertical component $\mathbf{v}_{0y}$, of magnitude $v_0 \sin\theta_0$.

Since the horizontal velocity component is constant, we have at any later time t,

$$v_x = v_{0x} = v_0 \cos\theta_0.$$

The vertical acceleration is $-g$, so the vertical velocity component at time t is

$$v_y = v_{0y} - gt = v_0 \sin\theta_0 - gt.$$

These components can be added vectorially to find the resultant velocity $\mathbf{v}$. Its magnitude is

$$v = \sqrt{v_x^2 + v_y^2},$$

and the angle θ it makes with the horizontal is given by

$$\tan\theta = v_y/v_x.$$

The velocity vector $\mathbf{v}$ is tangent to the trajectory, so its direction is the same as that of the trajectory.

The coordinates of the projectile at any time can now be found from the equations of motion with constant velocity, and with constant acceleration. The x-coordinate is

$$x = v_{0x}t = (v_0 \cos\theta_0)t$$

and the y-coordinate is

$$y = v_{0y}t - \tfrac{1}{2}gt^2 = (v_0 \sin\theta_0)t - \tfrac{1}{2}gt^2.$$

The two preceding equations give the equation of the trajectory in terms of the parameter t. The equation in terms of x and y can be obtained by eliminating t. This gives

$$y = (\tan\theta_0)x - \frac{g}{2v_0^2 \cos^2\theta_0} x^2. \qquad (6\text{–}6)$$

The quantities v_0, $\tan\theta_0$, $\cos\theta_0$, and g are constants, so the equation has the form

$$y = ax - bx^2,$$

which will be recognized as the equation of a *parabola*.

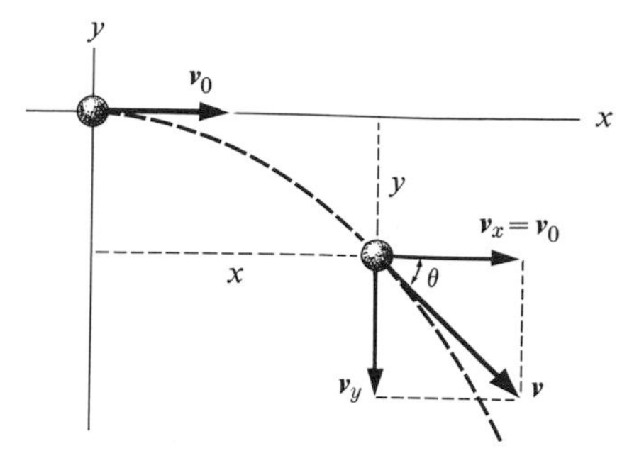

6–6 Trajectory of a body projected horizontally.

Example 1. A ball is projected horizontally with a velocity $\mathbf{v}_0$ of magnitude 8 ft s^{-1}. Find its position and velocity after $\frac{1}{4}$ s (see Fig. 6–6).

In this case, the departure angle is zero. The initial vertical velocity component is therefore zero. The horizontal velocity component equals the initial velocity and is constant.

The x- and y-coordinates, when $t = \frac{1}{4}$ s, are

$$x = v_x t = 8 \text{ ft s}^{-1} \times \tfrac{1}{4} \text{ s} = 2 \text{ ft},$$
$$y = -\tfrac{1}{2}gt^2 = -\tfrac{1}{2} \times 32 \text{ ft s}^{-2} \times (\tfrac{1}{4} \text{ s})^2 = -1 \text{ ft}.$$

The components of velocity are

$$v_x = v_0 = 8 \text{ ft s}^{-1},$$
$$v_y = -gt = -32 \text{ ft s}^{-2} \times \tfrac{1}{4} \text{ s} = -8 \text{ ft s}^{-1}.$$

The resultant velocity is

$$v = \sqrt{v_x^2 + v_y^2} = 8\sqrt{2} \text{ ft s}^{-1}.$$

The angle θ is

$$\theta = \tan^{-1}\frac{v_y}{v_x} = \tan^{-1}\frac{8 \text{ ft s}^{-1}}{-8 \text{ ft s}^{-1}} = -45°.$$

That is, the velocity is 45° *below* the horizontal.

Example 2. In Fig. 6–5, let $v_0 = 160$ ft s^{-1}, $\theta_0 = 53°$. Then

$$v_{0x} = v_0 \cos\theta_0 = 160 \text{ ft s}^{-1} \times 0.60 = 96 \text{ ft s}^{-1},$$
$$v_{0y} = v_0 \sin\theta_0 = 160 \text{ ft s}^{-1} \times 0.80 = 128 \text{ ft s}^{-1}.$$

(a) Find the position of the projectile, and the magnitude and direction of its velocity, when $t = 2.0$ s. (This corresponds to the first position of the projectile in Fig. 6–5, after starting its flight.) We have

$$x = 96 \text{ ft s}^{-1} \times 2.0 \text{ s} = 192 \text{ ft},$$
$$y = (128 \text{ ft s}^{-1} \times 2.0 \text{ s}) - (\tfrac{1}{2} \times 32 \text{ ft s}^{-2} \times (2.0 \text{ s})^2) = 192 \text{ ft},$$

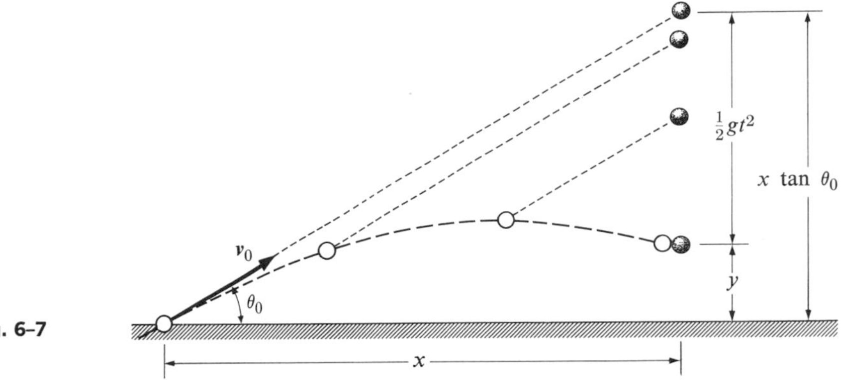

Fig. 6-7

Also

$$v_x = 96 \text{ ft s}^{-1},$$
$$v_y = 128 \text{ ft s}^{-1} - (32 \text{ ft s}^{-2} \times 2.0 \text{ s}) = 64 \text{ ft s}^{-1},$$
$$\theta = \tan^{-1} \frac{64 \text{ ft s}^{-1}}{96 \text{ ft s}^{-1}} = \tan^{-1} 0.667 = 33.5°.$$

(b) Find the time at which the projectile reaches the highest point of its flight, and find the elevation of this point.

At the highest point, the vertical velocity v_y is zero. If t_1 is the time at which this point is reached,

$$v_y = 0 = 128 \text{ ft s}^{-1} - (32 \text{ ft s}^{-2} \times t_1),$$
$$t_1 = 4 \text{ s}.$$

The elevation of the point is

$$y = (128 \text{ ft s}^{-1} \times 4 \text{ s}) - (\tfrac{1}{2} \times 32 \text{ ft s}^{-2} \times (4 \text{ s})^2) = 256 \text{ ft}.$$

(c) Find the *horizontal range R*, that is, the horizontal distance from the starting point to the point at which the projectile returns to its original elevation and at which, therefore, $y = 0$. Let t_2 be the time at which this point is reached. Then

$$y = 0 = (128 \text{ ft s}^{-1} \times t_2) - (\tfrac{1}{2} \times 32 \text{ ft s}^{-2} \times t_2^2).$$

This quadratic equation has two roots,

$$t_2 = 0 \quad \text{and} \quad t_2 = 8 \text{ s},$$

corresponding to the two points at which $y = 0$. Evidently the time desired is the second root, $t_2 = 8$ s, which is just twice the time to reach the highest point. The time of descent therefore equals the time of rise.

The horizontal range is

$$R = v_x t_2 = 96 \text{ ft s}^{-1} \times 8 \text{ s} = 768 \text{ ft}.$$

The vertical velocity at this point is

$$v_y = 128 \text{ ft s}^{-1} - (32 \text{ ft s}^{-2} \times 8 \text{ s}) = -128 \text{ ft s}^{-1}.$$

That is, the vertical velocity has the same magnitude as the initial vertical velocity, but the opposite direction. Since v_x is constant, the angle below the horizontal at this point equals the angle of departure.

(d) If unimpeded, the projectile continues to travel beyond its horizontal range. It is left as an exercise to compute the position and velocity at a time 10 s after the start, corresponding to the last position shown in Fig. 6-5. The results are:

$$x = 960 \text{ ft},$$
$$y = -320 \text{ ft},$$
$$v_x = 96 \text{ ft s}^{-1},$$
$$v_y = -192 \text{ ft s}^{-1}.$$

Example 3. Figure 6-7 illustrates an interesting experimental demonstration of the properties of projectile motion. A ball (shown by the open circle) is projected directly toward a second ball (the solid circle). The second ball is released from rest at the instant the first is projected, and the balls collide as shown regardless of the values of the initial velocity. To show that this happens, we note that the initial elevation of the second ball is $x \tan \theta_0$, and that in time t it falls a distance $\frac{1}{2}gt^2$. Its elevation at the instant of collision is therefore

$$y = x \tan \theta_0 - \tfrac{1}{2}gt^2,$$

which is the same as the elevation of the first ball as given by Eq. (6-6).

For any given initial velocity, there is one particular angle of departure for which the horizontal range is a maximum. To find this angle, write the general algebraic expression for the horizontal range R as

$$R = v_x t_2 = v_0 \cos\theta_0 \times \frac{2v_0 \sin\theta_0}{g}$$
$$= \frac{2v_0^2 \sin\theta_0 \cos\theta_0}{g} = \frac{v_0^2 \sin 2\theta_0}{g}. \qquad (6\text{-}7)$$

The maximum range is that for which $dR/d\theta_0 = 0$:

$$\frac{dR}{d\theta_0} = \frac{2v_0^2}{g} \cos 2\theta_0 = 0,$$
$$\cos 2\theta_0 = 0,$$
$$2\theta_0 = 90^\circ, \qquad \theta_0 = 45^\circ,$$

and the maximum horizontal range is attained with a departure angle of 45°.

From the standpoint of gunnery, what we usually want to know is what the departure angle should be for a given muzzle velocity v_0 in order to hit a target whose position is known. Let us assume target and gun are at the same elevation. Then, from Eq. (6-7),

$$\theta_0 = \tfrac{1}{2} \sin^{-1} \left(\frac{Rg}{v_0^2}\right).$$

Provided R is less than the maximum range, this equation has two solutions for values of θ_0 between 0° and 90°.

Thus if $R = 800$ ft, $g = 32$ ft s^{-2}, and $v_0 = 200$ ft s^{-1},

$$\theta_0 = \tfrac{1}{2} \sin^{-1} \left[\frac{800 \text{ ft} \times 32 \text{ ft s}^{-2}}{(200 \text{ ft s}^{-1})^2}\right]$$
$$= \tfrac{1}{2} \sin^{-1} 0.64.$$

But $\sin^{-1} 0.64 = 40^\circ$, or $180^\circ - 40^\circ$. Therefore

$$\theta_0 = 20^\circ \quad \text{or} \quad 70^\circ.$$

Either of these angles gives the same range. Of course both the time of flight and the maximum height reached are greater for the high-angle trajectory.

Figure 6-8 is a multiflash photograph of the trajectory of a ball, to which have been added x- and y-axes and the initial velocity vector. The horizontal distances between consecutive positions are all equal, showing that the horizontal velocity component is constant. The vertical distances first decrease and then increase, showing that the vertical motion is accelerated.

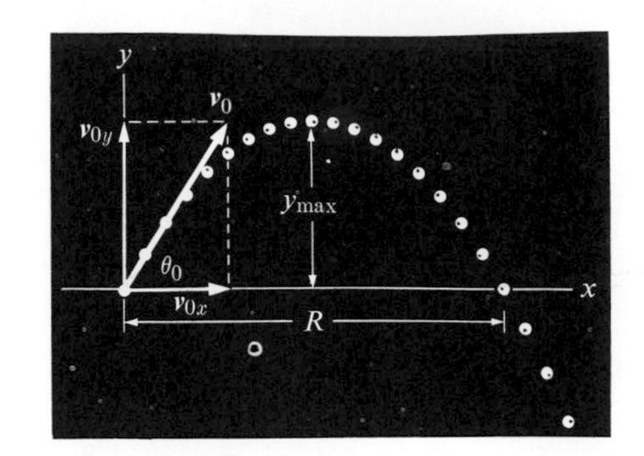

6-8 Trajectory of a body projected at an angle with the horizontal.

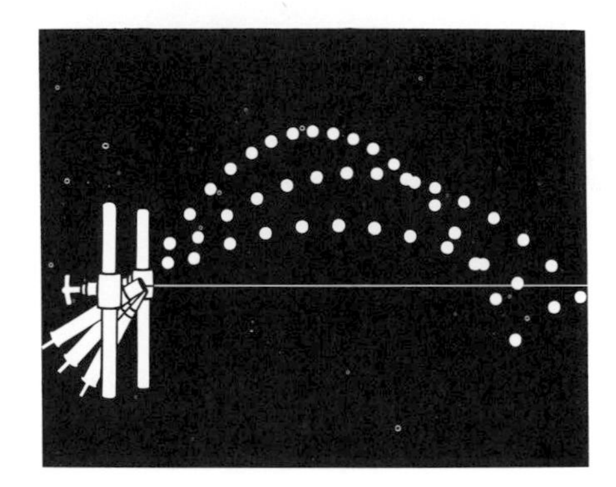

6-9 An angle of departure of 45° gives the maximum horizontal range.

Figure 6-9 is a composite photograph of the three trajectories of a ball projected from a spring gun with departure angles of 30°, 45°, and 60°. It will be seen that the horizontal ranges are (nearly) the same for the 30° and 60° angles and that both are less than the range when the angle is 45°. (The spring gun does not impart exactly the same initial velocity to the ball as the departure angle is altered.)

If the departure angle is *below* the horizontal, as for instance in the motion of a ball after rolling off a sloping roof or the trajectory of a bomb released from a dive bomber, exactly the same principles apply. The horizontal velocity component remains constant and equal to $v_0 \cos\theta_0$. The vertical motion is the same as that of a body projected *downward* with an initial velocity $v_0 \sin\theta_0$.

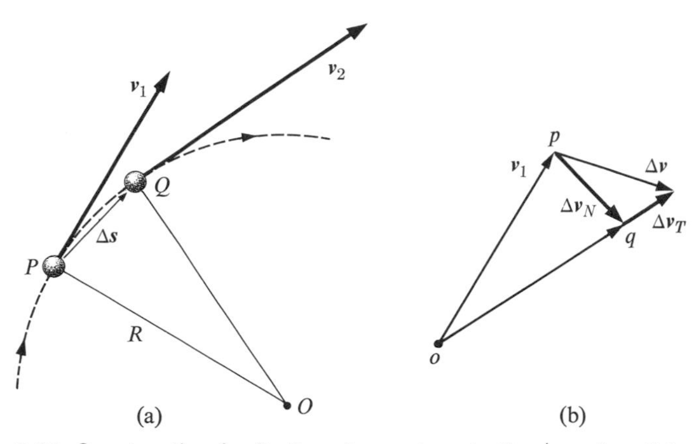

6–10 Construction for finding change in velocity, $\Delta \boldsymbol{v}$, of particle moving in a circle.

6–6 CIRCULAR MOTION

The acceleration of a particle moving in a curved path can be resolved into components normal and tangential to the path. There is a simple relation between the normal component of acceleration, the speed of the particle, and the radius of curvature of the path. We now derive this for the special case of motion in a circle.

Figure 6–10(a) represents a particle moving in a circular path of radius R with center at O. Vectors $\boldsymbol{v}_1$ and $\boldsymbol{v}_2$ represent its velocities at points P and Q. The vector change in velocity, $\Delta \boldsymbol{v}$, is obtained in Fig. 6–10(b), which is the same as Fig. 6–4. Vectors $\Delta \boldsymbol{v}_N$ and $\Delta \boldsymbol{v}_T$ are the normal and tangential components of $\Delta \boldsymbol{v}$.

The triangles OPQ and opq in Fig. 6–10(a) and (b) are similar, since both are isosceles triangles and their long sides are mutually perpendicular. Hence

$$\frac{\Delta v_N}{v_1} = \frac{\Delta s}{R}, \qquad \text{or} \qquad \Delta v_N = \frac{v_1}{R}\Delta s.$$

The magnitude of the average normal acceleration $\bar{a}_N$ is therefore

$$\bar{a}_N = \frac{\Delta v_N}{\Delta t} = \frac{v_1}{R}\frac{\Delta s}{\Delta t}.$$

The instantaneous normal acceleration a_N at point P is the limiting value of this expression, as point Q is taken closer and closer to point P:

$$a_N = \lim_{\Delta t \to 0} \frac{v_1}{R}\frac{\Delta s}{\Delta t} = \frac{v_1}{R}\lim_{\Delta t \to 0}\frac{\Delta s}{\Delta t}.$$

But the limiting value of $\Delta s/\Delta t$ is the speed v_1 at point P, and since P can be any point of the path we can drop the subscript from v_1 and let v represent the speed at any point. Then

$$\boxed{a_N = \frac{v^2}{R}.} \qquad (6\text{–}8)$$

The magnitude of the instantaneous normal acceleration is therefore equal to the square of the speed divided by the radius. The direction is inward along the radius, toward the center of the circle. Because of this it is called a *central*, a *centripetal*, or a *radial* acceleration. (The term "centripetal" means "seeking a center.")

The unit of radial acceleration is the same as that of an acceleration resulting from a change in the *magnitude* of a velocity. Thus if a particle travels with a speed of 4 ft s^{-1} in a circle of radius 2 ft, its radial acceleration is

$$a = \frac{(4 \text{ ft s}^{-1})^2}{2 \text{ ft}} = 8 \text{ ft s}^{-2}.$$

If the *speed* of the particle changes, it will also have a *tangential* component of acceleration, defined as

$$\boldsymbol{a}_T = \lim_{\Delta t \to 0} \frac{\Delta \boldsymbol{v}_T}{\Delta t}.$$

If the speed is constant, there is no tangential component of acceleration and the acceleration is purely normal, resulting from the continuous change in *direction* of the velocity. In general, there will be both tangential and normal components of acceleration.

Figure 6–11 shows the directions of the velocity and acceleration vectors at a number of points, for a particle revolving in a circle with a velocity of constant magnitude.

A *centrifuge* is a device for whirling an object with a high velocity. The consequent large radial acceleration is equivalent to increasing the value of g, and such processes as sedimentation, which would otherwise take place only slowly, can be greatly accelerated in this way. Very high speed centrifuges, called ultracentrifuges, have been operated at velocities as high as 180,000 rev min^{-1}, and small experimental units have been driven as fast as 1,300,000 rev min^{-1}.

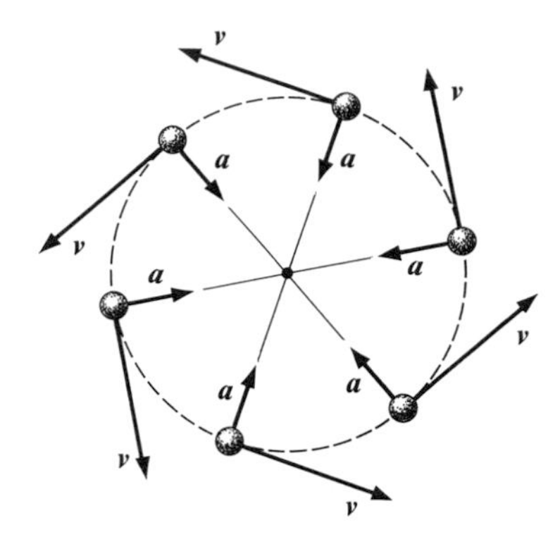

6-11 Velocity and acceleration vectors of a particle in uniform circular motion.

Example. The moon revolves about the earth in a circle (very nearly) of radius $R = 239{,}000$ mi or 12.6×10^8 ft, and requires 27.3 days or 23.4×10^5 s to make a complete revolution.

(a) What is the acceleration of the moon toward the earth?
The velocity of the moon is

$$v = \frac{2\pi R}{T} = \frac{2\pi \times 12.6 \times 10^8 \text{ ft}}{23.4 \times 10^5 \text{ s}} = 3360 \text{ ft s}^{-1}.$$

Its radial acceleration is therefore

$$a = \frac{v^2}{R} = \frac{(3360 \text{ ft s}^{-1})^2}{12.6 \times 10^8 \text{ ft}} = 0.00896 \text{ ft s}^{-2}.$$

(b) If the gravitational force exerted on a body by the earth is inversely proportional to the square of the distance from the earth's center, the acceleration produced by this force should vary in the same way. Therefore, if the acceleration of the moon is caused by the gravitational attraction of the earth, the ratio of the moon's acceleration to that of a falling body at the earth's surface should equal the ratio of the square of the earth's radius (3950 mi or 2.09×10^7 ft) to the square of the radius of the moon's orbit. Is this true?

The ratio of the two accelerations is

$$\frac{8.96 \times 10^{-3} \text{ ft s}^{-2}}{32.2 \text{ ft s}^{-2}} = 2.78 \times 10^{-4}.$$

The ratio of the squares of the distances is

$$\frac{(2.09 \times 10^7 \text{ ft})^2}{(12.6 \times 10^8 \text{ ft})^2} = 2.75 \times 10^{-4}.$$

The agreement is very close, although not exact because we have used average values.

It was the calculation above which Newton made first to justify his hypothesis that gravitation was truly *universal* and that the earth's pull extended out indefinitely into space. The numerical values available in Newton's time were not highly precise. While he did not obtain as close an agreement as that above, he states that he found his results to "answer pretty nearly," and he concluded that his hypothesis was verified.

6-7 CENTRIPETAL FORCE

Having obtained an expression for the radial acceleration of a particle revolving in a circle, we can now use Newton's second law to find the radial force on the particle. Since the magnitude of the radial acceleration equals v^2/R, and its direction is toward the center, the magnitude of the radial force on a particle of mass m is

$$F = m\frac{v^2}{R}. \tag{6-9}$$

The direction of this force is toward the center also, and it is called a *centripetal force*. (It is unfortunate that it has become common practice to characterize the force by the adjective "centripetal," since this seems to imply that there is some difference in nature between centripetal forces and other forces. This is not the case. Centripetal forces, like other forces, are pushes and pulls exerted by sticks and strings, or arise from the action of gravitational or other causes. The term "centripetal" refers to the *effect* of the force, that is, to the fact that it results in a change in the *direction* of the velocity of the body on which it acts, rather than a change in the *magnitude* of this velocity.)

Anyone who has ever tied an object to a cord and whirled it in a circle will realize the necessity of exerting this inward, centripetal force. If the cord breaks, the direction of the velocity ceases to change (unless other forces are acting) and the object flies off along a tangent to the circle.

Example 1. A small body of mass 200 g revolves uniformly in a circle on a horizontal frictionless surface, attached by a cord 20 cm long to a pin set in the surface. If the body makes two

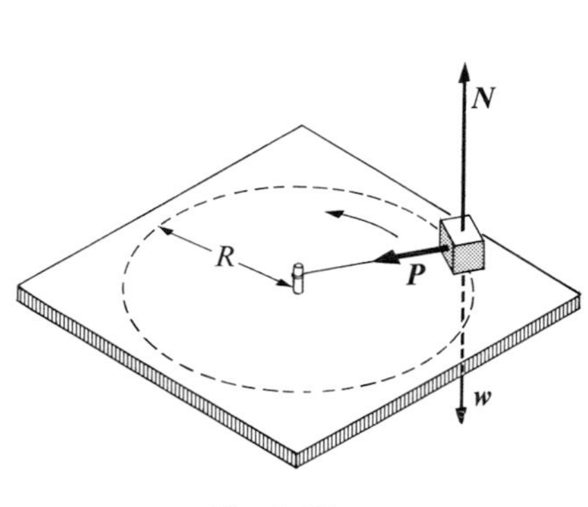

Fig. 6–12

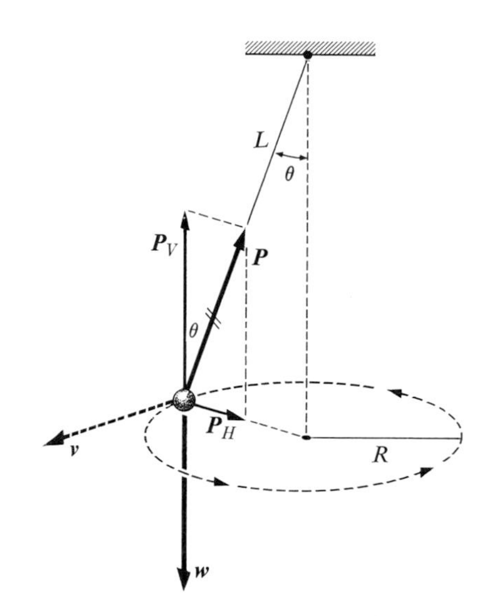

6–13 The conical pendulum.

complete revolutions per second, find the force $\boldsymbol{P}$ exerted on it by the cord. (See Fig. 6–12.)

The circumference of the circle is

$$2\pi \times 20\ \text{cm} = 40\pi\ \text{cm},$$

so the velocity is 80π cm s^{-1}. The magnitude of the centripetal acceleration is

$$a = \frac{v^2}{R} = \frac{(80\pi\ \text{cm s}^{-1})^2}{20\ \text{cm}} = 3150\ \text{cm s}^{-2}.$$

Since the body has no vertical acceleration, the forces $\boldsymbol{N}$ and $\boldsymbol{w}$ are equal and opposite and the force $\boldsymbol{P}$ is the resultant force. Therefore

$$\begin{aligned} P = ma &= 200\ \text{g} \times 3150\ \text{cm s}^{-2} \\ &= 6.3 \times 10^5\ \text{dyn}. \end{aligned}$$

Example 2. Figure 6–13 represents a small body of mass m revolving in a horizontal circle with velocity $\boldsymbol{v}$ of constant magnitude at the end of a cord of length L. As the body swings around its path, the cord sweeps over the surface of a cone. The cord makes an angle θ with the vertical, so the radius of the circle in which the body moves is $R = L \sin\theta$ and the magnitude of the velocity $\boldsymbol{v}$ equals $2\pi L \sin\theta/T$, where T is the time for one complete revolution.

The forces exerted on the body when in the position shown are its weight $\boldsymbol{w}$ and the tension $\boldsymbol{P}$ in the cord. (Note that the force diagram in Fig. 6–13 is exactly like that in Fig. 5–8(b). The only difference is that in this case the acceleration a is the *radial* acceleration, v^2/R.) Let $\boldsymbol{P}$ be resolved into a horizontal component $\boldsymbol{P}_H$ and a vertical component $\boldsymbol{P}_V$, of magnitudes $P \sin\theta$ and $P\cos\theta$, respectively. The body has no vertical acceleration, so the vertical forces $P\cos\theta$ and w are equal, and the resultant inward, radial, or centripetal force is the horizontal component $P\sin\theta$. Then

$$P\sin\theta = m\frac{v^2}{R}, \qquad P\cos\theta = w.$$

When the first of these equations is divided by the second, and w is replaced by mg, we get

$$\tan\theta = \frac{v^2}{Rg}. \tag{6–10}$$

When we make use of the relations $R = L\sin\theta$ and $v = 2\pi L \sin\theta/T$, Eq. (6–10) becomes

$$\cos\theta = \frac{gT^2}{4\pi^2 L}, \tag{6–11}$$

or

$$T = 2\pi\sqrt{L\cos\theta/g}. \tag{6–12}$$

Equation (6–11) indicates how the angle θ depends on the time of revolution T and the length L of the cord. For a given

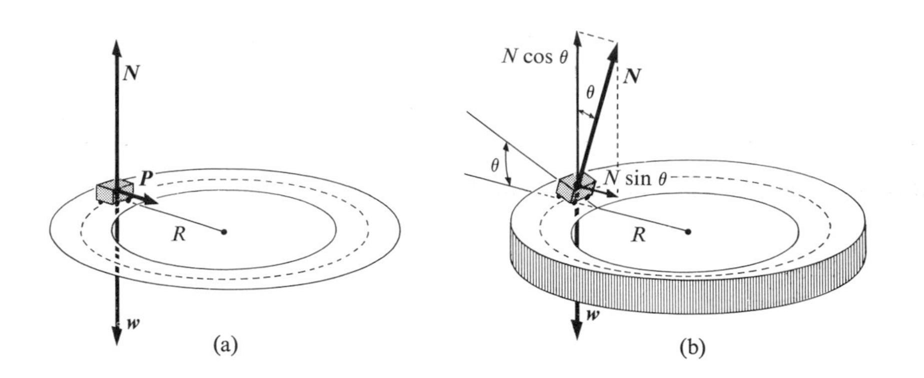

6-14 (a) Forces on a vehicle rounding a curve on a level track. (b) Forces when the track is banked.

length L, $\cos\theta$ *decreases* as the time is made shorter, and the angle θ *increases*. The angle never becomes 90°, however, since this requires that $T = 0$ or $v = \infty$.

Equation (6-12) is similar in form to the expression for the time of swing of a simple pendulum, which will be derived in Chapter 11. Because of this similarity, the present device is called a *conical pendulum*.

Some readers may wish to add to the forces shown in Fig. 6-13 an outward "centrifugal," force, to "keep the body out there," or to "keep it in equilibrium." ("Centrifugal" means "fleeing a center.") Let us examine this point of view. In the first place, to look for a force to "*keep* the body out there" is an example of faulty observation, because the body doesn't stay there. A moment later it will be at a different position on its circular path. At the instant shown it is moving in the direction of the velocity vector $\boldsymbol{v}$, and unless a resultant force acts on it, it will, according to Newton's first law, continue to move in this direction. If an outward force *were* acting on it, equal and opposite to the inward component of the force P, there would be no resultant inward force to deviate it sidewise from its present direction of motion.

Those who wish to add a force to "keep the body in equilibrium" forget that the term equilibrium refers to a state of rest, or of motion *in a straight line* with constant speed. Here, the body is *not* moving in a straight line, but in a circle. It is *not* in equilibrium, but has an acceleration toward the center of the circle and must be acted on by a resultant or *un*balanced force to produce this acceleration.

Example 3. Figure 6-14(a) represents an automobile or a railway car rounding a curve of radius R, on a level road or track. The forces acting on it are its weight $\boldsymbol{w}$, the normal force $\boldsymbol{N}$, and the centripetal force $\boldsymbol{P}$. The force $\boldsymbol{P}$ must be provided by friction, in the case of an automobile, or by a force exerted by the rails against the flanges on the wheels of a railway car.

In order not to have to rely on friction, or to reduce wear on the rails and flanges, the road or the track may be banked as shown in Fig. 6-14(b). The normal force $\boldsymbol{N}$ then has a vertical component of magnitude $N\cos\theta$, and a horizontal component of magnitude $N\sin\theta$ toward the center, which provides the centripetal force. The banking angle θ can be computed as follows. If $\boldsymbol{v}$ is the velocity and R the radius, then

$$N\sin\theta = \frac{mv^2}{R}.$$

Since there is no vertical acceleration, $N\cos\theta = w$.

Dividing the first equation by the second, and replacing w by mg, we get

$$\tan\theta = \frac{v^2}{Rg}.$$

The tangent of the angle of banking is proportional to the square of the speed and inversely proportional to the radius. For a given radius no one angle is correct for all speeds. Hence in the design of highways and railroads, curves are banked for the average speed of the traffic over them.

The same considerations apply to the correct banking angle of a plane when it makes a turn in level flight.

Note that the banking angle is given by the same expression as that for the angle with the vertical made by the cord of a conical pendulum. In fact, the force diagrams in Figs. 6-13 and 6-14 are identical.

6-8 MOTION IN A VERTICAL CIRCLE

Figure 6-15 represents a small body attached to a cord of length R and whirling in a vertical circle about a fixed point O to which the other end of the cord is attached. The motion, while circular, is not uniform, since the speed increases on the way down and decreases on the way up.

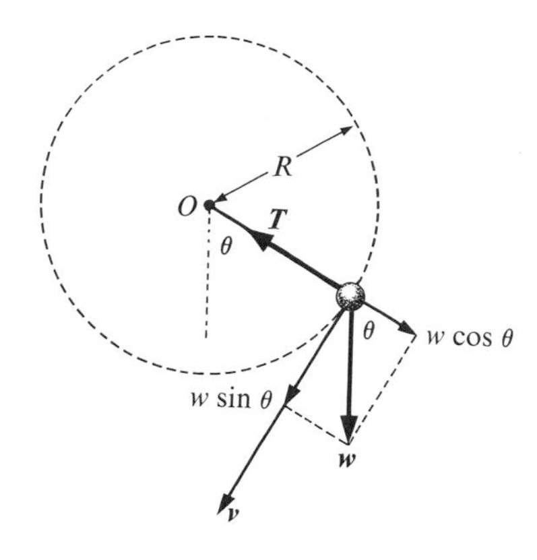

6–15 Forces on a body whirling in a vertical circle with center at O.

The forces on the body at any point are its weight $\boldsymbol{w} = m\boldsymbol{g}$ and the tension $\boldsymbol{T}$ in the cord. Let the weight be resolved into a normal component, of magnitude $w \cos \theta$, and a tangential component of magnitude $w \sin \theta$, as in Fig. 6–15. The resultant tangential and normal forces are then $F_T = w \sin \theta$ and $F_N = T - w \cos \theta$. The tangential acceleration, from Newton's second law, is

$$a_T = \frac{F_T}{m} = g \sin \theta,$$

and is the same as that of a body sliding on a frictionless inclined plane of slope angle θ. The radial acceleration, $a_N = v^2/R$, is

$$a_N = \frac{F_N}{m} = \frac{T - w \cos \theta}{m} = \frac{v^2}{R},$$

and the tension in the cord is therefore

$$T = m\left(\frac{v^2}{R} + g \cos \theta\right). \tag{6–13}$$

At the lowest point of the path, $\theta = 0$, $\sin \theta = 0$, $\cos \theta = 1$. Hence at this point $F_T = 0$, $a_T = 0$, and the acceleration is purely radial (upward). The magnitude of the tension, from Eq. (6–13), is

$$T = m\left(\frac{v^2}{R} + g\right).$$

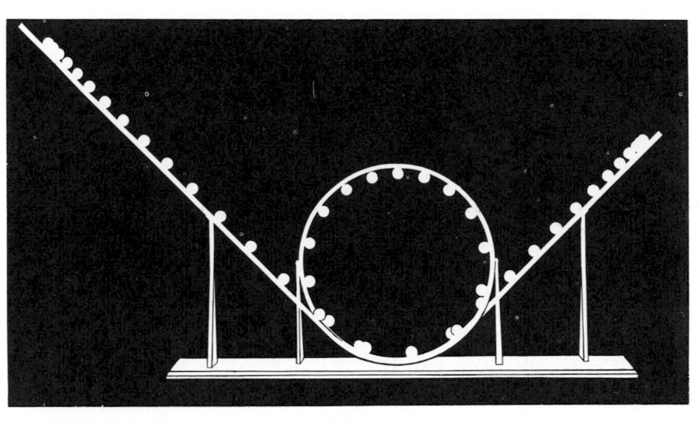

(a)

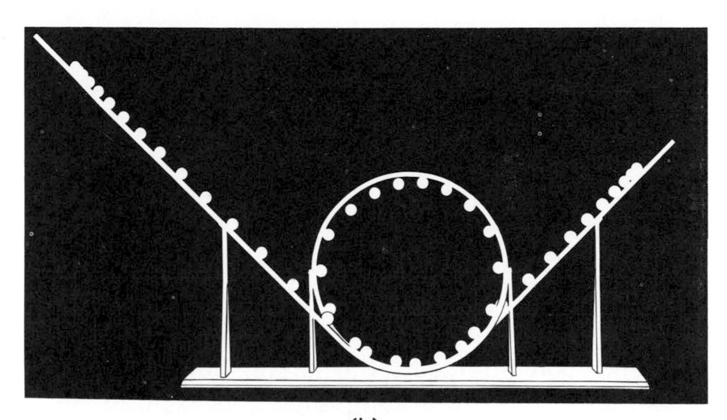

(b)

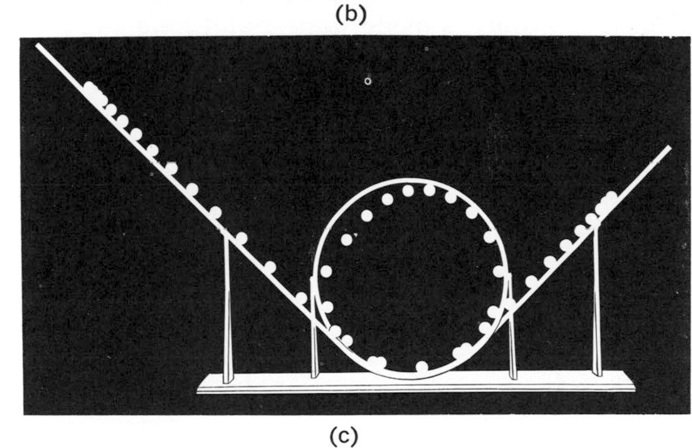

(c)

6–16 Multiflash photographs of a ball looping-the-loop in a vertical circle.

At the highest point, $\theta = 180°$, $\sin \theta = 0$, $\cos \theta = -1$, and the acceleration is again purely radial (downward). The tension is

$$T = m\left(\frac{v^2}{R} - g\right). \tag{6–14}$$

With motion of this sort, it is a familiar fact that there is a certain critical speed v_c at the highest point, below which the cord becomes slack. To find this speed, set $T = 0$ in Eq. (6-14):

$$0 = m\left(\frac{v_c^2}{R} - g\right), \qquad v_c = \sqrt{Rg}.$$

The multiflash photographs of Fig. 6-16 illustrate another case of motion in a vertical circle, a small ball "looping-the-loop" on the inside of a vertical circular track. The inward normal force exerted on the ball by the track takes the place of the tension $\boldsymbol{T}$ shown in Fig. 6-15.

In Fig. 6-16(a), the ball is released from an elevation such that its speed at the top of the track is greater than the critical speed, $\sqrt{Rg}$. In Fig. 6-16(b), the ball starts from a lower elevation and reaches the top of the circle with a speed such that its own weight is slightly larger than the requisite centripetal force. In other words, the track would have to pull *outward* to maintain the circular motion. Since this is impossible, the ball leaves the track and moves for a short distance in a parabola. This parabola soon intersects the circle, however, and the remainder of the trip is completed successfully. In Fig. 6-16(c), the start is made from a still lower elevation, the ball leaves the track sooner, and the parabolic path is clearly evident.

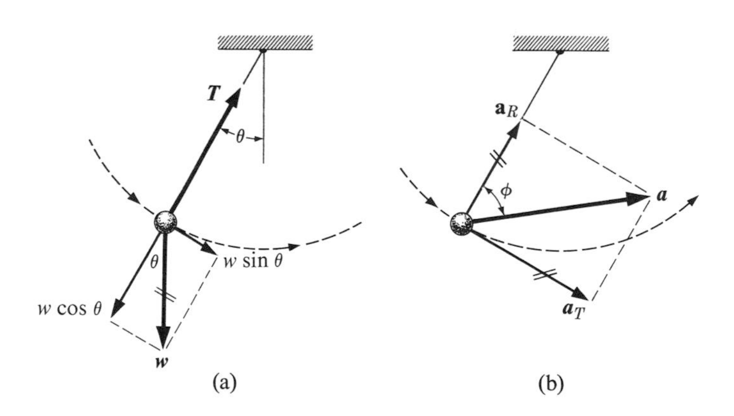

6-17 (a) Forces on a body swinging in a vertical circle. (b) The radial and tangential components of acceleration are combined to obtain the resultant acceleration $\boldsymbol{a}$.

Example. In Fig. 6-17, a small body of mass $m = 0.10$ kg swings in a vertical circle at the end of a cord of length $R = 1.0$ m. If its speed $v = 2.0$ m s^{-1} when the cord makes an angle $\theta = 30°$ with the vertical, find (a) the radial and tangential components of its acceleration at this instant, (b) the magnitude and direction of the resultant acceleration, and (c) the tension $\boldsymbol{T}$ in the cord.

(a) The radial component of acceleration is

$$a_R = \frac{v^2}{R} = \frac{(2.0 \text{ m s}^{-1})^2}{1.0 \text{ m}} = 4.0 \text{ m s}^{-2}.$$

The tangential component of acceleration is

$$a_T = g \sin\theta = 9.8 \text{ m s}^{-2} \times 0.50 = 4.9 \text{ m s}^{-2}.$$

(b) The magnitude of the resultant acceleration (see Fig. 6-17(b)) is

$$a = \sqrt{a_R^2 + a_T^2} = 6.3 \text{ m s}^{-2}.$$

The angle ϕ is

$$\phi = \tan^{-1}\frac{a_T}{a_R} = 51°.$$

(c) The tension in the cord is

$$T = m\left(\frac{v^2}{R} + g\cos\theta\right) = 1.3 \text{ N}.$$

Note that the magnitude of the tangential acceleration is not constant but is proportional to the sine of the angle θ. Hence the equations of motion with constant acceleration *cannot* be used to find the speed at other points of the path. We shall show in the next chapter, however, how the speed at any point can be found from *energy* considerations.

6-9 MOTION OF A SATELLITE

In discussing the trajectory of a projectile in Section 6-5, we assumed that the gravitational force on the projectile (its weight $\boldsymbol{w}$) had the same direction and magnitude at all points of the trajectory. Under these conditions the trajectory is a parabola.

In reality, the gravitational force is directed toward the center of the earth and is inversely proportional to the square of the distance from the earth's center, so that it is not constant in either magnitude or direction. It can be shown that under an inverse square force directed toward a fixed point the trajectory will be a *conic section* (ellipse, circle, parabola, or hyperbola).

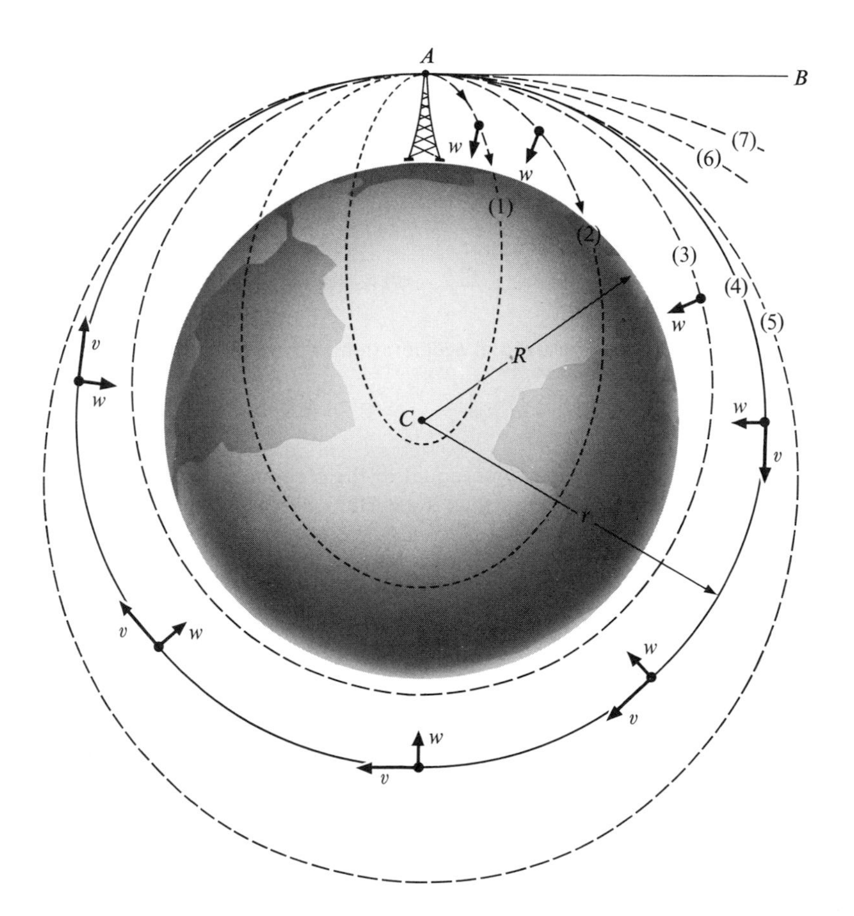

6-18 Trajectories of a body projected from point *A* in the direction *AB* with different initial velocities.

Suppose that a very tall tower could be constructed as in Fig. 6-18, and that a projectile were launched from point *A* at the top of the tower in the "horizontal" direction *AB*. If the initial velocity is not too great, the trajectory will be like that numbered (1), which is a portion of an ellipse with the earth's center *C* at one focus. (If the trajectory is so short that changes in the magnitude and direction of $\boldsymbol{w}$ can be neglected, the ellipse *approximates* a parabola.)

The trajectories numbered from (2) to (7) illustrate the effect of increasing the initial velocity. (Any effect of the earth's atmosphere is neglected.) Trajectory (2) is again a portion of an ellipse. Trajectory (3), which just misses the earth, is a *complete* ellipse and the projectile has become an earth satellite. Its velocity when it returns to point *A* is the same as its initial velocity, and in the absence of retarding forces it will repeat its motion indefinitely. (The earth's rotation will have moved the tower to a different point by the time the satellite returns to point *A*, but does not affect the orbit.)

Trajectory (4) is a special case in which the orbit is a circle. Trajectory (5) is again an ellipse, (6) is a parabola, and (7) is a hyperbola. Trajectories (6) and (7) are not closed orbits.

All man-made earth satellites have orbits like (3) or (5), but some are very nearly circles and for simplicity we shall consider circular orbits only. Let us calculate the velocity required for such an orbit, and the time of one revolution. The gravitational force on the satellite is the centripetal force retaining it in its orbit, and is equal to the product of the mass of the satellite and its radial acceleration.

Thus

$$w = F_g = G\frac{mm_E}{r^2} = m\frac{v^2}{r},$$

and from the last two terms,

$$v^2 = \frac{Gm_E}{r}, \qquad v = \sqrt{\frac{Gm_E}{r}}, \qquad (6\text{–}15)$$

The larger the radius r, the smaller the orbital velocity.

The satellite, like any projectile, is a freely falling body (the only force on it is its weight $\boldsymbol{w}$). Its radial acceleration v^2/r is therefore equal to the free-fall acceleration g *at the orbit*. That is,

$$w = mg = G\frac{mm_E}{r^2},$$

and

$$g = \frac{Gm_E}{r^2}, \qquad \text{or} \qquad \frac{Gm_E}{r} = rg.$$

Hence Eq. (6–15) can also be written as

$$v = \sqrt{rg}.$$

The free-fall acceleration is inversely proportional to the square of the distance from the earth's center. Then if g_R is the free-fall acceleration at the earth's surface, where r equals the earth's radius R,

$$\frac{g}{g_R} = \frac{R^2}{r^2}, \qquad rg = g_R\frac{R^2}{r},$$

and

$$v = R\sqrt{\frac{g_R}{r}}. \qquad (6\text{–}16)$$

The period T, or the time required for one complete revolution, equals the circumference of the orbit, $2\pi r$, divided by the velocity v:

$$T = \frac{2\pi r}{v} = \frac{2\pi r}{R\sqrt{g_R/r}} = \frac{2\pi}{R\sqrt{g_R}}r^{3/2}, \qquad (6\text{–}17)$$

and the larger the radius, the longer the period.

Example. An earth satellite revolves in a circular orbit at a height of 300 km (about 200 miles) above the earth's surface. (a) What is the velocity of the satellite, assuming the earth's radius to be 6400 km and g_R to be 9.80 m s^{-2}?

From Eq. (6–16),

$$v = R\sqrt{\frac{g_R}{r}} = 6.40 \times 10^6\ \text{m}\left(\frac{9.80\ \text{m s}^{-2}}{6.70 \times 10^6\ \text{m}}\right)^{1/2}$$
$$= 7740\ \text{m s}^{-1} = 25{,}400\ \text{ft s}^{-1} = 17{,}300\ \text{mi hr}^{-1}.$$

(b) What is the period T?

$$T = \frac{2\pi r}{v} = 90.6\ \text{min} = 1.51\ \text{hr}.$$

(c) What is the radial acceleration of the satellite?

$$a_R = \frac{v^2}{r} = 8.94\ \text{m s}^{-2}.$$

This is of course equal to the free-fall acceleration g at a height of 300 km above the earth.

A space vehicle in orbit is a freely falling body, with an acceleration a toward the earth's center equal to the value of g at its orbit. The apparent weight w' of the body is

$$w' = w - ma = mg - mg = 0.$$

It is in this sense that an astronaut in the vehicle is said to be "weightless" or in a state of "zero g." The vehicle is like the freely falling elevator discussed in Chapter 5, except that it has a large and constant tangential speed along its orbit.

6-10 EFFECT OF THE EARTH'S ROTATION ON g

Figure 6–19 is a cutaway view of our rotating earth, with three observers each holding a body of mass m, hanging from a string. Each body is attracted toward the earth's center with a force $F_g = Gmm_E/R^2$, and which we now designate by w_0. Let us consider the earth's *center* as the origin of an inertial reference system.

Except at the pole, each body is carried along by the earth's rotation and moves in a circle with center on the earth's axis. It therefore has a radial acceleration a_R, equal to v^2/r, toward the axis. The resultant or vector sum of the forces $\boldsymbol{T}$ and $\boldsymbol{w}_0$ must therefore be such as to provide the requisite radial acceleration. That is, in general vector form,

$$\boldsymbol{T} + \boldsymbol{w}_0 = m\boldsymbol{a}_R.$$

At some arbitrary latitude θ, $\boldsymbol{T}$ and $\boldsymbol{w}_0$ have directions as shown in the diagram, such that their resultant $\boldsymbol{F}$

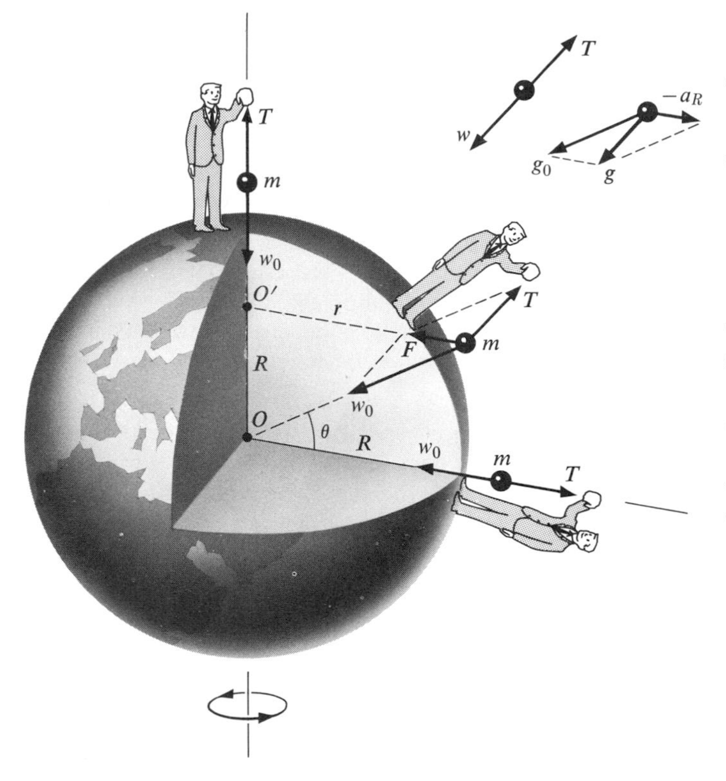

6–19 The resultant of the forces $\boldsymbol{T}$ and $\boldsymbol{w}_0$ is the centripetal force, equal to mv^2/r.

points toward the point O' and equals $m\boldsymbol{a}_R$. Thus we see that except at the pole and the equator, a plumb line does not point toward the earth's center.

At the pole, where $\boldsymbol{a}_R = 0$, the force $\boldsymbol{T}$ is equal and opposite to the true weight $\boldsymbol{w}_0$.

At the equator, where the direction of the radial acceleration is toward the earth's center O, $\boldsymbol{T}$ and $\boldsymbol{w}_0$ have the same action line but the magnitude of $\boldsymbol{w}_0$ is greater than that of $\boldsymbol{T}$.

Since each body appears in equilibrium to its respective observer, the *apparent* weight of each body, which we now represent by $\boldsymbol{w}$, is a force equal and opposite to $\boldsymbol{T}$, as shown in the inset diagram for the body at latitude θ. That is,

$$\boldsymbol{w} = -\boldsymbol{T} = \boldsymbol{w}_0 - m\boldsymbol{a}_R.$$

At an intermediate latitude, the apparent weight $\boldsymbol{w}$ differs in both magnitude and direction from the true weight $\boldsymbol{w}_0$.

At the pole, the apparent and true weights are equal. At the equator,

$$w = w_0 - ma_R,$$

where a_R is the radial acceleration at the equator. Dividing by m, we get

$$\frac{w}{m} = \frac{w_0}{m} - a_R,$$

where w/m is the observed acceleration g relative to the surface of the earth of a body falling at the equator, and w_0/m is the acceleration g_0 relative to the center of the earth. Thus, at the equator, $g = g_0 - a_R$.

Let us calculate the magnitude of the radial acceleration a_R at the equator. The equatorial velocity v, equal to the earth's circumference divided by the time of one rotation, is

$$v = \frac{2\pi \times 6.4 \times 10^6 \text{ m}}{8.64 \times 10^4 \text{ s}} = 465 \text{ m s}^{-1},$$

and hence

$$a_R = \frac{v^2}{R} = 0.034 \text{ m s}^{-2} = 3.4 \text{ cm s}^{-2}.$$

TABLE 6–1 VARIATIONS OF g WITH LATITUDE AND ELEVATION

Station	North latitude	Elevation, m	g, m s^{-2}	g, ft s^{-2}
Canal Zone	9°	0	9.78243	32.0944
Jamaica	18°	0	9.78591	32.1059
Bermuda	32°	0	9.79806	32.1548
Denver	40°	1638	9.79609	32.1393
Cambridge, Mass.	42°	0	9.80398	32.1652
Standard station			9.80665	32.1740
Greenland	70°	0	9.82534	32.2353

Thus if the free-fall acceleration g_0 at the equator is 9.880 m s^{-2}, the observed acceleration g is 9.766 m s^{-2}. Table 6–1 lists the measured free-fall acceleration g at a number of points. A part of the variation results from the fact that the earth is not spherical, that the observation points are at different elevations, or that there are local variations in earth density. It will be seen, however, that the change in the radial acceleration between equator and poles has the right magnitude to account for most of the differences.

Problems

6–1 A ball rolls off the edge of a table-top 4 ft above the floor, and strikes the floor at a point 6 ft horizontally from the edge of the table. (a) Find the time of flight. (b) Find the initial velocity. (c) Find the magnitude and direction of the velocity of the ball just before it strikes the floor. Draw a diagram to scale.

6–2 A block slides off a horizontal tabletop 4 ft high with a velocity of 12 ft s^{-1}. Find (a) the horizontal distance from the table at which the block strikes the floor, and (b) the horizontal and vertical components of its velocity when it reaches the floor.

6–3 A level-flight bomber, flying at 300 ft s^{-1}, releases a bomb at an elevation of 6400 ft. (a) How long before the bomb strikes the earth? (b) How far does it travel horizontally? (c) Find the horizontal and vertical components of its velocity when it strikes.

6–4 A block passes a point 10 ft from the edge of a table with a velocity of 12 ft s^{-1}. It slides off the edge of the table, which is 4 ft high, and strikes the floor 4 ft from the edge of the table. What was the coefficient of sliding friction between block and table?

6–5 A golf ball is driven horizontally from an elevated tee with a velocity of 80 ft s^{-1}. It strikes the fairway 2.5 s later. (a) How far has it fallen vertically? (b) How far has it traveled horizontally? (c) Find the horizontal and vertical components of its velocity, and the magnitude and direction of its resultant velocity, just before it strikes.

6–6 A level-flight bombing plane, flying at an altitude of 1024 ft with a velocity of 240 ft s^{-1}, is overtaking a motor torpedo boat traveling at 80 ft s^{-1} in the same direction as the plane. At what distance astern of the boat should a bomb be released in order to hit the boat?

6–7 A bomber is making a horizontal bombing run on a destroyer from an altitude of 25,600 ft. The magnitude of the velocity of the bomber is 300 mi hr^{-1}. (a) How much time is available for the destroyer to change its course after the bombs are released? (b) If the bomber is to be shot down before its bombs can reach the ship, what is the maximum angle that the line of sight from ship to bomber can make with the horizontal? Draw a diagram showing distances approximately to scale.

6–8 A ball is projected with an initial upward velocity component of 80 ft s^{-1} and a horizontal velocity component of 100 ft s^{-1}. (a) Find the position and velocity of the ball after 2 s, 3 s, 6 s. (b) How long a time is required to reach the highest point of the trajectory? (c) How high is this point? (d) How long a time is required for the ball to return to its original level? (e) How far has it traveled horizontally during this time? Show your results in a neat sketch, large enough to show all features clearly.

6–9 A spring gun projects a golf ball at an angle of 45° above the horizontal. The horizontal range is 32 ft. (a) What is the maximum height to which the ball rises? (b) For the same initial speed, what are the two angles of departure for which the range is 20 ft? (c) Sketch all three trajectories to scale in the same diagram.

6–10 Suppose the departure angle θ_0 in Fig. 6–7 is 15° and the distance $x = 5$ m. Where will the balls collide if the muzzle velocity of the first is (a) 20 m s^{-1}, (b) 5 m s^{-1}? Sketch both trajectories. (c) Will a collision take place if the departure angle is below horizontal?

6–11 If a baseball player can throw a ball a maximum distance of 200 ft over the ground, what is the maximum vertical height to which he can throw it? Assume the ball to have the same initial speed in each case.

6–12 A player kicks a football at an angle of 37° with the horizontal and with an initial velocity of 48 ft s^{-1}. A second player standing at a distance of 100 ft from the first in the direction of the kick starts running to meet the ball at the instant it is kicked. How fast must he run in order to catch the ball before it hits the ground?

6–13 A baseball leaves the bat at a height of 4 ft above the ground, traveling at an angle of 45° with the horizontal, and with a velocity such that the horizontal range would be 400 ft. At a distance of 360 ft from home plate is a fence 30 ft high. Will the ball be a home run?

6–14 (a) What must be the velocity of a projectile fired vertically upward to reach an altitude of 20,000 ft? (b) What velocity is required to reach the same height if the gun makes an angle of 45° with the vertical? (c) Compute the time required to reach the highest point in both trajectories. (d) How many feet would a plane traveling at 300 mi hr^{-1} move in this time?

6–15 The angle of elevation of an antiaircraft gun is 70° and the muzzle velocity is 2700 ft s^{-1}. For what time after firing should the fuse be set if the shell is to explode at an altitude of 5000 ft?

6–16 The Olympic Games record in the high jump is 2.16 m, in the broad jump it is 8.06 m, and in the shot-put it is 18.57 m. Suppose these records were made at a point where $g = 9.82$ m s^{-2}. Estimate the changes that might be expected at a point where $g = 9.78$ m s^{-2}. (Treat the changes as differentials.)

6–17 A trench mortar fires a projectile at an angle of 53° above the horizontal with a muzzle velocity of 200 ft s^{-1}. A tank is advancing directly toward the mortar on level ground at a speed of 10 ft s^{-1}. What should be the distance from mortar to tank at the instant the mortar is fired in order to score a hit?

6–18 The projectile of a trench mortar has a muzzle velocity of 300 ft s^{-1}. (a) Find the two angles of elevation to hit a target at the same level as the mortar and 300 yd distant. (b) Compute the maximum height of each trajectory, and (c) the time of flight of each. Make a neat sketch of the trajectories, approximately to scale.

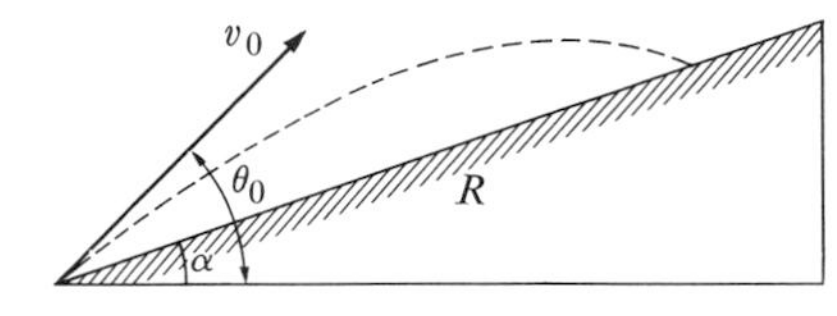

Fig. 6–20

6–19 A projectile is fired with an initial speed v_0 at a departure angle θ_0, from the foot of an inclined plane of slope angle α (Fig. 6–20). (a) What is the range R measured along the plane? (b) Show that the expression for R reduces to that for the horizontal range when $\alpha = 0$.

6–20 A bomber, diving at an angle of 53° with the vertical, releases a bomb at an altitude of 2400 ft. The bomb is observed to strike the ground 5 s after its release. (a) What was the velocity of the bomber, in feet per second? (b) How far did the bomb travel horizontally during its flight? (c) What were the horizontal and vertical components of its velocity just before striking?

6–21 A 15-lb stone is dropped from a cliff in a high wind. The wind exerts a steady horizontal 10-lb force on the stone as it falls. Is the path of the stone a straight line, a parabola, or some more complicated path? Explain.

6–22 A lump of ice slides 20 ft down a smooth sloping roof making an angle of 30° with the horizontal. The edge of the roof is 20 ft above a sidewalk which extends 10 ft out from the side of the building. Will the ice land on the sidewalk or in the street?

6–23 The second stage of a three-stage rocket designed to launch an earth satellite will exhaust its fuel at a height of about 130 mi above the earth's surface, when its speed will be approximately 11,000 mi hr^{-1}. The rocket will then "coast" to a height of about 300 mi, at which point its velocity will be horizontal. Assume a flat earth and a constant value of $\boldsymbol{g}$. (a) Find the direction of the trajectory at the 130-mi high point. (b) Find the horizontal distance traveled while the rocket rises from 130 mi to 300 mi. (c) Find the speed at the 300-mi high point. (d) At a height of 300 mi, the third stage will separate from the second and the latter will fall to the earth. How far will it travel horizontally while falling? (e) How long a time will be required?

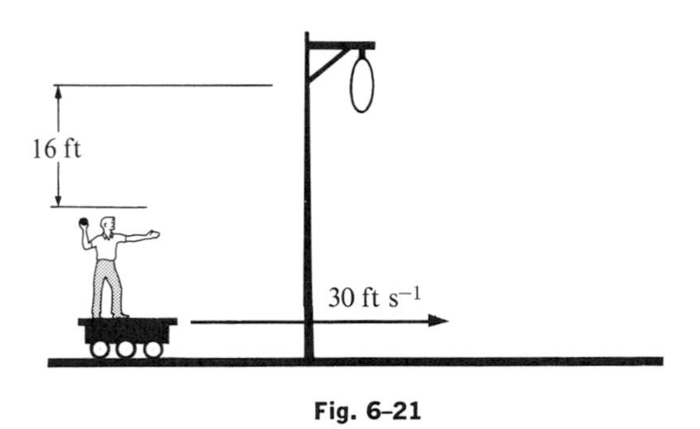

Fig. 6–21

6–24 A man is riding on a flatcar traveling with a constant velocity of 30 ft s^{-1} (Fig. 6–21). He wishes to throw a ball through a stationary hoop 16 ft above the height of his hands in such a manner that the ball will move horizontally as it passes through the hoop. He throws the ball with a velocity of 40 ft s^{-1} with respect to himself. (a) What must be the vertical component of the initial velocity of the ball? (b) How many seconds after he releases the ball will it pass through the hoop? (c) At what horizontal distance in front of the hoop must he release the ball?

6–25 At time $t = 0$ a body is moving east at 10 cm s^{-1}. At time $t = 2$ s it is moving 25° north of east at 14 cm s^{-1}. Find graphically the magnitude and direction of its change in velocity and the magnitude and direction of its average acceleration.

6–26 An automobile travels around a circular track 5000 ft in circumference at a constant speed of 100 ft s^{-1}. (a) Show in a diagram the velocity vectors of the automobile at the beginning and end of a time interval $\Delta t = 5$ s. Let 1 cm = 20 ft s^{-1}. (b) Find graphically the change in velocity, $\Delta \boldsymbol{v}$, in this time interval. (c) Find the magnitude of the average acceleration during this interval, $\Delta v/\Delta t$. (d) What is the radial acceleration, v^2/R? (e) If the track is 30 ft wide, what should be the elevation of the outer circumference above the inner circumference, for the speed above?

6–27 The radius of the earth's orbit around the sun (assumed circular) is 93×10^6 mi, and the earth travels around this orbit in 365 days. (a) What is the magnitude of the orbital velocity of the earth, in miles per hour? (b) What is the radial acceleration of the earth toward the sun, in feet per second squared?

6–28 A model of a helicopter rotor has four blades, each 5 ft long, and is rotated in a wind tunnel at 1500 rev min^{-1}. (a) What is the linear speed of the blade tip, in feet per second? (b) What is the radial acceleration of the blade tip, in terms of the acceleration of gravity, g? (c) A pressure measuring device weighing $\frac{1}{4}$ lb is mounted at the blade tip. Find the centripetal force on it, and compare with its weight.

6–29 A stone of mass 1 kg is attached to one end of a string 1 m long, of breaking strength 500 N, and is whirled in a horizontal circle on a frictionless tabletop. The other end of the string is kept fixed. Find the maximum velocity the stone can attain without breaking the string.

6–30 An unbanked circular highway curve on level ground makes a turn of 90°. The highway carries traffic at 60 mi hr^{-1}, and the centripetal force on a vehicle is not to exceed $\frac{1}{10}$ of its weight. What is the minimum length of the curve, in miles?

6–31 A coin placed on a 12-in. record will revolve with the record when it is brought up to a speed of 78 rev min^{-1}, provided the coin is not more than 2.5 in. from the axis. (a) What is the coefficient of static friction between coin and record? (b) How far from the axis can the coin be placed, without slipping, if the turntable rotates at 45 rev min^{-1}?

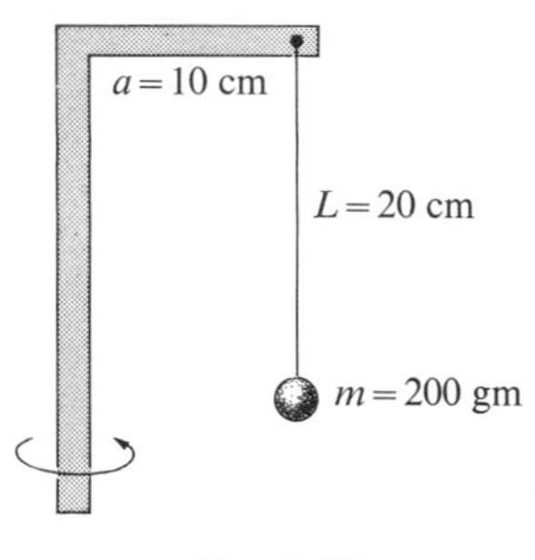

Fig. 6–22

6–32 (a) At how many revolutions per second must the apparatus of Fig. 6–22 rotate about the vertical axis in order that the cord shall make an angle of 45° with the vertical? (b) What is then the tension in the cord? (c) Find the angle θ which the cord makes with the vertical if the system is rotating at 1.5 rev s^{-1}. (Set up the general equation relating the angle θ to the number of revolutions per second, n, the lengths a and L, and the acceleration of gravity, g. Then find by trial the angle θ which satisfies this equation.)

6–33 The "Giant Swing" at a county fair consists of a vertical central shaft with a number of horizontal radial arms attached at its upper end. Each arm supports a seat suspended from a cable 15 ft long, the upper end of the cable being fastened to the arm at a point 12 ft from the central shaft. (a) Find the time of one revolution of the swing if the cable supporting a seat makes an angle of 30° with the vertical. (b) Does the angle depend on the weight of the passenger for a given rate of revolution?

6–34 The 8-lb block in Fig. 6–23 is attached to a vertical rod by means of two strings. When the system rotates about the axis of the rod, the strings are extended as shown in the diagram. (a) How many revolutions per minute must the system make in order that the tension in the upper cord shall be 15 lb? (b) What is then the tension in the lower cord?

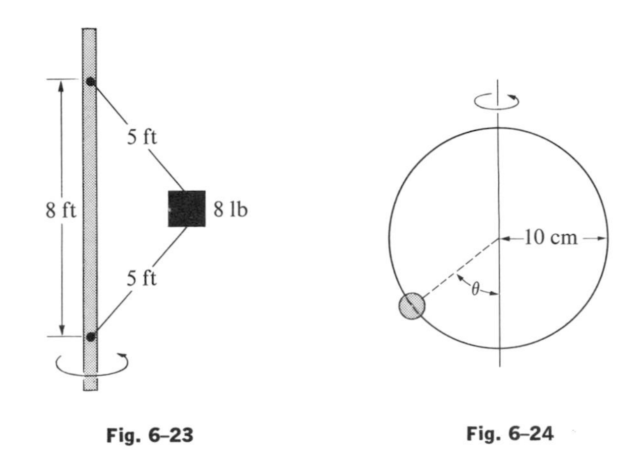

6–35 A bead can slide without friction on a circular hoop of radius 10 cm in a vertical plane. The hoop rotates at a constant rate of 2 rev s^{-1} about a vertical diameter, as in Fig. 6–24. (a) Find the angle θ at which the bead is in vertical equilibrium. (Of course it has a radial acceleration toward the axis.) (b) Is it possible for the bead to "ride" at the same elevation as the center of the hoop? (c) What will happen if the hoop rotates at 1 rev s^{-1}?

6–36 A curve of 600-ft radius on a level road is banked at the correct angle for a velocity of 30 mi hr^{-1}. If an automobile rounds this curve at 60 mi hr^{-1}, what is the minimum coefficient of friction between tires and road so that the automobile will not skid? Assume all forces to act at the center of gravity.

6–37 An airplane in level flight is said to make a *standard turn* when it makes a complete circular turn in 2 min. (a) What is the banking angle of a standard turn if the speed of the airplane is 400 ft s^{-1}? (b) What is the radius of the circle in which it turns? (c) What is the centripetal force on the airplane, expressed as a fraction (or multiple) of its weight?

6–38 The radius of the circular track in Fig. 6–16 is 40 cm and the mass of the ball is 100 gm. (a) Find the critical velocity at the highest point of the track. (b) If the actual velocity at the highest point of the track is twice the critical velocity, find the force exerted by the ball against the track.

6–39 The pilot of a dive bomber who has been diving at a velocity of 400 mi hr^{-1} pulls out of the dive by changing his course to a circle in a vertical plane. (a) What is the minimum radius of the circle in order that the acceleration at the lowest point shall not exceed "7*g*"? (b) How much does a 180-lb pilot apparently weigh at the lowest point of the pullout?

6–40 A cord is tied to a pail of water and the pail is swung in a vertical circle of radius 4 ft. What must be the minimum velocity of the pail at the highest point of the circle if no water is to spill from the pail?

6–41 The radius of a Ferris wheel is 15 ft and it makes one revolution in 10 s. (a) Find the difference between the apparent weight of a passenger at the highest and lowest points, expressed as a fraction of his weight. (That is, find the difference between the upward force exerted on the passenger by the seat at these two points.) (b) What would the time of one revolution be if his apparent weight at the highest point were zero? (c) What would then be his apparent weight at the lowest point? (d) What would happen, at this rate of revolution, if his seat belt broke at the highest point, and if he did not hang on to his seat?

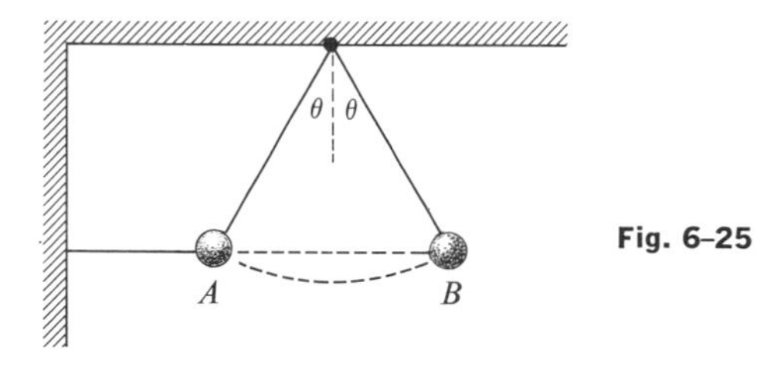

Fig. 6–25

6–42 A ball is held at rest in position *A* in Fig. 6–25 by two light cords. The horizontal cord is cut and the ball starts swinging as a pendulum. What is the ratio of the tension in the supporting cord, in position *B*, to that in position *A*?

6–43 The questions are often asked: "What keeps an earth satellite moving in its orbit?" and "What keeps the satellite up?" (a) What are your answers to these questions? (b) Are your answers applicable to the moon?

6–44 What is the period of revolution of a man-made satellite of mass *m* which is orbiting the earth in a circular path of radius 8000 km? (Mass of earth $= 5.98 \times 10^{24}$ kg.)

6–45 There exist several satellites moving in a circle in the earth's equatorial plane and at such a height above the earth's surface that they remain always above the same point. Find the height of such a satellite.

6–46 It is desired to launch a satellite 400 mi above the earth in a circular orbit. If suitable rockets are used to reach this elevation, what horizontal orbital velocity must be imparted to the satellite? The radius of the earth is 3950 mi.

6–47 An earth satellite rotates in a circular orbit of radius 4400 mi (about 400 mi above the earth's surface) with an orbital speed of 17,000 mi hr^{-1}. (a) Find the time of one revolution. (b) Find the acceleration of gravity at the orbit.

6–48 What would be the length of a day if the rate of rotation of the earth were such that $g = 0$ at the equator?

6–49 The weight of a man as determined by a spring balance at the equator is 180 lb. By how many ounces does this differ from the true force of gravitational attraction at the same point?

7 Work and Energy

7-1 INTRODUCTION

The curved line in Fig. 7-1 represents the trajectory, or path, of a particle of mass m moving in the xy-plane and acted on by a *resultant* force $\boldsymbol{F}$ which may vary in magnitude and direction from point to point of the path. Let us resolve the force into a component $\boldsymbol{F_S}$ along the path and a component $\boldsymbol{F_N}$ normal to the path.

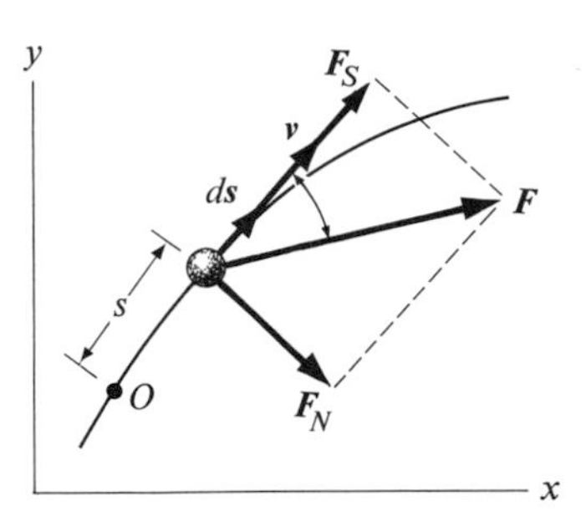

7-1 Path of a particle in the xy-plane.

The component $\boldsymbol{F_N}$, at right angles to the velocity $\boldsymbol{v}$, is a *centripetal* force, and its only effect is to change the *direction* of the velocity. The effect of the component $\boldsymbol{F_S}$ is to change the *magnitude* of the velocity.

Let s be the distance of the particle from some fixed point O, measured along the path. In general, the magnitude of $\boldsymbol{F_S}$ will be a function of s. From Newton's second law,

$$F_S = m\frac{dv}{dt}.$$

Since F_S is a function of s, we use the chain rule and write

$$\frac{dv}{dt} = \frac{dv}{ds}\frac{ds}{dt} = v\frac{dv}{ds}.$$

Then we have

$$F_S = mv\frac{dv}{ds},$$

$$F_S\,ds = mv\,dv.$$

If v_1 is the velocity when $s = s_1$, and v_2 the velocity when $s = s_2$, it follows that

$$\int_{s_1}^{s_2} F_S\,ds = \int_{v_1}^{v_2} mv\,dv. \tag{7-1}$$

The integral on the left is called the *work* W of the force $\boldsymbol{F}$, between the points s_1 and s_2:

$$W = \int_{s_1}^{s_2} F_S(s)\,ds.$$

This integral, known as the *line integral* of F_S, can be evaluated, of course, only when the component F_S is known as a function of s, or when both F_S and s are known as functions of another variable. The integral on the right side of Eq. (7-1), however, can always be evaluated:

$$\int_{v_1}^{v_2} mv\,dv = \tfrac{1}{2}mv_2^2 - \tfrac{1}{2}mv_1^2.$$

One-half the product of the mass of a particle and the square of the magnitude of its velocity is called the *kinetic energy* of the particle, E_k:

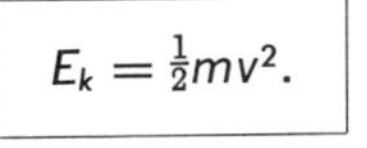

$$E_k = \tfrac{1}{2}mv^2.$$

Equation (7-1) can therefore be written

$$W = E_{k2} - E_{k1}. \tag{7-2}$$

That is, *the work of the resultant force exerted on a particle equals the change in kinetic energy of the particle.* This statement is known as the *work-energy* principle.

We now discuss these concepts of work and kinetic energy in more detail.

7-2 WORK

In everyday life, the word *work* is applied to any form of activity that requires the exertion of muscular or mental effort. In physics, the term is used in a very restricted sense. Work is done only when a force is exerted on a body while the body at the same time moves in such a way that the force has a component along the line of motion of its point of application. If the component of the force is in the *same direction* as the displacement, the work is *positive*. If *opposite* to the displacement, the work is negative. If the force is at *right angles* to the displacement, it has no component in the direction of the displacement and the work is *zero*.

Thus when a body is lifted, the work of the lifting force is positive; when a spring is stretched, the work of the stretching force is positive; when a gas is compressed in a cylinder, again the work of the compressing force is positive. On the other hand, the work of the gravitational force on a body being lifted is negative, since the (downward) gravitational force is opposite to the (upward) displacement. When a body slides on a fixed surface, the work of the frictional force exerted *on the body* is negative, since this force is always opposite to the displacement of the body. No work is done by the frictional force acting *on the fixed surface* because there is no motion of this surface. Also, although it would be considered "hard work" to hold a heavy object stationary at arm's length, no work would be done in the technical sense because there is no motion. Even if one were to walk along a level floor while carrying the object, no work would be done, since the (vertical) supporting force has no component in the direction of the (horizontal) motion. Similarly, the work of the normal force exerted on a body by a surface on which it moves is zero, as is the work of the centripetal force on a body moving in a circle.

In the mks system the unit of work is one *newton meter* (1 N m) which is called one *joule* (1 J). The unit of work in the cgs system is one *dyne centimeter* (1 dyn cm), called one *erg*. Since 1 m = 100 cm and 1 N = 10^5 dyn, it follows that

$$1 \text{ N m} = 10^7 \text{ dyn cm}, \qquad \text{or} \qquad 1 \text{ J} = 10^7 \text{ erg}.$$

In the engineering system, the unit of work is one *foot pound* (1 ft lb):

$$1 \text{ J} = 0.7376 \text{ ft lb}, \qquad 1 \text{ ft lb} = 1.356 \text{ J}.$$

The expression for the work of a force can be written in several ways. If θ is the angle between the force vector $\boldsymbol{F}$ and the infinitesimal displacement vector $d\boldsymbol{s}$, the component F_S is equal to $F \cos \theta$, so

$$W = \int_{s_1}^{s_2} F_S \, ds = \int_{s_1}^{s_2} F \cos \theta \, ds.$$

In Chapter 1 we discussed the *addition* and *subtraction* of vectors. The definition of work suggests a third process of vector algebra, namely, *scalar multiplication* of two vectors. In computing the work of a force, we multiply the magnitude of the vector $d\boldsymbol{s}$ by the magnitude of the component of another vector $\boldsymbol{F}$ in the direction of $d\boldsymbol{s}$. The product is called the *scalar product* or the *dot product* of the vectors. Thus if $\boldsymbol{A}$ and $\boldsymbol{B}$ are any two vectors, their scalar or dot product is defined as a scalar, equal to the product of the magnitudes of the vectors and the cosine of the angle θ between their positive directions:

$$\boldsymbol{A} \cdot \boldsymbol{B} = AB \cos \theta.$$

Evidently, $\boldsymbol{A} \cdot \boldsymbol{B} = \boldsymbol{B} \cdot \boldsymbol{A}$. In this new notation, the work of a force can be written as

$$W = \int_{s_1}^{s_2} \boldsymbol{F} \cdot d\boldsymbol{s}.$$

In the special case in which the force $\boldsymbol{F}$ is constant in magnitude and makes a constant angle with the direction of motion of its point of application, $F \cos \theta$ is constant and may be taken outside the integral sign. If in addition we measure displacements from the starting point of the motion and let $s_1 = 0$ and $s_2 = s$, the work of the force is

$$W = \int_0^s F \cos \theta \, ds = F \cos \theta \int_0^s ds = (F \cos \theta)s.$$

If the force is constant and in the same direction as the motion or opposite to the motion, the angle θ equals zero or 180°, $\cos \theta = \pm 1$, and

$$W = \pm Fs.$$

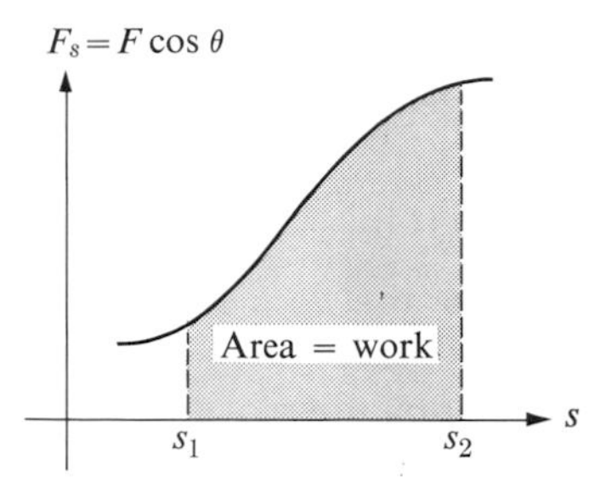

7-2 Work is equal to the area under the $F\cos\theta$-versus-s curve.

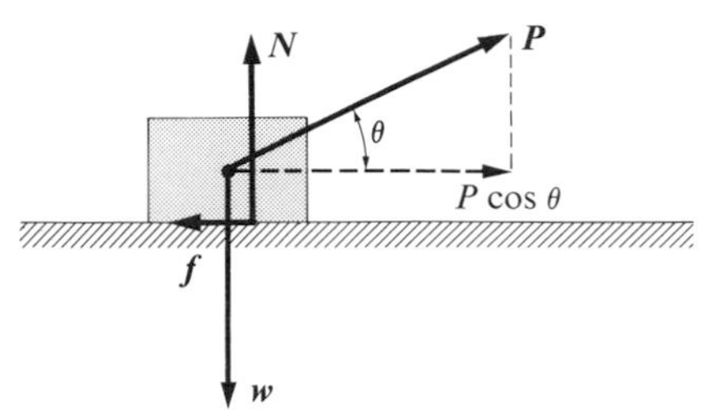

7-3 An object on a rough horizontal surface moving to the right under the action of a force $\boldsymbol{P}$ inclined at an angle θ.

That is, in this *very special case only*, we can say, "Work equals force times distance." It is important to remember, however, that the *general* definition of the work of a force is

$$W = \int_{s_1}^{s_2} \boldsymbol{F} \cdot d\boldsymbol{s} = \int_{s_1}^{s_2} F\cos\theta\, ds. \tag{7-3}$$

In the preceding discussion, the force $\boldsymbol{F}$ has represented the *resultant* of *all* external forces on a body. We often wish to consider the works of the separate forces that may act on a body. Each of these may be computed from the general definition of work in Eq. (7-3). Then, since work is a scalar quantity, the total work is the algebraic sum of the individual works. Note, however, that only the *total* work is equal to the change in kinetic energy of the body.

The work of a force can be represented graphically by plotting the component $F\cos\theta$ as a function of s, as in Fig. 7-2. The work in a displacement from s_1 to s_2 is then equal to the *area* under the curve between vertical lines drawn at s_1 and s_2.

Example 1. Figure 7-3 shows a box being dragged along a horizontal surface by a constant force $\boldsymbol{P}$ making a constant angle θ with the direction of motion. The other forces on the box are its weight $\boldsymbol{w}$, the normal upward force $\boldsymbol{N}$ exerted by the surface, and the friction force $\boldsymbol{f}$. What is the work of each force when the box moves a distance s along the surface to the right?

The component of $\boldsymbol{P}$ in the direction of motion is $P\cos\theta$. The work of the force $\boldsymbol{P}$ is therefore

$$W_P = (P\cos\theta)\cdot s.$$

The forces $\boldsymbol{w}$ and $\boldsymbol{N}$ are both at right angles to the displacement. Hence

$$W_w = 0, \qquad W_N = 0.$$

The friction force $\boldsymbol{f}$ is opposite to the displacement, so the work of the friction force is

$$W_f = -fs.$$

Since work is a scalar quantity, the total work W of all forces on the body is the algebraic (not the vector) sum of the individual works:

$$\begin{aligned} W &= W_P + W_w + W_N + W_f \\ &= (P\cos\theta)\cdot s + 0 + 0 - f\cdot s \\ &= (P\cos\theta - f)s. \end{aligned}$$

But $(P\cos\theta - f)$ is the *resultant* force on the body. Hence *the total work of all forces is equal to the work of the resultant force.*

Suppose that $w = 100$ lb, $P = 50$ lb, $f = 15$ lb, $\theta = 37°$, and $s = 20$ ft. Then

$$\begin{aligned} W_P &= (P\cos\theta)\cdot s = 50 \times 0.8 \times 20 = 800 \text{ ft lb}, \\ W_f &= -fs = -15 \times 20 = -300 \text{ ft lb}, \\ W &= W_P + W_f = 500 \text{ ft lb}. \end{aligned}$$

As a check, the total work may be expressed as

$$W = (P\cos\theta - f)\cdot s = (40 \text{ lb} - 15 \text{ lb}) \times 20 \text{ ft} = 500 \text{ ft lb}.$$

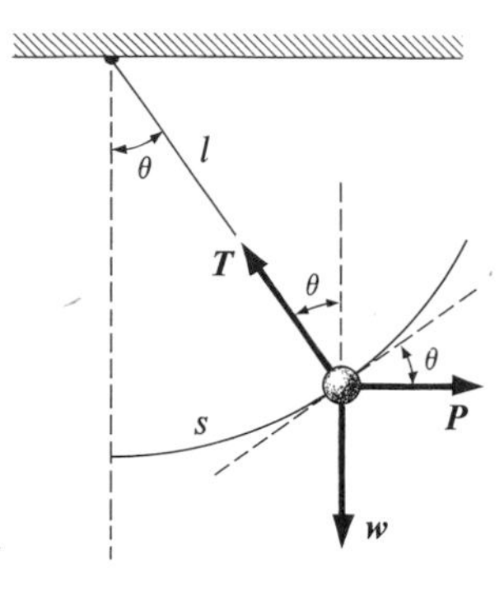

7-4 A variable horizontal force $\boldsymbol{P}$ acts on a small object while the displacement varies from zero to s.

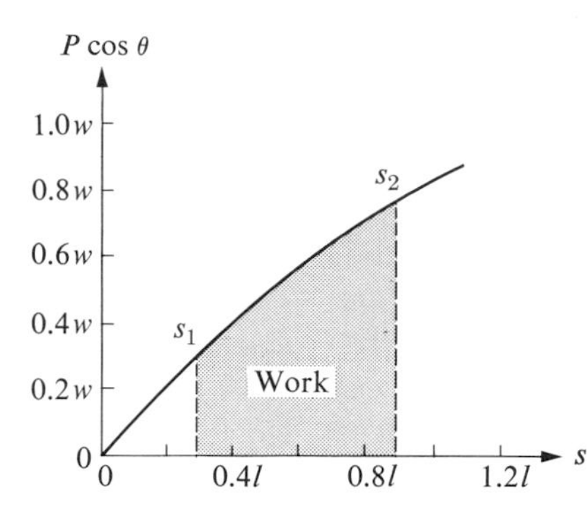

7-5 The work on a force $\boldsymbol{P}$ when its point of application moves from $s_1 = 0.3\ell$ to $s_2 = 0.9\ell$ is shaded area under the curve of $P \cos \theta$, plotted against s.

Example 2. A small object of weight $\boldsymbol{w}$ hangs from a string of length ℓ, as shown in Fig. 7-4. A *variable* horizontal force $\boldsymbol{P}$, which starts at zero and gradually increases, is used to pull the object very slowly (so that equilibrium exists at all times) until the string makes an angle θ with the vertical. Calculate the work of the force $\boldsymbol{P}$.

Since the object is in equilibrium, the sum of the horizontal forces equals zero, whence

$$P = T \sin \theta.$$

Equating the sum of the vertical forces to zero, we find

$$w = T \cos \theta.$$

Dividing these two equations, we get

$$P = w \tan \theta.$$

The point of application of $\boldsymbol{P}$ swings through the arc s. Since $s = \ell\theta$, $ds = \ell\, d\theta$ and

$$W = \int \boldsymbol{P} \cdot d\boldsymbol{s} = \int P \cos \theta\, ds$$

$$= \int_0^\theta w \tan \theta \cos \theta\, \ell\, d\theta = w\ell \int_0^\theta \sin \theta\, d\theta = w\ell(1 - \cos \theta). \tag{7-4}$$

A graph of $P \cos \theta$ $(= w \sin \theta)$ plotted against s $(= \ell\theta)$ is shown in Fig. 7-5. The work during any displacement is represented by the area under this curve. The shaded area is the work when the point of application of $\boldsymbol{P}$ moves from a position where $s_1 = 0.3\ell$ to a place where $s_2 = 0.9\ell$.

7-3 KINETIC ENERGY

Kinetic energy, like work, is a scalar quantity. The kinetic energy of a moving body depends only on the *magnitude* of its velocity (or its speed) and not on the direction in which it is moving or on the particular process by which it was set in motion.

It follows from the work-energy principle that the *change* in kinetic energy of a body depends only on the work $W = \int \boldsymbol{F} \cdot d\boldsymbol{s}$ and not on the individual values of F and s. That is, the force could have been large and the displacement small, or the reverse might have been true. If the mass m and the speeds v_1 and v_2 are known, the *work* of the resultant force can be found without any knowledge of the force and the displacement.

If the work W is *positive*, the final kinetic energy is greater than the initial kinetic energy and the kinetic energy *increases*. If the work is *negative*, the kinetic energy *decreases*. In the special case in which the work is *zero*, the kinetic energy remains *constant*.

In computing the kinetic energy of a body, consistent units must be used for m and v. In the mks system, m must be in kilograms and v in meters per second. In the cgs system m must be expressed in grams and v in centimeters per second. In the engineering system, m must be in slugs and v in feet per second. The corresponding units of kinetic energy are 1 kg m^2 s^{-2}, 1 g cm^2 s^{-2}, and 1 slug ft^2 s^{-2}. However, the unit of kinetic energy in any system is equal to the unit of work in that system, and kinetic energy is customarily expressed in joules, ergs, or foot pounds. That is,

$$1 \text{ kg m}^2\text{ s}^{-2} = 1 \text{ N m}^{-1}\text{ s}^2 \cdot \text{m}^2\text{ s}^{-2} = 1 \text{ N m} = 1 \text{ J}.$$

In the same way,

$$1 \text{ g cm}^2\text{ s}^{-2} = 1 \text{ erg}, \qquad 1 \text{ slug ft}^2\text{ s}^{-2} = 1 \text{ ft lb}.$$

Example. Refer again to the body in Fig. 7-3 and the numerical values given at the end of Example 1. The total work of the external forces was shown to be 500 ft lb. Hence the kinetic energy of the body increases by 500 ft lb. To verify this, suppose the initial speed v_1 is 4 ft s^{-1}. The initial kinetic energy is

$$E_{k1} = \tfrac{1}{2}mv_1^2 = \tfrac{1}{2}\tfrac{100}{32} \text{ slugs} \times 16 \text{ ft}^2\text{ s}^{-2} = 25 \text{ ft lb}.$$

To find the final kinetic energy we must first find the acceleration:

$$a = \frac{F}{m} = \frac{40 \text{ lb} - 15 \text{ lb}}{(100/32)\text{ slugs}} = 8 \text{ ft s}^{-2}.$$

Then

$$v_2^2 = v_1^2 + 2as = 16 \text{ ft}^2\text{ s}^{-2} + 2 \times 8 \text{ ft s}^{-2} \times 20 \text{ ft} = 336 \text{ ft}^2\text{ s}^{-2},$$

and

$$E_{k2} = \frac{1}{2}\frac{100}{32} \text{ slugs} \times 336 \text{ ft}^2\text{ s}^{-2} = 525 \text{ ft lb}.$$

The increase in kinetic energy is therefore 500 ft lb.

7-4 GRAVITATIONAL POTENTIAL ENERGY

Suppose a body of mass m (and of weight $\boldsymbol{w} = m\boldsymbol{g}$) moves vertically, as in Fig. 7-6(a), from a point where its center of gravity is at a height y_1 above an arbitrarily chosen plane (the *reference level*) to a point at a height y_2. For the present, we shall consider only displacements near the earth's surface, so that variations of gravitational force with distance from the earth's center can be neglected. The downward gravitational force on the body is then constant and equal to $\boldsymbol{w}$. Let $\boldsymbol{P}$ represent the resultant of all other forces acting on the body, and let W' be the work of these forces. The direction of the gravitational force $\boldsymbol{w}$ is opposite to the upward displacement, and the work of this force is

$$W_{\text{grav}} = -w(y_2 - y_1) = -(mgy_2 - mgy_1). \qquad (7\text{-}5)$$

[The reader should convince himself that the work of the gravitational force is given by $-mg(y_2 - y_1)$ whether the body moves up or down.]

Now suppose that the body starts at the same elevation y_1 but is moved up to the elevation y_2 along some arbitrary path, as in Fig. 7-6(b). Part (c) of the figure is an enlarged view of a small portion of the path. The work of the gravitational force is

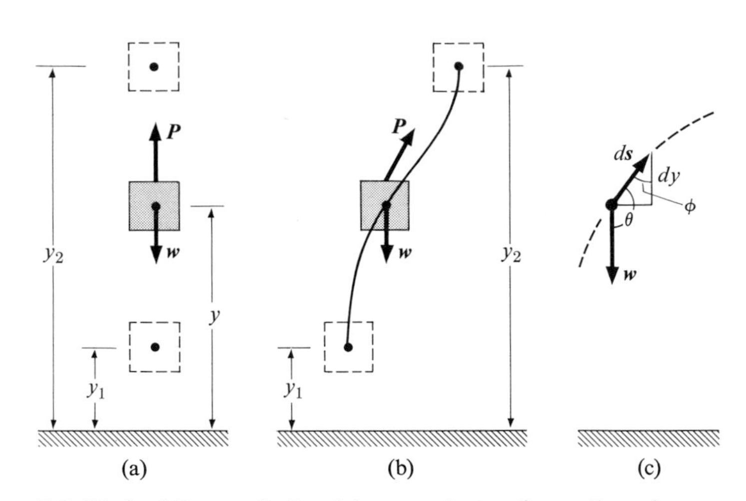

7-6 Work of the gravitational force $\boldsymbol{w}$ during the motion of an object from one point in a gravitational field to another.

$$W_{\text{grav}} = \int_{s_1}^{s_2} w \cos\theta \, ds.$$

Let ϕ represent the angle between ds and its vertical component dy. Then $dy = ds\cos\phi$, and since $\phi = 180° - \theta$,

$$\cos\phi = -\cos\theta, \qquad \cos\theta \, ds = -dy$$

and

$$W_{\text{grav}} = -\int_{y_1}^{y_2} w \, dy = -w(y_2 - y_1) = -(mgy_2 - mgy_1). \qquad (7\text{-}6)$$

The work of the gravitational force, therefore, depends only on the initial and final elevations and not on the path. If these points are at the *same* elevation the work is zero.

Since the *total* work equals the change in kinetic energy,

$$W' + W_{\text{grav}} = E_{k2} - E_{k1},$$
$$W' - (mgy_2 - mgy_1) = (\tfrac{1}{2}mv_2^2 - \tfrac{1}{2}mv_1^2).$$

The quantities $\tfrac{1}{2}mv_2^2$ and $\tfrac{1}{2}mv_1^2$ depend only on the final and initial *speeds;* the quantities mgy_2 and mgy_1

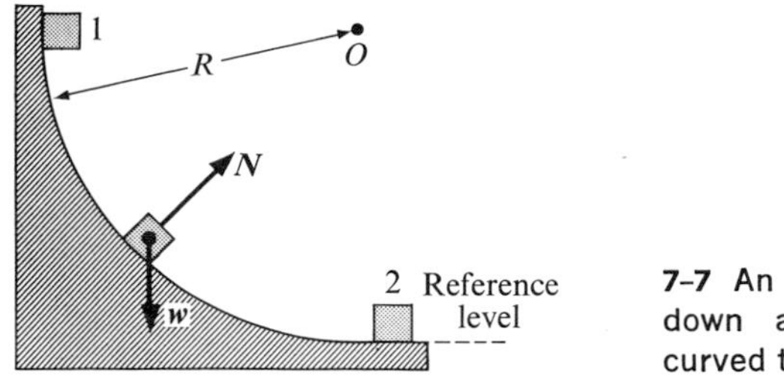

7-7 An object sliding down a frictionless curved track.

depend only on the final and initial *elevations*. Let us therefore rearrange this equation, transferring the quantities mgy_2 and mgy_1 from the "work" side of the equation to the "energy" side:

$$W' = (\tfrac{1}{2}mv_2^2 - \tfrac{1}{2}mv_1^2) + (mgy_2 - mgy_1). \quad (7\text{-}7)$$

The left side of Eq. (7-7) contains only the work of the force **P**. The terms on the right depend only on the final and initial states of the body (its speed and elevation) and not specifically on the way in which it moved. The quantity mgy, the product of the weight mg of the body and the height y of its center of gravity above the reference level, is called its *gravitational potential energy*, E_p.

$$E_P \text{ (gravitational)} = mgy. \quad (7\text{-}8)$$

The first expression in parentheses on the right of Eq. (7-7) is the change in kinetic energy of the body, and the second is the change in its gravitational potential energy.

Equation (7-7) can also be written

$$W' = (\tfrac{1}{2}mv_2^2 + mgy_2) - (\tfrac{1}{2}mv_1^2 + mgy_1). \quad (7\text{-}9)$$

The sum of the kinetic and potential energy of the body is called its *total mechanical energy*. The first expression in parentheses on the right of Eq. (7-9) is the final value of the total mechanical energy, and the second is the initial value. Hence, *the work of all forces acting on the body*, **with the exception of the gravitational force,** *equals the change in the total mechanical energy of the body*. If the work W' is positive, the mechanical energy increases. If W' is negative, the energy decreases.

In the special case in which the *only* force on the body is the gravitational force, the work W' is zero. Equation (7-9) can then be written

$$\tfrac{1}{2}mv_2^2 + mgy_2 = \tfrac{1}{2}mv_1^2 + mgy_1.$$

Under these conditions, then, the *total mechanical energy remains constant*, or is *conserved*. This is a special case of the *principle of the conservation of mechanical energy*.

Example 1. A man holds a ball of weight $w = \frac{1}{4}$ lb at rest in his hand. He then throws the ball vertically upward. In this process, his hand moves up 2 ft and the ball leaves his hand with an upward velocity of 48 ft s^{-1}. Discuss the motion of the ball from the work-energy standpoint.

First consider the throwing process. Take the reference level at the initial position of the ball. Then $E_{k1} = 0$, $E_{p1} = 0$. Take point 2 at the point where the ball leaves the thrower's hand. Then

$$E_{p2} = mgy_2 = \tfrac{1}{4}\text{ lb} \times 2\text{ ft} = 0.5\text{ ft lb},$$
$$E_{k2} = \tfrac{1}{2}mv_2^2 = \tfrac{1}{2} \times (\tfrac{1}{4}/32)\text{ slug} \times (48\text{ ft s}^{-1})^2 = 9\text{ ft lb}.$$

Let P represent the upward force exerted on the ball by the man in the throwing process. The work W' is then the work of this force, and is equal to the sum of the changes in kinetic and potential energy of the ball.

The kinetic energy of the ball increases by 9 ft lb and its potential energy by 0.5 ft lb. The work W' of the upward force P is therefore 9.5 ft lb.

If the force P is constant, the work of this force is given by

$$W' = P(y_2 - y_1),$$

and the force P is then

$$P = \frac{W'}{y_2 - y_1} = \frac{9.5\text{ ft lb}}{2\text{ ft}} = 4.75\text{ lb}.$$

However, the *work* of the force P is 9.5 ft lb whether or not the force is constant.

Now consider the flight of the ball after it leaves the thrower's hand. In the absence of air resistance, the only force on the ball is then its weight $\mathbf{w} = m\mathbf{g}$. Hence the total mechanical energy of the ball remains constant. The calculations will be simplified if we take a new reference level at the point where the ball leaves the thrower's hand. Calling this point 1, we have

$$E_{k1} = 9\text{ ft lb}, \qquad E_{p1} = 0,$$
$$E_k + E_p = 9\text{ ft lb},$$

and the total mechanical energy at any point of the path equals 9 ft lb.

Suppose we wish to find the speed of the ball at a height of 20 ft above the reference level. Its potential energy at this elevation is 5 ft lb. (Why?) Its kinetic energy is therefore 4 ft lb. To find its speed, we have

$$\tfrac{1}{2}mv^2 = E_k, \qquad v = \pm\sqrt{2E_k/m} = \pm 32 \text{ ft s}^{-1}.$$

The significance of the $\pm$ sign is that the ball passes this point *twice*, once on the way up and again on the way down. Its *potential* energy at this point is the same whether it is moving up or down. Hence its kinetic energy is the same and its *speed* is the same. The algebraic sign of the speed is + when the ball is moving up and − when it is moving down.

Next, let us find the height of the highest point reached. At this point, $v = 0$ and $E_k = 0$. Therefore $E_p = 9$ ft lb, and the ball rises to a height of 36 ft above the point where it leaves the thrower's hand.

Finally, suppose we were asked to find the speed at a point 40 ft above the reference level. The potential energy at this point would be 10 ft lb. But the total energy is only 9 ft lb, so the ball never reaches a height of 40 ft.

Example 2. A body slides down a curved track which is one quadrant of a circle of radius R, as in Fig. 7-7. If it starts from rest and there is no friction, find its speed at the bottom of the track. The motion of this body is exactly the same as that of a body attached to one end of a string of length R, the other end of which is held at point O.

The equations of motion with constant acceleration cannot be used, since the acceleration decreases during the motion. (The slope angle of the track becomes smaller and smaller as the body descends.) However, if there is no friction, the only force on the body in addition to its weight is the normal force $\boldsymbol{N}$ exerted on it by the track. The work of this force is zero, so $W' = 0$ and mechanical energy is conserved. Take point 1 at the starting point and point 2 at the bottom of the track. Take the reference level at point 2. Then $y_1 = R$, $y_2 = 0$, and

$$E_{k2} + E_{p2} = E_{k1} + E_{p1},$$
$$\tfrac{1}{2}mv_2^2 + 0 = 0 + mgR,$$
$$v_2 = \pm\sqrt{2gR}.$$

The speed is therefore the same as if the body had fallen *vertically* through a height R. (What is now the significance of the $\pm$ sign?)

As a numerical example, let $R = 1$ m. Then

$$v = \pm\sqrt{2 \times 9.8 \text{ (m s}^{-2}) \times 1 \text{ m}} = \pm 4.43 \text{ m s}^{-1}.$$

Example 3. Suppose a body of mass 0.5 kg slides down a track of radius $R = 1$ m, like that in Fig. 7-7, but its speed at the bottom is only 3 m s^{-1}. What was the work of the frictional force acting on the body?

In this case, $W' = W_f$, and

$$\begin{aligned} W_f &= (\tfrac{1}{2}mv_2^2 - \tfrac{1}{2}mv_1^2) + (mgy_2 - mgy_1) \\ &= (\tfrac{1}{2} \times 0.5 \text{ kg} \times 9 \text{ m}^2 \text{ s}^{-2} - 0) \\ &\quad + (0 - 0.5 \text{ kg} \times 9.8 \text{ m s}^{-2} \times 1 \text{ m}) \\ &= 2.25\text{J} - 4.9 \text{ J} = -2.65 \text{ J}. \end{aligned}$$

The frictional work was therefore −2.65 J, and the total mechanical energy *decreased* by 2.65 J. The mechanical energy of a body is *not* conserved when friction forces act on it.

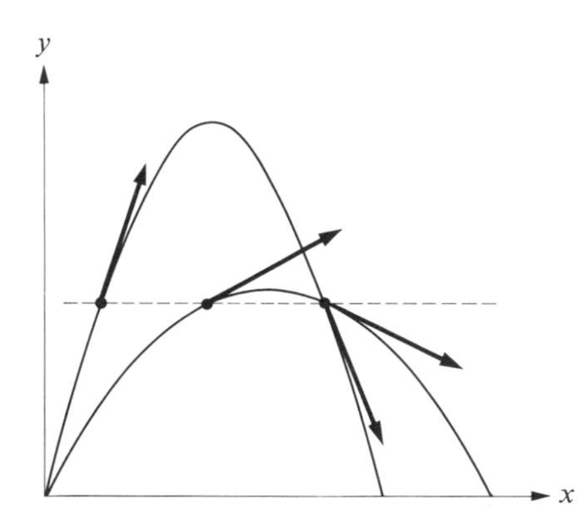

7-8 For the same initial speed, the speed is the same at all points at the same elevation.

Example 4. In the absence of air resistance, the only force on a projectile is its weight, and the mechanical energy of the projectile remains constant. Figure 7-8 shows two trajectories of a projectile with the same initial speed (and hence the same total energy) but with different angles of departure. At all points at the same elevation the potential energy is the same; hence the kinetic energy is the same and the speed is the same.

Example 5. A small object of weight $\boldsymbol{w}$ hangs from a string of length ℓ, as shown in Fig. 7-9. A *variable* horizontal force $\boldsymbol{P}$ which starts at zero and gradually increases is used to pull the object very slowly until the string makes an angle θ with the vertical. Calculate the work of the force $\boldsymbol{P}$.

The sum of the works W' of all the forces other than the gravitational force must equal the change of kinetic energy plus the change of gravitational potential energy. Hence

$$W' = W_P + W_T = \Delta E_k + \Delta E_p.$$

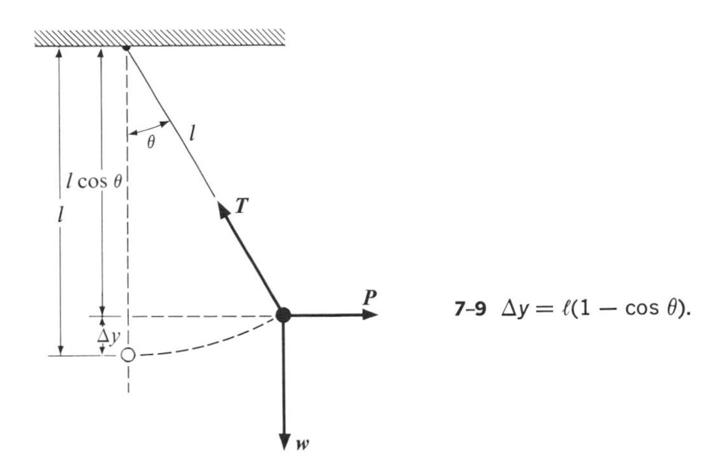

7-9 $\Delta y = \ell(1 - \cos\theta)$.

Since $\boldsymbol{T}$ is perpendicular to the path of its point of application, $W_T = 0$; and since the body was pulled very slowly at all times, the change of kinetic energy is also zero. Hence

$$W_P = \Delta E_p = w\Delta y,$$

where Δy is the distance that the object has been raised. From Fig. 7-9, Δy is seen to be $\ell(1 - \cos\theta)$. Therefore

$$W_P = w\ell(1 - \cos\theta).$$

The reader is referred to the example at the end of Section 7-2, where this same problem was solved by performing the integration in the equation $W_P = \int P \cos\theta \, ds$. The two answers, of course, are identical, but how much more simply the energy principle leads to the final result!

It has been assumed thus far in this section that the changes in elevation were small enough so that the gravitational force on a body could be considered constant. We now consider the more general case. The force $\boldsymbol{w}$ in Fig. 7-6 is the force of gravitational attraction exerted on the body by the earth, and the general expression for the magnitude of this force is

$$\frac{G\, mm_E}{r^2},$$

where m_E is the mass of the earth and r is the distance from the earth's center. When r increases from r_1 to r_2, the work of the gravitational force is

$$W_{\text{grav}} = -Gmm_E \int_{r_1}^{r_2} \frac{dr}{r^2} = \frac{Gmm_E}{r_2} - \frac{Gmm_E}{r_1}.$$

Setting the total work equal to the change in kinetic energy and rearranging terms, we get, instead of Eq. (7-9),

$$\boxed{W' = \left(\tfrac{1}{2}mv_2^2 - \frac{Gmm_E}{r_2}\right) - \left(\tfrac{1}{2}mv_1^2 - \frac{Gmm_E}{r_1}\right).}$$

The quantity $-G(mm_E/r)$ is therefore the general expression for the gravitational potential energy of a body attracted by the earth:

$$\boxed{E_p \text{ (gravitational)} = -G\frac{mm_E}{r}.} \qquad (7\text{-}10)$$

The total mechanical energy of the body, the sum of its kinetic energy and potential energy, is

$$E = E_k + E_p = \tfrac{1}{2}mv^2 - G\frac{mm_E}{r}.$$

If the only force on the body is the gravitational force, then $W' = 0$ and the total mechanical energy remains constant, or is conserved.

Example. Use the principle of conservation of mechanical energy to find the velocity with which a body must be projected vertically upward, in the absence of air resistance, (a) to rise to a height above the earth's surface equal to the earth's radius, R, and (b) to escape from the earth.

(a) Let v_1 be the initial velocity. Then

$$r_1 = R, \qquad r_2 = 2R, \qquad v_2 = 0,$$

and

$$\tfrac{1}{2}mv_1^2 - G\frac{mm_E}{R} = 0 - G\frac{mm_E}{2R},$$

or

$$v_1^2 = \frac{Gm_E}{R},$$

in agreement with the result obtained in Example 8 at the end of Section 5-6.

(b) When v_1 is the escape velocity, $r_1 = R$, $r_2 = \infty$, $v_2 = 0$. Then

$$\tfrac{1}{2}mv_1^2 - G\frac{mm_E}{R} = 0 \quad \text{or} \quad v_1^2 = \frac{2Gm_E}{R}.$$

It may be puzzling at first sight that the general expression for gravitational potential energy should contain a minus sign. The reason for this lies in the choice of a reference state or reference level in which the potential energy is considered zero. If we set $E_p = 0$ in Eq. (7-10) and solve for r, we get

$$r = \infty.$$

That is, *the gravitational potential energy of a body is now considered to be zero when the body is at an infinite distance from the earth.* Since the potential energy *decreases* as the body approaches the earth, it must be *negative* at any finite distance from the earth. The *change* in potential energy of a body as it moves from one point to another is the same, whatever the choice of reference level, and it is only changes in potential energy that are significant.

Finally, we show that the general expression for the change in potential energy reduces to Eq. (7-6) for small changes in elevation near the earth's surface. Thus the increase in potential energy of a body of mass m, when its distance from the earth's center increases from r_1 to r_2, is

$$\begin{aligned} E_{p2} - E_{p1} &= -G\frac{mm_E}{r_2} - \left(-G\frac{mm_E}{r_1}\right) \\ &= Gmm_E\left(\frac{1}{r_1} - \frac{1}{r_2}\right) \\ &= Gmm_E\left(\frac{r_2 - r_1}{r_1 r_2}\right). \end{aligned}$$

If points 1 and 2 are at elevations y_1 and y_2 above the earth's surface, then

$$r_2 - r_1 = y_2 - y_1.$$

Furthermore, if y_1 and y_2 are both small compared with the earth's radius R, the product $r_1 r_2$ is very nearly equal to R^2. The preceding equation then reduces to

$$E_{p2} - E_{p1} = \frac{Gmm_E}{R^2}(y_2 - y_1).$$

But

$$\frac{Gm_E}{R^2} = g,$$

where g is the acceleration of gravity at the earth's surface, so

$$E_{p2} - E_{p1} = mg(y_2 - y_1),$$

which is the same as the expression derived earlier when variations in g were neglected.

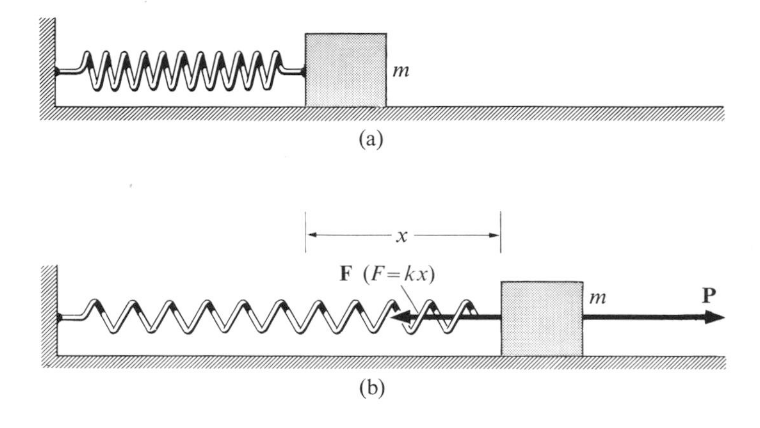

7-10 When an applied force $\boldsymbol{P}$ produces an extension x of a spring, an elastic restoring force $\boldsymbol{F}$ is created within the spring, where $F = kx$.

7-5 ELASTIC POTENTIAL ENERGY

Figure 7-10 shows a body of mass m on a level surface. One end of a spring is attached to the body and the other end of the spring is fixed. Take the origin of coordinates at the position of the body when the spring is unstretched (Fig. 7-10a). An outside agent exerts a force $\boldsymbol{P}$ of sufficient magnitude to cause the spring to stretch. As soon as the slightest extension takes place, a force $\boldsymbol{F}$ is created within the spring which acts in a direction opposite that of increasing x, and therefore opposite that of $\boldsymbol{P}$. The force $\boldsymbol{F}$ is called an *elastic force*. If the force $\boldsymbol{P}$ is reduced or made zero, the elastic force will restore the spring to its original unstretched condition. It may therefore be referred to as a *restoring force*. The subject of elastic restoring forces was first studied by Robert Hooke, in 1678, who observed that if the extension x of a spring was not so great as to permanently distort the spring, *the elastic force is directly proportional to the extension*, or

$$F = kx, \tag{7-11}$$

a relation known as *Hooke's law*. The constant of proportionality k is called the *force constant* or the *stiffness coefficient*.

The work of the elastic force W_{el}, during any process in which the spring is extended from a value x_1 to x_2, is

$$W_{el} = \int \mathbf{F} \cdot d\mathbf{s} = \int_{x_1}^{x_2} F \cos\theta \, dx.$$

Since the direction of F is opposite the direction of dx, $\cos\theta = -1$, whence

$$W_{el} = -\int_{x_1}^{x_2} kx \, dx,$$

or

$$W_{el} = -(\tfrac{1}{2}kx_2^2 - \tfrac{1}{2}kx_1^2).$$

Let W' stand for the work of the applied force $\mathbf{P}$. Setting the total work equal to the change in kinetic energy of the body, we have

$$W' + W_{el} = \Delta E_k,$$

$$W' - (\tfrac{1}{2}kx_2^2 - \tfrac{1}{2}kx_1^2) = (\tfrac{1}{2}mv_2^2 - \tfrac{1}{2}mv_1^2).$$

The quantities $\frac{1}{2}kx_2^2$ and $\frac{1}{2}kx_1^2$ depend only on the initial and final positions of the body and not specifically on the way in which it moved. Let us therefore transfer them from the "work" side of the equation to the "energy" side. Then

$$\boxed{W' = (\tfrac{1}{2}mv_2^2 - \tfrac{1}{2}mv_1^2) + (\tfrac{1}{2}kx_2^2 - \tfrac{1}{2}kx_1^2).} \tag{7-12}$$

The quantity $\frac{1}{2}kx^2$, one-half the product of the force constant and the square of the coordinate of the body, is called the *elastic potential energy* of the body, E_p. (The symbol E_p is used for any form of potential energy.)

$$\boxed{E_p \text{ (elastic)} = \tfrac{1}{2}kx^2.} \tag{7-13}$$

Hence the work W' of the force $\mathbf{P}$ equals the sum of the change in the kinetic energy of the body and the change in its elastic potential energy.

Equation (7-12) can also be written

$$W' = (\tfrac{1}{2}mv_2^2 + \tfrac{1}{2}kx_2^2) - (\tfrac{1}{2}mv_1^2 + \tfrac{1}{2}kx_1^2).$$

The sum of the kinetic and potential energies of the body is its total mechanical energy and *the work of all forces acting on the body*, **with the exception of the elastic force**, *equals the change in the total mechanical energy of the body*.

If the work W' is positive, the mechanical energy increases. If W' is negative, it decreases. In the special case in which W' is zero, the mechanical energy remains constant or is *conserved*.

Example 1. Let the force constant k of the spring in Fig. 7-10 be 24 N m^{-1}, and let the mass of the body be 4 kg. The body is initially at rest, and the spring is initially unstretched. Suppose that a constant force $\mathbf{P}$ of 10 N is exerted on the body, and that there is no friction. What will be the speed of the body when it has moved 0.5 m?

The equations of motion with constant acceleration cannot be used, since the resultant force on the body varies as the spring is stretched. However, the speed can be found from energy considerations:

$$W' = \Delta E_k + \Delta E_p,$$

$$10 \text{ N} \times 0.5 \text{ m} = (\tfrac{1}{2} \times 4 \text{ kg} \times v_2^2 - 0) + (\tfrac{1}{2} \times 24 \text{ N m}^{-1} \times 0.25 \text{ m}^2 - 0),$$

$$v_2 = 1 \text{ m s}^{-1}.$$

Example 2. Suppose the force $\mathbf{P}$ ceases to act when the body has moved 0.5 m. How much farther will the body move before coming to rest?

The elastic force is now the only force, and mechanical energy is conserved. The kinetic energy is $\frac{1}{2}mv^2 = 2$ J, and the potential energy is $\frac{1}{2}kx^2 = 3$ J. The total energy is therefore 5 J (equal to the work of the force $\mathbf{P}$). When the body comes to rest, its kinetic energy is zero and its potential energy is therefore 5 J. Hence

$$\tfrac{1}{2}kx_{max}^2 = 5 \text{ J}, \qquad x_{max} = 0.645 \text{ m}.$$

7-6 CONSERVATIVE AND DISSIPATIVE FORCES

When an object is moved from any position above a zero reference level to any other position, the work of the gravitational force is found to be independent of the path and equal to the difference between the final and the initial values of a function called the *gravitational po-*

tential energy. If the gravitational force alone acts on the object, the total mechanical energy (the sum of the kinetic and gravitational potential energies) is constant or conserved, and therefore the gravitational force is called a *conservative force.* Thus, if the object is ascending, the work of the gravitational force is accomplished at the expense of the kinetic energy. If, however, the object is descending, the work of the gravitational force serves to increase the kinetic energy, or, in other words, this work is *completely recovered.* Complete recoverability is an important aspect of the work of a conservative force.

When an object attached to a spring is moved from one value of the spring extension to any other value, the work of the elastic force is also independent of the path and equal to the difference between the final and initial values of a function called the *elastic potential energy.* If the elastic force alone acts on the object, the sum of the kinetic and elastic potential energies is conserved, and therefore the elastic force is also a conservative force. If the object moves so as to increase the extension of the spring, the work of the elastic force is accomplished at the expense of the kinetic energy. If, however, the extension is decreasing, then the work of the elastic force serves to increase the kinetic energy so that this work is completely recovered also.

To summarize, we see that the work of a conservative force has the following properties:

1) It is independent of the path.
2) It is equal to the difference between the final and initial values of an energy function.
3) It is completely recoverable.

Contrast a conservative force with a friction force exerted on a moving object by a fixed surface. The work of the friction force *does depend* on the path; the longer the path, the greater the work. There is *no* function the difference of two values of which equals the work of the friction force. When we slide an object on a rough fixed surface back to its original position, the friction force reverses, and instead of recovering the work done in the first displacement, we must again do work on the return trip. In other words, frictional work is *not* completely recoverable. When the friction force acts alone, the total mechanical energy is *not* conserved. The friction force is therefore called a *nonconservative* or a *dissipative* force. *The mechanical energy of a body is conserved only when no dissipative forces act on it.*

We find that when friction forces act on a moving body, another form of energy is involved. The more general principle of conservation of energy includes this other form of energy, along with kinetic and potential energy, and when it is included the *total* energy of any system remains constant. We shall study this general conservation principle more fully in a later chapter.

Example. Example 3 in Section 7-4 illustrates the motion of a body acted on by a dissipative friction force. The initial mechanical energy of the body is its initial potential energy of 4.9 J. Its final mechanical energy is its final kinetic energy of 2.25 J. The frictional work W_f is -2.65 J. A quantity of energy equivalent to 2.65 J is developed as the body slides down the track. The sum of this energy and the final mechanical energy equals the initial mechanical energy, and the total energy of the system is conserved.

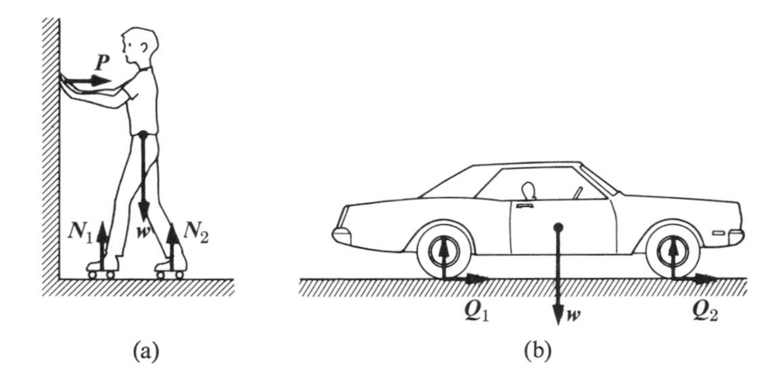

7-11 **(a) External forces acting on a man who is pushing against a wall. The work of these forces is zero.** (b) External forces on an automobile. The work of these forces is zero. In both cases, the work of the internal force is responsible for the increase in kinetic energy.

7-7 INTERNAL WORK

Figure 7-11(a) shows a man standing on frictionless roller skates on a level surface, facing a rigid wall. Suppose that he sets himself in motion backward by pushing against the wall. The external forces *on the man* are his weight $\boldsymbol{w}$, the upward forces $\boldsymbol{N}_1$ and $\boldsymbol{N}_2$ exerted by the ground, and the horizontal force $\boldsymbol{P}$ exerted by the wall. (The latter is the reaction to the force with which the man pushes against the wall.) The works of $\boldsymbol{w}$ and of $\boldsymbol{N}$ are zero because they are perpendicular to the motion. The force $\boldsymbol{P}$ is the unbalanced horizontal force which imparts to the system a horizontal acceleration. *The work of* $\boldsymbol{P}$*, however, is zero because there is no motion of its point of application.* We are therefore confronted with a curious situation in which a force is responsible for acceleration, but its work, being zero, is not equal to the increase in kinetic energy of the system!

It is at this point that the concept of *internal work* must be introduced. Although internal forces play no role in accelerating the system, their points of application may move in such a way that work is done. Thus, in this case, an *internal* muscular force is exerted within the man away from the wall. (Think of his lengthening arm as an expanding spring.) Since the point of application of this force moves in the same direction, the work W_i of the internal force is not zero. This is the work responsible for the increase in kinetic energy.

If therefore both external and internal forces act on the particles of a system, the total work W of all forces, external and internal, is the sum of the works W_e and W_i and is equal to the change in the total kinetic energy of the system:

$$W = W_e + W_i = \Delta E_k.$$

The same principle holds in the case of an accelerated automobile. The portions of the rubber tires that are in contact with the rough roadway push back on the ground and the reactions to these forces, designated by $\boldsymbol{Q}_1$ and $\boldsymbol{Q}_2$ in Fig. 7-11(b), are the external horizontal forces acting *on* the automobile responsible for imparting the horizontal acceleration to the system. Since, however, the portions of the tires momentarily in contact with the road are at rest with respect to the road, the works of the forces $\boldsymbol{Q}_1$ and $\boldsymbol{Q}_2$ are zero. As a result of the expanding gases in the cylinders of the automobile engine there are many internal forces, some of which do work. Again in this case, it is the work W_i of the internal forces which is equal to the increase in kinetic energy.

7-8 INTERNAL POTENTIAL ENERGY

If any of the *external* forces on the particles of a system are conservative (gravitational or elastic), the work of these forces can be transferred to the energy side of the work-energy equation and called the change in *external potential energy of the system.* In many instances, the *internal* forces depend only on the distances between *pairs* of particles. The internal work then depends only on the initial and final distances between the particles, and not on the particular way in which they moved. The work of these internal forces can then also be transferred to the energy side of the work-energy equation and called the change in *internal potential energy of the system.* Internal potential energy is a property of the system as a whole and cannot be assigned to any specific particle.

Let W' represent the works of all external and internal forces that have *not* been transferred to the energy side of the work-energy equation and called changes in external and internal potential energy. Let E_p^e and E_p^i represent the external and internal potential energies. The work-energy equation then takes the form

$$W' = \Delta E_k + \Delta E_p^e + \Delta E_p^i, \tag{7-14}$$

where the total mechanical energy now includes the kinetic energy of the system and both its external and internal potential energy. If the work W' is zero, the total mechanical energy is conserved.

Example 1. Consider a system consisting of two particles in "outer space," very far from all other matter. No external forces act on the system, no external work is done when the bodies move, and the system has no external potential energy. A gravitational force of attraction acts between the bodies, which depends only on the distance between them. We can therefore say that the system has an internal potential energy. Suppose the particles start from rest and accelerate toward each other. The total kinetic energy of the system increases and its internal potential energy decreases. The sum of its kinetic energy and internal potential energy remains constant.

Energy considerations *alone* do not suffice to tell us how much kinetic energy is gained by each body separately. This can be determined, however, from *momentum* considerations, as will be explained in the next chapter.

Example 2. In Section 7-5 we computed the elastic potential energy of a body acted on by a force exerted by a spring. This force was an *external* force acting on the body, and the elastic potential energy $\frac{1}{2}kx^2$ should properly be called the *external* elastic potential energy *of the body.*

Let us now consider the spring itself. The spring is a *system* composed of an enormous number of molecules which exert internal forces on one another. When the spring is stretched, the distances between its molecules change. The intermolecular forces depend only on the distances between pairs of molecules, so that a stretched spring has an internal elastic potential energy. To calculate this, suppose the spring is slowly stretched from its no-load length by equal and opposite forces applied at its ends. The work of these forces was shown to equal $\frac{1}{2}kx^2$, where x is the elongation of the spring above its no-load length. Let us retain this work on the left side of the work-energy equation so that it becomes the work W' in Eq. (7-14). There is no change in the kinetic energy of the spring and no change in its external potential energy. If we call the internal potential energy zero when $x = 0$, the *change* in internal potential energy in the stretching process equals the final potential energy E_p^i. Then

$$W' = \Delta E_p^i = E_p^i, \qquad E_p^i = \tfrac{1}{2}kx^2.$$

Hence the *internal elastic potential energy of a stretched spring* is equal to $\frac{1}{2}kx^2$. Note that the internal potential energy can be calculated without any detailed information regarding the intermolecular forces.

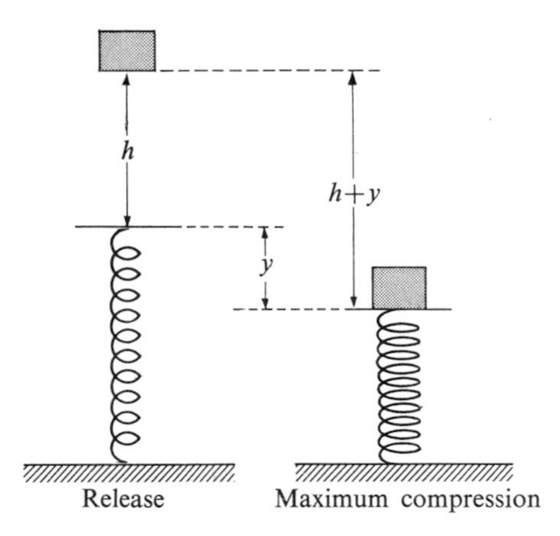

7-12 The total fall of the block is $h + y$.

Example 3. A block of mass m, initially at rest, is dropped from a height h onto a spring whose force constant is k. Find the maximum distance y that the spring will be compressed. (See Fig. 7-12.)

This is a process for which the principle of the conservation of mechanical energy holds. At the moment of release, the kinetic energy is zero. At the moment when maximum compression occurs, there is also no kinetic energy. Hence, the loss of gravitational potential energy of the block equals the gain of elastic potential energy of the spring. As shown in Fig. 7-12, the total fall of the block is $h + y$, whence

$$mg(h + y) = \tfrac{1}{2}ky^2, \qquad \text{or} \qquad y^2 - \frac{2mg}{k}y - \frac{2mgh}{k} = 0.$$

Therefore

$$y = \frac{1}{2}\left[\frac{2mg}{k} \pm \sqrt{\left(\frac{2mg}{k}\right)^2 + \frac{8mgh}{k}}\right].$$

7-9 POWER

The time element is not involved in the definition of work. The same amount of work is done in raising a given weight through a given height whether the work is done in 1 s, or 1 hr, or 1 yr. In many instances, however, it is necessary to consider the *rate* at which work is done as well as the total amount of work accomplished. The rate at which work is done by a working agent is called the *power* developed by that agent.

If a quantity of work ΔW is done in a time interval Δt, the average power $\overline{P}$ is defined as

$$\text{Average power} = \frac{\text{work done}}{\text{time interval}},$$

$$\overline{P} = \frac{\Delta W}{\Delta t}.$$

The instantaneous power P is the limiting value of this quotient as Δt approaches zero:

$$\boxed{P = \lim_{\Delta t \to 0} \frac{\Delta W}{\Delta t} = \frac{dW}{dt}.} \tag{7-15}$$

The mks unit of power is one joule per second (1 J s^{-1}), which is called one *watt* (1 W). Since this is a rather small unit, the kilowatt (1 kW = 10^3 W) and the megawatt (1 MW = 10^6 W) are commonly used. The cgs power unit is one erg per second (1 erg s^{-1}). No single term is assigned to this unit.

In the engineering system, where work is expressed in foot pounds and time in seconds, the unit of power is one foot pound per second. Since this unit is inconveniently small, a larger unit called the *horsepower* (hp) is in common use: 1 hp = 550 ft lb s^{-1} = 33,000 ft lb min^{-1}. That is, a 1-hp motor running at full load is doing 33,000 ft lb of work every minute it runs.

A common misconception is that there is something inherently *electrical* about a watt or a kilowatt. This is not the case. It is true that electrical power is usually expressed in watts or kilowatts, but the power consumption of an incandescent lamp could equally well be expressed in horsepower, or an automobile engine rated in kilowatts.

From the relations between the newton, pound, meter, and foot, we can show that 1 hp = 746 W = 0.746 kW, or about $\frac{3}{4}$ of a kilowatt, a useful figure to remember.

Having defined two units of power, the horsepower and the kilowatt, we may use these in turn to define two new units of work, the *horsepower hour* (hp hr) and the *kilowatt hour* (kwh).

One horsepower hour is the work done in one hour by an agent working at the constant rate of one horsepower.

Since such an agent does 33,000 ft lb of work each minute, the work done in 1 hr is 60 × 33,000 = 1,980,000 ft lb:

$$1 \text{ hp hr} = 1.98 \times 10^6 \text{ ft lb.}$$

One kilowatt hour is the work done in one hour by an agent working at the constant rate of one kilowatt.

Since such an agent does 1000 J of work each second, the work done in 1 hr is 3600 × 1000 = 3,600,000 J:

$$1 \text{ kwh} = 3.6 \times 10^6 \text{ J} = 3.6 \text{ MJ.}$$

Note that the horsepower hour and the kilowatt hour are units of *work*, not power.

One aspect of work or energy which may be pointed out here is that although it is an abstract physical quantity, it nevertheless has a monetary value. A pound of force or a foot per second of velocity are not things which are bought and sold as such, but a foot pound or a kilowatt hour of energy are quantities offered for sale at a definite market rate. In the form of electrical energy, a kilowatt hour can be purchased at a price varying from a few tenths of a cent to a few cents, depending on the locality and the quantity purchased. In the form of heat, 778 ft lb (one Btu) costs about a thousandth of a cent.

7-10 POWER AND VELOCITY

Suppose that a force $\boldsymbol{F}$ is exerted on a particle while the particle moves a distance Δs along its path. If F_S is the magnitude of the component of $\boldsymbol{F}$ tangent to the path, then the work of $\boldsymbol{F}$ is given by $\Delta W = F_S\,\Delta s$, and the average power is

$$\overline{P} = \frac{\Delta W}{\Delta t} = F_S \frac{\Delta s}{\Delta t} = F_S \overline{v}.$$

The instantaneous power is therefore

$$\boxed{P = F_S v,} \tag{7-16}$$

where v is the instantaneous velocity. Another way of writing Eq. (7-16) is in terms of the scalar product

$$\boxed{P = \boldsymbol{F} \cdot \boldsymbol{v}.} \tag{7-17}$$

Example. The engine of a jet aircraft develops a thrust of 3000 lb. What horsepower does it develop at a velocity of 600 mi hr^{-1} = 880 ft s^{-1}?

$$P = F_S v = 3000 \text{ lb} \times 880 \text{ ft s}^{-1} = 2{,}640{,}000 \text{ ft lb s}^{-1}$$
$$= \frac{2.64 \times 10^6 \text{ ft lb s}^{-1}}{550 \text{ ft lb s}^{-1} \text{ hp}^{-1}} = 4800 \text{ hp.}$$

7-11 MASS AND ENERGY

While the mass of a body can ordinarily be considered constant, there is ample experimental evidence that it is actually a function of the velocity of the body, increasing with increasing velocity according to the relation

$$m = \frac{m_0}{\sqrt{1 - v^2/c^2}}, \tag{7-18}$$

where m_0 is the "rest mass" of the body, c is the velocity of light, and v the velocity of the body.

This equation was predicted by Lorentz and Einstein on theoretical grounds based on relativity considerations, and it has been directly verified by experiments on rapidly moving electrons and ions. The increase in mass is not appreciable until the velocity approaches that of light, and therefore it ordinarily escapes detection.

Figure 7-13 is a graph of the ratio m/m_0 plotted as a function of the ratio v/c. For a rifle bullet, v/c is about 3×10^{-6}, and evidently even at this speed the mass equals the rest mass for all practical purposes. On the other hand, an electron accelerated by a potential

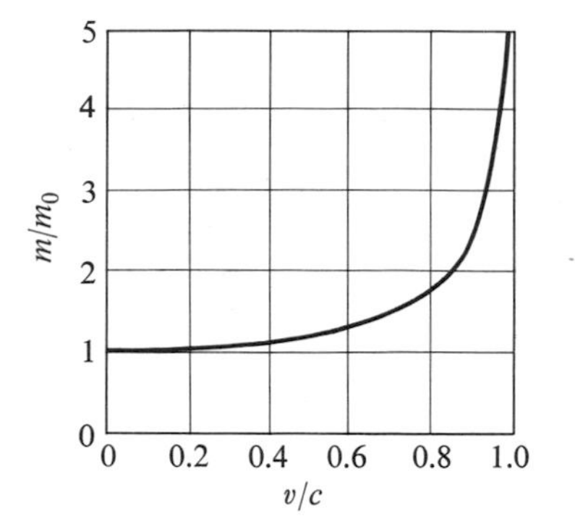

7-13 Graph of the ratio m/m_0 plotted as a function of v/c.

difference as small as 100,000 volts already has a speed of about 0.6c, so it is evident that the increase of mass with velocity is of the greatest importance when dealing with atomic and subatomic particles that have been accelerated through millions of volts.

It was stated at the beginning of Chapter 5 that when a force F acts on a particle, the ratio of force to acceleration is a constant m, the mass of the particle:

$$\frac{\mathbf{F}}{d\mathbf{v}/dt} = \text{constant} = m.$$

We now see that the statement above is only an approximation, although a very good approximation, when the velocity of the particle relative to an observer is small compared with the velocity of light, c. The mass is in reality not constant, but increases with increasing velocity.

The relativistic form of the second law, instead of being

$$\mathbf{F} = m\frac{d\mathbf{v}}{dt}, \qquad \text{is} \qquad \mathbf{F} = \frac{d}{dt}(m\mathbf{v}).$$

The product $m\mathbf{v}$ of a particle is called its momentum and is discussed fully in the next chapter.The relativistic form of the second law may therefore be stated as follows: *the resultant force on a particle equals the rate of change of momentum of the particle.*

If the relative velocity is small, then $m = m_0 =$ constant (very nearly) and

$$\mathbf{F} = m_0\frac{d\mathbf{v}}{dt}, \tag{7-19}$$

so that for small velocities the relativistic equation reduces to the classical form of the second law. In general, however, we have

$$\mathbf{F} = m\frac{d\mathbf{v}}{dt} + \mathbf{v}\frac{dm}{dt}, \tag{7-20}$$

and the effect of a force is not only to change the *velocity* v of a particle, but to change its *mass* m as well.

The term v^2 in Eq. (7-18) refers to the square of the relative *speed* of the particle. If the force is at *right angles* to the velocity, as will be the case when a particle moves in a circle, it produces no change in speed, so that $dm/dt = 0$ and

$$\mathbf{F} = m\frac{d\mathbf{v}}{dt},$$

which has the same *form* as Eq. (7-19), although if the speed of the particle is large the mass m is not equal to the mass m_0.

One of the experimental methods of measuring the masses of electrons and ions is to whirl them in a circular path in a magnetic field such as that of a cyclotron. The force on the particle, its velocity, and the radius of its circular path can all be measured and the mass computed from its definition as the ratio of the force to the radial acceleration. In fact, it is by measurements of this sort that the variation of mass with speed has been verified.

Now consider the motion of a particle acted on by a force in the *same direction* as the velocity, so that the speed and the mass both increase. We wish to compute the work of the force. If ds is a small displacement along the path, then

$$v = \frac{ds}{dt}, \qquad dt = \frac{ds}{v},$$

and Eq. (7-20) can be written as

$$F\,ds = mv\,dv + v^2\,dm. \tag{7-21}$$

Next, we square both sides of Eq. (7-18) and write the result as

$$v^2 = c^2\left(1 - \frac{m_0^2}{m^2}\right).$$

Now take differentials of both sides and multiply by $m/2$:

$$mv\,dv = \frac{m_0c^2}{m^2}\,dm.$$

When these expressions for v^2 and $mv\,dv$ are inserted in Eq. (7-21), we get the simple relation

$$F\,ds = c^2\,dm.$$

We now *define* the kinetic energy of the particle as the work of the accelerating force:

$$E_k = \int F\,ds = \int_{m_0}^{m} c^2\,dm = (m - m_0)c^2. \quad (7\text{-}22)$$

The kinetic energy therefore equals the increase in mass over the rest mass, multiplied by the square of the velocity of light. The preceding equation is one form of the famous Einstein *mass-energy relation*. It appears very different from the familiar expression $E_k = \frac{1}{2}mv^2$. However, all relativistic equations reduce to those of Newtonian mechanics when v is small compared with c. If this is the case, then v^2/c^2 is very small compared with unity, and

$$\left(1 - \frac{v^2}{c^2}\right)^{-1/2} \approx 1 + \frac{1}{2}\frac{v^2}{c^2},$$

as may be seen by expanding the quantity on the left by the binomial theorem and neglecting the terms in v^4/c^4 and higher powers. Then

$$m = \frac{m_0}{\sqrt{1 - v^2/c^2}} \approx m_0\left(1 + \frac{1}{2}\frac{v^2}{c^2}\right)$$

and

$$(m - m_0)c^2 \approx \tfrac{1}{2}mv^2.$$

Equation (7-22) can also be written as

$$m = m_0 + \frac{E_k}{c^2}. \quad (7\text{-}23)$$

The increase in mass of a particle in relative motion, over its rest mass, equals its kinetic energy divided by c^2. This relation is not restricted to *kinetic* energy, however, but includes *potential* energy as well. Thus the mass of a *system* of particles increases not only when the kinetic energies of the particles increase, but when the internal potential energy of the system increases. The general form of Eq. (7-23) is therefore

$$m = m_0 + \frac{E_k}{c^2} + \frac{E_p}{c^2}. \quad (7\text{-}24)$$

When the kinetic and potential energies are both zero, the mass equals m_0, so that the term "rest mass" implies not only that the system is at rest, but also that it has no internal potential energy. The mass of a stretched spring, at rest in the laboratory, is greater than its mass in an unstretched state by the elastic potential energy E_p (divided by c^2). The value of c^2 is so large, however, that the increase in mass is quite undetectable even by the most sensitive mass-measuring instruments. When it comes to the potential energies of nuclei, the situation is different. The energies associated with the binding forces that hold the nucleons together are relatively so great that the masses of nuclei can be measurably different from the sums of the masses of their component parts when they are separated from one another.

Equation (7-24) can be written

$$mc^2 = m_0c^2 + E_k + E_p. \quad (7\text{-}25)$$

Each term in this equation can now be interpreted as an *energy*. This product mc^2 is the *total* energy E:

$$E = mc^2. \quad (7\text{-}26)$$

The product m_0c^2 is the *rest energy* E_0:

$$E_0 = m_0c^2. \quad (7\text{-}27)$$

Then we can write

$$E = E_0 + E_k + E_p. \quad (7\text{-}28)$$

Thus the concepts of *mass* and *energy* become simply two aspects of a single entity that can be called *mass-energy*. Even when the kinetic and potential energies are zero, a system (or a particle) has a rest energy E_0, associated with its rest mass m_0.

One frequently reads that in the processes of nuclear fission and nuclear fusion, "mass is converted to energy." Such statements give the impression that the mass of a system *decreases* in a nuclear reaction while at the same time its energy *increases*. This is not the case. Let us consider as an example the fission of a uranium nucleus. Before the fission process occurs, the mass of the nucleus is the sum of the rest masses of the fission products and their internal *potential* energy (divided by c^2). Their kinetic energy is zero. Immediately after fission takes place, and before the fission products have made any collisions with the molecules of the surrounding material, the products have a large kinetic energy but no potential energy. Their total mass is now the sum of their rest masses and their *kinetic* energy (divided by c^2). But the kinetic energy equals the original potential energy, so the mass of the system is equal to the original mass of the uranium nucleus. The "source" of the kinetic energy cannot be a decrease in mass of the system because no such decrease has occurred. Rather, the process consists of a conversion of internal potential energy to kinetic energy. Energy and mass are *both* conserved, and the principles of conservation of mass and conservation of energy become two aspects of a *single* principle of *conservation of mass-energy*.

Consider next the collision processes between the fission fragments and the molecules of the surrounding material. Immediately after the fission process, the fragments are traveling at very high speeds and their masses are greater than their rest masses. As they collide with the surrounding molecules, their kinetic energies decrease and their masses decrease also. When they have come to rest, their combined mass is *less* than that of the original nucleus, and in this sense the mass *has* decreased. But in the process of colliding with the surrounding molecules, the kinetic energies and velocities of these molecules are *increased* and their masses are increased also. The increase in energy of these molecules is just equal to the decrease in energy of the fission fragments, and the increase in mass of the molecules is just equal to the decrease in mass of the fragments. Again, there has been no change in either the total mass or the total energy of the *entire system*. Mass is conserved and energy is conserved also.

It is true, of course, that the mass *of an individual particle* such as an electron or proton can no longer be considered constant. But in any process in which the energy and mass of a particle increase, both the energy and mass of some other particle or particles decrease by exactly the same amounts, and both the total energy and the total mass of the system remain constant.

Problems

7-1 The locomotive of a freight train exerts a constant force of 6 tons on the train while drawing it at 50 mi hr^{-1} on a level track. How many foot-pounds of work are done in a distance of 1 mile?

7-2 An 80-lb block is pushed a distance of 20 ft along a level floor at constant speed by a force at an angle of 30° below the horizontal. The coefficient of friction between block and floor is 0.25. How many foot pounds of work are done?

7-3 A horse is towing a canal boat, the towrope making an angle of 10° with the towpath. If the tension in the rope is 100 lb, how many foot pounds of work are done while moving 100 ft along the towpath?

7-4 A block is pushed 4 ft along a fixed horizontal surface by a horizontal force of 10 lb. The opposing force of friction is 2 lb. (a) How much work is done by the 10-lb force? (b) What is the work of the friction force?

7-5 A body is attracted toward the origin with a force given by $F = -6x^3$, where F is in pounds and x in feet. (a) What force is required to hold the body at point a, 1 ft from the origin? (b) At point b, 2 ft from the origin? (c) How much work must be done to move the body from point a to point b?

7-6 The force exerted by a gas in a cylinder on a piston whose area is A is given by $F = pA$, where p is the force per unit area, or *pressure*. The work W in a displacement of the piston from x_1 to x_2 is

$$W = \int_{x_1}^{x_2} F\,dx = \int_{x_1}^{x_2} pA\,dx = \int_{V_1}^{V_2} p\,dV,$$

where dV is the accompanying infinitesimal change of volume of the gas. (a) During an expansion of a gas at constant temperature (isothermal) the pressure depends on the volume according to the relation

$$p = \frac{nRT}{V},$$

where n and R are constants and T is the constant temperature. Calculate the work in expanding isothermally from volume V_1 to volume V_2. (b) During an expansion of a gas at constant entropy (adiabatic) the pressure depends on the volume according to the relation

$$p = \frac{K}{V^{\gamma}},$$

where K and γ are constants. Calculate the work in expanding adiabatically from V_1 to V_2.

7-7 (a) Compute the kinetic energy of an 1800-lb automobile traveling at 30 mi hr^{-1}. (b) How many times as great is the kinetic energy if the velocity is doubled?

7-8 Compute the kinetic energy, in ergs and in joules, of a 2-g rifle bullet traveling at 500 m s^{-1}.

7-9 An electron strikes the screen of a cathode-ray tube with a velocity of 10^9 cm s^{-1}. Compute its kinetic energy in ergs. The mass of an electron is 9×10^{-28} g.

7-10 What is the potential energy of a 1600-lb elevator at the top of the Empire State building, 1248 ft above street level? Assume the potential energy at street level to be zero.

7-11 What is the increase in potential energy of a 1-kg body when lifted from the floor to a table 1 m high?

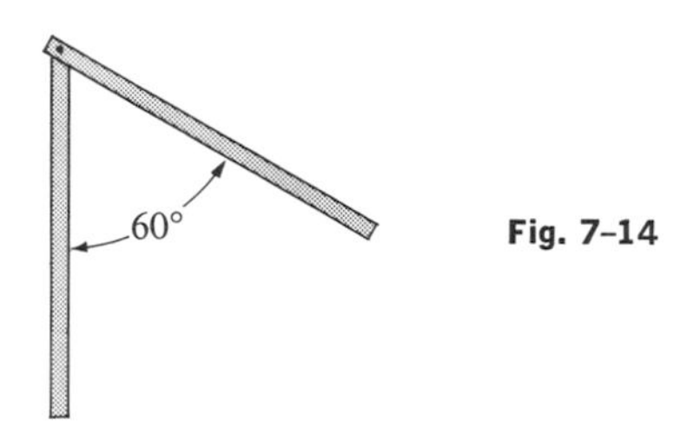

Fig. 7-14

7-12 A meter stick whose mass is 300 g is pivoted at one end as in Fig. 7-14 and displaced through an angle of 60°. What is the increase in its potential energy?

7-13 The force in pounds required to stretch a certain spring a distance of x ft beyond its unstretched length is given by $F = 10x$. (a) What force will stretch the spring 6 in.? 1 ft? 2 ft? (b) How much work is required to stretch the spring 6 in.? 1 ft? 2 ft?

7-14 The scale of a certain spring balance reads from zero to 400 lb and is 8 in. long. (a) What is the potential energy of the spring when it is stretched 8 in.? 4 in.? (b) When a 50-lb weight hangs from the spring?

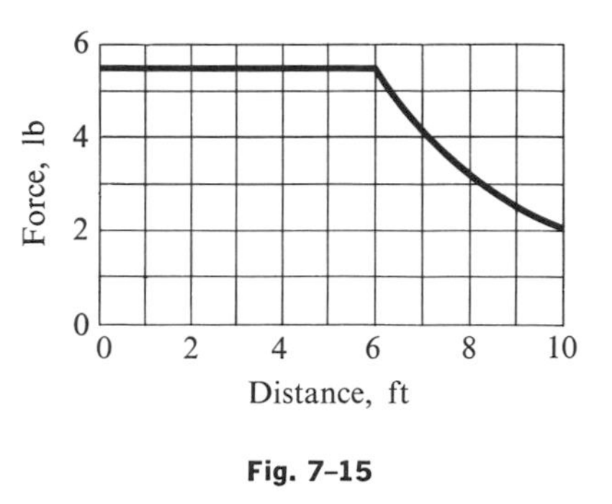

Fig. 7-15

7-15 A body moves a distance of 10 ft under the action of a force which has the constant value of 5.5 lb for the first 6 ft and then decreases to a value of 2 lb, as shown by the graph in Fig. 7-15. (a) How much work is done in the first 6 ft of the motion? (b) How much work is done in the last 4 ft?

7-16 A block weighing 16 lb is pushed 20 ft along a horizontal frictionless surface by a horizontal force of 8 lb. The block starts from rest. (a) How much work is done? What becomes of this work? (b) Check your answer by computing the acceleration of the block, its final velocity, and its kinetic energy.

7-17 In the preceding problem, suppose the block had an initial velocity of 10 ft s^{-1}, other quantities remaining the same. (a) How much work is done? (b) Check by computing the final velocity and the increase in kinetic energy.

7-18 A 16-lb block is lifted vertically at a constant velocity of 10 ft s^{-1} through a height of 20 ft. (a) How great a force is required? (b) How much work is done? What becomes of this work?

7-19 A 25-lb block is pushed 100 ft up the sloping surface of a plane inclined at an angle of 37° to the horizontal by a constant force F of 32.5 lb acting parallel to the plane. The coefficient of friction between the block and plane is 0.25. (a) What is the work of the force F? (b) Compute the increase in kinetic energy of the block. (c) Compute the increase in potential energy of the block. (d) Compute the work done against friction. What becomes of this work? (e) What can you say about the sum of (b), (c), and (d)?

7-20 A man weighing 150 lb sits on a platform suspended from a movable pulley and raises himself by a rope passing over a fixed pulley (Fig. 7-16). Assuming no friction losses, find (a) the force he must exert, (b) the increase in his energy when he raises himself 2 ft. Answer part (b) by calculating his increase in potential energy, and also by computing the product of the

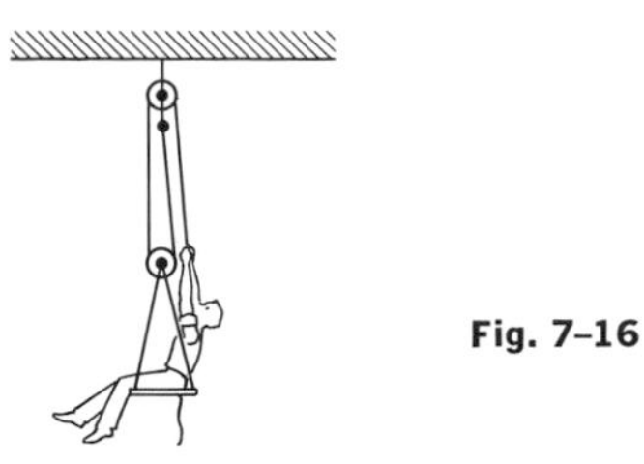

Fig. 7-16

force on the rope and the length of rope passing through his hands.

7-21 A barrel weighing 250 lb is suspended by a rope 30 ft long. (a) What horizontal force is necessary to hold the barrel sideways 5 ft from the vertical? (b) How much work is done in moving it to this position?

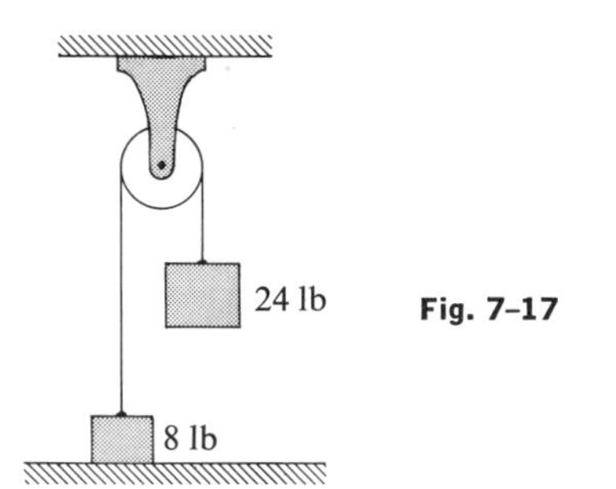

Fig. 7-17

7-22 The system in Fig. 7-17 is released from rest with the 24-lb block 8 ft above the floor. Use the principle of conservation of energy to find the velocity with which the block strikes the floor. Neglect friction and inertia of the pulley.

7-23 The spring of a spring gun has a force constant of 3 lb in^{-1}. It is compressed 2 in. and a ball weighing 0.02 lb is placed in the barrel against the compressed spring. (a) Compute the maximum velocity with which the ball leaves the gun when released. (b) Determine the maximum velocity if a resisting force of 2.25 lb acts on the ball.

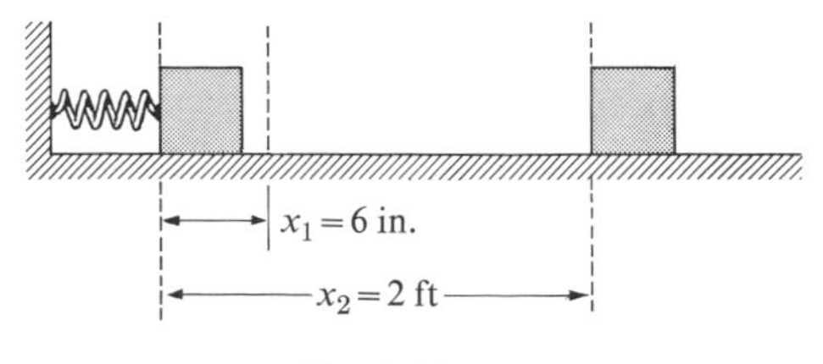

Fig. 7-18

7-24 A block weighing 2 lb is forced against a horizontal spring of negligible mass, compressing the spring an amount $x_1 = 6$ in. When released, the block moves on a horizontal tabletop a distance $x_2 = 2$ ft before coming to rest. The force constant k is 8 lb ft^{-1} (Fig. 7-18). What is the coefficient of friction, μ, between the block and the table?

7-25 A 2-kg block is dropped from a height of 40 cm onto a spring whose force constant k is 1960 N m^{-1}. Find the maximum distance the spring will be compressed.

7-26 A 16-lb projectile is fired from a gun with a muzzle velocity of 800 ft s^{-1} at an angle of departure of 45°. The angle is then increased to 90° and a similar projectile is fired with the same muzzle velocity. (a) Find the maximum heights attained by the projectiles. (b) Show that the total energy at the top of the trajectory is the same in the two cases. (c) Using the energy principle, find the height attained by a similar projectile if fired at an angle of 30°.

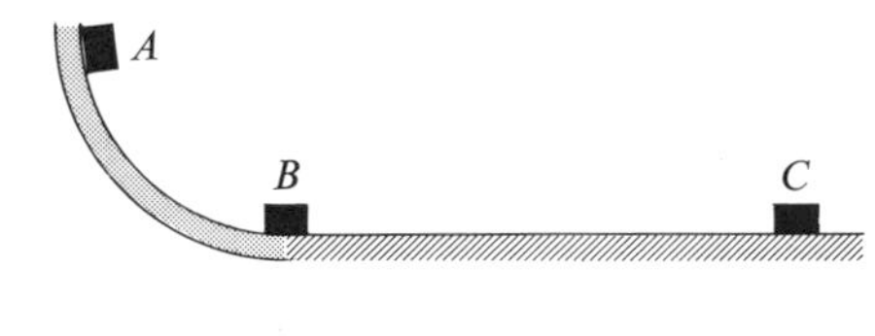

Fig. 7-19

7-27 A block weighing 2 lb is released from rest at point A on a track which is one quadrant of a circle of radius 4 ft (Fig. 7-19). It slides down the track and reaches point B with a velocity of 12 ft s^{-1}. From point B it slides on a level surface a distance of 9 ft to point C, where it comes to rest. (a) What was the coefficient of sliding friction on the horizontal surface? (b) How much work was done against friction as the body slid down the circular arc from A to B?

7-28 A small sphere of mass m is fastened to a weightless string of length 2 ft to form a pendulum. The pendulum is swinging so as to make a maximum angle of 60° with the vertical. (a) What is the velocity of the sphere when it passes through the vertical position? (b) What is the instantaneous acceleration when the pendulum is at its maximum deflection?

7-29 (a) A ball is tied to a cord and set in rotation in a vertical circle. Prove that the tension in the cord at the lowest point exceeds that at the highest point by six times the weight of the ball. (b) A particle (Fig. 7-20) at the end of a string is originally held in position 1 and then allowed to swing freely until it impinges against the stop S. Prove that position 2 is at the same level as 1 regardless of the position of the stop.

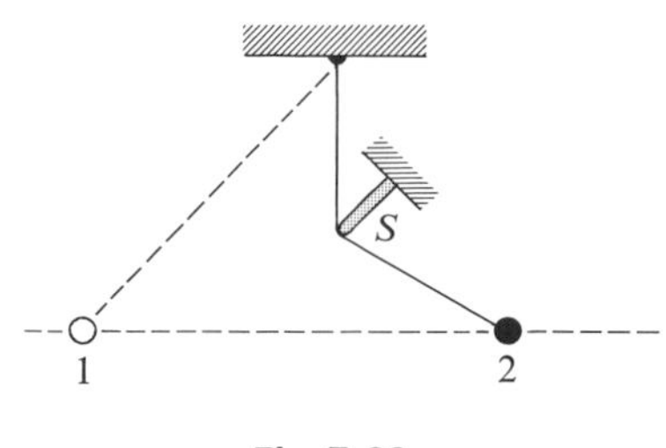

Fig. 7-20

7-30 A small body of mass m slides without friction around the loop-the-loop apparatus shown in Fig. 7-21. It starts from rest at point A at a height $3R$ above the bottom of the loop. When it reaches point B at the end of a horizontal diameter of the loop, compute (a) its radial acceleration, and (b) its tangential acceleration.

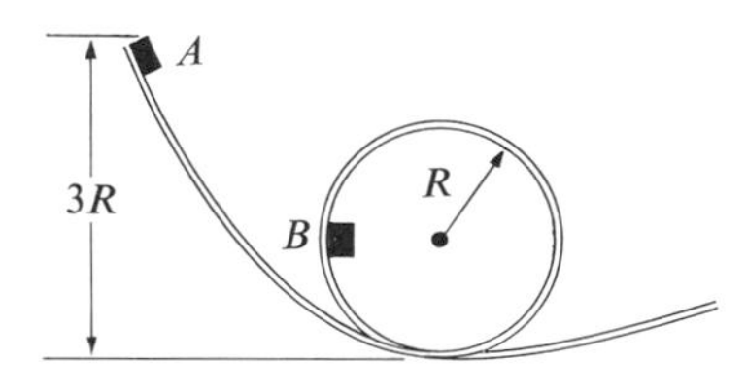

Fig. 7-21

7-31 A meter stick, pivoted about a horizontal axis through its center, has a body of mass 2 kg attached to one end and a body of mass 1 kg attached to the other. The mass of the meter stick can be neglected. The system is released from rest with the stick horizontal. What is the velocity of each body as the stick swings through a vertical position?

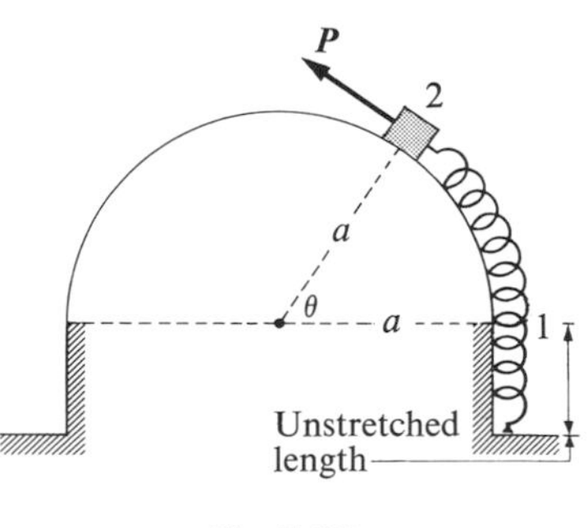

Fig. 7-22

7-32 A variable force $\boldsymbol{P}$ is maintained tangent to a frictionless cylindrical surface of radius a, as shown in Fig. 7-22. By slowly varying this force, a block of weight w is moved and the spring to which it is attached is stretched from position 1 to position 2. The spring, of force constant k, is unstretched in position 1. Calculate the work of the force $\boldsymbol{P}$, (a) by integration, (b) by use of the energy principle.

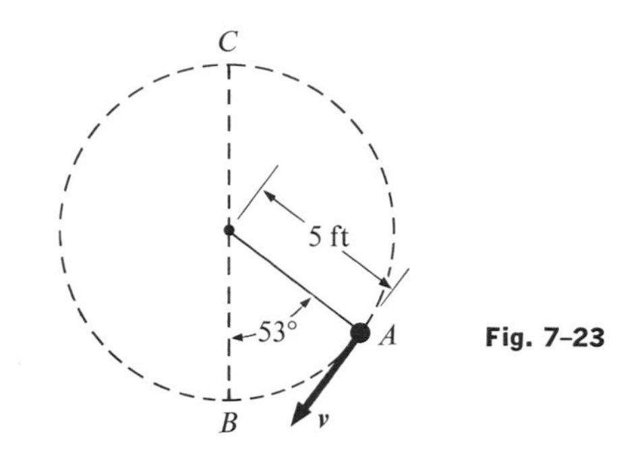

Fig. 7-23

7-33 A small 4-lb body is fastened to a weightless string 5 ft long to form a pendulum, as shown in Fig. 7-23. The body is pulled aside until the string makes an angle of 53° with the vertical. (a) With what tangential speed v must the body be started from point A so that it will reach C, the highest point, with a tangential speed of 10 ft s^{-1}? (b) With what speed does it pass through the lowest point B when started from A with the speed v? (c) What is the tension in the cord as the body passes through B? (d) If the body is started from A with the tangential speed v of part (a) in the direction opposite to that shown, with what speed will it then arrive at C?

7-34 The potential energy of a diatomic molecule, as given by Lennard-Jones, is the following function of the distance r between the atoms:

$$E_p(r) = \epsilon_0 \left[\left(\frac{r_0}{r} \right)^{12} - 2 \left(\frac{r_0}{r} \right)^{6} \right].$$

Prove that (a) r_0 is the intermolecular separation when the potential energy is a minimum, (b) the minimum potential energy is $-\epsilon_0$, and (c) the intermolecular separation when $E_p(r) = 0$ is equal to $r_0/\sqrt[6]{2}$. (d) Sketch the graph of $E_p(r)$.

7-35 A particle originally at rest at the origin is constrained to move along the x-axis. Its potential energy is a function of x, $E_p(x)$, and its total energy is a constant E. Prove that the time t to go to a point where the coordinate is x is

$$t = \int_0^x \frac{dx}{\sqrt{(2/m)[E - E_p(x)]}}.$$

7-36 An earth satellite of mass m revolves in a circle at a height above the earth equal to twice the earth's radius, $2R$. In terms of m, R, the gravitational constant G, and the mass of the earth m_E, what are the values of (a) the kinetic energy of the satellite, (b) its gravitational potential energy, and (c) its total mechanical energy?

7–37 A bicycle rider sits still on his seat and pedals so that acceleration takes place. The center of gravity of the system composed of bicycle and rider remains at a constant level. (a) What is the accelerating force? (b) What is the work of the accelerating force? (c) What is responsible for the increase in kinetic energy?

7–38 The hammer of a pile driver weighs 1000 lb and must be lifted a vertical distance of 6 feet in 3 s. What horsepower engine is required?

7–39 A ski tow is to be operated on a 37° slope 800 ft long. The rope is to move at 8 mi hr^{-1} and power must be provided for 80 riders at one time, each weighing, on an average, 150 lb. Estimate the horsepower required to operate the tow.

7–40 (a) If energy costs 5 cents per kwh, how much is one horsepower hour worth? (b) How many foot pounds can be purchased for 1 cent?

7–41 Compute the monetary value of the kinetic energy of the projectile of a 14-in. naval gun, at the rate of 2 cents kwh^{-1}. The projectile weighs 1400 lb and its muzzle velocity is 2800 ft s^{-1}.

7–42 At 5 cents kwh^{-1}, what does it cost to operate a 10-hp motor for 8 hr?

7–43 The engine of an automobile develops 20 hp when the automobile is traveling at 30 mi hr^{-1}. (a) What is the resisting force in pounds? (b) If the resisting force is proportional to the velocity, what horsepower will drive the car at 15 mi hr^{-1}? At 60 mi hr^{-1}?

7–44 The engine of a motorboat delivers 40 hp to the propeller while the boat is making 20 mi hr^{-1}. What would be the tension in the towline if the boat were being towed at the same speed?

7–45 A man whose mass is 70 kg walks up to the third floor of a building. This is a vertical height of 12 m above the street level. (a) How many joules of work has he done? (b) By how much has he increased his potential energy? (c) If he climbs the stairs in 20 s, what was his rate of working, in horsepower?

7–46 A pump is required to lift 200 gallons of water per minute from a well 20 ft deep and eject it with a speed of 30 ft s^{-1}. (a) How much work is done per minute in lifting the water? (b) How much in giving it kinetic energy? (c) What horsepower engine is needed?

7–47 A 4800-lb elevator starts from rest and is pulled upward with a constant acceleration of 10 ft s^{-2}. (a) Find the tension in the supporting cable. (b) What is the velocity of the elevator after it has risen 45 ft? (c) Find the kinetic energy of the elevator 3 s after it starts. (d) How much is its potential energy increased in the first 3 s? (e) What horsepower is required when the elevator is traveling 22 ft s^{-1}?

7–48 An automobile weighing 2000 lb has a speed of 100 ft s^{-1} on a horizontal road when the engine is developing 50 hp. What is its speed, with the same horsepower, if the road rises 1 ft in 20 ft? Assume all friction forces to be constant.

7–49 (a) If 20 hp are required to drive a 2400-lb automobile at 30 mi hr^{-1} on a level road, what is the retarding force of friction, windage, etc.? (b) What power is necessary to drive the car at 30 mi hr^{-1} up a 10% grade, i.e., one rising 10 ft vertically in 100 ft horizontally? (c) What power is necessary to drive the car at 30 mi hr^{-1} *down* a 2% grade? (d) Down what grade would the car coast at 30 mi hr^{-1}?

7–50 (a) Calculate the distance from the surface of the moon to the point P where the gravitational fields of the moon and of the earth cancel.

b) If the potential energy of a 1-kg object is zero at an infinite distance from the moon and the earth, what is its potential energy at point P?

c) Assume that the Apollo 8 space capsule acquired a speed of 12,000 m s^{-1} while still close to the earth and then coasted on an orbit designed to bring it close to the moon. What would be its speed at point P?

d) What would be its speed almost at the surface of the moon?

e) At what speed would a space vehicle execute a circular orbit a short distance from the surface of the moon?

f) What would be the period of the motion in part (e)?

[*Note:* The distance between centers of the moon and of the earth is 3.84×10^8 m; the radius of the moon is 1.74×10^6 m; the radius of the earth is 6.37×10^6 m; the mass of the moon is 7.34×10^{22} kg; the mass of the earth is 5.98×10^{24} kg; the gravitational constant is 6.67×10^{-11} N m^2 kg^{-2}.]

7–51 An atomic bomb containing 20 kg of plutonium explodes. The rest mass of the products of the explosion is less than the original rest mass by one ten-thousandth of the original rest mass. (a) How much energy is released in the explosion? (b) If the explosion takes place in 1 μs, what is the average power developed by the bomb? (c) How much water could the released energy lift to a height of 1 mile?

7–52 Construct a right triangle in which one of the angles is α, where $\sin \alpha = v/c$, and v is the speed of a particle and c is the speed of light. If the base of the triangle equals the rest energy, $E_0 = m_0c^2$, where m_0 is the rest mass, show that (a) the hypotenuse is $E = mc^2$, (b) the vertical side of the triangle equals the product of c and the momentum mv. (c) Describe a simple graphical procedure for finding the kinetic energy E_k.

8 Impulse and Momentum

8-1 IMPULSE AND MOMENTUM

In the preceding chapter it was shown how the concepts of work and energy are developed from Newton's laws of motion. We shall next see how two similar concepts, those of *impulse* and *momentum*, also arise from these laws.

Let us again consider a particle of mass m moving in the xy-plane, as in Fig. 8-1, and acted on by a resultant force $\boldsymbol{F}$ that may vary in magnitude and direction. If we limit the velocity of the particle to the nonrelativistic range where the mass has the constant value m, Newton's second law states that at every instant

$$\boldsymbol{F} = m\frac{d\boldsymbol{v}}{dt},$$

or

$$\boldsymbol{F}\,dt = m\,d\boldsymbol{v}.$$

If $\boldsymbol{v}_1$ is the velocity when $t = t_1$, and $\boldsymbol{v}_2$ the velocity when $t = t_2$, it follows that

$$\int_{t_1}^{t_2} \boldsymbol{F}\,dt = \int_{\boldsymbol{v}_1}^{\boldsymbol{v}_2} m\,d\boldsymbol{v}. \tag{8-1}$$

The integral on the left is called the *impulse of the force* $\boldsymbol{F}$ in the time interval $t_2 - t_1$, and is a *vector quantity:*

$$\text{Impulse} = \int_{t_1}^{t_2} \boldsymbol{F}\,dt.$$

This integral can be evaluated, of course, only when the force is known as a function of the time. The integral on the right, however, always yields the result

$$\int_{\boldsymbol{v}_1}^{\boldsymbol{v}_2} m\,d\boldsymbol{v} = m\boldsymbol{v}_2 - m\boldsymbol{v}_1.$$

The product of the mass of a particle and its velocity is called the *linear momentum* of the particle, and is also a *vector* quantity:

$$\text{Linear momentum} = m\boldsymbol{v}.$$

(We refer to this product as *linear* momentum to distinguish it from a similar quantity called *angular* momentum, which will be discussed later. When there is no opportunity for confusion, we shall speak of the linear momentum simply as the momentum.)

Equation (8-1) is now written as

$$\int_{t_1}^{t_2} \boldsymbol{F}\,dt = m\boldsymbol{v}_2 - m\boldsymbol{v}_1, \tag{8-2}$$

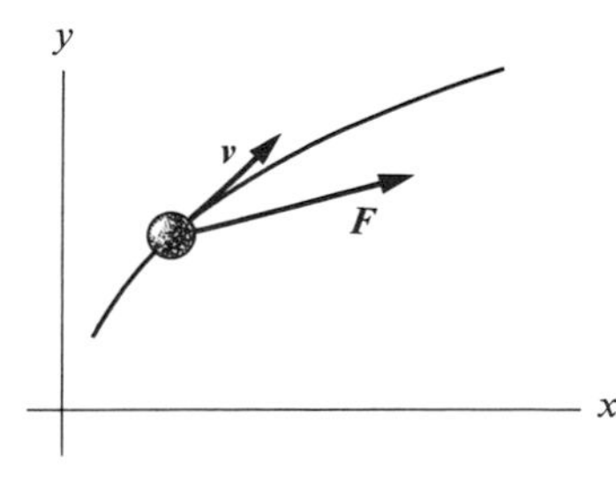

8-1 Particle moving in the xy-plane.

expressing the important fact that: *The* **vector impulse** *of the resultant force on a particle, in any time interval, is equal in magnitude and direction to the* **vector change** *in momentum of the particle.* This is known as the **impulse-momentum** principle.

The impulse-momentum principle finds its chief application in connection with forces of short duration, such as those arising in collisions or explosions. Such forces are called *impulsive forces.*

The unit of impulse, in any system, equals the product of the units of force and time in that system. Thus in the mks system the unit is one newton second (1 N s), in the cgs system it is one dyne second (1 dyn s), and in the engineering system it is one pound second (1 lb s).

The unit of momentum in the mks system is one kilogram meter per second (1 kg m s^{-1}), in the cgs system it is one gram centimeter per second (1 g cm s^{-1}), and in the engineering system it is one slug foot per second (1 slug ft s^{-1}). Since

$$1 \text{ kg m s}^{-1} = (1 \text{ kg m s}^{-2})\text{s} = 1 \text{ N s},$$

it follows that the unit of momentum in any system equals the unit of impulse in that system.

In contrast to work and energy, which are scalars, impulse and momentum are vector quantities, and therefore Eq. (8–2), like every vector equation, is equivalent (for forces and velocities in the xy-plane) to the two scalar equations

$$\int_{t_1}^{t_2} F_x\, dt = mv_{x2} - mv_{x1},$$
$$\int_{t_1}^{t_2} F_y\, dt = mv_{y2} - mv_{y1}. \qquad (8\text{–}3)$$

For the special case of a force that is constant in magnitude and direction, we can take $\boldsymbol{F}$ outside the integral sign in Eq. (8–2), and if we let $t_1 = 0$ and $t_2 = t$, we have

$$\boldsymbol{F}t = m\boldsymbol{v}_2 - m\boldsymbol{v}_1. \qquad (8\text{–}4)$$

That is, the impulse of a *constant* force equals the product of the force and the time interval during which it acts. The *vector change* in momentum produced by such a force, $(m\boldsymbol{v}_2 - m\boldsymbol{v}_1)$, is in the same direction as the force.

If the force and the velocities $\boldsymbol{v}_1$ and $\boldsymbol{v}_2$ are in the same direction, Eq. (8–4) reduces to the scalar equation

$$Ft = mv_2 - mv_1.$$

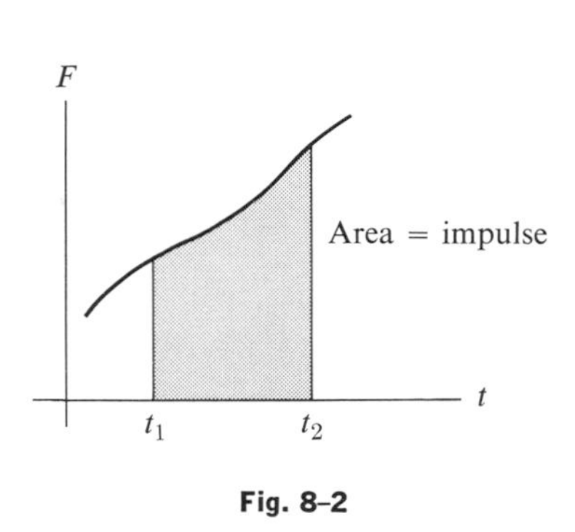

Fig. 8–2

The impulse of any force component, or any force whose direction is constant, can be represented graphically by plotting the force vertically and the time horizontally, as in Fig. 8–2. The *area* under the curve, between vertical lines at t_1 and t_2, is equal to the impulse of the force in this time interval.

If the impulse of a force is *positive*, the momentum of the body on which it acts *increases* algebraically. If the impulse is *negative*, the momentum *decreases*. If the impulse is *zero*, there is no change in momentum.

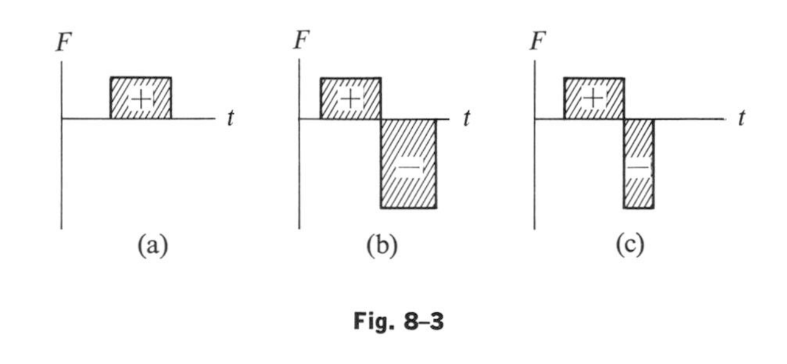

Fig. 8–3

Example 1. Consider the changes in momentum produced by the following forces: (a) A body moving on the x-axis is acted on for 2 s by a constant force of 10 N toward the right. (b) The body is acted on for 2 s by a constant force of 10 N toward the right and then for 2 s by a constant force of 20 N toward the left. (c) The body is acted on for 2 s by a constant force of 10 N toward the right, then for 1 s by a constant force of 20 N toward the left. The three forces are shown graphically in **Fig. 8–3.**

(a) The impulse of the force is +10 N × 2 s = +20 N s. Hence the momentum of *any* body on which the force acts increases by 20 kg m s^{-1}. This change is the same whatever the mass of the body and whatever the magnitude and direction of its initial velocity.

Suppose the mass of the body is 2 kg and that it is initially at rest. Its *final* momentum then equals its *change* in momentum and its final *velocity* is 10 m s^{-1} toward the right. (The reader should verify by computing the acceleration.)

Had the body been initially moving toward the right at 5 m s^{-1}, its initial momentum would have been 10 kg m s^{-1}, its final momentum 30 kg m s^{-1}, and its final velocity 15 m s^{-1} toward the right.

Had the body been moving initially toward the *left* at 5 m s^{-1}, its initial momentum would have been −10 kg m s^{-1}, its final momentum +10 kg m s^{-1}, and its final velocity 5 m s^{-1} toward the right. That is, the constant force of 10 N toward the right would first have brought the body to rest and then given it a velocity in the direction opposite to its initial velocity.

(b) The impulse of this force is (+10 N × 2 s − 20 N × 2 s) = −20 N s. The momentum of *any* body on which it acts is decreased by 20 kg m s^{-1}. The reader should examine various possibilities, as in the preceding example.

(c) The impulse of this force is (+10 N × 2 s − 20 N × 1 s) = 0. Hence the momentum of any body on which it acts is not changed. Of course, the momentum of the body is *increased* during the first 2 s but it is *decreased* by an equal amount in the next second. As an exercise, describe the motion of a body of mass 2 kg, moving initially to the left at 5 m s^{-1}, and acted on by this force. It will help to construct a graph of velocity versus time.

Example 2. A ball of mass 0.4 kg is thrown against a brick wall. When it strikes the wall it is moving horizontally to the left at 30 m s^{-1}, and it rebounds horizontally to the right at 20 m s^{-1}. Find the impulse of the force exerted on the ball by the wall.

The initial momentum of the ball is 0.4 kg × −30 m s^{-1} = −12 kg m s^{-1}. The final momentum is +8 kg m s^{-1}. The *change* in momentum is

$$
\begin{aligned}
mv_2 - mv_1 &= 8 \text{ kg m s}^{-1} - (-12 \text{ kg m s}^{-1}) \\
&= 20 \text{ kg m s}^{-1}.
\end{aligned}
$$

Hence, the impulse of the force exerted on the ball was 20 N s. Since the impulse is *positive*, the force must be toward the right.

Note that the *force* exerted on the ball cannot be found without further information regarding the collision. The general nature of the force-time graph is shown by one of the curves in Fig. 8–4. The force is zero before impact, rises to a maximum, and decreases to zero when the ball leaves the wall. If the ball is

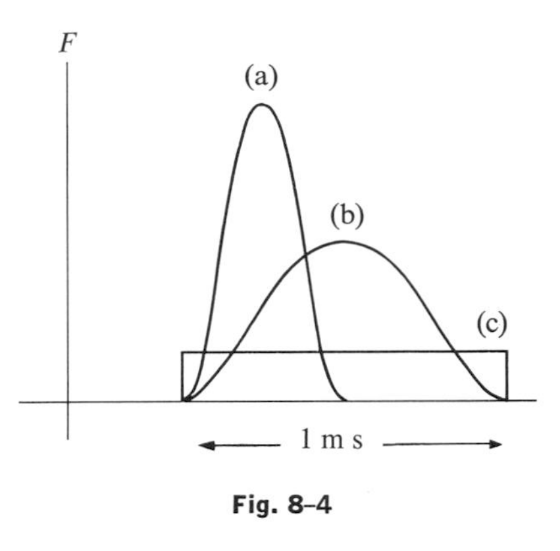

Fig. 8–4

relatively rigid, like a baseball, the time of collision is small and the maximum force is large, as in curve (a). If the ball is more yielding, like a tennis ball, the collision time is larger and the maximum force is less, as in curve (b). In any event, the *area* under the force-time graph must equal 20 N s.

For an idealized case in which the force is constant and the collision time is 1 ms (10^{-3} s), as represented by the horizontal straight line, the force is 20,000 N.

Explain why one "eases off" when catching a fast ball.

8–2 CONSERVATION OF LINEAR MOMENTUM

Whenever there is a force of interaction between two particles, the momentum of each particle is changed as a result of the force exerted on it by the other. (The force may be gravitational, electric, magnetic, or of any other origin.) Furthermore, since by Newton's *third* law the force on one particle is always equal in magnitude and opposite in direction to that on the other, the impulses of the forces are equal in magnitude and opposite in direction. It follows that *the vector change in momentum of either particle,* in any time interval, *is equal in magnitude and opposite in direction to the vector change in momentum of the other.* The *net change in momentum of the* **system** (the two particles together) *is therefore zero.*

The pair of action-reaction forces are *internal* forces of the system and we conclude that *the total momentum of a system of bodies cannot be changed by internal forces between the bodies.* Hence if the *only* forces acting on the particles of a system are *internal* forces (that is, if there are no *external* forces), the total momentum of the system remains constant in magnitude and direction. This is the *principle of* **conservation of linear momentum:** *When* **no resultant external force** *acts on a system, the* **total momentum of the system remains constant in magnitude and direction.**

The principle of conservation of momentum is one of the most fundamental and important principles of mechanics. Note that it is more general than the principle of conservation of mechanical energy; mechanical energy is conserved *only* when the internal forces are *conservative*. The principle of conservation of momentum holds whatever the nature of the internal forces.

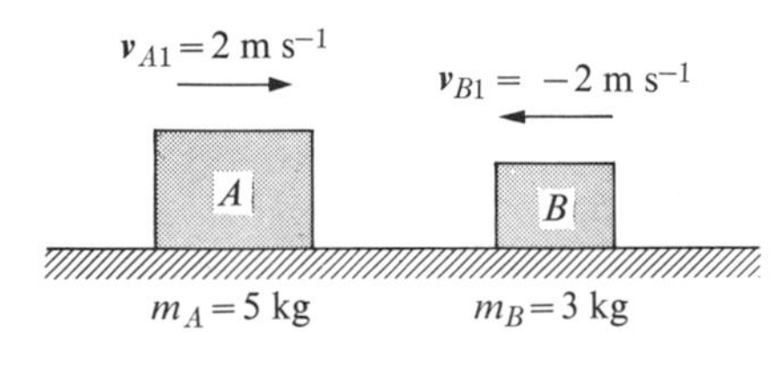

Fig. 8-5

Example 1. Figure 8-5 shows a body A of mass m_A moving toward the right on a level frictionless surface with a velocity $\mathbf{v}_{A1}$. It collides with a second body B of mass m_B moving toward the left with a velocity $\mathbf{v}_{B1}$. Since there is no friction and the resultant vertical force on the system is zero, the only forces on the bodies are the internal action-reaction forces which they exert on each other in the collision process, and the momentum of the system remains constant in magnitude and direction.

Let $\mathbf{v}_{A2}$ and $\mathbf{v}_{B2}$ represent the velocities of A and B after the collision. Then

$$m_A\mathbf{v}_{A1} + m_B\mathbf{v}_{B1} = m_A\mathbf{v}_{A2} + m_B\mathbf{v}_{B2}. \tag{8-5}$$

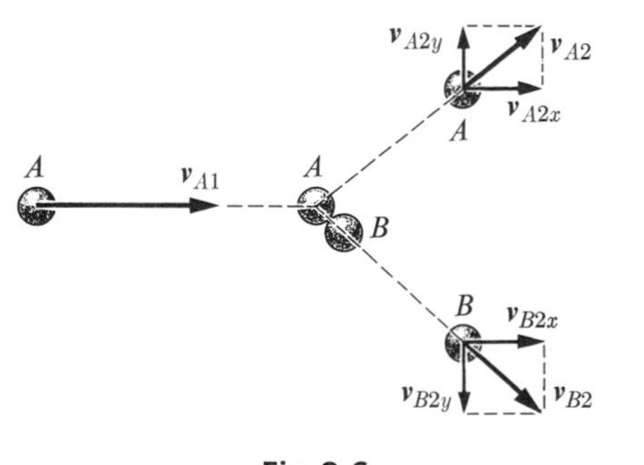

Fig. 8-6

Example 2. In Fig. 8-6, body A of mass m_A is initially moving toward the right with a velocity $\mathbf{v}_{A1}$. It collides with body B, initially at rest, after which the bodies separate and move with velocities $\mathbf{v}_{A2}$ and $\mathbf{v}_{B2}$. No forces act on the system except those in the collision process.

This example illustrates the *vector* nature of momentum, that is, the x- and y-components of momentum are *both* conserved. Let us take the x-axis in the direction of $\mathbf{v}_{A1}$. The initial x-momentum is m_Av_{A1} and the initial y-momentum is zero.

The final x-momentum of the system is

$$m_Av_{A2x} + m_Bv_{B2x}$$

and the final y-momentum is

$$m_Av_{A2y} - m_Bv_{B2y}.$$

Therefore

$$m_Av_{A2x} + m_Bv_{B2x} = m_Av_{A1}, \tag{8-6}$$

$$m_Av_{A2y} - m_Bv_{B2y} = 0. \tag{8-7}$$

8-3 ELASTIC AND INELASTIC COLLISIONS

Suppose we are given the masses and initial velocities of two colliding bodies, and we wish to compute their velocities after the collision. If the collision is "head-on," as in Fig. 8-5, Eq. (8-5) provides one equation for the velocities $\mathbf{v}_{A2}$ and $\mathbf{v}_{B2}$. If it is of the type shown in Fig. 8-6, Eqs. (8-6) and (8-7) provide two equations for the four velocity components v_{A2x}, v_{A2y}, v_{B2x} and v_{B2y}. Hence momentum considerations *alone* do not suffice to completely determine the final velocities; we must have more information about the collision process.

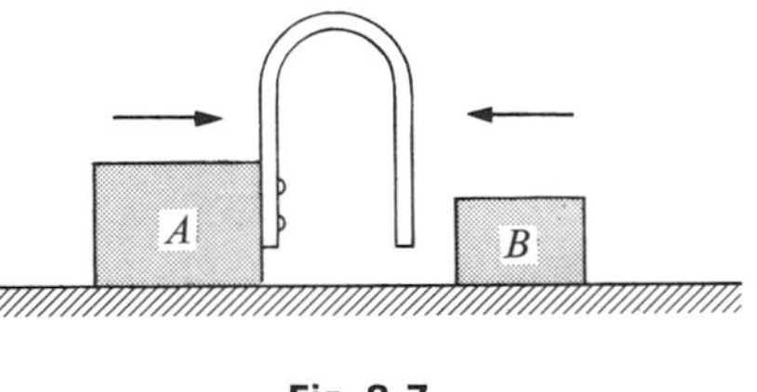

Fig. 8-7

If the forces of interaction between the bodies are *conservative*, the total kinetic energy is the same before and after the collision and the collision is said to be *completely elastic*. Such a collision is closely approximated if one end of an inverted U-shaped steel spring is attached to one of the bodies, as in Fig. 8-7. When the bodies collide the spring is momentarily compressed and some of the original kinetic energy is momentarily converted to elastic potential energy. The spring then expands and when the bodies separate this potential energy is reconverted to kinetic energy.

At the opposite extreme from a completely elastic collision is one in which the colliding bodies stick together and move as a unit after the collision. Such a collision is called a *completely inelastic* collision, and will result if the bodies in Fig. 8-5 are provided with a coupling mechanism like that between two freight cars, or if the bodies in Fig. 8-7 are locked together at the instant when their velocities become equal and the spring is fully compressed.

8-4 INELASTIC COLLISIONS

For the special case of a completely inelastic collision between two bodies A and B, we have, from the definition of such a collision,

$$\mathbf{v}_{A2} = \mathbf{v}_{B2} = \mathbf{v}_2.$$

When this is combined with the principle of conservation of momentum, we obtain

$$m_A \mathbf{v}_{A1} + m_B \mathbf{v}_{B1} = (m_A + m_B)\mathbf{v}_2, \tag{8-8}$$

and the final velocity can be computed if the initial velocities and the masses are known.

The kinetic energy of the system before collision is $E_{k1} = \frac{1}{2}m_A v_{A1}^2 + \frac{1}{2}m_B v_{B1}^2$. The final kinetic energy is $E_{k2} = \frac{1}{2}(m_A + m_B)v_2^2$.

For the special case in which body B is initially at rest, $v_{B1} = 0$ and the ratio of the final to the initial kinetic energy is

$$\frac{E_{k2}}{E_{k1}} = \frac{(m_A + m_B)v_2^2}{m_A v_{A1}^2}.$$

Inserting the expression for v_2 from Eq. (8-8), we find that this reduces to

$$\frac{E_{k2}}{E_{k1}} = \frac{m_A}{m_A + m_B}.$$

The right side is necessarily less than unity, so *the total kinetic energy decreases in an inelastic collision.*

Example 1. Suppose the collision in Fig. 8-5 is completely inelastic and that the masses and velocities have the values shown. The velocity after the collision is then

$$v_2 = \frac{m_A v_{A1} + m_B v_{B1}}{m_A + m_B} = 0.5 \text{ m s}^{-1}.$$

Since v_2 is positive, the system moves to the right after the collision.

The kinetic energy of body A before the collision is

$$\tfrac{1}{2}m_A v_{A1}^2 = 10 \text{ J},$$

and that of body B is

$$\tfrac{1}{2}m_B v_{B1}^2 = 6 \text{ J}.$$

The total kinetic energy before collision is therefore 16 J.

Note that the kinetic energy of body B is positive, although its velocity v_{B1} and its momentum mv_{B1} are both negative. The kinetic energy after the collision is $\frac{1}{2}(m_A + m_B)v_2^2 = 1$ J.

Hence, far from remaining constant, the final kinetic energy is only 1/16 of the original, and 15/16 is "lost" in the collision. If the bodies couple together like two freight cars, most of this energy is converted to elastic waves which are eventually absorbed.

If there is a spring between the bodies, as in Fig. 8-7, and the bodies are locked together when their velocities become equal, the energy is trapped as potential energy in the compressed spring. If all these forms of energy are taken into account, the *total* energy of the system is conserved although its *kinetic* energy is not. However, *momentum is always conserved* in a collision, whether or not the collision is elastic.

Example 2. The *ballistic pendulum* is a device for measuring the velocity of a bullet. The bullet is allowed to make a completely inelastic collision with a body of much greater mass. The momentum of the system immediately after the collision equals the original momentum of the bullet, but since the velocity is very much smaller, it can be determined more easily. Although the ballistic pendulum has now been superseded by other devices, it is still an important laboratory experiment for illustrating the concepts of momentum and energy.

In Fig. 8-8, the pendulum, consisting perhaps of a large wooden block of mass m', hangs vertically by two cords. A bullet of mass m, traveling with a velocity v, strikes the pendulum and remains embedded in it. If the collision time is very small compared with the time of swing of the pendulum, the supporting cords remain practically vertical during this time. Hence no external horizontal forces act on the system during the collision, and the horizontal momentum is conserved. Then if V represents the velocity of bullet and block immediately after the collision,

$$mv = (m + m')V, \qquad v = \frac{m + m'}{m} V.$$

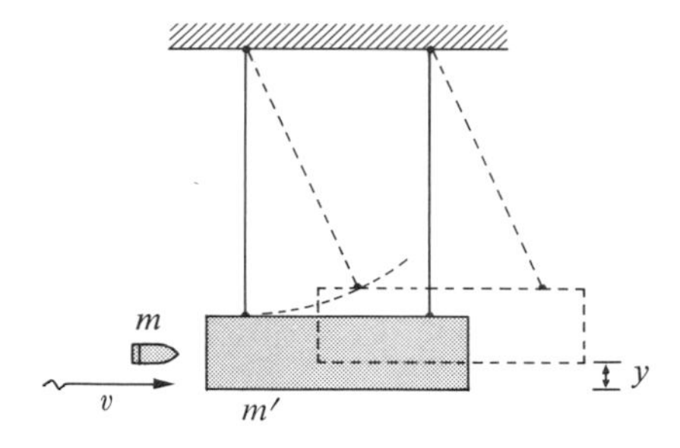

8-8 The ballistic pendulum.

The kinetic energy of the system, immediately after the collision, is $E_k = \frac{1}{2}(m + m')V^2$.

The pendulum now swings to the right and upward until its kinetic energy is converted to gravitational potential energy. (Small frictional effects can be neglected.) Hence

$$\tfrac{1}{2}(m + m')V^2 = (m + m')gy,$$
$$V = \sqrt{2gy},$$

and

$$v = \frac{m + m'}{m}\sqrt{2gy}.$$

By measuring m, m', and y, the original velocity v of the bullet can be computed.

It is important to remember that kinetic energy is not conserved *in the collision*. The ratio of the kinetic energy of bullet and pendulum, after the collision, to the original kinetic energy of the bullet, is

$$\frac{\frac{1}{2}(m + m')V^2}{\frac{1}{2}mv^2} = \frac{m}{m + m'}.$$

Thus if $m' = 1000$ g and $m = 1$ g, only about 0.1% of the original energy remains as kinetic energy; 99.9% is converted to internal energy.

8-5 ELASTIC COLLISIONS

Consider next a perfectly elastic "head-on" or *central* collision between two bodies A and B. The bodies separate after the collision and have different velocities, $\mathbf{v}_{A2}$ and $\mathbf{v}_{B2}$. Since kinetic energy and momentum are *both* conserved, we have:

Conservation of kinetic energy:

$$\tfrac{1}{2}m_A v_{A1}^2 + \tfrac{1}{2}m_B v_{B1}^2 = \tfrac{1}{2}m_A v_{A2}^2 + \tfrac{1}{2}m_B v_{B2}^2.$$

Conservation of momentum:

$$m_A v_{A1} + m_B v_{B1} = m_A v_{A2} + m_B v_{B2}.$$

Hence if the masses and initial velocities are known, we have two independent equations from which the final velocities can be found. Simultaneous solution of these equations gives

$$(v_{B2} - v_{A2}) = -(v_{B1} - v_{A1}), \qquad (8\text{–}9)$$

$$v_{A2} = \frac{2m_B v_{B1} + v_{A1}(m_A - m_B)}{m_A + m_B}, \qquad (8\text{–}10)$$

$$v_{B2} = \frac{2m_A v_{A1} - v_{B1}(m_A - m_B)}{m_A + m_B}. \qquad (8\text{–}11)$$

The difference $(v_{B2} - v_{A2})$ is the velocity of B relative to A after the collision, while $(v_{B1} - v_{A1})$ is its relative velocity before the collision. Thus, from Eq. (8–9), *the relative velocity of two particles in a central, completely elastic collision is unchanged in magnitude but reversed in direction.*

For the special case in which body B is at rest before the collision, $v_{B1} = 0$ and Eqs. (8–10) and (8–11) simplify to

$$v_{A2} = \frac{m_A - m_B}{m_A + m_B}v_{A1}, \qquad v_{B2} = \frac{2m_A}{m_A + m_B}v_{A1}.$$

If the masses of A and B are equal, $v_{A2} = 0$ and $v_{B2} = v_{A1}$. That is, the first body comes to rest and the second moves off with a velocity equal to the original velocity of the first. Both the momentum and kinetic energy of the first are completely transferred to the second.

When the masses are unequal, the kinetic energies after the collision are

$$(E_{k2})_A = \tfrac{1}{2}m_A v_{A2}^2 = \left(\frac{m_A - m_B}{m_A + m_B}\right)^2 (E_{k1})_A,$$

$$(E_{k2})_B = \tfrac{1}{2}m_B v_{B2}^2 = \frac{4m_A m_B}{(m_A + m_B)^2}(E_{k1})_A.$$

It is of interest to derive the expression for the fractional decrease in kinetic energy of body A, that is, the ratio of its decrease in kinetic energy to its original kinetic energy. Since, in an elastic collision, the energy lost by A equals the energy gained by B, this ratio is

$$\frac{(E_{k2})_B}{(E_{k1})_A} = \frac{4m_A m_B}{(m_A + m_B)^2} = 4\frac{m_A}{m_B}\frac{1}{[1 + (m_A/m_B)]^2}.$$

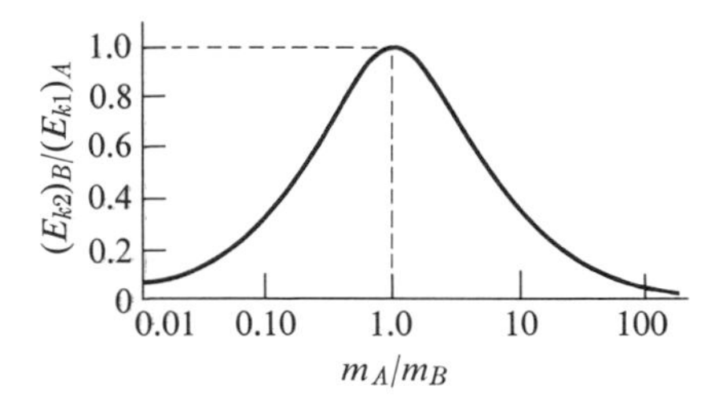

8-9 Fractional loss in energy in a head-on elastic collision, plotted as a function of the ratio of the masses of the colliding bodies.

The fractional decrease in kinetic energy of A is plotted in Fig. 8-9 as a function of the ratio m_A/m_B (note the logarithmic scale). The energy loss approaches zero as m_A/m_B approaches zero (a very small body colliding with a very large one) and also as m_A/m_B approaches infinity (a very large body colliding with a very small one). In the former case, the first body rebounds with practically its original velocity. In the latter, the first body continues to move with very nearly its original velocity. The maximum fractional energy decrease occurs when $m_A = m_B$, and for this ratio of masses the fractional energy loss equals unity, as shown above.

This fact has an important bearing on the problem of slowing down the rapidly moving neutrons resulting from certain nuclear reactions. As neutrons pass through matter, they make occasional elastic collisions with nuclei. In a head-on collision, the greatest loss of kinetic energy results when the mass of the nucleus equals that of the neutron, that is, when the nucleus is that of ordinary hydrogen. The greater the nuclear mass, the less of the neutron's energy is transferred to the nucleus.

Example. Suppose the collision illustrated in Fig. 8-5 is completely elastic. What are the velocities of A and B after the collision?

From the principle of conservation of momentum,

$$5\text{ kg} \times 2\text{ m s}^{-1} + 3\text{ kg} \times (-2\text{ m s}^{-1}) = (5\text{ kg} \times v_{A2}) + (3\text{ kg} \times v_{B2}),$$

$$5v_{A2} + 3v_{B2} = 4\text{ m s}^{-1}.$$

Since the collision is completely elastic,

$$v_{B2} - v_{A2} = -(v_{B1} - v_{A1}) = 4\text{ m s}^{-1}.$$

When these equations are solved simultaneously, we get

$$v_{A2} = -1\text{ m s}^{-1}, \qquad v_{B2} = 3\text{ m s}^{-1}.$$

Both bodies therefore reverse their directions of motion, A traveling to the left at 1 m s^{-1} and B to the right at 3 m s^{-1}.

The total kinetic energy after the collision is

$$\tfrac{1}{2} \cdot 5\text{ kg} \cdot (-1\text{ m s}^{-1})^2 + \tfrac{1}{2} \cdot 3\text{ kg} \cdot (3\text{ m s}^{-1})^2 = 16\text{ J},$$

which equals the total kinetic energy before the collision.

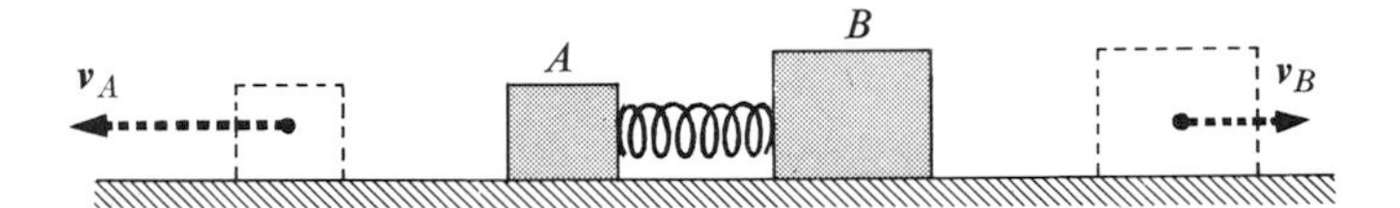

8-10 Conservation of momentum in recoil.

8-6 RECOIL

Figure 8-10 shows two blocks A and B between which there is a compressed spring. When the system is released from rest, the spring exerts equal and opposite forces on the blocks until it has expanded to its natural unstressed length. It then drops to the surface, while the blocks continue to move. The original momentum of the system is zero, and if frictional forces can be neglected, the resultant external force on the system is zero. The momentum of the system therefore remains constant and equal to zero. Then if $\mathbf{v}_A$ and $\mathbf{v}_B$ are the velocities acquired by A and B, we have

$$m_A v_A + m_B v_B = 0, \qquad \frac{v_A}{v_B} = -\frac{m_B}{m_A}.$$

The velocities are of opposite sign and their magnitudes are inversely proportional to the corresponding masses.

The original kinetic energy of the system is also zero. The final kinetic energy is

$$E_k = \tfrac{1}{2}m_A v_A^2 + \tfrac{1}{2}m_B v_B^2.$$

The source of this energy is the original elastic potential energy of the system. The ratio of kinetic energies is

$$\frac{\frac{1}{2}m_A v_A^2}{\frac{1}{2}m_B v_B^2} = \frac{m_A}{m_B}\left(\frac{v_A}{v_B}\right)^2 = \frac{m_B}{m_A}.$$

Thus, although the momenta are equal in magnitude, the kinetic energies are inversely proportional to the corresponding masses, the body of smaller mass receiving the larger share of the original potential energy. The reason is that the change in *momentum* of a body equals the *impulse* of the force acting on it ($\int \boldsymbol{F}\, dt$), while the change in *kinetic energy* equals the *work* of the force ($\int \boldsymbol{F}\, ds$). The forces on the two bodies are equal in magnitude and act for equal times, so they produce equal and opposite changes in momentum. The points of application of the forces, however, do not move through equal distances (except when $m_A = m_B$), since the acceleration, velocity, and displacement of the smaller body are greater than those of the larger. Hence more work is done on the body of smaller mass.

Considerations like those above also apply to the firing of a rifle. The initial momentum of the system is zero. When the rifle is fired, the bullet and the powder gases acquire a forward momentum, and the rifle (together with any system to which it is attached) acquires a rearward momentum of the same magnitude. The ratio of velocities cannot be expressed as simply as in the case discussed above. The bullet travels with a definite velocity, but different portions of the powder gases have different velocities. Because of the relatively large mass of the rifle compared with that of the bullet and powder charge, the velocity and kinetic energy of the rifle are much smaller than those of the bullet and powder gases.

In the processes of radioactive decay and nuclear fission, a nucleus splits into two or more parts which fly off in different directions. Although we do not yet understand the nature of the forces which act in these processes, it has been verified by many experiments that the principle of conservation of momentum applies to them also. The total (vector) momentum of the products equals the original momentum of the system. The source of the kinetic energy of the products is the so-called *binding energy* of the original nucleus, analogous to the chemical energy of the powder charge in a rifle.

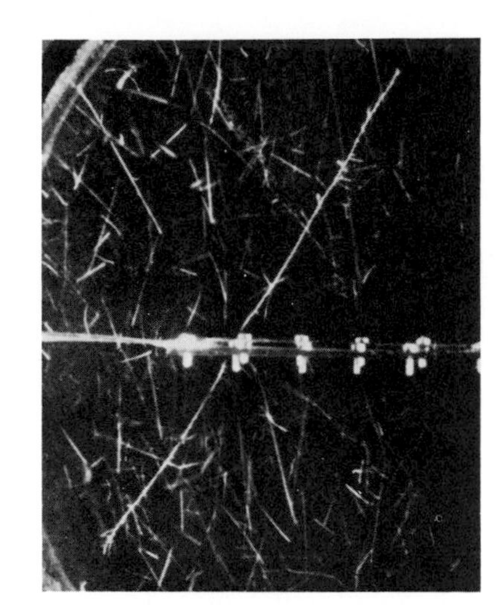

8-11 Cloud-chamber photograph of the fission of an atom of uranium. [From Bøggild, Brostrøm, and Lauritsen, *Phys. Rev.* **59**, 275, 1941.]

Figure 8-11 shows the explosive break-up (fission) of one uranium atom into two fragments of approximately equal mass. The atom was at rest in the thin foil that stretches horizontally in the picture, then it was "triggered off" by a passing neutron, and the two fission products recoiled from each other in opposite directions. While they cannot, of course, be directly observed, the two fragments give themselves away indirectly by the thin streaks of fog or clouds which can be made to condense along their paths.* It is evident that the fragments are slowed down to a stop after a few centimeters in the medium through which they have to plow, but from such data as the length and density of the tracks we can deduce the speeds of separation. From the law of conservation of momentum, applied to the fission fragments, we can then obtain directly the ratio of their masses, an important aid in the identification of the nuclear reaction.

In this simplified account we are neglecting the contribution to momentum of minor fission products not apparent in this picture, and of the incident neutron. In the "background" can be seen the result of collisions between other incident neutrons and the atmosphere in the cloud chamber.

---

* The equipment for this purpose, a most important tool in contemporary research, is called a *cloud chamber.* It originated in a design devised in 1895 by the British Nobel Prize physicist C. T. R. Wilson.

8–7 PRINCIPLES OF ROCKET PROPULSION

A rocket is propelled by the ejection of a portion of its mass to the rear. The forward force on the rocket is the reaction to the backward force on the ejected material, and as more material is ejected, the mass of the rocket decreases. The problem is best handled by impulse-momentum considerations. In order not to bring in too many complicating factors, we shall consider a rocket fired vertically upward, and neglect air resistance and variations in $\mathbf{g}$.

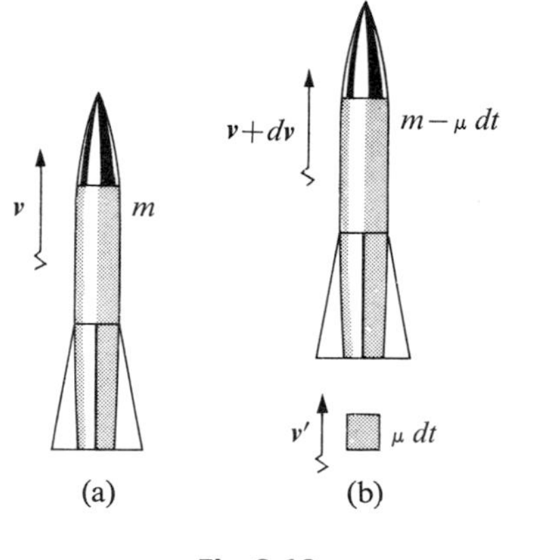

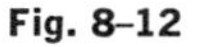
Fig. 8–12

Figure 8–12(a) represents the rocket at a time t after take-off, when its mass is m and its upward velocity is $\mathbf{v}$. In part (b), at a time $t + dt$, the velocity of the rocket has increased to $\mathbf{v} + d\mathbf{v}$. Let μ represent the mass ejected per unit time. The mass of the rocket is then $m - \mu\, dt$ and that of the ejected material, represented by the small rectangle, is $\mu\, dt$. Let $\mathbf{v}_r$ represent the velocity of the rocket relative to that of the ejected material. The velocity $\mathbf{v}'$ of the latter is then

$$\mathbf{v}' = \mathbf{v} - \mathbf{v}_r. \tag{8–12}$$

The only external force on the system is its weight $m\mathbf{g}$. The impulse of this force in time dt, taking the upward direction as positive, is $-mg\, dt$, and from the impulse-momentum theorem this equals the change in momentum of the system. The initial momentum is mv. The final momentum of the rocket is $(m - \mu t)(v + dv)$, and that of the ejected material is $v'\mu\, dt$. Then

$$-mg\, dt = [(m - \mu\, dt)(v + dv) + v'\mu\, dt] - mv.$$

Now expand the right side of this equation, eliminate v' (by Eq. 8–12) and neglect the relatively small quantity $\mu\, dt\, dv$. (The result is therefore valid only so long as m is very large compared with $\mu\, dt$.) This leads to the equation

$$m\, dv = v_r \mu\, dt - mg\, dt. \tag{8–13}$$

The change in mass of the rocket in time dt is $dm = -\mu\, dt$. Therefore

$$dv = -v_r \frac{dm}{m} - g\, dt,$$

which integrates to

$$v = -v_r \ln m - gt + C.$$

Let m_0 and v_0 be the mass and velocity at time $t = 0$. Then

$$v_0 = -v_r \ln m_0 + C,$$

and

$$v = v_0 - gt + v_r \ln \frac{m_0}{m}. \tag{8–14}$$

The first two terms on the right are the same as those in Eq. (4–18). The third term represents the excess velocity over that of a projectile fired vertically upward with initial velocity v_0. It is evident that in order to attain a high velocity v, the relative velocity v_r and the mass ratio m_0/m must be large.

Example. If a single-stage rocket, fired vertically from rest at the earth's surface, burns its fuel in a time of 30 s, and the relative velocity $v_r = 3000\ \mathrm{m\ s^{-1}}$, what must be the mass ratio m_0/m for a final velocity v of $8\ \mathrm{km\ s^{-1}}$ (about equal to the orbital velocity of an earth satellite)?

$$\ln \frac{m_0}{m} = \frac{v + gt}{v_r} = 2.76, \qquad \frac{m_0}{m} = 16.$$

Problems

8–1 (a) What is the momentum of a 10-ton truck whose velocity is 30 mi hr^{-1}? What velocity must a 5-ton truck attain in order to have (b) the same momentum, (c) the same kinetic energy?

8–2 A baseball weighs $5\frac{1}{2}$ oz. (a) If the velocity of a pitched ball is 80 ft s^{-1}, and after being batted it is 120 ft s^{-1} in the opposite direction, find the change in momentum of the ball and the impulse of the blow. (b) If the ball remains in contact with the bat for 0.002 s, find the average force of the blow.

8–3 A bullet having a mass of 0.05 kg, moving with a velocity of 400 m s^{-1}, penetrates a distance of 0.1 m in a wooden block firmly attached to the earth. Assume the decelerating force constant. Compute (a) the deceleration of the bullet, (b) the decelerating force, (c) the time of deceleration, (d) the impulse of the collision. Compare the answer to part (d) with the initial momentum of the bullet.

8–4 A bullet of mass 2 g emerges from the muzzle of a gun with a velocity of 300 m s^{-1}. The resultant force on the bullet, while it is in the gun barrel, is given by

$$F = 400 - \frac{4 \times 10^5}{3}\,t,$$

where F is in newtons and t in seconds. (a) Construct a graph of F versus t. (b) Compute the time required for the bullet to travel the length of the barrel.

8–5 A box, initially sliding on the floor of a room, is eventually brought to rest by friction. Is the momentum of the box conserved? If not, does this process contradict the principle of conservation of momentum? What becomes of the original momentum of the box?

8–6 Compare the damage to an automobile (and its occupants) in the following circumstances: (a) The automobile makes a completely inelastic head-on collision with an identical automobile traveling with the same speed in the opposite direction, and (b) it makes a completely inelastic head-on collision with a vertical rock cliff. (c) Which would be worse (for the occupants of a light car), to collide head-on with a truck traveling in the opposite direction with a momentum of equal magnitude, or to collide head-on with a truck having the same kinetic energy?

8–7 (a) An empty freight car weighing 10 tons rolls at 3 ft s^{-1} along a level track and collides with a loaded car weighing 20 tons, standing at rest with brakes released. If the cars couple together, find their speed after the collision. (b) Find the decrease in kinetic energy as a result of the collision. (c) With what speed should the loaded car be rolling toward the empty car, in order that both shall be brought to rest by the collision?

8–8 When a bullet of mass 20 g strikes a ballistic pendulum of mass 10 kg, the center of gravity of the pendulum is observed to rise a vertical distance of 7 cm. The bullet remains embedded in the pendulum. (a) Calculate the original velocity of the bullet. (b) What fraction of the original kinetic energy of the bullet remains as kinetic energy of the system immediately after the collision? (c) What fraction of the original momentum remains as momentum of the system?

8–9 A bullet weighing 0.01 lb is shot through a 2-lb wooden block suspended on a string 5 ft long. The center of gravity of the block is observed to rise a distance of 0.0192 ft. Find the speed of the bullet as it emerges from the block if the initial speed is 1000 ft s^{-1}.

8–10 When a bullet of mass 10 g strikes a ballistic pendulum of mass 2 kg, the center of gravity of the pendulum is observed to rise a vertical distance of 10 cm. The bullet remains embedded in the pendulum. Calculate the velocity of the bullet.

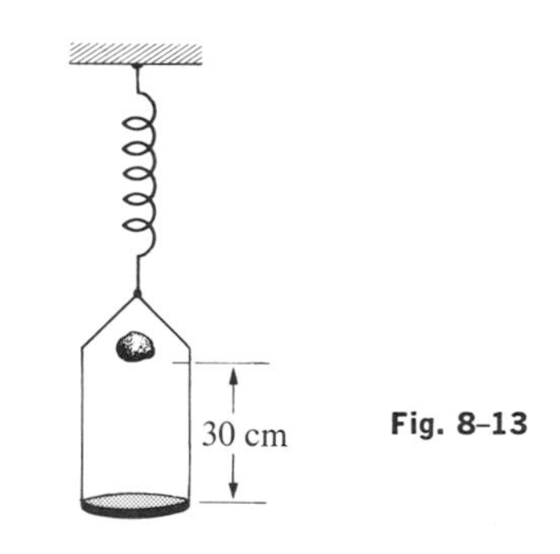

Fig. 8–13

8–11 A frame of mass 200 g, when suspended from a certain coil spring, is found to stretch the spring 10 cm. A lump of putty of mass 200 g is dropped from rest onto the frame from a height of 30 cm (Fig. 8–13). Find the maximum distance the frame moves downward.

8–12 A bullet of mass 2 g, traveling in a horizontal direction with a velocity of 500 m s^{-1}, is fired into a wooden block of mass 1 kg, initially at rest on a level surface. The bullet passes through the block and emerges with its velocity reduced to 100 m s^{-1}. The block slides a distance of 20 cm along the surface from its initial position. (a) What was the coefficient of sliding friction between block and surface? (b) What was the decrease in kinetic energy of the bullet? (c) What was the kinetic energy of the block at the instant after the bullet passed through it?

8–13 A rifle bullet weighing 0.02 lb is fired with a velocity of 2500 ft s^{-1} into a ballistic pendulum weighing 10 lb and suspended from a cord 3 ft long. Compute (a) the vertical height through which the pendulum rises, (b) the initial kinetic energy of the bullet, (c) the kinetic energy of bullet and pendulum after the bullet is embedded in the pendulum.

8–14 A 5-gm bullet is fired horizontally into a 3-kg wooden block resting on a horizontal surface. The coefficient of sliding friction between block and surface is 0.20. The bullet remains embedded in the block, which is observed to slide 25 cm along the surface. What was the velocity of the bullet?

8–15 A bullet of mass 2 g, traveling at 500 m s^{-1}, is fired into a ballistic pendulum of mass 1 kg suspended from a cord 1 m long. The bullet penetrates the pendulum and emerges with a velocity of 100 m s^{-1}. Through what vertical height will the pendulum rise?

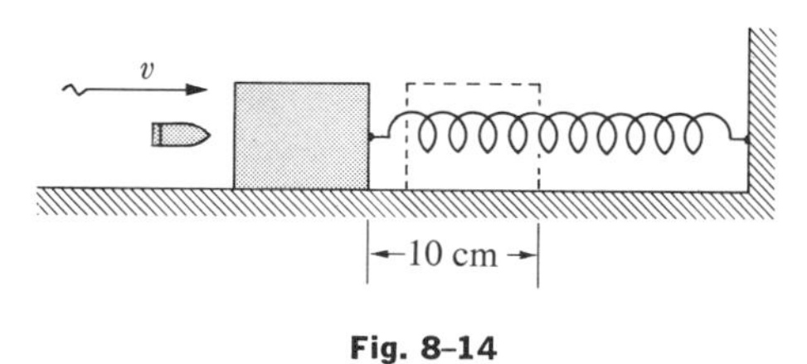

Fig. 8–14

8–16 A rifle bullet of mass 10 g strikes and embeds itself in a block of mass 990 g which rests on a horizontal frictionless surface and is attached to a coil spring, as shown in Fig. 8–14. The impact compresses the spring 10 cm. Calibration of the spring shows that a force of 100,000 dyn is required to compress the spring 1 cm. (a) Find the maximum potential energy of the spring. (b) Find the velocity of the block just after the impact. (c) What was the initial velocity of the bullet?

8–17 A 4000-lb automobile going eastward on Chestnut Street at 40 mi hr^{-1} collides with a truck weighing 4 tons which is going southward across Chestnut Street at 15 mi hr^{-1}. If they become coupled on collision, what is the magnitude and direction of their velocity immediately after colliding?

8–18 On a frictionless table, a 3-kg block moving 4 m s^{-1} to the right collides with an 8-kg block moving 1.5 m s^{-1} to the left. (a) If the two blocks stick together, what is the final velocity? (b) If the two blocks make a completely elastic head-on collision, what are their final velocities? (c) How much mechanical energy is dissipated in the collision of part (a)?

8–19 Two blocks of mass 300 g and 200 g are moving toward each other along a horizontal frictionless surface with velocities of 50 cm s^{-1} and 100 cm s^{-1}, respectively. (a) If the blocks collide and stick together, find their final velocity. (b) Find the loss of kinetic energy during the collision. (c) Find the final velocity of each block if the collision is completely elastic.

8–20 (a) Prove that when a moving body makes a perfectly inelastic collision with a second body of equal mass, initially at rest, one-half of the original kinetic energy is "lost." (b) Prove that when a very heavy particle makes a perfectly elastic collision with a very light particle that is at rest, the light one goes off with twice the velocity of the heavy one.

8–21 A 10-g block slides with a velocity of 20 cm s^{-1} on a smooth level surface and makes a head-on collision with a 30-g block moving in the opposite direction with a velocity of 10 cm s^{-1}. If the collision is perfectly elastic, find the velocity of each block after the collision.

8–22 A block of mass 200 g, sliding with a velocity of 12 cm s^{-1} on a smooth, level surface, makes a perfectly elastic head-on collision with a block of mass m, initially at rest. After the collision, the velocity of the 200-g block is 4 cm s^{-1} in the same direction as its initial velocity. Find (a) the mass m, and (b) its velocity after the collision.

8–23 A body of mass 600 g is initially at rest. It is struck by a second body of mass 400 g initially moving with a velocity of 125 cm s^{-1} toward the right along the x-axis. After the collision the 400-g body has a velocity of 100 cm s^{-1} at an angle of 37° above the x-axis in the first quadrant. Both bodies move on a horizontal frictionless plane. (a) What is the magnitude and direction of the velocity of the 600-g body after the collision? (b) What is the loss of kinetic energy during the collision?

8–24 A small steel ball moving with speed v_0 in the positive x-direction makes a perfectly elastic, noncentral collision with an identical ball originally at rest. After impact, the first ball moves with speed v_1 in the first quadrant at an angle θ_1 with the x-axis, and the second with speed v_2 in the fourth quadrant at an angle θ_2 with the x-axis. (a) Write the equations expressing conservation of linear momentum in the x-direction, and in the y-direction. (b) Square these equations and add them. (c) At this point, introduce the fact that the collision is perfectly elastic. (d) Prove that $\theta_1 + \theta_2 = \pi/2$.

8–25 A jet of liquid of cross-sectional area A and density ρ moves with speed v_J in the positive x-direction and impinges against a perfectly smooth blade B which deflects the stream at right angles but does not slow it down, as shown in Fig. 8–15. If the blade is *stationary*, prove that (a) the rate of arrival of mass at the blade is $\Delta m/\Delta t = \rho A v_J$. (b) If the impulse momentum

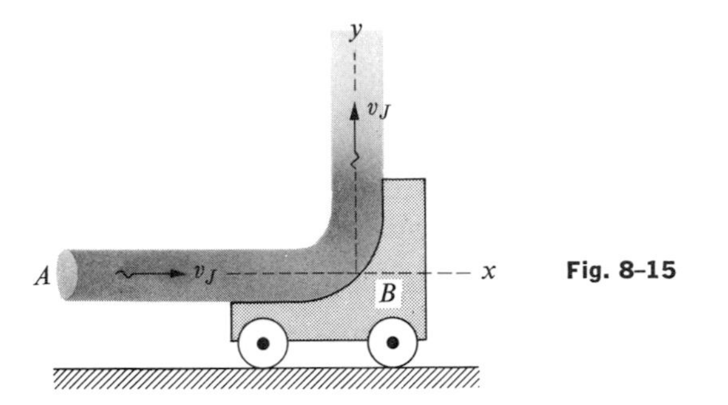

Fig. 8-15

theorem is applied to a small mass Δm, the x-component of the force acting on this mass for the time interval Δt is given by

$$F_x = -\frac{\Delta m}{\Delta t} v_J.$$

(c) The *steady* force exerted *on* the blade in the x-direction is

$$F_x = \rho A v_J^2.$$

If the blade moves to the right with a speed v_B ($v_B < v_J$), derive the equations for (d) the rate of arrival of mass at the moving blade, (e) the force F_x on the blade, and (f) the power delivered to the blade.

8-26 A stone whose mass is 100 g rests on a horizontal frictionless surface. A bullet of mass 2.5 g, traveling horizontally at 400 m s^{-1}, strikes the stone and rebounds horizontally at right angles to its original direction with a speed of 300 m s^{-1}. (a) Compute the magnitude and direction of the velocity of the stone after it is struck. (b) Is the collision perfectly elastic?

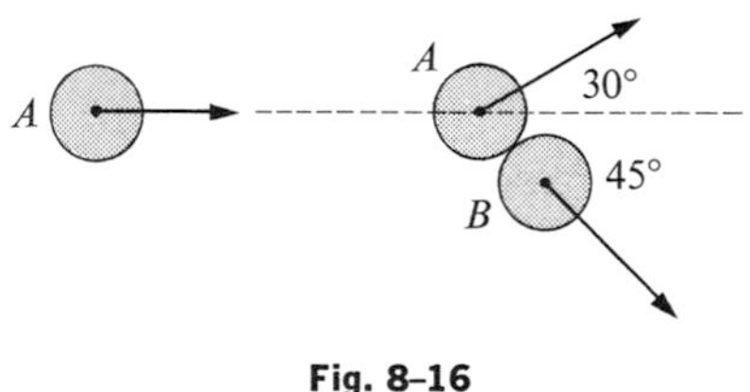

Fig. 8-16

8-27 A hockey puck B rests on a smooth ice surface and is struck by a second puck, A which was originally traveling at 80 ft s^{-1} and which is deflected 30° from its original direction (Fig. 8-16). Puck B acquires a velocity at 45° with the original velocity of A. (a) Compute the speed of each puck after the collision. (b) Is the collision perfectly elastic? If not, what fraction of the original kinetic energy of puck A is "lost"?

8-28 Imagine that a ball of mass 200 g rolls back and forth between opposite sides of a billiard table 1 m wide, with a velocity that remains constant in magnitude but reverses in direction at each collision with the cushions. The magnitude of the velocity is 4 m s^{-1}. (a) What is the change in momentum of the ball at each collision? (b) How many collisions per unit time are made by the ball with one of the cushions? (c) What is the average time rate of change of momentum of the ball as a result of these collisions? (d) What is the average force exerted by the ball on a cushion? (e) Sketch a graph of the force exerted on the ball as a function of time, for a time interval of 5 s. [*Note:* This problem illustrates how one computes the average force exerted by the molecules of a gas on the walls of the containing vessel.]

8-29 A projectile is fired at an angle of departure of 60° and with a muzzle velocity of 1200 ft s^{-1}. At the highest point of its trajectory the projectile explodes into two fragments of equal mass, one of which falls vertically. (a) How far from the point of firing does the other fragment strike if the terrain is level? (b) How much energy was released during the explosion?

8-30 A railroad handcar is moving along straight frictionless tracks. In each of the following cases the car initially has a total weight (car and contents) of 500 lb and is traveling with a velocity of 10 ft s^{-1}. Find the final velocity of the car in each of the three cases. (a) A 50-lb weight is thrown sideways out of the car with a velocity of 8 ft s^{-1} relative to the car. (b) A 50-lb weight is thrown backwards out of the car with a velocity of 10 ft s^{-1} relative to the car. (c) A 50-lb weight is thrown into the car with a velocity of 12 ft s^{-1} relative to the ground and opposite in direction to the velocity of the car.

8-31 A bullet weighing 0.02 lb is fired with a muzzle velocity of 2700 ft s^{-1} from a rifle weighing 7.5 lb. (a) Compute the recoil velocity of the rifle, assuming it free to recoil. (b) Find the ratio of the kinetic energy of the bullet to that of the rifle.

8-32 A 75-mm gun fires a projectile weighing 16 lb with a muzzle velocity of 1900 ft s^{-1}. By how many miles per hour is the velocity of a plane mounting such a gun decreased when a projectile is fired directly ahead? The plane weighs 32,000 lb.

8-33 The projectile of a 16-in. seacoast gun weighs 2400 lb, travels a distance of 38 ft in the bore of the gun, and has a muzzle velocity of 2250 ft s^{-1}. The gun weighs 300,000 lb. (a) Compute the initial recoil velocity of the gun, assuming it free to recoil. (b) Find the ratio of the kinetic energy of the projectile to that of the recoiling gun.

8-34 Block A in Fig. 8-17 has a mass of 1 kg and block B has a mass of 2 kg. The blocks are forced together, compressing a spring S between them, and the system is released from rest on a level frictionless surface. The spring is not fastened to either block and drops to the surface after it has expanded. Block B acquires a speed of 0.5 m s^{-1}. How much potential energy was stored in the compressed spring?

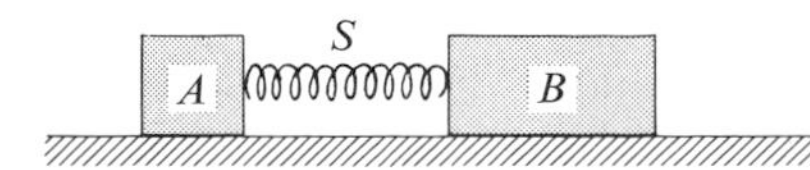

Fig. 8–17

8–35 An open-topped freight car weighing 10 tons is coasting without friction along a level track. It is raining very hard, with the rain falling vertically downward. The car is originally empty and moving with a velocity of 2 ft s^{-1}. What is the velocity of the car after it has traveled long enough to collect one ton of rain water?

8–36 A neutron of mass 1.67×10^{-24} g, moving with a velocity of 2×10^{6} cm s^{-1}, makes a head-on collision with a boron nucleus of mass 17.0×10^{-24} g, originally at rest. (a) If the collision is completely inelastic, what is the final kinetic energy of the system, expressed as a fraction of the original kinetic energy? (b) If the collision is perfectly elastic, what fraction of its original kinetic energy does the neutron transfer to the boron nucleus?

8–37 A nucleus, originally at rest, decays radioactively by emitting an electron of momentum 9.22×10^{-16} g cm s^{-1}, and at right angles to the direction of the electron a neutrino with momentum 5.33×10^{-16} g cm s^{-1}. (a) In what direction does the residual nucleus recoil? (b) What is its momentum? (c) If the mass of the residual nucleus is 3.90×10^{-22} g, what is its kinetic energy?

8–38 A 160-lb man standing on ice throws a 6-oz ball horizontally with a speed of 80 ft s^{-1}. (a) With what speed and in what direction will the man begin to move? (b) If the man throws 4 such balls every 3 s, what is the average force acting on him? [*Hint:* average force equals average rate of change of momentum.]

8–39 Find the average recoil force on a machine gun firing 120 shots per minute. The weight of each bullet is 0.025 lb, and the muzzle velocity is 2700 ft s^{-1}.

8–40 A rifleman, who together with his rifle weighs 160 lb, stands on roller skates and fires 10 shots horizontally from an automatic rifle. Each bullet weighs 0.0257 lb (180 grains) and has a muzzle velocity of 2500 ft s^{-1}. (a) If the rifleman moves back without friction, what is his velocity at the end of the 10 shots? (b) If the shots were fired in 10 s, what was the average force exerted on him? (c) Compare his kinetic energy with that of the 10 bullets.

8–41 A rocket burns 50 g of fuel per second, ejecting it as a gas with a velocity of 500,000 cm s^{-1}. (a) What force does this gas exert on the rocket? Give the result in dynes and newtons. (b) Would the rocket operate in free space? (c) If it would operate in free space, how would you steer it? Could you brake it?

8–42 This problem illustrates the advantage of using a multistage rather than a single-stage rocket. Suppose that the first stage of a two-stage rocket has a total weight of 12 tons, of which 9 tons is fuel. The total weight of the second stage is 1 ton, of which 0.75 ton is fuel. Assume that the relative velocity v_r of ejected material is constant, and neglect any effect of gravity. (The latter effect is small during the firing period if the rate of fuel consumption is large, as shown by the example in Section 8–7.)

(a) Suppose that the entire fuel supply carried by the two-stage rocket were utilized in a single-stage rocket of the same total weight of 13 tons. What would be the velocity of the rocket, starting from rest, when its fuel was exhausted?

(b) What is the velocity when the fuel of the first stage is exhausted, if the first stage carries the second stage with it to this point? This velocity then becomes the initial velocity of the second stage.

(c) What is the final velocity of the second stage?

(d) Sputnik I's velocity was about 18,000 mi hr^{-1} or 8 km s^{-1}. What value of v_r would be required to give the second stage of the above rocket a velocity of this magnitude?

8–43 (a) Is it possible for a rocket to acquire a velocity greater than the relative velocity of the material ejected? (b) Divide each term of Eq. (8–13) by dt and give the physical significance of each of the resulting terms.

8–44 (a) Show that the acceleration of a rocket fired vertically upward is given by

$$a = -\frac{v_r}{m}\frac{dm}{dt} - g.$$

(b) Suppose that the rate of ejection of mass by the rocket is constant, that is, the rate of decrease of mass is $dm/dt = -km_0$, where k is a positive constant and m_0 is the initial mass. What is the numerical value of k, and in what units is it expressed, for the rocket in the example at the end of Section 8–7?

(c) Show that the mass at any time t is given by the equation

$$m = m_0(1 - kt).$$

(d) Show that the acceleration at any time is equal to

$$\frac{v_r k}{1 - kt} - g.$$

(e) Find the initial acceleration of the rocket in the example at the end of Section 8–7, in terms of the acceleration of gravity, g.

(f) Find the acceleration 15 s after the motion starts.

(g) Sketch the acceleration-time graph.

9 Rotation

9-1 INTRODUCTION

The most general type of motion which a body can undergo is a combination of *translation* and *rotation*. Thus far we have considered only the special case of translational motion, along a straight line or along a curve. We next discuss motion of rotation about a fixed axis, that is, motion of rotation without translation. We shall see that many of the equations describing rotation about a fixed axis are exactly analogous to those encountered in rectilinear motion. If the axis is *not* fixed, the problem becomes much more complicated, and we shall not attempt to give a complete discussion of the general case of translation plus rotation.

9-2 ANGULAR VELOCITY

Figure 9-1 represents a rigid body of arbitrary shape rotating about a fixed axis through point O and perpendicular to the plane of the diagram. Line OP is a line fixed with respect to the body and rotating with it. The position of the entire body is evidently completely specified by the angle θ which the line OP makes with some reference line fixed in space, such as Ox. The motion of the body is therefore analogous to the rectilinear motion of a particle whose position is completely specified by a single coordinate such as x or y. The equations of motion are greatly simplified if the angle θ is expressed in *radians*.

One radian (1 rad) is the angle subtended at the center of a circle by an arc of length equal to the radius of the circle (Fig. 9-2a). Since the radius is contained 2π times ($2\pi = 6.28\ldots$) in the circumference, there are 2π or $6.28\ldots$ rad in one complete revolution or 360°. Hence

$$1 \text{ rad} = \frac{360}{2\pi} = 57.3 \text{ degrees}$$

$$360° = 2\pi \text{ rad} = 6.28 \text{ rad}$$

$$180° = \pi \text{ rad} = 3.14 \text{ rad}$$

$$90° = \pi/2 \text{ rad} = 1.57 \text{ rad}$$

$$60° = \pi/3 \text{ rad} = 1.05 \text{ rad}$$

and so on.

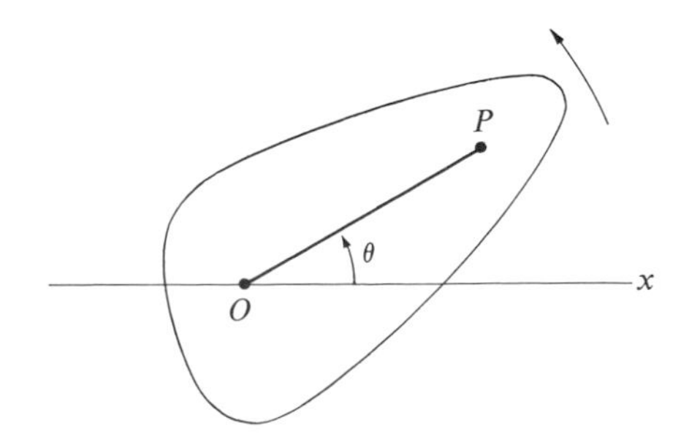

9-1 Body rotating about a fixed axis through point O.

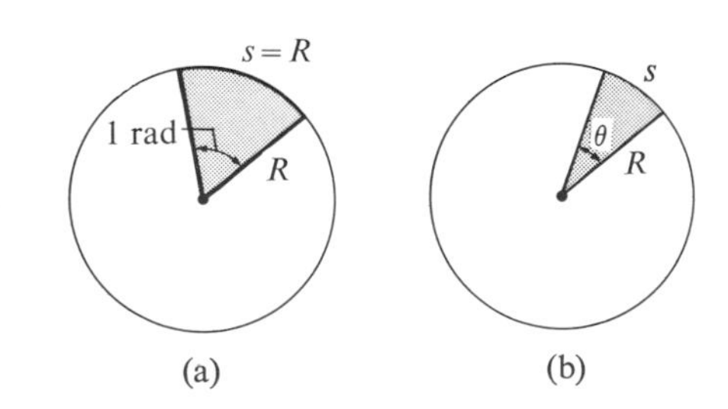

9-2 An angle θ in radians is defined as the ratio of the arc s to the radius R.

In general (Fig. 9-2b), if θ represents any arbitrary angle subtended by an arc of length s on the circumference of a circle of radius R, then θ (in radians) is equal to the length of the arc s divided by the radius R:

$$\theta = \frac{s}{R}, \qquad s = R\theta. \qquad (9\text{-}1)$$

An angle in radians, being defined as the ratio of a length to a length, is a pure number.

In Fig. 9-3, a reference line OP in a rotating body makes an angle θ_1 with the reference line Ox, at a time t_1. At a later time t_2 the angle has increased to θ_2. The *average angular velocity* of the body, $\bar{\omega}$, in the time interval between t_1 and t_2, is defined as the ratio of the *angular displacement* $\theta_2 - \theta_1$, or $\Delta\theta$, to the elapsed time $t_2 - t_1$ or Δt:

$$\bar{\omega} = \frac{\Delta\theta}{\Delta t}.$$

The *instantaneous angular velocity* ω is defined as the limit approached by this ratio as Δt approaches zero:

$$\omega = \lim_{\Delta t \to 0} \frac{\Delta\theta}{\Delta t} = \frac{d\theta}{dt}. \qquad (9\text{-}2)$$

Since the body is rigid, *all* lines in it rotate through the same angle in the same time, and the angular velocity is characteristic of the body as a whole. If the angle θ is in radians, the unit of angular velocity is one *radian per second* (1 rad s^{-1}). Other units, such as the revolution per minute (rev min^{-1}), are in common use.

9-3 ANGULAR ACCELERATION

If the angular velocity of a body changes by $\Delta\omega$ in a time interval Δt, it is said to have an angular acceleration. The *average angular acceleration* $\bar{\alpha}$ is defined as

$$\bar{\alpha} = \frac{\Delta\omega}{\Delta t},$$

and the *instantaneous angular acceleration* α is defined as the limit of this ratio when Δt approaches zero:

$$\alpha = \lim_{\Delta t \to 0} \frac{\Delta\omega}{\Delta t} = \frac{d\omega}{dt}. \qquad (9\text{-}3)$$

The unit of angular acceleration is 1 rad s^{-2}.

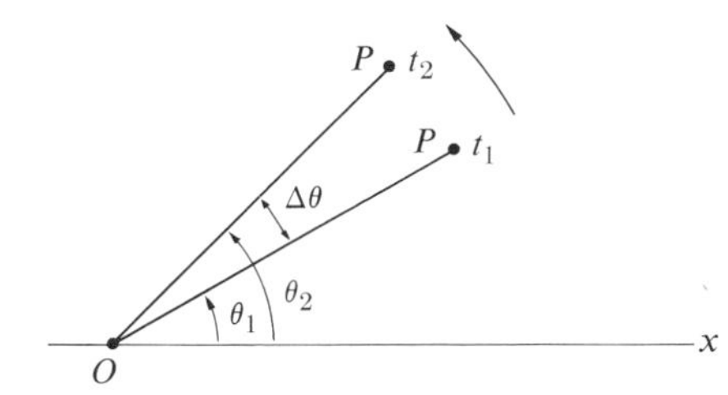

9-3 Angular displacement $\Delta\theta$ of a rotating body.

Since $\omega = d\theta/dt$, the angular acceleration can be written

$$\alpha = \frac{d}{dt}\frac{d\theta}{dt} = \frac{d^2\theta}{dt^2}.$$

Also, by the chain rule,

$$\alpha = \frac{d\omega}{d\theta}\frac{d\theta}{dt} = \omega\frac{d\omega}{d\theta}. \qquad (9\text{-}4)$$

Angular velocity and angular acceleration are exactly analogous to linear velocity and acceleration.

Many of the equations describing the motion of a rigid body can be written more compactly if we introduce the concepts of *vector angular velocity*, $\boldsymbol{\omega}$, and *vector angular acceleration* $\boldsymbol{\alpha}$. The vector angular velocity $\boldsymbol{\omega}$ is defined as a vector of magnitude ω, pointing in the direction of advance of a right-hand screw which is turned in the direction of rotation of the body. For a rigid body rotating about a fixed axis, the vector $\boldsymbol{\omega}$ is parallel to the axis, as shown in Fig. 9-4.

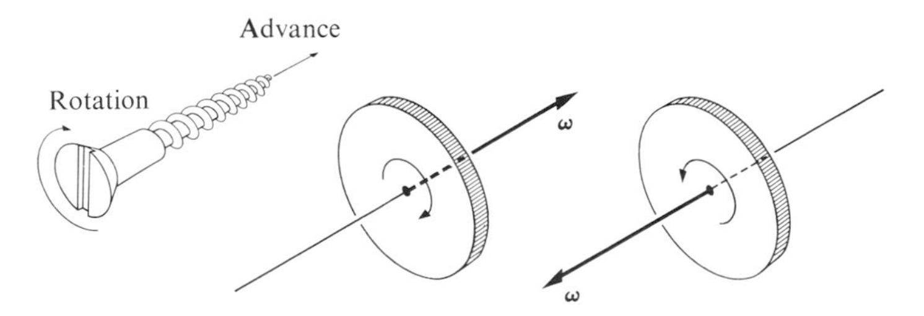

9-4 Vector angular velocity $\boldsymbol{\omega}$ of a rotating body.

Similarly, the vector angular acceleration $\boldsymbol{\alpha}$ is defined as a vector of magnitude α, having at any instant the direction of the vector change in angular velocity,

$d\boldsymbol{\omega}$. For rotation about a fixed axis, where $\boldsymbol{\omega}$ is always parallel to the axis, the vector $\boldsymbol{\alpha}$ is parallel to the axis also.

9-4 ROTATION WITH CONSTANT ANGULAR ACCELERATION

The simplest type of accelerated rotational motion is that in which the angular acceleration is constant. When this is the case, the expressions for the angular velocity and angular coordinate can readily be found by integration. We have

$$d\omega/dt = \alpha = \text{constant},$$

$$\int d\omega = \int \alpha\, dt,$$

$$\omega = \alpha t + C_1.$$

If ω_0 is the angular velocity when $t = 0$, the integration constant $C_1 = \omega_0$ and

$$\boxed{\omega = \omega_0 + \alpha t.} \tag{9-5}$$

Then, since $\omega = d\theta/dt$,

$$\int d\theta = \int \omega_0\, dt + \int \alpha t\, dt, \qquad \theta = \omega_0 t + \tfrac{1}{2}\alpha t^2 + C_2.$$

In general, the integration constant C_2 is the value of θ when $t = 0$, say θ_0. If $\theta_0 = 0$, then

$$\boxed{\theta = \omega_0 t + \tfrac{1}{2}\alpha t^2.} \tag{9-6}$$

If we write the angular acceleration as

$$\alpha = \omega \frac{d\omega}{d\theta},$$

then

$$\int \alpha\, d\theta = \int \omega\, d\omega + C_3, \qquad \alpha\theta = \tfrac{1}{2}\omega^2 + C_3.$$

If the angle θ is zero when $t = 0$, and if the initial angular velocity is ω_0, then $C_3 = -\tfrac{1}{2}\omega_0^2$ and

$$\boxed{\omega^2 = \omega_0^2 + 2\alpha\theta.} \tag{9-7}$$

Equations (9-5), (9-6), and (9-7) are exactly analogous to the corresponding equations for linear motion with constant acceleration:

$$v = v_0 + at,$$

$$x = v_0 t + \tfrac{1}{2}at^2,$$

$$v^2 = v_0^2 + 2ax.$$

Example. The angular velocity of a body is 4 rad s^{-1} at time $t = 0$, and its angular acceleration is constant and equal to 2 rad s^{-2}. A line OP in the body is horizontal at time $t = 0$. (a) What angle does this line make with the horizontal at time $t = 3$ s? (b) What is the angular velocity at this time?

(a) $\theta = \omega_0 t + \tfrac{1}{2}\alpha t^2$

$= 4 \text{ rad s}^{-1} \times 3 \text{ s} + \tfrac{1}{2} \times 2 \text{ rad s}^{-2} \times (3 \text{ s})^2$

$= 21 \text{ rad} = 3.34 \text{ rev}.$

(b) $\omega = \omega_0 + \alpha t$

$= 4 \text{ rad s}^{-1} + 2 \text{ rad s}^{-2} \times 3 \text{ s} = 10 \text{ rad s}^{-1}.$

Alternatively, from Eq. (9-7),

$\omega^2 = \omega_0^2 + 2\alpha\theta$

$= (4 \text{ rad s}^{-1})^2 + 2 \times 2 \text{ rad s}^{-2} \times 21 \text{ rad} = 100 \text{ rad}^2 \text{ s}^{-2},$

$\omega = 10 \text{ rad s}^{-1}.$

9-5 RELATION BETWEEN ANGULAR AND LINEAR VELOCITY AND ACCELERATION

In Section 6-2 we discussed the linear velocity and acceleration of a *particle* revolving in a circle. When a *rigid body* rotates about a fixed axis, every point in the body moves in a circle whose center is on the axis and which lies in a plane perpendicular to the axis. There are some useful and simple relations between the *angular* velocity and acceleration of the rotating body and the *linear* velocity and acceleration of points within it.

Let r be the distance from the axis to some point P in the body, so that the point moves in a circle of radius r, as in Fig. 9-5. When the radius makes an angle θ with the reference axis, the distance s to the point P, measured along the circular path, is

$$\boxed{s = r\theta,} \tag{9-8}$$

if θ is in radians.

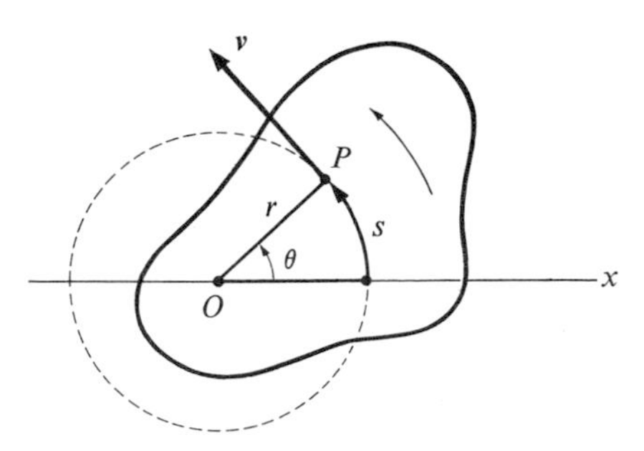

9-5 The distance s moved through by point P equals $r\theta$.

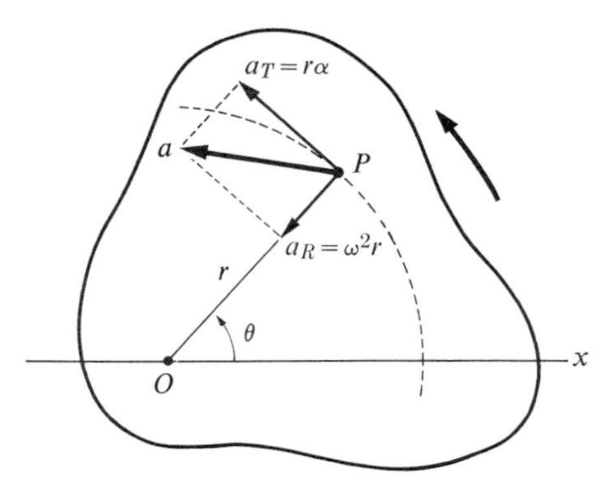

9-6 Nonuniform rotation about a fixed axis through point O. The tangential component of acceleration of point P equals $r\alpha$; the radial component equals $\omega^2 r$.

Differentiating both sides of this equation with respect to t, we have, since r is constant,

$$\frac{ds}{dt} = r\frac{d\theta}{dt}.$$

But ds/dt is the magnitude of the linear velocity v of point P, and $d\theta/dt$ is the angular velocity ω of the rotating body. Hence

$$v = r\omega \tag{9-9}$$

and the magnitude v of the linear velocity equals the product of the angular velocity ω and the distance r of the point from the axis.

Differentiating Eq. (9-9) with respect to t gives

$$\frac{dv}{dt} = r\frac{d\omega}{dt}.$$

But dv/dt is the magnitude of the tangential component of acceleration a_T of point P, and $d\omega/dt$ is the angular acceleration α of the rotating body, so

$$a_T = r\alpha \tag{9-10}$$

and the tangential component of acceleration equals the product of the angular acceleration and the distance from the axis.

The *radial* component of acceleration v^2/r of the point P can also be expressed in terms of the angular velocity:

$$a_R = \frac{v^2}{r} = \omega^2 r = \omega v. \tag{9-11}$$

The tangential and radial components of acceleration of any arbitrary point P in a rotating body are shown in Fig. 9-6.

9-6 TORQUE AND ANGULAR ACCELERATION. MOMENT OF INERTIA

We are now ready to consider the *dynamics* of rotation about a fixed axis, that is, the relation between the forces on a pivoted body and its angular acceleration.

Figure 9-7 represents a rigid body pivoted about a fixed axis through point O, perpendicular to the plane of the diagram.

The solid circle represents one of the particles of the body, of mass m_i. The particle is acted on by an external force $\boldsymbol{F}_i$, and by an internal force $\boldsymbol{f}_i$, the resultant of the forces exerted on it by all of the other particles of the body. It will suffice to consider only the case in which the forces $\boldsymbol{F}_i$ and $\boldsymbol{f}_i$ lie in a plane perpendicular to the axis. From Newton's second law,

$$\boldsymbol{F}_i + \boldsymbol{f}_i = m_i\boldsymbol{a}_i.$$

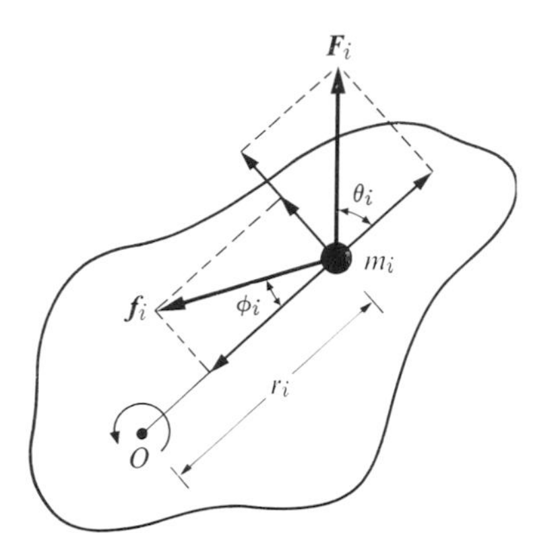

9-7 An external force $\boldsymbol{F}_i$ and an internal force $\boldsymbol{f}_i$ acting on a particle of mass m_i in a rigid body rotating about a fixed axis.

Let us resolve the forces and accelerations into radial and tangential components. Then

$$F_i \cos\theta_i + f_i \cos\phi_i = m_i a_{iR} = m_i r_i \omega^2,$$
$$F_i \sin\theta_i + f_i \sin\phi_i = m_i a_{iT} = m_i r_i \alpha.$$

The first of these equations does not concern us further. When both sides of the second are multiplied by the distance r_i of the particle from the axis, we get

$$F_i r_i \sin\theta_i + f_i r_i \sin\phi_i = m_i r_i^2 \alpha. \qquad (9\text{-}12)$$

The first term on the left is the moment Γ_i of the external force about the axis, and the second is the moment of the internal force.

Equations corresponding to Eq. (9–12) can be written for every particle of the body. When these equations are added, the moments of the *internal* forces will cancel, since the resultant moment of every internal action-reaction force pair is zero. The sum of the left sides of the equations is then simply the resultant moment Γ of the *external* forces about the axis, or

$$\Gamma = \sum\Gamma_i = \sum F_i r_i \sin\theta_i.$$

Since the body is rigid, all particles have the same angular acceleration α and therefore

$$\Gamma = (m_1 r_1^2 + m_2 r_2^2 + \cdots)\alpha = (\sum m_i r_i^2)\alpha. \qquad (9\text{-}13)$$

The sum $\sum m_i r_i^2$ is called the *moment of inertia* of the body about the axis through point O, and is represented by I:

$$\boxed{I = \sum m_i r_i^2.} \qquad (9\text{-}14)$$

Equation (9–13) then becomes

$$\boxed{\Gamma = I\alpha = I\frac{d\omega}{dt}.} \qquad (9\text{-}15)$$

That is, *when a rigid body is pivoted about a fixed axis, the resultant external torque about the axis equals the product of the moment of inertia of the body about the axis, and the angular acceleration.*

Thus the *angular* acceleration of a rigid body about a fixed axis is given by an equation having exactly the same form as that for the *linear* acceleration of a particle:

$$F = ma = m\frac{dv}{dt}.$$

The resultant torque Γ about the axis corresponds to the resultant force F, the angular acceleration α corresponds to the linear acceleration a, and the moment of inertia I about the axis corresponds to the mass m.

The concept of moment of inertia can be thought of either as the sum of the product of the mass of each particle in a rigid body and the square of its distance from the axis, $I = \sum m_i r_i^2$, or as the ratio of the resultant torque to the angular acceleration, $I = \Gamma/\alpha$.

The *vector moment* $\boldsymbol{\Gamma}$ of a force about an axis is a vector of magnitude Γ, parallel to the axis and pointing in a direction given by the right-hand screw rule. Similarly, the *vector angular velocity* and *acceleration*, for a body rotating about a fixed axis, are also parallel to the axis. Equation (9–15) can therefore be written as a *vector equation*:

$$\boldsymbol{\Gamma} = I\boldsymbol{\alpha} = I\frac{d\boldsymbol{\omega}}{dt}, \qquad (9\text{-}16)$$

in which form it implies not only the scalar equation, Eq. (9–15), but also that the vectors $\boldsymbol{\Gamma}$ and $\boldsymbol{\alpha} = d\boldsymbol{\omega}/dt$ are in the same direction. Thus in Fig. 9–7 the vectors $\boldsymbol{\Gamma}$ and $\boldsymbol{\alpha}$ both point toward the reader.

TABLE 9-1 ANALOGY BETWEEN TRANSLATIONAL AND ROTATIONAL QUANTITIES

Concept	Translation	Rotation	Comments
Displacement	s	θ	$s = r\theta$
Velocity	$v = ds/dt$	$\omega = d\theta/dt$	$v = r\omega$
Acceleration	$a = dv/dt$	$\alpha = d\omega/dt$	$a_T = r\alpha$
Resultant force, moment	F	Γ	$\Gamma = Fr$
Equilibrium	$F = 0$	$\Gamma = 0$	
Acceleration constant	$v = v_0 + at$ $s = v_0t + \frac{1}{2}at^2$ $v^2 = v_0^2 + 2as$	$\omega = \omega_0 + \alpha t$ $\theta = \omega_0 t + \frac{1}{2}\alpha t^2$ $\omega^2 = \omega_0^2 + 2\alpha\theta$	
Mass, moment of inertia	m	I	$I = \sum m_i r_i^2$
Newton's second law	$F = ma$	$\Gamma = I\alpha$	
Work	$W = \int F\, ds$	$W = \int \Gamma\, d\theta$	
Power	$P = Fv$	$P = \Gamma\omega$	
Potential energy	$E_p = mgy$		
Kinetic energy	$E_k = \frac{1}{2}mv^2$	$E_k = \frac{1}{2}I\omega^2$	
Impulse	$\int F\, dt$	$\int \Gamma\, dt$	
Momentum	mv	$L = I\omega$	

Equation (9-16) is exactly analogous to the vector equation for the linear motion of a particle:

$$\mathbf{F} = m\mathbf{a} = m\frac{d\mathbf{v}}{dt}.$$

The analogy between translational and rotational quantities is displayed in Table 9-1.

Example. A wheel of radius R, mass m_2, and moment of inertia I is mounted on an axle supported in fixed bearings, as in Fig. 9-8. A light flexible cord is wrapped around the rim of the wheel and carries a body of mass m_1. Friction in the bearings can be neglected. Discuss the motion of the system.

We must consider the resultant *force* on the suspended body and the resultant *torque* on the wheel. Let $\boldsymbol{T}$ represent the tension in the cord and $\boldsymbol{P}$ the upward force exerted on the shaft of the wheel by the bearings.

The resultant force on the suspended body is $w_1 - T$, and from Newton's second law for linear motion,

$$w_1 - T = m_1 a.$$

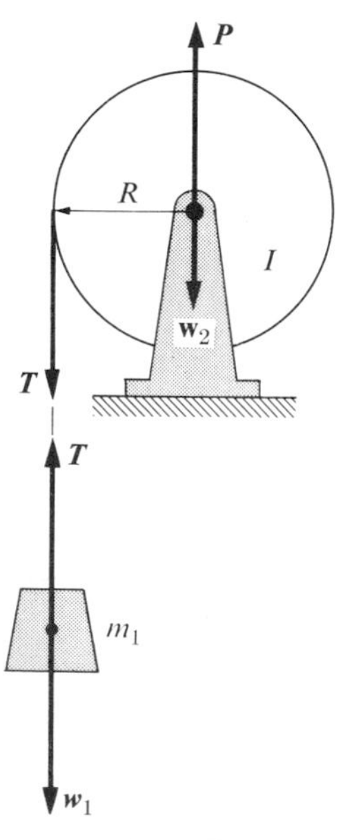

Fig. 9-8

[We have taken the downward direction as positive in order that a positive (counterclockwise) angular displacement of the wheel shall correspond to a positive linear displacement of the suspended body.]

The forces $\boldsymbol{P}$ and $\boldsymbol{w}_2$ have no moment about the axis of the wheel. The resultant torque on the wheel, about the axis, is TR, and from Newton's second law for rotation,

$$TR = I\alpha.$$

Since the linear acceleration of the suspended body equals the tangential acceleration of the rim of the wheel, we have

$$a = R\alpha.$$

Simultaneous solution of these equations gives

$$a = g\frac{1}{1 + (I/m_1R^2)}.$$

If the system starts from rest, the linear speed v of the suspended body, after descending a distance y (the acceleration is constant) is given by

$$v^2 = 2ay = 2\left[g\frac{1}{1 + (I/m_1R^2)}\right]y.$$

9-7 CALCULATION OF MOMENTS OF INERTIA

The moment of inertia of a body about an axis can be found *experimentally* by pivoting the body about the axis, applying a measured torque Γ to the body, and measuring the resulting angular acceleration α. The moment of inertia is then given by

$$I = \frac{\Gamma}{\alpha}.$$

A more convenient experimental method is described in Section 11-7.

The moment of inertia can be *calculated* from the defining equation $I = \sum m_ir_i^2$, for any system consisting of discrete point masses.

Example. Three small bodies, which can be considered as particles, are connected by light rigid rods, as in Fig. 9-9. What is the moment of inertia of the system (a) about an axis through point A, perpendicular to the plane of the diagram, and (b) about an axis coinciding with the rod BC?

(a) The particle at point A lies on the axis. Its distance *from* the axis is zero and it contributes nothing to the moment of inertia.

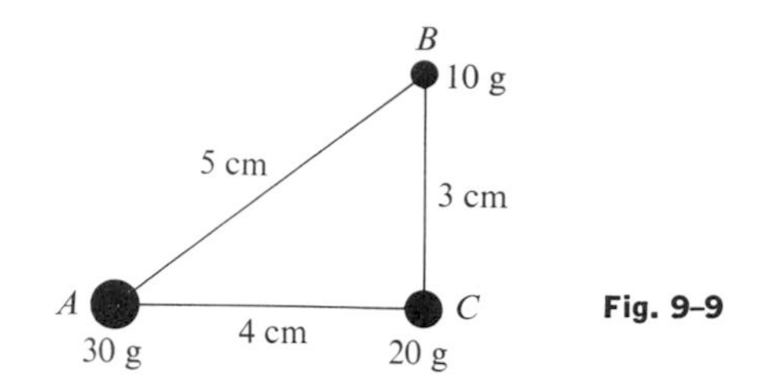

Fig. 9-9

Therefore

$$I = \sum m_ir_i^2 = 10\text{ g} \times (5\text{ cm})^2 + 20\text{ g} \times (4\text{ cm})^2$$
$$= 570\text{ g cm}^2.$$

(b) The particles at B and C both lie on the axis. The moment of inertia is

$$I = \sum m_ir_i^2 = 30\text{ g} \times (4\text{ cm})^2$$
$$= 480\text{ g cm}^2.$$

This illustrates the important fact that the moment of inertia of a body, unlike its mass, is not a unique property of the body but depends on the axis about which it is computed.

For a body which is not composed of discrete point masses but is a continuous distribution of matter, the summation expressed in the definition of moment of inertia, $I = \sum m_ir_i^2$, must be evaluated by the methods of calculus. The body is imagined to be subdivided into volume elements, each of mass Δm. Let r be the distance from any element to the axis of rotation. If each mass Δm is multiplied by the square of its distance from the axis and all the products $r^2\,\Delta m$ summed over the whole body, the moment of inertia is

$$I = \lim_{\Delta m \to 0} \sum r^2\,\Delta m = \int r^2\,dm. \tag{9-17}$$

If dV and dm represent the volume and mass, respectively, of an element, the density ρ is defined by the relation $dm = \rho\,dV$. Equation (9-17) may therefore be written

$$I = \int r^2\rho\,dV.$$

If the density of a body is the same at all points, the body is said to be *uniform* or *homogeneous*, in which case

$$I = \rho\int r^2\,dV.$$

In using this equation, any convenient volume element may be chosen, *provided that all points within the element are the same distance r from the axis.*

The evaluation of integrals of this type may present considerable difficulty if the body is irregular, but for bodies of simple shape the integration can be carried out relatively easily. Some examples are given below.

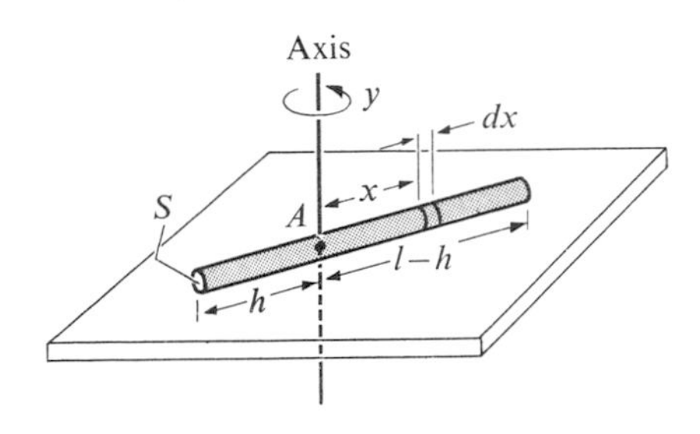

Fig. 9–10

Example 1. *Uniform slender rod, axis perpendicular to length.* Figure 9–10 shows a slender uniform rod of mass m and length ℓ. We wish to compute its moment of inertia about an axis through A, at an arbitrary distance h from one end. Select as an element of volume a short section of length dx and cross-sectional area S, at a distance x from point A. Then

$$dm = \rho\, dV = \rho S\, dx = \frac{\rho S \ell}{\ell}\, dx = \frac{m}{\ell}\, dx.$$

Now

$$I_A = \int x^2\, dm = \frac{m}{\ell}\int_{-h}^{\ell-h} x^2\, dx$$

$$= \frac{m}{\ell}\frac{x^3}{3}\Big]_{-h}^{\ell-h} = \tfrac{1}{3}m(\ell^2 - 3\ell h + 3h^2).$$

From this general expression, we can find the moment of inertia about'an axis through any point on the rod. For example, if the axis is at the left end, $h = 0$ and

$$I = \tfrac{1}{3}m\ell^2.$$

If the axis is at the right end, $h = \ell$ and

$$I = \tfrac{1}{3}m\ell^2,$$

as would be expected. If the axis passes through the center,

$$h = \ell/2 \quad \text{and} \quad I = \tfrac{1}{12}m\ell^2.$$

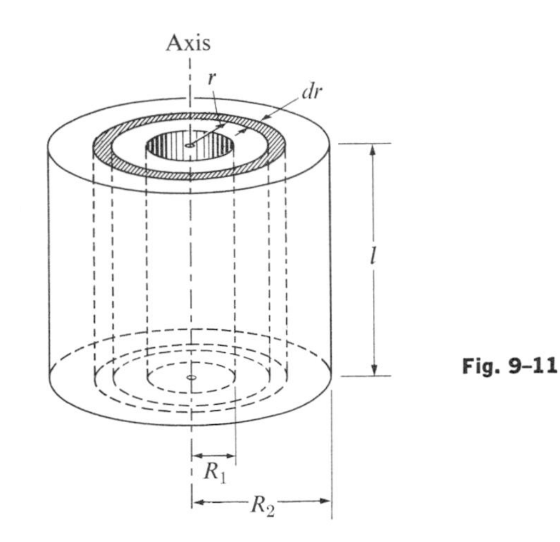

Fig. 9–11

Example 2. *Hollow or solid cylinder, axis of symmetry.* Figure 9–11 shows a hollow cylinder of length ℓ and inner and outer radii R_1 and R_2. We choose as the most convenient volume element the thin cylindrical shell of radius r, thickness dr, and length ℓ. If ρ is the density of the material, that is, the mass per unit volume, then

$$dm = \rho\, dV = \rho(2\pi r\, dr) \times \ell.$$

The moment of inertia is given by

$$I = \int r^2\, dm = 2\pi\ell \int_{R_1}^{R_2} \rho r^3\, dr.$$

If the body were of nonuniform density, one would have to know ρ as a function of r before the integration could be carried out. For a homogeneous solid, however, ρ is constant, and

$$I = 2\pi\ell\rho \int_{R_1}^{R_2} r^3\, dr = \frac{\pi\ell\rho}{2}(R_2^4 - R_1^4).$$

The mass m of the entire cylinder is the product of its density and its volume. The volume is given by

$$\pi\ell(R_2^2 - R_1^2).$$

Hence

$$m = \pi\ell\rho(R_2^2 - R_1^2),$$

and the moment of inertia is therefore

$$I = \tfrac{1}{2}m(R_1^2 + R_2^2). \qquad (9\text{–}18)$$

If the cylinder is solid, $R_1 = 0$, and if we let R represent the outer radius,

$$I = \tfrac{1}{2}mR^2. \qquad (9\text{–}19)$$

If the cylinder is very thin-walled (like a stovepipe), $R_1 = R_2$ (very nearly) and if R represents this common radius,

$$I = mR^2.$$

Note that the moment of inertia of a cylinder about an axis coinciding with its axis of symmetry does not depend on the length ℓ. Two hollow cylinders of the same inner and outer radii, one of wood and one of brass, but having the same mass m, have equal moments of inertia even though the length of the former is much greater. Moment of inertia depends only on the *radial* distribution of mass, not on its distribution along the axis. Thus Eq. (9–18) holds for a very short cylinder like a washer, and Eq. (9–19) for a thin disk.

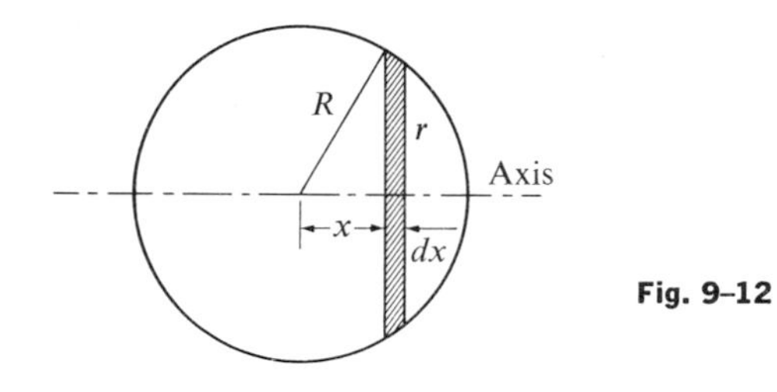

Fig. 9–12

Example 3. *Uniform sphere, axis through center.* Divide the sphere into thin disks, as indicated in Fig. 9–12. The radius r of the disk shown is

$$r = \sqrt{R^2 - x^2}.$$

Its volume is

$$dV = \pi r^2\, dx = \pi(R^2 - x^2)\, dx$$

and its mass is

$$dm = \rho\, dV.$$

Hence from Eq. (9–19) its moment of inertia is

$$dI = \frac{\pi\rho}{2}(R^2 - x^2)^2\, dx,$$

and for the whole sphere,

$$I = 2 \times \frac{\pi\rho}{2}\int_0^R (R^2 - x^2)^2\, dx,$$

since by symmetry the right hemisphere has the same moment of inertia as the left. Carrying out the integration, we get

$$I = \frac{8\pi\rho}{15}R^5.$$

The mass m of the sphere is

$$m = \rho V = \frac{4\pi\rho R^3}{3}.$$

Hence

$$I = \tfrac{2}{5}mR^2.$$

The moments of inertia of a few simple but important bodies are shown in Fig. 9–13 for convenience.

Whatever the shape of a body, it is always possible to find a radial distance from any given axis at which the mass of the body could be concentrated without altering the moment of inertia of the body about that axis. This distance is called the *radius of gyration* of the body about the given axis, and is represented by k.

If the mass m of the body actually were concentrated at this distance, the moment of inertia would be that of a particle of mass m at a distance k from an axis, or mk^2. Since this equals the moment of inertia I, then

$$mk^2 = I, \qquad k = \sqrt{I/m}. \qquad (9\text{–}20)$$

Example. What is the radius of gyration of a slender rod of mass m and length ℓ about an axis perpendicular to its length and passing through the center?

The moment of inertia about an axis through the center is

$$I_0 = \tfrac{1}{12}m\ell^2.$$

Hence

$$k_0 = \sqrt{\frac{\frac{1}{12}m\ell^2}{m}} = \frac{\ell}{2\sqrt{3}} = 0.289\ell.$$

The radius of gyration, like the moment of inertia, depends on the location of the axis.

Note carefully that, in general, the mass of a body *cannot* be considered as concentrated at its center of mass for the purpose of computing its moment of inertia. For example, when a rod is pivoted about its center, the distance from the axis to the center of mass is zero, although the radius of gyration is $\ell/2\sqrt{3}$.

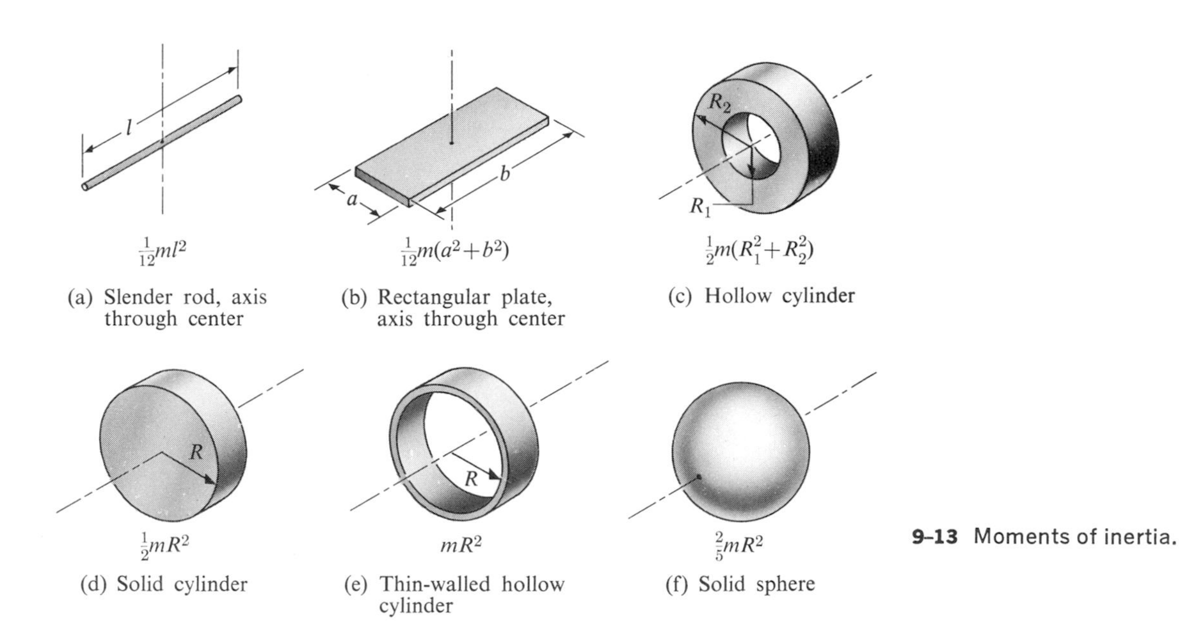

$\frac{1}{12}ml^2$
(a) Slender rod, axis through center

$\frac{1}{12}m(a^2+b^2)$
(b) Rectangular plate, axis through center

$\frac{1}{2}m(R_1^2+R_2^2)$
(c) Hollow cylinder

$\frac{1}{2}mR^2$
(d) Solid cylinder

mR^2
(e) Thin-walled hollow cylinder

$\frac{2}{5}mR^2$
(f) Solid sphere

9-13 Moments of inertia.

9-8 KINETIC ENERGY, WORK, AND POWER

When a rigid body rotates about a fixed axis, the velocity v_i of a particle at a perpendicular distance r_i from the axis equals $r_i\omega$, where ω is the angular velocity. The kinetic energy of the particle is then

$$\tfrac{1}{2}m_iv_i^2 = \tfrac{1}{2}m_ir_i^2\omega^2,$$

and the total kinetic energy of the body is

$$E_k = \sum \tfrac{1}{2}m_ir_i^2\omega^2 = \tfrac{1}{2}(\sum m_ir_i^2)\omega^2.$$

But $\sum m_ir_i^2$ equals the moment of inertia I about the axis, so

$$\boxed{E_k = \tfrac{1}{2}I\omega^2.} \tag{9-21}$$

Thus the kinetic energy of a rigid body rotating about a fixed axis is given by an expression exactly analogous to that for the kinetic energy of a particle in linear motion, the moment of inertia I corresponding to the mass m and the angular velocity ω corresponding to the linear velocity v. (See Table 9-1.)

Example. Let us consider the motion of the system in Fig. 9-8 from the energy standpoint. Looking at the system as a whole, the external forces are the forces $\boldsymbol{P}$ and $\boldsymbol{w}_2$, which do no work and the force $\boldsymbol{w}_1$, which is conservative. We can therefore apply the principle of conservation of energy, setting the decrease in potential energy of the body, as it descends a distance y, equal to the sum of the increase in *translational* kinetic energy of the body and the increase in *rotational* kinetic energy of the wheel:

$$m_1gy = \tfrac{1}{2}m_1v^2 + \tfrac{1}{2}I\omega^2.$$

But

$$v = \omega R,$$

and again we find that

$$v^2 = 2\left[g\,\frac{1}{1+(I/m_1R^2)}\right]y.$$

In Fig. 9-14, an external force $\boldsymbol{F}$ is applied at point P of a rigid body rotating about a fixed axis through O, perpendicular to the plane of the diagram. As the body rotates through a small angle $d\theta$, point P moves a distance $ds = r\,d\theta$ and the work done by the force $\boldsymbol{F}$ is

$$W = \int F_s\,ds = \int F_sr\,d\theta.$$

But F_sr is the moment Γ of the force about the axis, so

$$W = \int_{\theta_1}^{\theta_2} \Gamma\,d\theta. \tag{9-22}$$

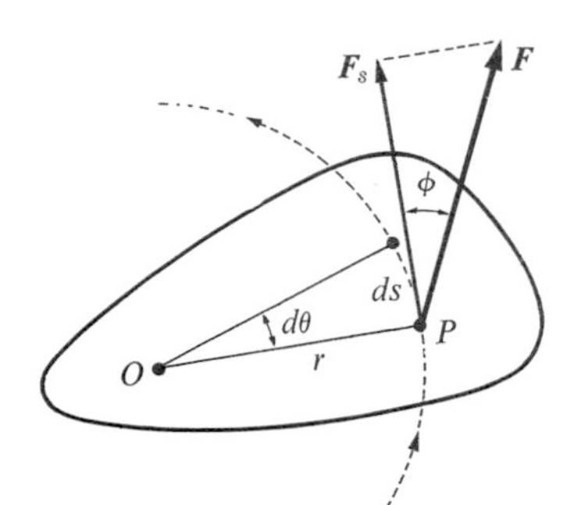

9-14 The work done by the force $\boldsymbol{F}$ in an angular displacement $d\theta$.

If more than one force acts on the body, the total work equals the work of the resultant moment.

From Eq. (9-15),

$$\Gamma = I\alpha = I\omega \frac{d\omega}{d\theta}.$$

Hence

$$\Gamma \, d\theta = I\omega \, d\omega,$$

and

$$W = \int_{\theta_1}^{\theta_2} \Gamma \, d\theta = \int_{\omega_1}^{\omega_2} I\omega \, d\omega = \tfrac{1}{2}I\omega_2^2 - \tfrac{1}{2}I\omega_1^2. \tag{9-23}$$

That is, *the work of the resultant moment equals the increase in kinetic energy*, in analogy with the work-energy equation for linear motion.

The power developed by the force Γ in Fig. 9-14, if $\boldsymbol{v}$ is the velocity of its point of application, is

$$P = F_S v = F_S r\omega,$$

and since $F_S r = \Gamma$,

$$\boxed{P = \Gamma\omega,} \tag{9-24}$$

the rotational analog of $P = F_S v$. The power output of an engine therefore equals the product of torque and angular velocity. The transmission of an automobile engine is often called a "torque converter." An automobile engine can deliver full power only when running at its rated angular velocity and delivering the corresponding torque. When accelerating, or climbing a steep hill, the torque converter permits the engine to run at rated speed, torque, and power, while the same power is transmitted to the drive shaft at a larger torque and a smaller angular velocity.

Example. The manufacturers of an automobile state that its engine develops 345 hp and a torque of 475 lb ft. What is the corresponding angular velocity?

$$\omega = \frac{P}{\Gamma} = \frac{345 \times 550 \text{ ft lb s}^{-1}}{475 \text{ lb ft}} = 400 \text{ rad s}^{-1} \approx 3800 \text{ rev min}^{-1}.$$

9-9 ANGULAR MOMENTUM

The equation

$$\Gamma = I\alpha$$

can be written as

$$\Gamma = I\frac{d\omega}{dt} = \frac{d}{dt}(I\omega). \tag{9-25}$$

The product of moment of inertia I and angular velocity ω is the rotational analog of the product of mass m and linear velocity v. The latter product is the linear momentum, and by analogy we call the product $I\omega$ the *angular momentum* L:

$$\boxed{L = I\omega.} \tag{9-26}$$

(Although in this special case of a rigid body rotating about a fixed axis the angular momentum is equal to $I\omega$, this is not the general definition of this quantity.)

Equation (9-25) can now be written

$$\Gamma = \frac{dL}{dt}, \tag{9-27}$$

or, *the resultant external torque is equal to the rate of change of angular momentum*, just as the resultant external force equals the rate of change of linear momentum.

Multiplying by dt and integrating, we get

$$\int_0^t \Gamma \, dt = L - L_0. \tag{9-28}$$

The integral $\int \Gamma\, dt$ is called the *angular impulse* of the torque and is analogous to the impulse of a force, $\int F\, dt$. The preceding equation is therefore the analog of the impulse-momentum principle in linear motion. That is, *the resultant angular impulse of the torque on a body is equal to the change in angular momentum of the body.*

Thus far we have considered only problems of rotation about a fixed axis, in which all of the forces were in a plane perpendicular to the axis. When the motion of a body does not take place about a fixed axis it is necessary to make use of the *vector moment of a force.*

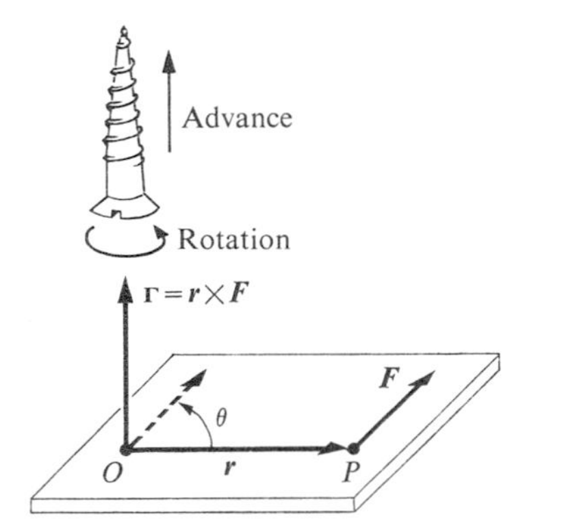

9-15 (a) Vector $\boldsymbol{\Gamma}$ is the vector moment of the force $\boldsymbol{F}$ about the point O: $\boldsymbol{\Gamma} = \boldsymbol{r} \times \boldsymbol{F}$.

In Fig. 9-15, a force $\boldsymbol{F}$, lying in the horizontal plane, is applied at the point P. Point O is at a perpendicular distance r from point P, and $\boldsymbol{r}$ is the vector from O to P. There may or may not be an axis through point O. If there *were* an axis through O, perpendicular to the plane of $\boldsymbol{r}$ and $\boldsymbol{F}$, the moment Γ of the force $\boldsymbol{F}$ about this axis would be

$$\Gamma = rF.$$

We define the *vector moment* of the force $\boldsymbol{F}$, *about the* **point** O, as a vector $\boldsymbol{\Gamma}$ whose magnitude equals the moment rF and whose direction is perpendicular to the plane of $\boldsymbol{r}$ and $\boldsymbol{F}$, as shown. The sense of $\boldsymbol{\Gamma}$ along the axis is specified by the *right-hand screw rule:* Rotate the first vector ($\boldsymbol{r}$) through the smaller angle (θ) that will bring it into parallelism with the second vector ($\boldsymbol{F}$), as shown by the broken line in the diagram. The vector $\boldsymbol{\Gamma}$ then points in the direction of *advance* of a right-hand screw when it is *rotated* through the same angle θ. The vector moment of $\boldsymbol{F}$ about point O has the same magnitude and direction whatever the direction of an actual axis through O, or even if there is no actual axis through this point.

Since $\boldsymbol{r}$ and $\boldsymbol{F}$ are vectors, the definition above suggests that we define a new product of two vectors called the *vector product* or *cross product*. The vector product of any two vectors $\boldsymbol{A}$ and $\boldsymbol{B}$ is written $\boldsymbol{A} \times \boldsymbol{B}$ and is defined as a *vector* of magnitude $AB \sin \theta$, where θ is the angle between $\boldsymbol{A}$ and $\boldsymbol{B}$. Then if $\boldsymbol{C}$ represents this product,

$$\boldsymbol{C} = \boldsymbol{A} \times \boldsymbol{B}, \qquad C = AB \sin \theta.$$

The direction and sense of the vector product is specified by the right-hand screw rule given above. The vector moment of the force $\boldsymbol{F}$ about the point O can therefore be written in general as

$$\boxed{\boldsymbol{\Gamma} = \boldsymbol{r} \times \boldsymbol{F},}$$

where

$$\Gamma = rF \sin \theta.$$

For the special case in Fig. 9-15, $\theta = 90°$ and $\sin \theta = 1$.

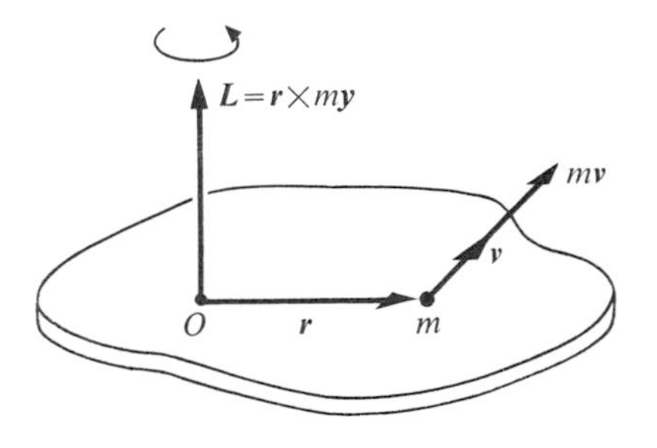

9-16 Vector $\boldsymbol{L}$ represents the moment of momentum, or angular momentum, of a particle of mass m about point O: $\boldsymbol{L} = \boldsymbol{r} \times m\boldsymbol{v}$.

Figure 9-16 shows a horizontal flat plate rotating about a fixed vertical axis through point O with an angular velocity ω. A particle of the plate of mass m, at a distance r from the axis, has a linear velocity $v = r\omega$ and a linear momentum mv. The angular momentum L

of the particle is defined as the product of the distance r and the linear momentum:

$$L = rmv.$$

Since angular momentum is defined in the same way as the moment of a force, it is also called the *moment of momentum*.

The *vector angular momentum* of the particle is defined as the *vector product* of $\boldsymbol{r}$ and $m\boldsymbol{v}$:

$$\boldsymbol{L} = \boldsymbol{r} \times m\boldsymbol{v}.$$

By the right-hand screw rule, the vector $\boldsymbol{L}$ points along the axis of rotation, as shown.

The vector angular momentum of any body is the *vector sum* of the angular momenta of the particles composing it. However, since in this case the angular momentum vectors of all particles are in the same direction, the *vector* sum becomes an *arithmetic* sum and the magnitude of the angular momentum of the plate is

$$L = \sum rmv.$$

But $v = r\omega$, and ω has the same value for all particles. Hence

$$L = (\sum mr^2)\omega = I\omega,$$

and this definition of angular momentum leads to the same expression for L as that previously stated.

Now suppose that an external torque Γ is exerted on the plate. In a small time interval Δt the angular momentum of the plate changes by ΔL, where $\Delta L = \Gamma\,\Delta t$.

Since the *vectors* $\Delta\boldsymbol{L}$ and $\boldsymbol{\Gamma}$ are both along the axis of rotation, this can be written as a *vector* equation:

$$\Delta\boldsymbol{L} = \boldsymbol{\Gamma}\,\Delta t. \tag{9-29}$$

An analysis of the general case, where $\boldsymbol{L}$ and $\boldsymbol{\Gamma}$ may not be in the same direction, shows that the result above is always true. That is, *the* **vector change** *in angular momentum of a body, $\Delta\boldsymbol{L}$, is equal in magnitude and direction to the vector impulse $\boldsymbol{\Gamma}\,\Delta t$ of the resultant external torque on the body.*

Let us now apply these vector concepts of torque and angular momentum to a specific example. In Fig. 9-17, a disk is mounted on a shaft through its center, the shaft being supported by fixed bearings. If the disk is rotating as indicated, its angular momentum vector $\boldsymbol{L}$ points toward the right, along the axis.

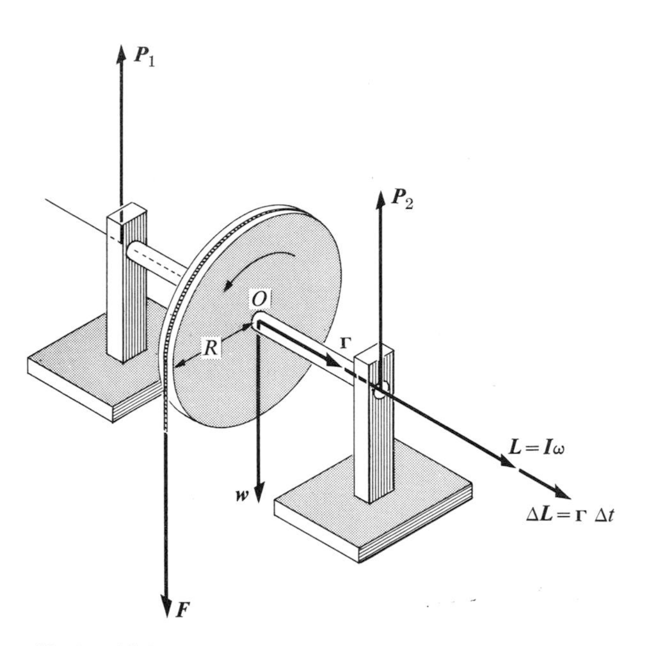

9-17 Vector $\Delta\boldsymbol{L}$ is the change in angular momentum produced in time Δt by the moment $\boldsymbol{\Gamma}$ of the force $\boldsymbol{F}$. Vectors $\Delta\boldsymbol{L}$ and $\boldsymbol{\Gamma}$ are in the same direction.

Suppose a cord is wrapped around the rim of the disk and a force $\boldsymbol{F}$ is exerted on the cord. The magnitude of the resultant torque on the disk is $\Gamma = FR$, and the torque vector $\boldsymbol{\Gamma}$ also points along the axis. In a time Δt, the torque produces a vector change $\Delta\boldsymbol{L}$ in the angular momentum, equal to $\boldsymbol{\Gamma}\,\Delta t$ and having the same direction as $\boldsymbol{\Gamma}$. When this change is added vectorially to the original angular momentum $\boldsymbol{L}$, the resultant is a vector of length $\boldsymbol{L} + \Delta\boldsymbol{L}$, in the same direction as $\boldsymbol{L}$. In other words, the *magnitude* of the angular momentum is increased, its *direction* remaining the same. An increase in the magnitude of the angular momentum simply means that the body rotates more rapidly.

The lengthy argument above appears at first to be nothing more than a difficult way of solving an easy problem in rotation about a fixed axis. However, the vector nature of torque and angular momentum are essential to an understanding of the gyroscope, to be discussed in the next section.

If the resultant external torque on a body is zero, then from Eq. (9–29) $\Delta \boldsymbol{L}$ is zero and *the angular momentum vector remains constant in magnitude and direction* (rotational analog of Newton's first law). This is the principle of *conservation of angular momentum*, and it ranks with the principles of conservation of linear momentum and conservation of energy as one of the most fundamental relations of mechanics.

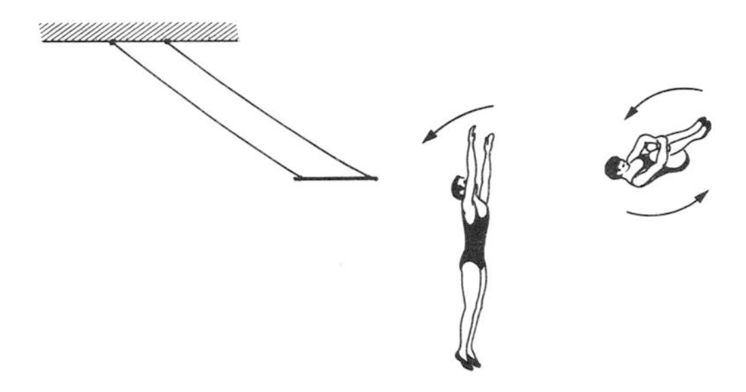

9–18 Conservation of angular momentum.

A circus acrobat, a diver, or a skater performing a pirouette on the toe of one skate, all take advantage of the principle. Suppose an acrobat has just left a swing, as in Fig. 9–18, with arms and legs extended and with a small counterclockwise angular momentum. When he pulls his arms and legs in, his moment of inertia I becomes much smaller. Since his angular momentum $I\omega$ remains constant and I decreases, his angular velocity ω increases.

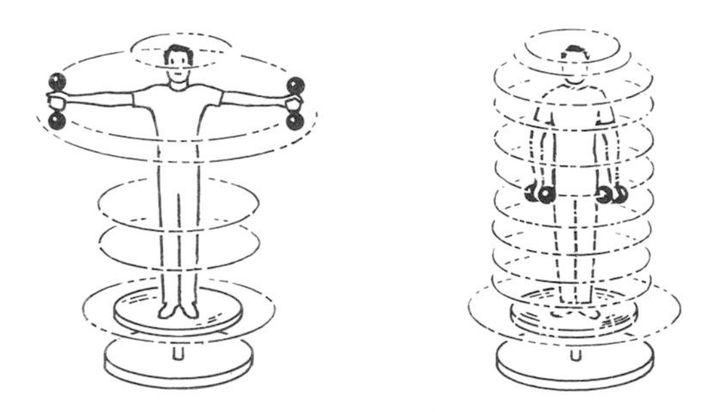

9–19 Conservation of angular momentum about a fixed axis.

Example. A man stands at the center of a turntable, holding his arms extended horizontally with a 10-lb weight in each hand, as shown in Fig. 9–19. He is set rotating about a vertical axis with an angular velocity of one revolution in 2 s. Find his new angular velocity if he drops his hands to his sides. The moment of inertia of the man may be assumed constant and equal to 4 slug ft^2. The original distance of the weights from the axis is 3 ft, and their final distance is 6 in.

If friction in the turntable is neglected, no external torques act about a vertical axis and the angular momentum about this axis is constant. That is,

$$I\omega = (I\omega)_0 = I_0\omega_0,$$

where I and ω are the final moment of inertia and angular velocity, and I_0 and ω_0 are the initial values of these quantities:

$$I = I_{\text{man}} + I_{\text{weights}},$$
$$I = 4 + 2(\tfrac{10}{32})(\tfrac{1}{2})^2 = 4.16 \text{ slug ft}^2,$$
$$I_0 = 4 + 2(\tfrac{10}{32})(3)^2 = 9.63 \text{ slug ft}^2,$$
$$\omega_0 = \pi \text{ rad s}^{-1},$$
$$\omega = \omega_0 \frac{I_0}{I} = 2.31\pi \text{ rad s}^{-1}.$$

That is, the angular velocity is more than doubled.

9–10 ROTATION ABOUT A MOVING AXIS. THE TOP AND THE GYROSCOPE

Figure 9–20 illustrates the usual mounting of a toy gyroscope, more properly called a top, since the fixed point O is not at the center of mass. The top is spinning about its axis of symmetry, and if the axis is initially set in motion in the direction shown, with the proper angular velocity, the system continues to rotate uniformly about the pivot at O, the spin axis remaining horizontal.

If the axis of the top were fixed in space, its angular momentum would equal the product of its moment of inertia about the axis and its angular velocity about the axis, and would point along the axis. Because the axis itself is rotating, the angular momentum vector no longer lies on the axis. However, if the angular velocity *of* the axis is small compared with the angular velocity *about* the axis, the component of angular momentum arising from the former effect is small, and we shall neglect it. The angular momentum vector $\boldsymbol{L}$, about the fixed point O, can then be drawn along the axis as shown, and as the top rotates about O its angular momentum vector rotates with it.

The upward force $\boldsymbol{P}$ at the pivot has no moment about O. The resultant external moment is that due to

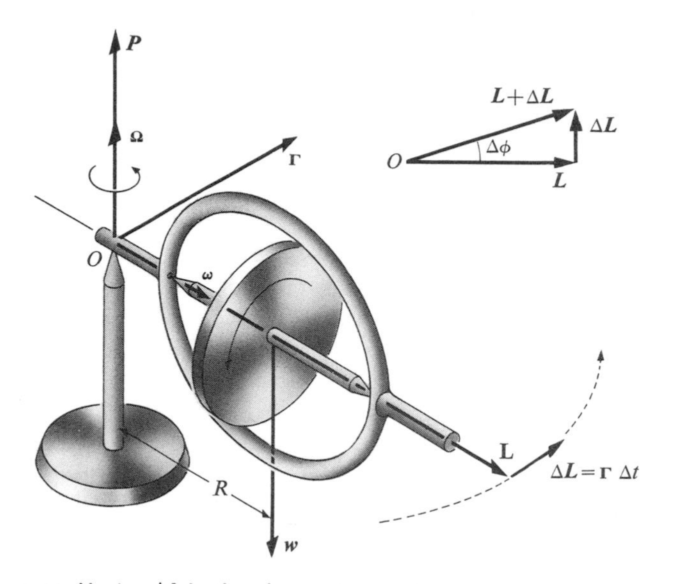

9-20 Vector $\Delta \boldsymbol{L}$ is the change in angular momentum produced in time Δt by the moment $\boldsymbol{\Gamma}$ of the force $\boldsymbol{w}$. Vectors $\Delta \boldsymbol{L}$ and $\boldsymbol{\Gamma}$ are in the same direction. (Compare with Fig. 9-17.)

the weight $\boldsymbol{w}$ and its magnitude is

$$\Gamma = wR.$$

The direction of $\boldsymbol{\Gamma}$ is perpendicular to the axis, as shown. In a time Δt (compare with the analysis of Fig. 9-17) this torque produces a change $\Delta \boldsymbol{L}$ in the angular momentum, having the same direction as $\boldsymbol{\Gamma}$ and given by

$$\Delta \boldsymbol{L} = \boldsymbol{\Gamma} \, \Delta t.$$

The angular momentum $\boldsymbol{L} + \Delta \boldsymbol{L}$, after a time Δt, is the vector sum of $\boldsymbol{L}$ and $\Delta \boldsymbol{L}$. Since $\Delta \boldsymbol{L}$ is perpendicular to $\boldsymbol{L}$, the new angular momentum vector has the same *magnitude* as the old but a different *direction*. The tip of the angular momentum vector moves as shown, and as time goes on it swings around a horizontal circle. But since the angular momentum vector lies along the gyroscope axis, the axis turns also, rotating in a horizontal plane about the point O. This motion of the axis of rotation is called *precession*.

The angle $\Delta\phi$ turned through by the vector $\boldsymbol{L}$ in time Δt (see the small inset diagram) is

$$\Delta\phi = \frac{\Delta L}{L}.$$

The *angular velocity of precession*, Ω, is

$$\Omega = \lim_{\Delta t \to 0} \frac{\Delta\phi}{\Delta t} = \frac{1}{L}\frac{dL}{dt}.$$

But

$$\frac{dL}{dt} = \Gamma,$$

so

$$\Omega = \frac{\Gamma}{L}, \qquad \Gamma = \Omega L. \tag{9-30}$$

The angular velocity of precession is therefore inversely proportional to the angular momentum. If this is large, the precessional angular velocity will be small.

The *vector* angular velocity of precession, $\boldsymbol{\Omega}$, by the usual right-hand screw rule, points upward as shown. Equation (9-30) can therefore be written as a vector product:

$$\boldsymbol{\Gamma} = \boldsymbol{\Omega} \times \boldsymbol{L}.$$

To within the approximation that the magnitude of the angular momentum is equal to $I\omega$, we can write

$$\Gamma = I\omega\Omega. \tag{9-31}$$

It will be noted from Fig. 9-20 that the change in direction of the vector $\boldsymbol{L}$ is such as to swing it toward the direction of the torque vector $\boldsymbol{\Gamma}$. This is always the case, and we say that "the angular momentum vector chases the torque vector." In this particular system, the torque vector also turns as the momentum vector turns, so the latter never catches up with the former. In other arrangements, the angular momentum vector eventually becomes aligned with the torque vector and the precessional motion then ceases.

From a purely kinematic point of view, precessional motion of a top is the rotational analog of uniform circular motion of a particle, as pointed out by Benfield in 1958. The analogy is illustrated in Fig. 9-21, where the diagram and calculation for uniform circular motion are shown on the left and the corresponding diagram and calculation for precessional motion of a top are displayed on the right.

Why doesn't the top in Fig. 9-20 fall? The answer is that the upward force $\boldsymbol{P}$ exerted on it by the pivot is just equal in magnitude to its weight $\boldsymbol{w}$, so that the resultant

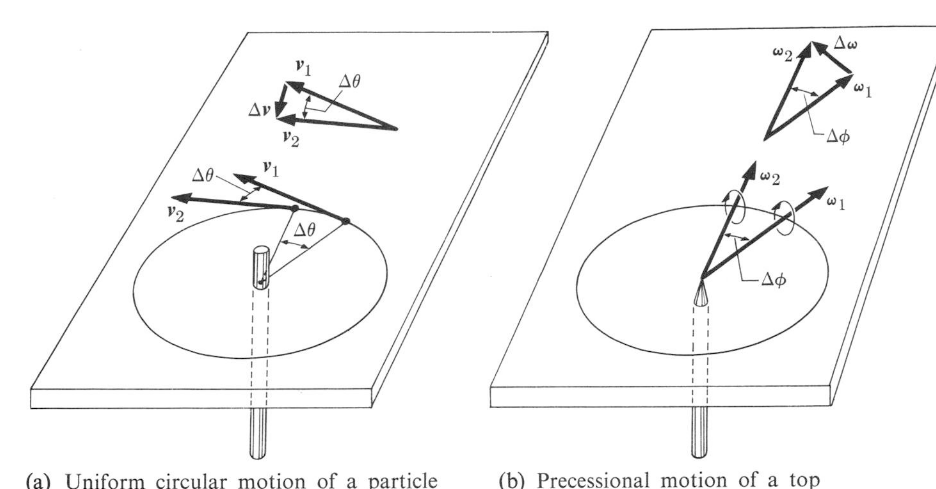

(a) Uniform circular motion of a particle (b) Precessional motion of a top

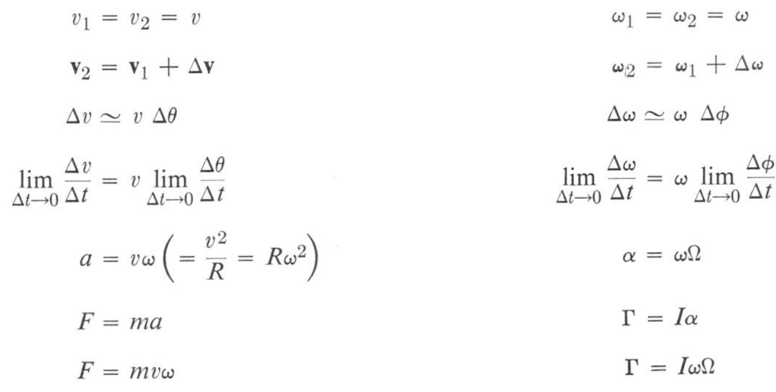

9–21 Analogy between uniform circular motion of a particle and precessional motion of a top.

vertical *force* is zero and the vertical acceleration of the center of gravity is zero. In other words, the vertical component of its linear momentum remains zero, since there is no resultant vertical force. The two forces $\boldsymbol{P}$ and $\boldsymbol{w}$ constitute a *couple* of moment $\Gamma = wR$, so the resultant moment is *not* zero and the angular momentum changes.

If the top were not rotating, it would have no angular momentum $\boldsymbol{L}$ to start with. Its angular momentum $\Delta\boldsymbol{L}$ after a time Δt would be that acquired from the couple acting on it and would be in the same direction as the moment $\boldsymbol{\Gamma}$ of this couple. In other words, the top would rotate about an axis through O in the direction of the vector $\boldsymbol{\Gamma}$. But if the top is originally rotating, the change in its angular momentum produced by the couple adds vectorially to the large angular momentum it already has, and since $\Delta\boldsymbol{L}$ is horizontal and perpendicular to $\boldsymbol{L}$, the result is a motion of precession with both the angular momentum vector and the axis remaining horizontal.

To understand why the vertical force P should equal w, we must look further into the way in which the precessional motion in Fig. 9–20 originated. If the frame of the top is initially held at rest, say by supporting the projecting portion of the frame opposite O with one's finger, the upward forces exerted by the finger and by the pivot are each equal to $w/2$. If the finger is suddenly removed, the upward force at O, at the first instant, is still $w/2$. The resultant vertical *force* is not zero, and the center of gravity has an initial downward acceleration. At the same time, precessional motion begins, although with a smaller angular velocity than that in the final steady state. The result of this motion is to cause the end of the frame at O to press down on the pivot with a greater force, so that the upward force at O increases and eventually becomes greater than w. When this happens, the center of gravity starts to accelerate upward. The process repeats itself and the motion consists of a precession together with an up-and-down oscillation of the axis, called *nutation*.

To start the top off with pure precession, it is necessary to give the outer end of the axis a push in the direction in which it would normally precess. This causes the end of the frame at O to bear down on the pivot so that the upward force at O increases. When this force equals w, the vertical forces are in equilibrium, the outer end can be released, and precession goes on as in Fig. 9–20.

Figure 9–22 is a photograph of a gyroscope mounted in gimbals. Except when two or more of the gimbal rings lie in the same plane, the outer frame can be turned in any direction in space without exerting a torque on the gyro wheel, except for the small frictional torques in the pivots. If the outer frame is fixed, the center of gravity of the gyro remains fixed whatever the orientation of its axis.

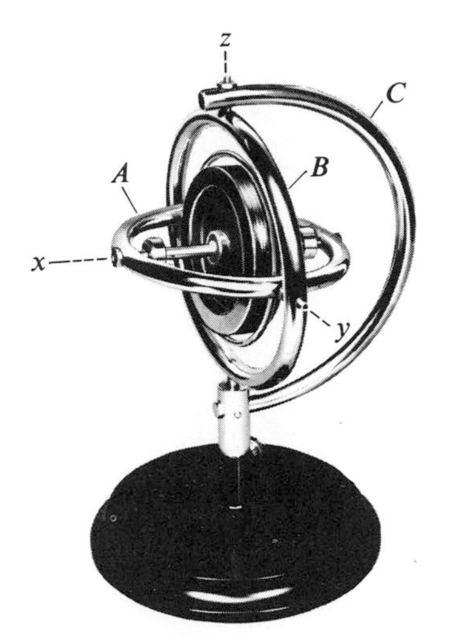

9–22 Photograph of a gimbal-mounted gyro. [Courtesy of Sperry Gyroscope Co.]

Problems

9–1 An electric motor running at 1800 rev min^{-1} has on its shaft three pulleys, of diameters 2, 4, and 6 in. respectively. Find the linear velocity of the surface of each pulley in ft s^{-1}. The pulleys may be connected by a belt to a similar set on a countershaft, the 2 in. to the 6 in., the 4 in. to the 4 in., and the 6 in. to the 2 in. Find the three possible angular velocities of the countershaft in revolutions per minute.

9–2 The angular velocity of a flywheel decreases uniformly from 1000 rev min^{-1} to 400 rev min^{-1} in 5 s. Find the angular acceleration and the number of revolutions made by the wheel in the 5-s interval. How many more seconds are required for the wheel to come to rest?

9–3 A flywheel requires 3 s to rotate through 234 rad. Its angular velocity at the end of this time is 108 rad s^{-1}. Find its constant angular acceleration.

9–4 A flywheel whose angular acceleration is constant and equal to 2 rad s^{-2}, rotates through an angle of 100 rad in 5 s. How long had it been in motion at the beginning of the 5-s interval if it started from rest?

9–5 (a) Distinguish clearly between tangential and radial acceleration. (b) A flywheel rotates with constant angular velocity. Does a point on its rim have a tangential acceleration? a radial acceleration? (c) A flywheel is rotating with constant angular acceleration. Does a point on its rim have a tangential acceleration? a radial acceleration? Are these accelerations constant in magnitude?

9–6 A wheel 30 in. in diameter is rotating about a fixed axis with an initial angular velocity of 2 rev s^{-1}. The acceleration is 3 rev s^{-2}. (a) Compute the angular velocity after 6 s. (b) Through what angle has the wheel turned in this time interval? (c) What is the tangential velocity of a point on the rim of the wheel at $t = 6$ s? (d) What is the resultant acceleration of a point on the rim of the wheel at $t = 6$ s?

9–7 A wheel having a diameter of 1 ft starts from rest and accelerates uniformly to an angular velocity of 900 rev min^{-1} in 5 s. (a) Find the position at the end of 1 s of a point originally at the top of the wheel. (b) Compute and show in a diagram the magnitude and direction of the acceleration at the end of 1 s.

9-8 A flywheel of radius 30 cm starts from rest and accelerates with a constant angular acceleration of 0.50 rad s^{-2}. Compute the tangential acceleration, the radial acceleration, and the resultant acceleration, of a point on its rim (a) at the start, (b) after it has turned through 120°, (c) after it has turned through 240°.

9-9 A wheel starts from rest and accelerates uniformly to an angular velocity of 900 rev min^{-1} in 20 s. At the end of 1 s, (a) find the angle through which the wheel has rotated, and (b) compute and show in a diagram the magnitude and direction of the tangential and radial components of acceleration of a point 6 in. from the axis.

9-10 An automobile engine is idling at 500 rev min^{-1}. When the accelerator is depressed, the angular velocity increases to 3000 rev min^{-1} in 5 s. Assume a constant angular acceleration. (a) What are the initial and final angular velocities, expressed in radians per second? (b) What was the angular acceleration, in radians per second squared? (c) How many revolutions did the engine make during the acceleration period? (d) The flywheel of the engine is 18 in. in diameter. What is the linear speed of a point at its rim when the angular speed is 3000 rev min^{-1}? (e) What was the tangential acceleration of the point during the acceleration period? (f) What is the radial acceleration of the point when the angular speed is 3000 rev min^{-1}?

9-11 Find the required angular velocity of an ultracentrifuge, in revolutions per minute, in order that the radial acceleration of a point 1 cm from the axis shall equal 300,000g (i.e., 300,000 times the acceleration due to gravity).

9-12 (a) Prove that when a body starts from rest and rotates about a fixed axis with constant angular acceleration, the radial acceleration of a point in the body is directly proportional to its angular displacement. (b) Through what angle will the body have turned when the resultant acceleration makes an angle of 60° with the radial acceleration?

9-13 A particle moves in the xy-plane according to the law

$$x = R \cos \omega t, \qquad y = R \sin \omega t,$$

where x and y are the coordinates of the body, t is the time, and R and ω are constants. (a) Eliminate t between these equations to find the equation of the curve in which the body moves. [*Hint:* Square each equation.] What is this curve? (b) Differentiate the original equations to find the x- and y-components of the velocity of the particle. Combine these expressions to obtain the magnitude and direction of the resultant velocity. (c) Differentiate again to obtain the magnitude and direction of the resultant acceleration. (This problem illustrates an alternative method of deriving the expressions for the speed and radial acceleration of a particle moving in a circle.)

9-14 Find the moment of inertia of a rod 4 cm in diameter and 2 m long, of mass 8 kg, (a) about an axis perpendicular to the rod and passing through its center, (b) about an axis perpendicular to the rod and passing through one end, (c) about a longitudinal axis through the center of the rod.

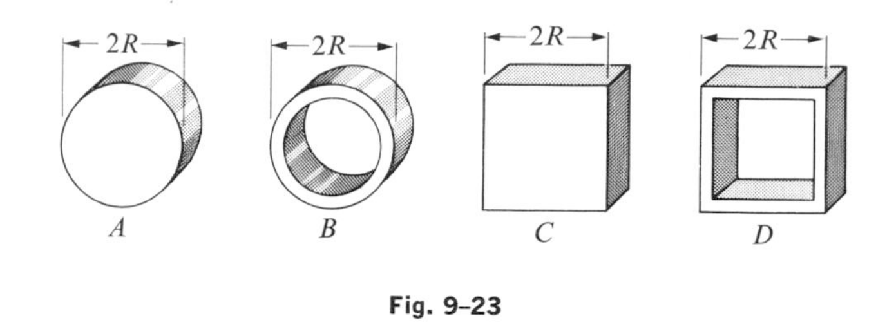

Fig. 9-23

9-15 The four bodies shown in Fig. 9-23 have equal masses m. Body A is a solid cylinder of radius R. Body B is a hollow thin cylinder of radius R. Body C is a solid square with length of side $= 2R$. Body D is the same size as C, but hollow (i.e., made up of four thin walls). The bodies have axes of rotation perpendicular to the page and through the center of gravity of each body. (a) Which body has the smallest moment of inertia? (b) Which body has the largest moment of inertia?

9-16 Small blocks, each of mass m, are clamped at the ends and at the center of a light rigid rod of length ℓ. Compute the moment of inertia and the radius of gyration of the system about an axis perpendicular to the rod and passing through a point one-quarter of the length from one end. Neglect the moment of inertia of the rod.

9-17 The radius of the earth is 4000 mi and its mass is 4×10^{23} slugs (approximately). Find (a) its moment of inertia about an axis through its center, and (b) its radius of gyration in miles. Assume the density to be uniform.

9-18 A flywheel consists of a solid disk 1 ft in diameter and 1 in. thick, and two projecting hubs 4 in. in diameter and 3 in. long. If the material of which it is constructed weighs 480 lb ft^{-3}, find (a) its moment of inertia, and (b) its radius of gyration about the axis of rotation.

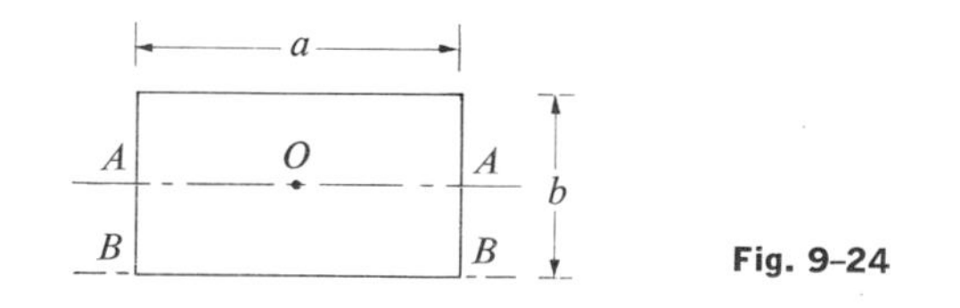

Fig. 9-24

9-19 The uniform thin rectangular plate in Fig. 9-24 has a length a, a width b, and a mass m. Find its moment of inertia (a) about the axis AA through its center O, (b) about the axis BB at one edge.

9-20 (a) Prove that the moment of inertia of the thin flat plate in Fig. 9-25, about the z-axis, equals the sum of its moments of inertia about the x- and y-axes. (b) Given that the moment of inertia of a disk about an axis through its center and perpendicular to its plane is $mR^2/2$, use the relation above to find its moment of inertia about a diameter. (c) Derive the result of part (b) by direct integration of the defining equation, $I = \int r^2\, dm$. (d) What is the moment of inertia of a disk about an axis tangent to its edge?

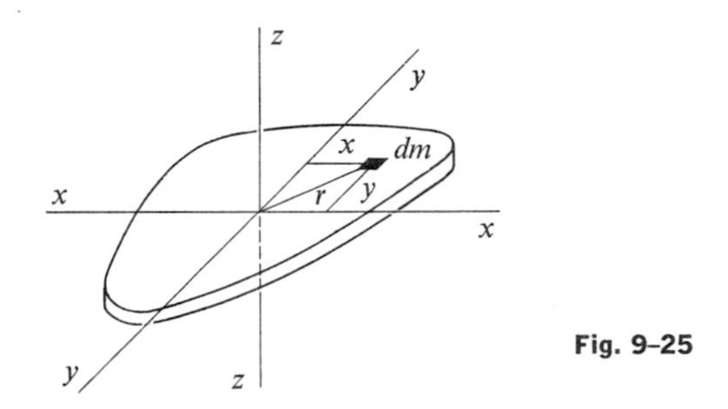

Fig. 9-25

9-21 A grindstone 3 ft in diameter, weighing 96 lb, is rotating at 900 rev min^{-1}. A tool is pressed normally against the rim with a force of 45 lb, and the grindstone comes to rest in 10 s. Find the coefficient of friction between the tool and the grindstone. Neglect friction in the bearings.

9-22 A 128-lb grindstone is 3 ft in diameter and has a radius of gyration of $\frac{3}{4}$ ft. A tool is pressed down on the rim with a normal force of 10 lb. The coefficient of sliding friction between the tool and stone is 0.6 and there is a constant friction torque of 3 lb ft between the axle of the stone and its bearings. (a) How much force must be applied normally at the end of a crank handle 15 in. long to bring the stone from rest to 120 rev min^{-1} in 9 s? (b) After attaining a speed of 120 rev min^{-1}, what must the normal force at the end of the handle become to maintain a constant speed of 120 rev min^{-1}? (c) How long will it take the grindstone to come from 120 rev min^{-1} to rest if it is acted on by the axle friction alone?

9-23 A constant torque of 20 N m is exerted on a pivoted wheel for 10 s, during which time the angular velocity of the wheel increases from zero to 100 rev min^{-1}. The external torque is then removed and the wheel is brought to rest by friction in its bearings in 100 s. Compute (a) the moment of inertia of the wheel, (b) the friction torque, and (c) the total number of revolutions made by the wheel.

9-24 A cord is wrapped around the rim of a flywheel 2 ft in radius, and a steady pull of 10 lb is exerted on the cord, as in

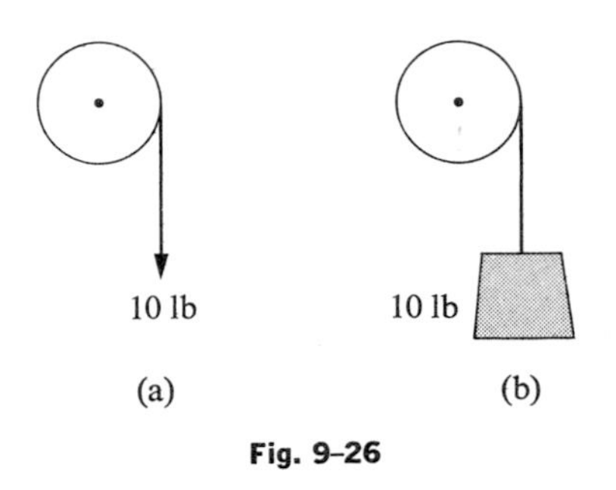

Fig. 9-26

Fig. 9-26(a). The wheel is mounted in frictionless bearings on a horizontal shaft through its center. The moment of inertia of the wheel is 2 slug ft^2. (a) Compute the angular acceleration of the wheel. (b) Show that the work done in unwinding 20 ft of cord equals the gain in kinetic energy of the wheel. (c) If a 10-lb weight hangs from the cord as in Fig. 9-26(b), compute the angular acceleration of the wheel. Why is this not the same as in part (a)?

9-25 A solid cylinder of mass 15 kg, 30 cm in diameter, is pivoted about a horizontal axis through its center, and a rope wrapped around the surface of the cylinder carries at its end a block of mass 8 kg. (a) How far does the block descend in 5 s, starting from rest? (b) What is the tension in the rope? (c) What is the force exerted on the cylinder by its bearings?

9-26 A bucket of water weighing 64 lb is suspended by a rope wrapped around a windlass in the form of a solid cylinder 1 ft in diameter, also weighing 64 lb. The bucket is released from rest at the top of a well and falls 64 ft to the water. (a) What is the tension in the rope while the bucket is falling? (b) With what velocity does the bucket strike the water? (c) What was the time of fall? Neglect the weight of the rope.

9-27 A 16-lb block rests on a horizontal frictionless surface. A cord attached to the block passes over a pulley, whose diameter is 6 in., to a hanging block which also weighs 16 lb. The system is released from rest, and the blocks are observed to move 16 ft in 2 s. (a) What was the moment of inertia of the pulley? (b) What was the tension in each part of the cord?

9-28 Figure 9-27 represents an Atwood's machine. Find the linear accelerations of blocks A and B, the angular acceleration of the wheel C, and the tension in each side of the cord (a) if the surface of the wheel is frictionless, (b) if there is no slipping between the cord and the surface of the wheel. Let the weights of blocks A and B be 8 lb and 4 lb, respectively, the moment of inertia of the wheel about its axis be 0.125 slug ft^2, and the radius of the wheel be 0.5 ft.

9-29 A flywheel 3 ft in diameter is pivoted on a horizontal axis. A rope is wrapped around the outside of the flywheel and a steady

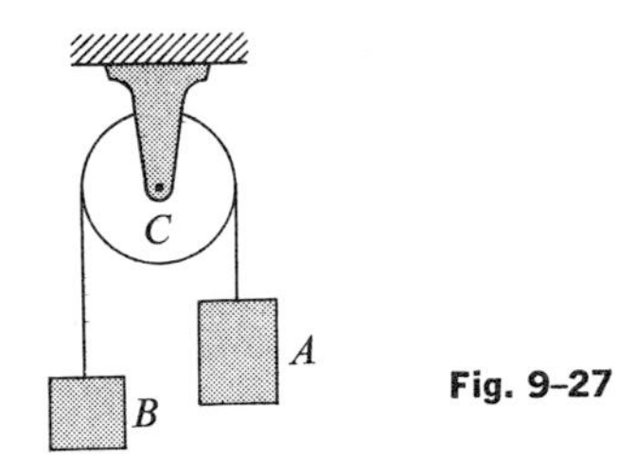

Fig. 9–27

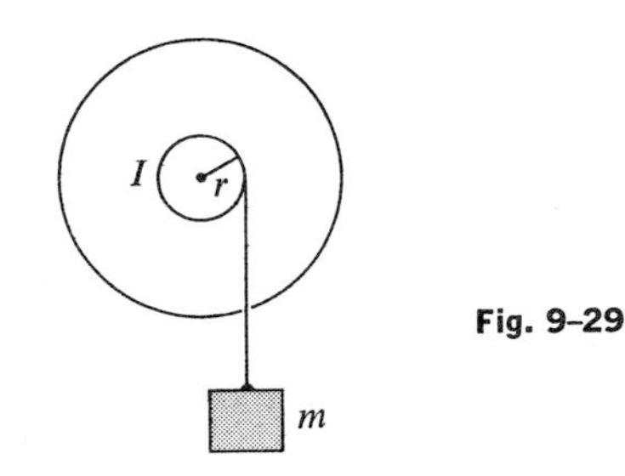

Fig. 9–29

pull of 10 lb is exerted on the rope. It is found that 24 ft of rope are unwound in 4 s. (a) What was the angular acceleration of the flywheel? (b) What was its final angular velocity? (c) What was its final kinetic energy? (d) What is its moment of inertia?

9–30 A light rigid rod 100 cm long has a small block of mass 50 g attached at one end. The other end is pivoted, and the rod rotates in a vertical circle. At a certain instant the rod makes an angle of 53° with the vertical, and the tangential speed of the block is 400 cm s^{-1}. (a) What are the horizontal and vertical components of the velocity of the block? (b) What is the moment of inertia of the system? (c) What is the radial acceleration of the block? (d) What is the tangential acceleration of the block? (e) What is the tension or compression in the rod?

9–31 (a) Compute the torque developed by an airplane engine whose output is 2000 hp at an angular velocity of 2400 rev min^{-1}. (b) If a drum 18 in. in diameter were attached to the motor shaft, and the power output of the motor were used to raise a weight hanging from a rope wrapped around the shaft, how large a weight could be lifted? (c) With what velocity would it rise?

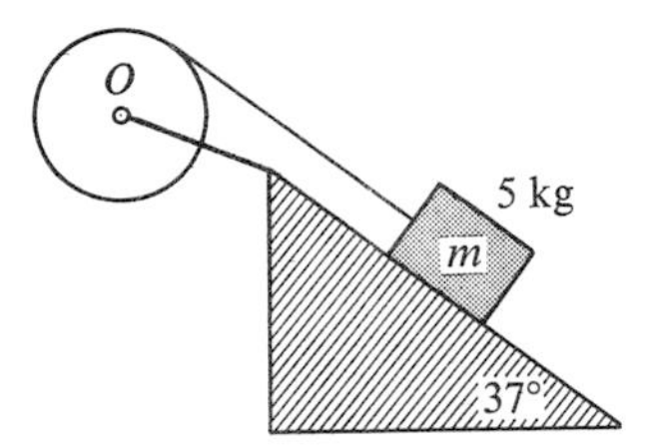

Fig. 9–28

9–32 A block of mass $m = 5$ kg slides down a surface inclined 37° to the horizontal, as shown in Fig. 9–28. The coefficient of sliding friction is 0.25. A string attached to the block is wrapped around a flywheel on a fixed axis at O. The flywheel has a mass $M = 20$ kg, an outer radius $R = 0.2$ m, and a radius of gyration with respect to the axis $k_O = 0.1$ m. (a) What is the acceleration of the block down the plane? (b) What is the tension in the string?

9–33 A body of mass m is attached to a light cord wound around the shaft of a wheel, as in Fig. 9–29. The radius of the shaft is r, and the shaft is supported in fixed frictionless bearings. When released from rest, the body descends a distance of 175 cm in 5 s. Find the moment of inertia of the wheel and shaft, in terms of m and r.

9–34 A disk of mass m and radius R is pivoted about a horizontal axis through its center, and a small body of mass m is attached to the rim of the disk. If the disk is released from rest with the small body at the end of a horizontal radius, find the angular velocity when the small body is at the bottom.

9–35 The flywheel of a gasoline engine is required to give up 380 ft lb of kinetic energy while its angular velocity decreases from 600 rev min^{-1} to 540 rev min^{-1}. What moment of inertia is required?

9–36 The flywheel of a punch press has a moment of inertia of 15 slug ft^2 and it runs at 300 rev min^{-1}. The flywheel supplies all the energy needed in a quick punching operation. (a) Find the speed in revolutions per minute to which the flywheel will be reduced by a sudden punching operation requiring 4500 ft lb of work. (b) What must be the constant power supply to the flywheel in horsepower to bring it back to its initial speed in 5 s?

9–37 A magazine article described a passenger bus in Zurich, Switzerland, which derived its motive power from the energy stored in a large flywheel. The wheel was brought up to speed periodically, when the bus stopped at a station, by an electric motor which could then be attached to the electric power lines. The flywheel was a solid cylinder of mass 1000 kg, diameter 180 cm, and its top speed was 3000 rev min^{-1}. (a) At this speed, what is the kinetic energy of the flywheel? (b) If the average power required to operate the bus is 25 hp, how long can it operate between stops?

9–38 A grindstone in the form of a solid cylinder has a radius of 2 ft and weighs 96 lb. (a) What torque will bring it from rest to an angular velocity of 300 rev min^{-1} in 10 s? (b) What is its kinetic energy when rotating at 300 rev min^{-1}?

9–39 The flywheel of a motor weighs 640 lb and has a radius of gyration of 4 ft. The motor develops a constant torque of 1280 lb ft, and the flywheel starts from rest. (a) What is the angular acceleration of the flywheel? (b) What will be its angular velocity after making 4 revolutions? (c) How much work is done by the motor during the first 4 revolutions?

9–40 The flywheel of a stationary engine has a moment of inertia of 20 slugs ft^2. (a) What constant torque is required to bring it up to an angular velocity of 900 rev min^{-1} in 10 s, starting from rest? (b) What is its final kinetic energy?

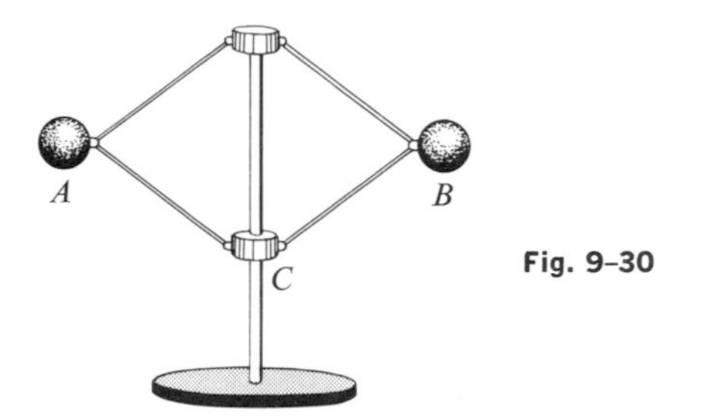

Fig. 9–30

9–41 In Fig. 9–30, the steel balls *A* and *B* have a mass of 500 g each, and are rotating about the vertical axis with an angular velocity of 4 rad s^{-1} at a distance of 15 cm from the axis. Collar *C* is now forced down until the balls are at a distance of 5 cm from the axis. How much work must be done to move the collar down?

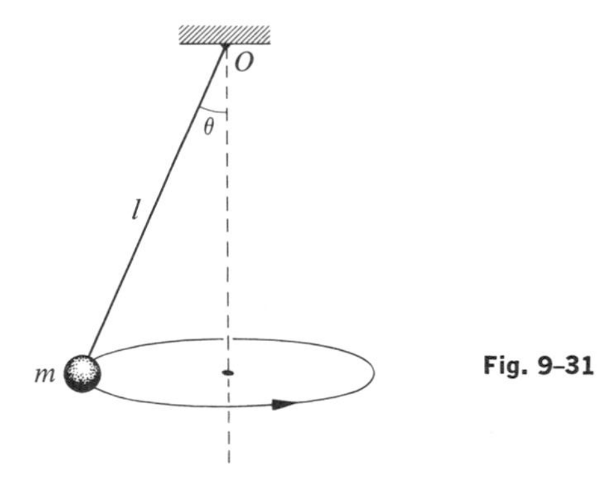

Fig. 9–31

9–42 A particle of mass *m*, attached to a cord of length ℓ, rotates as a conical pendulum, as shown in Fig. 9–31. What is the axial angular momentum of the particle about a fixed vertical axis through point *O*?

9–43 A man sits on a piano stool holding a pair of dumbbells at a distance of 3 ft from the axis of rotation of the stool. He is given an angular velocity of 2 rad s^{-1}, after which he pulls the dumbbells in until they are but 1 ft distant from the axis. The moment of inertia of the man about the axis of rotation is 3 slugs ft^2 and may be considered constant. The dumbbells weigh 16 lb each and may be considered point masses. Neglect friction. (a) What is the initial angular momentum of the system? (b) What is the angular velocity of the system after the dumbbells are pulled in toward the axis? (c) Compute the kinetic energy of the system before and after the dumbbells are pulled in. Account for the difference, if any.

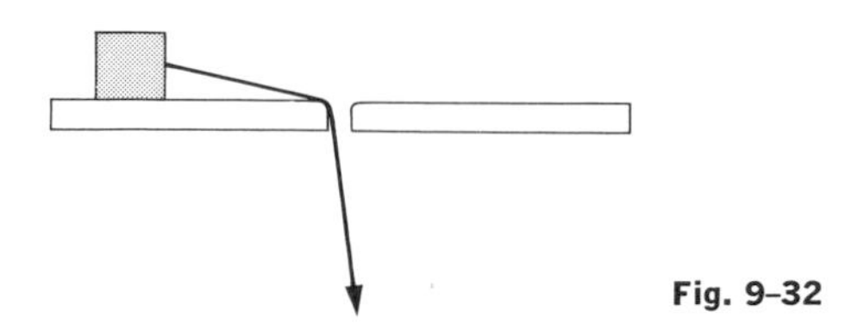

Fig. 9–32

9–44 A block of mass 50 g is attached to a cord passing through a hole in a horizontal frictionless surface as in Fig. 9–32. The block is originally revolving at a distance of 20 cm from the hole with an angular velocity of 3 rad s^{-1}. The cord is then pulled from below, shortening the radius of the circle in which the block revolves to 10 cm. The block may be considered a point mass. (a) What is the new angular velocity? (b) Find the change in kinetic energy of the block.

9–45 A small block weighing 8 lb is attached to a cord passing through a hole in a horizontal frictionless surface. The block is originally revolving in a circle of radius 2 ft about the hole with a tangential velocity of 12 ft s^{-1}. The cord is then pulled slowly from below, shortening the radius of the circle in which the block revolves. The breaking strength of the cord is 144 lb. What will be the radius of the circle when the cord breaks?

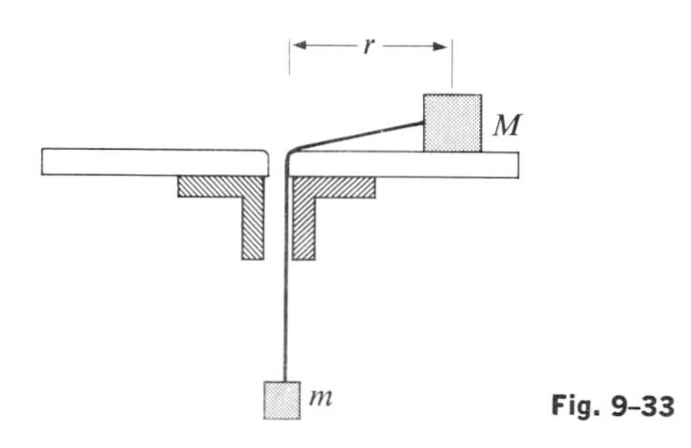

Fig. 9–33

9–46 A block of mass *M* rests on a turntable which is rotating at constant angular velocity ω. A smooth cord runs from the block through a hole in the center of the table down to a hanging block of mass *m*. The coefficient of friction between the first block and the turntable is μ. (See Fig. 9–33.) Find the largest and smallest values of the radius *r* for which the first block will remain at rest relative to the turntable.

9–47 A uniform rod of mass 30 g and 20 cm long rotates in a horizontal plane about a fixed vertical axis through its center. Two small bodies, each of mass 20 g, are mounted so that they can slide along the rod. They are initially held by catches at positions 5 cm on each side of the center of the rod, and the system is rotating at 15 rev min^{-1}. Without otherwise changing the system, the catches are released and the masses slide outward along the rod and fly off at the ends. (a) What is the angular velocity of the system at the instant when the small masses reach the ends of the rod? (b) What is the angular velocity of the rod after the small masses leave it?

9–48 A turntable rotates about a fixed vertical axis, making one revolution in 10 s. The moment of inertia of the turntable about this axis is 720 slugs ft^2. A man weighing 160 lb, initially standing at the center of the turntable, runs out along a radius. What is the angular velocity of the turntable when the man is 6 ft from the center?

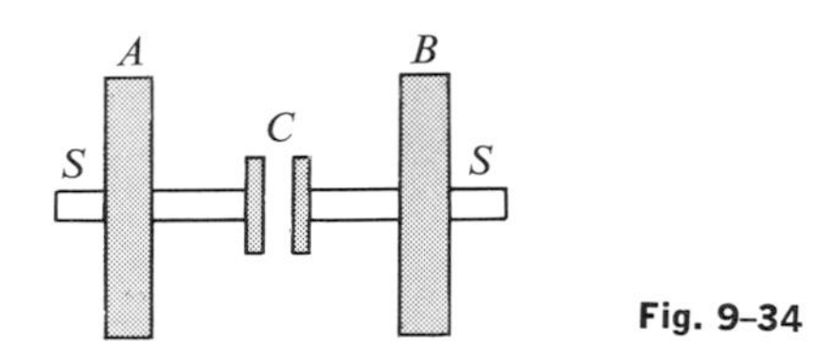

Fig. 9–34

9–49 Disks A and B are mounted on a shaft SS and may be connected or disconnected by a clutch C, as in Fig. 9–34. The moment of inertia of disk A is one-half that of disk B. With the clutch disconnected, A is brought up to an angular velocity ω_0. The accelerating torque is then removed from A and it is coupled to disk B by the clutch. Bearing friction may be neglected. It is found that 3000 ft lb of heat are developed in the clutch when the connection is made. What was the original kinetic energy of disk A?

9–50 A man weighing 160 lb stands at the rim of a turntable of radius 10 ft and moment of inertia 2500 slugs ft^2, mounted on a vertical frictionless shaft at its center. The whole system is initially at rest. The man now walks along the outer edge of the turntable with a velocity of 2 ft s^{-1}, relative to the earth. (a) With what angular velocity and in what direction does the turntable rotate? (b) Through what angle will it have rotated when the man reaches his initial position on the turntable? (c) Through what angle will it have rotated when he reaches his initial position relative to the earth?

9–51 A man weighing 160 lb runs around the edge of a horizontal turntable mounted on a vertical frictionless axis through its center. The velocity of the man, relative to the earth, is 4 ft s^{-1}. The turntable is rotating in the opposite direction with an angular velocity of 0.2 rad s^{-1}. The radius of the turntable is 8 ft and its moment of inertia about the axis of rotation is 320 slug ft^2. Find the final angular velocity of the system if the man comes to rest, relative to the turntable.

9–52 Two flywheels, A and B, are mounted on shafts which can be connected or disengaged by a friction clutch C. (Fig. 9–34.) The moment of inertia of wheel A is 4 slugs ft^2. With the clutch disengaged, wheel A is brought up to an angular velocity of 600 rev min^{-1}. Wheel B is initially at rest. The clutch is now engaged, accelerating B and decelerating A until both wheels have the same angular velocity. The final angular velocity of the system is 400 rev min^{-1}. (a) What was the moment of inertia of wheel B? (b) How much energy was lost in the process? Neglect all bearing friction.

9–53 The stabilizing gyroscope of a ship weighs 50 tons, its radius of gyration is 5 ft, and it rotates about a vertical axis with an angular velocity of 900 rev min^{-1}. (a) How long a time is required to bring it up to speed, starting from rest, with a constant power input of 100 hp? (b) Find the torque needed to cause the axis to precess in a vertical fore-and-aft plane at the rate of 1 deg s^{-1}.

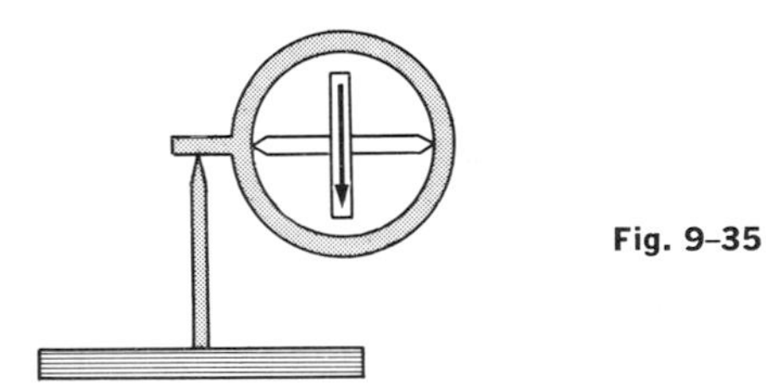

Fig. 9–35

9–54 The mass of the rotor of a toy gyroscope is 150 g and its moment of inertia about its axis is 1500 g cm^2. The mass of the frame is 30 g. The gyroscope is supported on a single pivot as in Fig. 9–35, with its center of gravity distant 4 cm horizontally from the pivot, and is precessing in a horizontal plane at the rate of one revolution in 6 s. (a) Find the upward force exerted by the pivot. (b) Find the angular velocity with which the rotor is spinning about its axis, expressed in revolutions per minute. (c) Copy the diagram, and show by vectors the angular momentum of the rotor and the torque acting on it.

9–55 The moment of inertia of the front wheel of a bicycle is 0.25 slug ft^2, its radius is 15 in., and the forward speed of the bicycle is 20 ft s^{-1}. With what angular velocity must the front wheel be turned about a vertical axis to counteract the capsizing torque due to a weight of 120 lb, applied 1 in. horizontally to the right or left of the line of contact of wheels and ground? [Bicycle riders: compare with experience and see if your answer seems reasonable.]

9–56 The rotor of a small control gyro can be accelerated from rest to an angular velocity of 50,000 rev min^{-1} in 0.2 s. The moment of inertia of the rotor about its spin axis is 385 g cm^2. (a) What is the angular acceleration (assumed constant) as the rotor is brought up to speed? (b) What torque is needed to produce this angular acceleration?

Suppose the data above refer to the gyro in Fig. 9–22, and that the rotor is spinning clockwise when viewed from the left along the x-axis. (c) Which way will the spin axis turn if ring B is forced to rotate clockwise, as seen from above, about the z-axis? (d) Suppose that rings A and B are locked in place. What torque must be exerted about the y-axis to rotate the outer frame about this axis at 1 deg s^{-1}?

9–57 A demonstration gyro wheel is constructed by removing the tire from a bicycle wheel 30 in. in diameter, wrapping lead wire around the rim, and taping it in place. The shaft projects 10 in. at each side of the wheel and a man holds the ends of the shaft in his hands. The weight of the system is 10 lb and its entire mass may be assumed to be located at its rim. The shaft is horizontal and the wheel is spinning about the shaft at 5 rev s^{-1}. Find the magnitude and direction of the force each hand exerts on the shaft under the following conditions: (a) The shaft is at rest. (b) The shaft is rotating in a horizontal plane about its center at 0.04 rev s^{-1}. (c) The shaft is rotating in a horizontal plane about its center at 0.20 rev s^{-1}. (d) At what rate must the shaft rotate in order that it may be supported at one end only?

9–58 A man stands on a turntable free to rotate without friction about a vertical axis. The moment of inertia of man and turntable together about this axis is 1.2 slug ft^2. The man holds the shaft of the bicycle wheel in Problem 9–57 in one hand, with the shaft vertical. The entire system is initially at rest. (a) The man grasps the rim of the wheel with his free hand and sets the wheel in rotation, in a clockwise direction as viewed from above, with an angular velocity of 5 rev s^{-1}. What is the angular velocity of man and turntable, in magnitude and direction? (b) The man now tips the shaft of the wheel downward until it is horizontal. What is the new angular velocity of man and turntable? (c) What is the kinetic energy of the entire system in part (a)? (d) What is the kinetic energy in part (b)? (e) Explain why the answers to (c) and (d) are not the same.

10 Elasticity

10-1 STRESS

The preceding chapter dealt with the motion of a "rigid" body, a convenient mathematical abstraction, since every real substance yields to some extent under the influence of applied forces. Ultimately, the change in shape or volume of a body when outside forces act on it is determined by the forces between its molecules. Although molecular theory is at present not sufficiently advanced to enable one to calculate the elastic properties of, say, a block of copper starting from the properties of a copper atom, the study of the solid state is an active subject in many research laboratories and our knowledge of it is steadily increasing. In this chapter we shall, however, confine ourselves to quantities that are directly measurable, and not attempt any molecular explanation of the observed behavior.

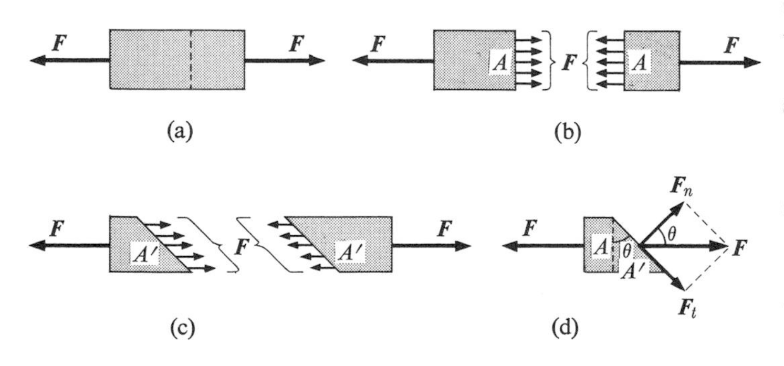

10-1 (a) A bar in tension. (b) The stress at a perpendicular section equals F/A. (c) and (d) The stress at an inclined section can be resolved into a normal stress F_n/A', and a tangential or shearing stress F_t/A'.

Figure 10-1(a) shows a bar of uniform cross-sectional area A subjected to equal and opposite pulls $\boldsymbol{F}$ at its ends. The bar is said to be in *tension*. Consider a section through the bar at right angles to its length, as indicated by the dotted line. Since every portion of the bar is in equilibrium, that portion at the right of the section must be pulling on the portion at the left with a force $\boldsymbol{F}$, and vice versa. If the section is not too near the ends of the bar, these pulls are uniformly distributed over the cross-sectional area A, as indicated by the short arrows in Fig. 10-1(b). We define the *stress* S at the section as the ratio of the force F to the area A:

$$\text{Stress} = \frac{F}{A}. \qquad (10\text{-}1)$$

The stress is called a *tensile* stress, meaning that each portion *pulls* on the other, and it is also a *normal* stress because the distributed force is perpendicular to the area. Units of stress are 1 newton per square meter (1 N m^{-2}), 1 dyne per square centimeter (1 dyn cm^{-2}), and 1 pound per square foot (1 lb ft^{-2}). The hybrid unit 1 lb in^{-2} is also commonly used.

Consider next a section through the bar in some arbitrary direction, as in Fig. 10-1(c). The resultant force exerted on the portion at either side of this section, by the portion at the other, is equal and opposite to the force F at the end of the section. Now, however, the force is distributed over a larger area A' and is not at right angles to the area. If we represent the resultant of the distributed forces by a single vector of magnitude F, as in Fig. 10-1(d), this vector can be resolved into a component F_n normal to the area A', and a component F_t

tangent to the area. The *normal* stress is defined, as before, as the ratio of the component F_n to the area A'. The ratio of the component F_t to the area A' is called the *tangential* stress or, more commonly, the *shearing* stress at the section:

$$\text{Normal stress} = \frac{F_n}{A'},$$

$$\text{Tangential (shearing) stress} = \frac{F_t}{A'}. \qquad (10\text{-}2)$$

Stress is not a vector quantity since, unlike a force, we cannot assign to it a specific direction. The *force* acting on the portion of the body on a specified side of a section has a definite direction. Stress is one of a class of physical quantities called *tensors*.

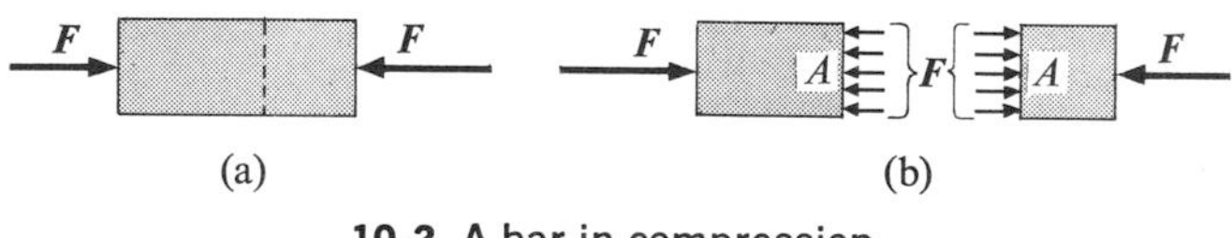

10-2 A bar in compression.

A bar subjected to pushes at its ends, as in Fig. 10-2, is said to be in *compression*. The stress on the dotted section, illustrated in part (b), is also a normal stress but is now a *compressive* stress, since each portion pushes on the other. It should be evident that if we take a section in some arbitrary direction it will be subject to both a tangential (shearing) and a normal stress, the latter now being a compression.

As another example of a body under stress, consider the block of square cross section in Fig. 10-3(a), acted on by two equal and opposite couples produced by the pairs of forces $\boldsymbol{F}_x$ and $\boldsymbol{F}_y$ distributed over its surfaces. The block is in equilibrium, and any portion of it is in equilibrium also. Thus the distributed forces over the diagonal face in part (b) must have a resultant $\boldsymbol{F}$ whose components are equal to $\boldsymbol{F}_x$ and $\boldsymbol{F}_y$. The stress at this section is therefore a pure compression, although the stresses at the right face and the bottom are both shearing stresses. Similarly, we see from Fig. 10-3(c) that the other diagonal is in pure tension.

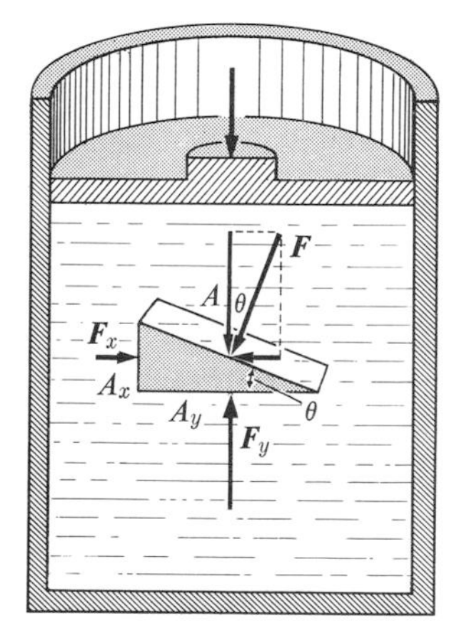

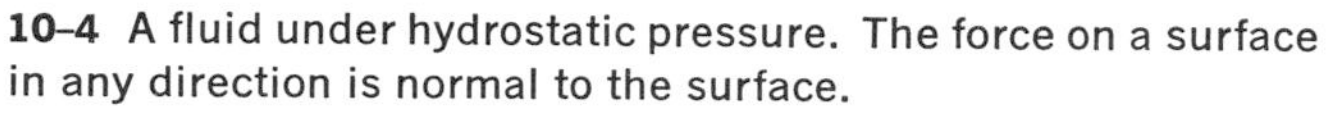
10-4 A fluid under hydrostatic pressure. The force on a surface in any direction is normal to the surface.

Consider next a fluid under pressure. The term "fluid" means a substance that can flow; hence the term applies to both liquids and gases. If there is a shearing stress at any point in a fluid, the fluid slips sidewise so long as the stress is maintained. Hence in a fluid at rest, the shearing stress is everywhere zero. Figure 10-4 represents a fluid in a cylinder provided with a piston, on which is exerted a downward force. The triangle is a side view of a wedge-shaped portion of the fluid. If for

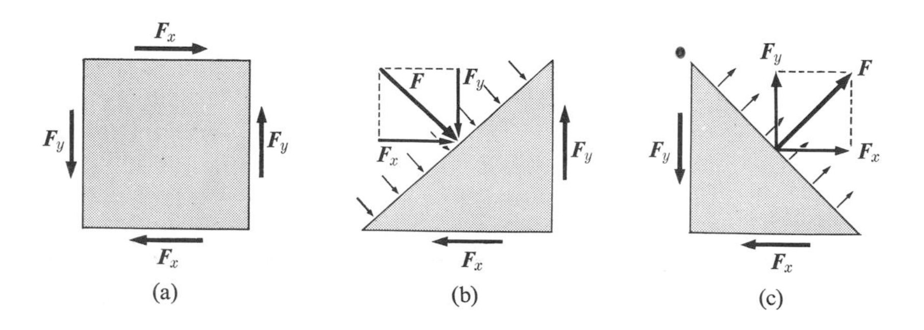

10-3 (a) A body in shear. The stress on one diagonal, part (b), is a pure compression; that on the other, part (c), is a pure tension.

the moment we neglect the weight of the fluid, the only forces on this portion are those exerted by the rest of the fluid, and since these forces can have no shearing (or tangential) component, they must be normal to the surfaces of the wedge. Let $\mathbf{F}_x$, $\mathbf{F}_y$, and $\mathbf{F}$ represent the forces against the three faces. Since the fluid is in equilibrium, it follows that

$$F \sin \theta = F_x, \qquad F \cos \theta = F_y.$$

Also,

$$A \sin \theta = A_x, \qquad A \cos \theta = A_y.$$

Dividing the upper equations by the lower, we find

$$\frac{F}{A} = \frac{F_x}{A_x} = \frac{F_y}{A_y}.$$

Hence the force per unit area is the *same*, regardless of the direction of the section, and is always a compression. Any one of the preceding ratios defines the *hydrostatic pressure p* in the fluid,

$$p = \frac{F}{A}, \qquad F = pA. \tag{10-3}$$

Units of pressure are 1 N m^{-2}, 1 dyn cm^{-2}, or 1 lb ft^{-2}. Like other types of stress, pressure is not a vector quantity and no direction can be assigned to it. The *force* against any area within (or bounding) a fluid at rest and under pressure is normal to the area, regardless of the orientation of the area. This is what is meant by the common statement that "the pressure in a fluid is the same in all directions."

The stress within a solid can also be a hydrostatic pressure, provided the stress at all points of the surface of the solid is of this nature. That is, the force per unit area must be the same at *all* points of the surface, and the force must be normal to the surface and directed inward. This is not the case in Fig. 10-2, where forces are applied at the ends of the bar only, but it is automatically the case if a solid is immersed in a fluid under pressure.

10-2 STRAIN

The term *strain* refers to the relative change in dimensions or shape of a body which is subjected to stress. Associated with each type of stress which we have described in the preceding section is a corresponding type of strain.

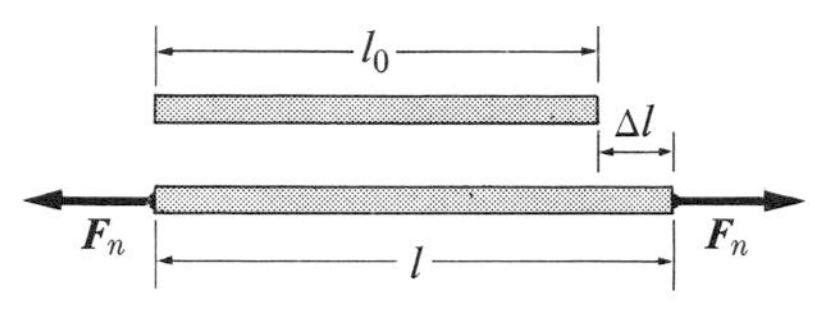

10-5 The longitudinal strain is defined as $\Delta\ell/\ell_0$.

Figure 10-5 shows a bar whose natural length is ℓ_0 and which elongates to a length ℓ when equal and opposite pulls are exerted at its ends. The elongation, of course, does not occur at the ends only; every element of the bar stretches in the same proportion as does the bar as a whole. The *tensile strain* in the bar is defined as the ratio of the increase in length to the original length:

$$\text{Tensile strain} = \frac{\ell - \ell_0}{\ell_0} = \frac{\Delta\ell}{\ell_0}. \tag{10-4}$$

The *compressive strain* of a bar in compression is defined in the same way, as the ratio of the decrease in length to the original length.

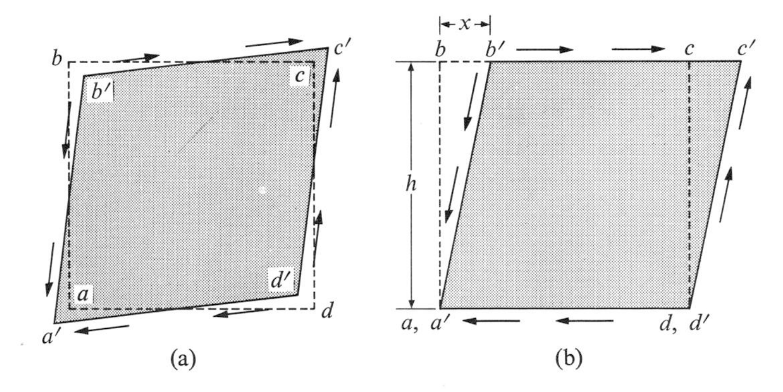

10-6 Change in shape of a block in shear. The shearing strain is defined as x/h.

Figure 10-6(a) illustrates the nature of the deformation when shearing stresses act on the faces of a block, as in Fig. 10-3. The dotted outline *abcd* represents the unstressed block, and the full lines *a′b′c′d′* represent the block under stress. The centers of the stressed and unstressed block coincide in part (a). In part (b), the edges

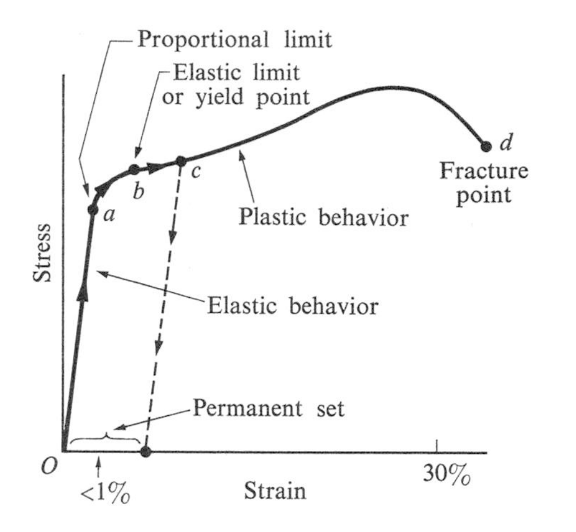

10-7 Typical stress-strain diagram for a ductile metal under tension.

ad and *a'd'* coincide. The lengths of the faces under shear remain very nearly constant, while all dimensions parallel to the diagonal *ac* increase in length, and those parallel to the diagonal *bd* decrease in length. Note that this is to be expected in view of the nature of the corresponding internal stresses (see Fig. 10-3). This type of strain is called a *shearing strain*, and is defined as the ratio of the displacement x of corner b to the transverse dimension h:

$$\text{Shearing strain} = x/h. \tag{10-5}$$

Like other types of strain, shearing strain is a pure number.

The strain produced by a hydrostatic pressure, called a *volume strain*, is defined as the ratio of the change in volume, ΔV, to the original volume V. It also is a pure number:

$$\text{Volume strain} = \frac{\Delta V}{V}. \tag{10-6}$$

10-3 ELASTICITY AND PLASTICITY

The relation between each of the three kinds of stress and its corresponding strain plays an important role in the branch of physics called the *theory of elasticity*, or its engineering counterpart, *strength of materials*. When any stress is plotted against the appropriate strain, the resulting stress-strain diagram is found to have several different shapes, depending on the kind of material. Two of the most important materials of present-day science and technology are metal and vulcanized rubber.

Even among metals there are wide variations. A typical stress-strain diagram for a ductile metal is shown in Fig. 10-7. The stress is a simple tensile stress and the strain is the percentage elongation. During the first portion of the curve (up to a strain of less than 1%), the stress and strain are proportional until the point a, the *proportional limit*, is reached. The proportional relation between stress and strain in this region is called *Hooke's law*. From a to b stress and strain are not proportional, but nevertheless, if the load is removed at any point between O and b, the curve will be retraced and the material will be restored to its original length. In the region Ob, the material is said to be *elastic* or to exhibit *elastic behavior* and the point b is called the *elastic limit*, or the *yield point*.

If the material is loaded further, the strain increases rapidly, but when the load is removed at some point beyond b, say at c, the material does not come back to its original length but traverses the dashed line in Fig. 10-7. The length at zero stress is now greater than the original length and the material is said to have a *permanent set*. Further increase of load beyond c produces a large increase in strain until a point d is reached at which *fracture* takes place. From b to d the metal is said to undergo *plastic flow* or *plastic deformation*, during which slipping takes place within the metal along the planes of maximum shearing stress. If large plastic deformation takes place between the elastic limit and the fracture point, the metal is said to be *ductile*. If, however, fracture occurs soon after the elastic limit, the metal is said to be *brittle*.

Figure 10-8 shows a stress-strain curve for a typical sample of vulcanized rubber which has been stretched to over seven times its original length. During no portion of this curve is the stress proportional to the strain. The substance, however, is elastic, in the sense that when the load is removed, the rubber is restored to its original length. On decreasing the load, the stress-strain curve is *not* retraced but follows the dashed curve of Fig. 10-8. The lack of coincidence of the curves for increasing and decreasing stress is known as *elastic hysteresis* (accent on the *third* syllable). An analogous phenomenon ob-

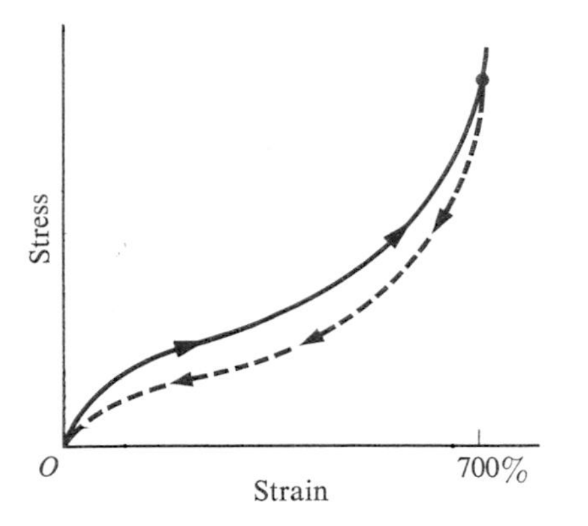

10–8 Typical stress-strain diagram for vulcanized rubber, showing elastic hysteresis.

served with magnetic materials is called *magnetic hysteresis*. It can be shown that the area bounded by the two curves, that is, the area of the *hysteresis loop*, is equal to the energy dissipated within the elastic or magnetic material. The large elastic hysteresis of some types of rubber makes these materials very valuable as vibration absorbers. If a block of such material is placed between a piece of vibrating machinery and, let us say, the floor, elastic hysteresis takes place during every cycle of vibration. Mechanical energy is converted to a form known as internal energy which evidences itself by a rise of temperature. As a result, only a small amount of energy of vibration is transmitted to the floor.

10–4 ELASTIC MODULUS

The stress required to produce a given strain depends on the nature of the material under stress. The ratio of stress to strain, or the *stress per unit strain*, is called an *elastic modulus* of the material. The larger the elastic modulus, the greater the stress needed for a given strain.

Consider first longitudinal (tensile or compressive) stresses and strains. Experiment shows that up to the proportional limit, a given longitudinal stress produces a strain of the same magnitude whether the stress is a tension or a compression. Hence the ratio of tensile stress to tensile strain, for a given material, equals the ratio of compressive stress to compressive strain. This ratio is called the *stretch modulus* or *Young's modulus* of the material and will be denoted by Y:

$$Y = \frac{\text{tensile stress}}{\text{tensile strain}} = \frac{\text{compressive stress}}{\text{compressive strain}}$$

$$= \frac{F_n/A}{\Delta\ell/\ell_0} = \frac{\ell_0}{A}\frac{F_n}{\Delta\ell}. \tag{10–7}$$

If the proportional limit is not exceeded, the ratio of stress to strain is constant and Hooke's law is therefore equivalent to the statement that *within the proportional limit, the elastic modulus of a given material is constant*, depending only on the nature of the material.

Since a strain is a pure number, the units of Young's modulus are the same as those of stress, namely, force per unit area. Tabulated values are usually in pounds per square inch or dynes per square centimeter. Some typical values are listed in Table 10–1.

TABLE 10–1 APPROXIMATE ELASTIC MODULI

Material	Young's modulus, Y		Shear modulus, S		Bulk modulus, B	
	10^{12} dyn cm^{-2}	10^{6} lb in^{-2}	10^{12} dyn cm^{-2}	10^{6} lb in^{-2}	10^{12} dyn cm^{-2}	10^{6} lb in^{-2}
Aluminum	0.70	10	0.24	3.4	0.70	10
Brass	0.91	13	0.36	5.1	0.61	8.5
Copper	1.1	16	0.42	6.0	1.4	20
Glass	0.55	7.8	0.23	3.3	0.37	5.2
Iron	0.91	13	0.70	10	1.0	14
Lead	0.16	2.3	0.056	0.8	0.077	1.1
Nickel	2.1	30	0.77	11	2.6	34
Steel	2.0	29	0.84	12	1.6	23
Tungsten	3.6	51	1.5	21	2.0	29

If the relation between stress and strain is not linear, an elastic modulus can be defined more generally as the limiting ratio of a small *change* in stress to the *change* in strain produced by it. Thus if the force F_n in Fig. 10-5 is increased by dF_n, and as a result the length of the bar increases by $d\ell$, the stretch modulus is defined as

$$Y = \frac{dF_n/A}{d\ell/\ell} = \frac{\ell}{A}\frac{dF_n}{dl}. \tag{10-8}$$

This is equivalent to defining the modulus at any point as the *slope* of a curve in a stress-strain diagram. Within the Hooke's law region, the two definitions are equivalent.

The *shear modulus* S of a material, within the Hooke's law region, is defined as the ratio of a shearing stress to the shearing strain it produces:

$$S = \frac{\text{shearing stress}}{\text{shearing strain}} = \frac{F_t/A}{x/h} = \frac{h}{A}\frac{F_t}{x}. \tag{10-9}$$

(Refer to Fig. 10-6 for the meaning of x and of h.)

The shear modulus of a material is also expressed as force per unit area. For most materials it is one-half to one-third as great as Young's modulus. The shear modulus is also called the *modulus of rigidity* or the *torsion modulus.*

The more general definition of shear modulus is

$$S = \frac{dF_t/A}{dx/h} = \frac{h}{A}\frac{dF_t}{dx}, \tag{10-10}$$

where dx is the increase in x when the shearing force increases by dF_t.

The shear modulus has a significance for *solid* materials only. A liquid or gas will flow under the influence of a shearing stress and will not *permanently* support such a stress.

The modulus relating a hydrostatic pressure to the volume strain it produces is called the *bulk modulus* and we shall represent it by B. The general definition of bulk modulus is the (negative) ratio of a change in pressure to the change in volume strain produced by it:

$$B = -\frac{dp}{dV/V} = -V\frac{dp}{dV}. \tag{10-11}$$

The minus sign is included in the definition of B because an *increase* of pressure always causes a *decrease* in volume. That is, if dp is positive, dV is negative. By including a minus sign in its definition, we make the bulk modulus itself a positive quantity.

The change in volume of a *solid* or *liquid* under pressure is so small that the volume V, in Eq. (10-11) can be considered constant. Provided the pressure is not too great, the ratio dp/dV is constant also, the bulk modulus is constant, and we can replace dp and dV by finite changes in pressure and volume. The volume of a *gas*, however, changes markedly with pressure, and the general definition of B must be used for gases.

The reciprocal of the bulk modulus is called the *compressibility* k. From its definition,

$$k = \frac{1}{B} = -\frac{dV/V}{dp} = -\frac{1}{V}\frac{dV}{dp}. \tag{10-12}$$

The compressibility of a material thus equals the *fractional decrease in volume, $-dV/V$, per unit increase dp in pressure.*

TABLE 10-2 COMPRESSIBILITIES OF LIQUIDS

Liquid	Compressibility, k		
	$(\text{N m}^{-2})^{-1}$	$(\text{lb in}^{-2})^{-1}$	atm^{-1}
Carbon disulfide	64×10^{-11}	45×10^{-7}	66×10^{-6}
Ethyl alcohol	110	78	115
Glycerine	21	15	22
Mercury	3.7	2.6	3.8
Water	49	34	50

The units of a bulk modulus are the same as those of pressure, and the units of compressibility are those of a *reciprocal pressure.* Thus the statement that the compressibility of water (see Table 10-2) is $50 \times 10^{-6}\ \text{atm}^{-1}$ means that the volume decreases by 50 one-millionths of the original volume for each atmosphere increase in pressure. (1 atm = 14.7 lb in^{-2}.)

Example 1. In an experiment to measure Young's modulus, a load of 1000 lb hanging from a steel wire 8 ft long, of cross section 0.025 in^2, was found to stretch the wire 0.010 ft above its no-load length. What were the stress, the strain, and the value of Young's modulus for the steel of which the wire was composed?

$$\text{Stress} = \frac{F_n}{A} = \frac{1000 \text{ lb}}{0.025 \text{ in}^2} = 40{,}000 \text{ lb in}^{-2}.$$

$$\text{Strain} = \frac{\Delta\ell}{\ell_0} = \frac{0.010 \text{ ft}}{8 \text{ ft}} = 0.00125.$$

$$Y = \frac{\text{stress}}{\text{strain}} = \frac{40{,}000 \text{ lb in}^{-2}}{0.00125} = 32 \times 10^6 \text{ lb in}^{-2}.$$

Example 2. Suppose the object in Fig. 10-6 is a brass plate 2 ft square and $\frac{1}{4}$ in. thick. How large a force F must be exerted on each of its edges if the displacement x in Fig. 10-6(b) is 0.01 in.? The shear modulus of brass is 5×10^6 lb in^{-2}.

The shearing stress on each edge is

$$\text{Shearing stress} = \frac{F_t}{A} = \frac{F}{24 \times \frac{1}{4} \text{ in}^2} = \frac{F}{6} \text{ in}^{-2}.$$

The shearing strain is

$$\text{Shearing strain} = \frac{x}{h} = \frac{0.01 \text{ in.}}{24 \text{ in.}} = 4.17 \times 10^{-4}.$$

$$\text{Shear modulus } S = \frac{\text{stress}}{\text{strain}},$$

$$5 \times 10^6 \text{ lb in}^{-2} = \frac{F/6}{4.17 \times 10^{-4}} \text{ in}^{-2}$$

$$F = 12{,}500 \text{ lb}.$$

Example 3. The volume of oil contained in a certain hydraulic press is 5 ft^3. Find the decrease in volume of the oil when subjected to a pressure of 2000 lb in^{-2}. The compressibility of the oil is 20×10^{-6} atm^{-1}.

The volume decreases by 20 parts per million for a pressure increase of 1 atm. Since 2000 lb in^{-2} = 136 atm, the volume decrease is $136 \times 20 = 2720$ parts per million. Since the original volume is 5 ft^3, the actual decrease is

$$\frac{2720}{1{,}000{,}000} \times 5 \text{ ft}^3 = 0.0136 \text{ ft}^3 = 23.5 \text{ in}^3.$$

Or, from Eq. (10-12), replacing dV and dp with ΔV and Δp,

$$\begin{aligned} \Delta V &= -kV\,\Delta p \\ &= -20 \times 10^{-6} \text{ atm}^{-1} \times 5 \text{ ft}^3 \times 136 \text{ atm} \\ &= -0.0136 \text{ ft}^3. \end{aligned}$$

10-5 THE FORCE CONSTANT

The various elastic moduli are quantities which describe the elastic properties of a particular *material* and do not directly indicate how much a given rod, cable, or spring constructed of the material will distort under load. If Eq. (10-7) is solved for F_n, one obtains

$$F_n = \frac{YA}{\ell_0}\Delta\ell$$

or, if YA/ℓ_0 is replaced by a single constant k, and the elongation $\Delta\ell$ is represented by x, then

$$F_n = kx. \qquad (10\text{-}13)$$

In other words, the elongation of a body in tension above its no-load length is directly proportional to the stretching force. Hooke's law was originally stated in this form, rather than in terms of stress and strain.

When a helical wire spring is stretched, the stress in the wire is practically a pure shear. The elongation of the spring as a whole is directly proportional to the stretching force. That is, an equation of the form $F = kx$ still applies, the constant k depending on the shear modulus of the wire, its radius, the radius of the coils, and the number of coils.

The constant k, or the ratio of the force to the elongation, is called the *force constant* or the *stiffness* of the spring, and is expressed in pounds per foot, newtons per meter, or dynes per centimeter. It is equal numerically to the force required to produce unit elongation.

The ratio of the elongation to the force, or the elongation per unit force, is called the *compliance* of the spring. The compliance equals the reciprocal of the force constant and is expressed in feet per pound, meters per newton, or centimeters per dyne. It is numerically equal to the elongation produced by unit force.

Problems

10–1 A steel wire 10 ft long and 0.1 in^2 in cross section is found to stretch 0.01 ft under a tension of 2500 lb. What is Young's modulus for this steel?

10–2 The elastic limit of a steel elevator cable is 40,000 lb in^{-2}. Find the maximum upward acceleration which can be given a 2-ton elevator when supported by a cable whose cross section is $\frac{1}{2}$ in^2 if the stress is not to exceed $\frac{1}{4}$ of the elastic limit.

10–3 A copper wire 12 ft long and 0.036 in. in diameter was given the test below. A load of 4.5 lb was originally hung from the wire to keep it taut. The position of the lower end of the wire was read on a scale.

Added load, lb	Scale reading, in.
0	3.02
2	3.04
4	3.06
6	3.08
8	3.10
10	3.12
12	3.14
14	3.65

(a) Make a graph of these values, plotting the increase in length horizontally and the added load vertically. (b) Calculate the value of Young's modulus. (c) What was the stress at the proportional limit?

10–4 A steel wire has the following properties:

Length = 10 ft
Cross section = 0.01 in^2
Young's modulus = 30,000,000 lb in^{-2}
Shear modulus = 10,000,000 lb in^{-2}
Proportional limit = 60,000 lb in^{-2}
Breaking stress = 120,000 lb in^{-2}

The wire is fastened at its upper end and hangs vertically. (a) How great a load can be supported without exceeding the proportional limit? (b) How much will the wire stretch under this load? (c) What is the maximum load that can be supported?

10–5 (a) What is the maximum load that can be supported by an aluminum wire 0.05 in. in diameter without exceeding the proportional limit of 14,000 lb in^{-2}? (b) If the wire was originally 20 ft long, how much will it elongate under this load?

10–6 A 10-lb weight hangs on a vertical steel wire 2 ft long and 0.001 in^2 in cross section. Hanging from the bottom of this weight is a similar steel wire which supports a 5-lb weight. Compute (a) the longitudinal strain, and (b) the elongation of each wire.

10–7 A 32-lb weight, fastened to the end of a steel wire of unstretched length 2 ft, is whirled in a vertical circle with an angular velocity of 2 rev s^{-1} at the bottom of the circle. The cross section of the wire is 0.01 in^2. Calculate the elongation of the wire when the weight is at the lowest point of its path.

10–8 A copper wire 320 in. long and a steel wire 160 in. long, each of cross section 0.1 in^2, are fastened end-to-end and stretched with a tension of 100 lb. (a) What is the change in length of each wire? (b) What is the elastic potential energy of the system?

10–9 A copper rod of length 3 ft and cross-sectional area 0.5 in^2 is fastened end-to-end to a steel rod of length L and cross-sectional area 0.2 in^2. The compound rod is subjected to equal and opposite pulls of magnitude 6000 lb at its ends. (a) Find the length L of the steel rod if the elongations of the two rods are equal. (b) What is the stress in each rod? (c) What is the strain in each rod?

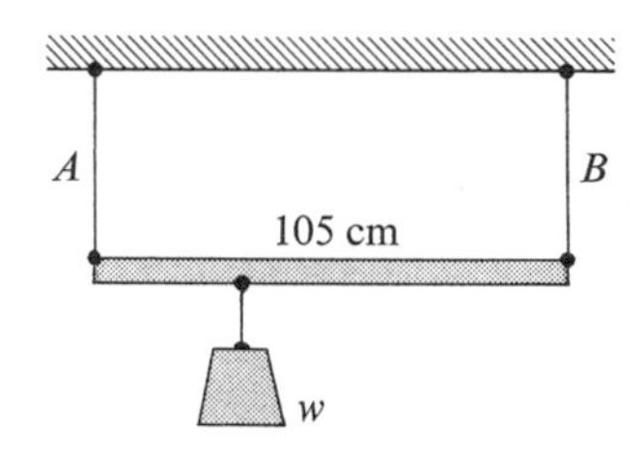

10–10 A rod 105 cm long, whose weight is negligible, is supported at its ends by wires A and B of equal length. The cross section of A is 1 mm^2, that of B is 2 mm^2. Young's modulus for wire A is 30×10^6 lb in^{-2} and for B it is 20×10^6 lb in^{-2}. At what point along the bar should a weight w be suspended in order to produce (a) equal stresses in A and B, (b) equal strains in A and B? (Fig. 10–9.)

10–11 A bar of length L, cross-sectional area A, and Young's modulus Y, is subjected to a tension F. Represent the stress in the bar by Q and the strain by P. Derive the expression for the elastic potential energy, per unit volume, of the bar in terms of Q and P.

10–12 The compressibility of sodium is to be measured by observing the displacement of the piston in Fig. 10–4 when a force is applied. The sodium is immersed in an oil which fills the cylinder below the piston. Assume that the piston and walls of the cylinder are perfectly rigid, that there is no friction, and no oil leak. Compute the compressibility of the sodium in terms of the applied force F, the piston displacement x, the piston area A, the initial volume of the oil V_0, the initial volume of the sodium v_0, and the compressibility of the oil k_0.

10–13 Two strips of metal are riveted together at their ends by four rivets, each of diameter 0.25 in. What is the maximum tension that can be exerted by the riveted strip if the shearing stress on the rivets is not to exceed 10,000 lb in^{-2}? Assume each rivet to carry one-quarter of the load.

10–14 Find the weight per cubic foot of ocean water at a depth where the pressure is 4700 lb ft^{-2}. The weight at the surface is 64 lb ft^{-3}.

10–15 Compute the compressibility of steel, in reciprocal atmospheres, and compare with that of water. Which material is the more readily compressed?

10–16 A steel post 6 in. in diameter and 10 ft long is placed vertically and is required to support a load of 20,000 lb. (a) What is the stress in the post? (b) What is the strain in the post? (c) What is the change in length of the post?

10–17 A hollow cylindrical steel column 10 ft high shortens 0.01 in. under a compression load of 72,000 lb. If the inner radius of the cylinder is 0.80 of the outer one, what is the outer radius?

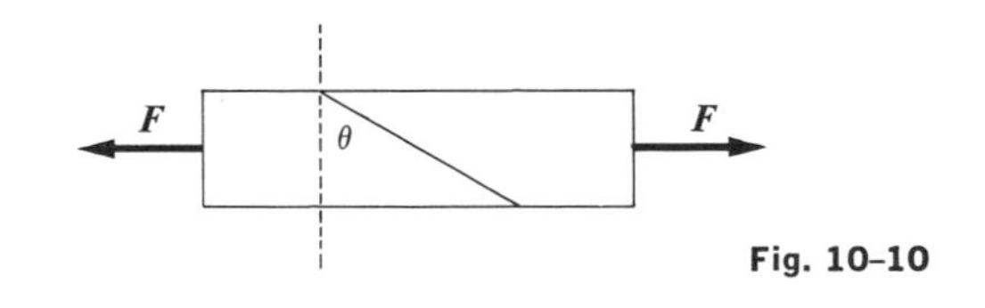

Fig. 10–10

10–18 A bar of cross section A is subjected to equal and opposite tensile forces $\mathbf{F}$ at its ends. Consider a plane through the bar making an angle θ with a plane at right angles to the bar (Fig. 10–10). (a) What is the tensile (normal) stress at this plane, in terms of F, A, θ? (b) What is the shearing (tangential) stress at the plane, in terms of F, A, and θ? (c) For what value of θ is the tensile stress a maximum? (d) For what value of θ is the shearing stress a maximum?

10–19 Suppose the block in Fig. 10–3 is rectangular instead of square, but is in equilibrium under the action of shearing stresses on those outside faces perpendicular to the plane of the diagram. (Note that then $F_x \neq F_y$.) (a) Show that the shearing stress is the same on all outside faces perpendicular to the plane of the diagram. (b) Show that on all sections perpendicular to the plane of the diagram and making an angle of 45° with an end face, the stress is still a pure tension or compression.

11 Harmonic Motion

11-1 INTRODUCTION

When a body executes a to-and-fro motion about some fixed point, its motion is said to be *oscillatory*. In this chapter we consider a special type of oscillatory motion called *harmonic motion*. Motion of this sort is closely approximated by that of a body suspended from a spring, by a pendulum swinging with a small amplitude, and by the balance wheel of a watch. The vibrations of the strings and air columns of musical instruments either are harmonic or are a superposition of harmonic motions. Modern atomic theory leads us to believe that the molecules of a solid body oscillate with nearly harmonic motion about their fixed lattice positions, although of course their motion cannot be directly observed.

In every form of wave motion, the particles of the medium in which the wave is traveling oscillate with harmonic motion or with a superposition of such motions. This is true even for light waves and radiowaves in empty space, except that instead of material particles the quantities that oscillate are the electric and magnetic intensities associated with the wave. As a final example, the equations describing the behavior of an electrical circuit in which there is an alternating current have the same form as those for the harmonic motion of a material body. It can be seen that a study of harmonic motion will lay the foundation for future work in many different fields of physics.

11-2 ELASTIC RESTORING FORCES

It has been shown in Chapter 10 that when a body is caused to change its shape, the distorting force is proportional to the amount of the change, provided the proportional limit of elasticity is not exceeded. The change may be in the nature of an increase in length, as of a rubber band or a coil spring, or a decrease in length, or a bending as of a flat spring, or a twisting of a rod about its axis, or of many other forms. The term "force" is to be interpreted liberally as the force, or torque, or pressure, or whatever may be producing the distortion. If we restrict the discussion to the case of a push or a pull, where the distortion is simply the displacement of the point of application of the force, the force and displacement are related by Hooke's law,

$$F = kx,$$

where k is a proportionality constant called the *force constant* and x is the displacement from the equilibrium position.

In this equation, F stands for the force which must be exerted *on* an elastic body to produce the displacement x. The force with which the elastic body pulls back on an object to which it is attached is called the *restoring force* and is equal to $-kx$.

11-3 DEFINITIONS

To fix our ideas, suppose that a flat strip of steel such as a hacksaw blade is clamped vertically in a vise and a small body is attached to its upper end, as in Fig. 11-1. Assume that the strip is sufficiently long and the displacement sufficiently small so that the motion is essentially along a straight line. The mass of the strip itself is negligible.

Let the top of the strip be pulled to the right a distance A, as in Fig. 11-1, and released. The attached body is then acted on by a restoring force exerted by the steel strip and directed toward the equilibrium position

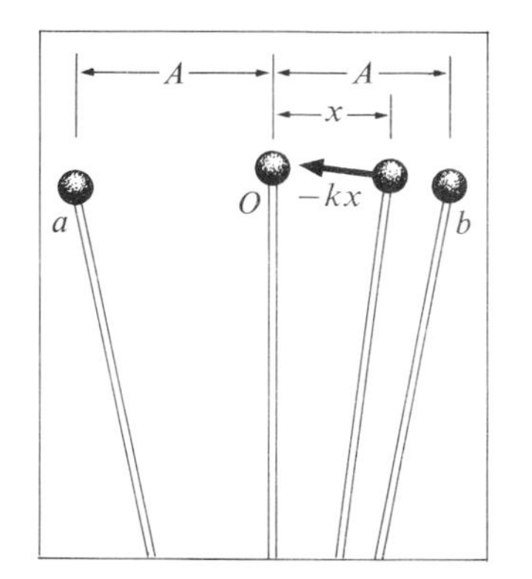

11-1 Motion under an elastic restoring force.

O. It therefore accelerates in the direction of this force, and moves in toward the center with increasing speed. The *rate* of increase (i.e., the acceleration) is not constant, however, since the accelerating force becomes smaller as the body approaches the center.

When the body reaches the center the restoring force has decreased to zero, but because of the velocity which has been acquired, the body "overshoots" the equilibrium position and continues to move toward the left. As soon as the equilibrium position is passed, the restoring force again comes into play, directed now toward the right. The body therefore decelerates, and at a rate which increases with increasing distance from O. It will therefore be brought to rest at some point to the left of O, and repeat its motion in the opposite direction.

Both experiment and theory show that the motion will be confined to a range $\pm A$ on either side of the equilibrium position, each to-and-fro movement taking place in the same length of time. If there were no loss of energy by friction, the motion would continue indefinitely once it had been started. This type of motion, under the influence of an elastic restoring force and in the absence of all friction, is called *simple harmonic motion*, often abbreviated SHM.

Any sort of motion which repeats itself in equal intervals of time is called *periodic*, and if the motion is back and forth over the same path it is also called *oscillatory*.

A *complete vibration* or *complete cycle* means one round trip, say from a to b and back to a, or from O to b to O to a and back to O.

The *periodic time*, or simply the *period* of the motion, represented by T, is the time required for one complete vibration.

The *frequency*, f, is the number of complete vibrations per unit time. Evidently the frequency is the reciprocal of the period, or $T = 1/f$. The mks unit, one cycle per second, is called one *hertz* (1 Hz).

The *coordinate*, x, at any instant, is the distance away from the equilibrium position or center of the path at that instant.

The *amplitude*, A, is the maximum coordinate. The total range of the motion is therefore $2A$.

11-4 EQUATIONS OF SIMPLE HARMONIC MOTION

We now wish to find expressions for the coordinate, velocity, and acceleration of a body moving with simple harmonic motion, just as we found those for a body moving with constant acceleration. It must be emphasized that the equations of motion with *constant* acceleration cannot be applied, since the acceleration is continually changing.

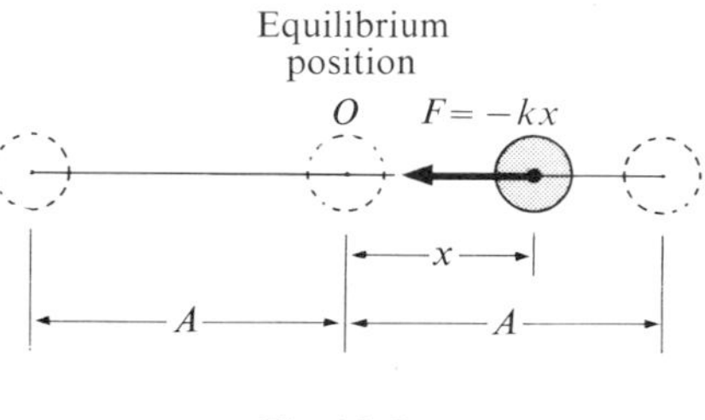

Fig. 11-2.

Figure 11-2 represents the vibrating body of Fig. 11-1 at some instant when its coordinate is x. The resultant force on it is simply the elastic restoring force, $-kx$, and from Newton's second law,

$$F = -kx = ma = mv\frac{dv}{dx}.$$

From the second and fourth terms, we have

$$mv\frac{dv}{dx} + kx = 0. \qquad (11\text{-}1)$$

Therefore

$$\int mv\,dv + \int kx\,dx = 0,$$

which integrates to

$$\tfrac{1}{2}mv^2 + \tfrac{1}{2}kx^2 = C_1. \qquad (11\text{-}2)$$

The first term on the left is the kinetic energy E_k of the body and the second is its elastic potential energy E_p. Equation (11-2) therefore states that the total energy of the system is constant, and the integration constant C_1 equals the total energy E (this result is only to be expected, since the system is conservative):

$$E_k + E_p = E.$$

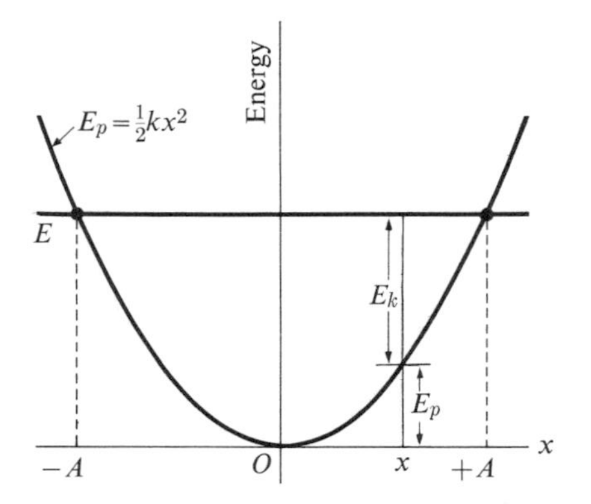

11-3 Relation between total energy E, potential energy E_p, and kinetic energy E_k, for a body oscillating with SHM.

The significance of this relation can be brought out by the graph shown in Fig. 11-3, in which energy is plotted vertically and the coordinate x horizontally. First, the curve representing the potential energy, $E_p = \frac{1}{2}kx^2$, is constructed. (This curve is a parabola.) Next, a horizontal line is drawn at a height equal to the total energy E. We see at once that the motion is restricted to values of x lying between the points at which the horizontal line intersects the parabola, since if x were outside this range the potential energy would exceed the total energy, which is impossible. The motion of the vibrating body is analogous to that of a particle released at a height E on a frictionless track shaped like the potential energy curve, and the motion is said to take place in a "potential energy well."

If a vertical line is constructed at any value of x within the permitted range, the length of the segment between the x-axis and the parabola represents the potential energy E_p at that value of x, and the length of the segment between the parabola and the horizontal line at height E represents the corresponding kinetic energy E_k. At the endpoints, therefore, the energy is all potential and at the midpoint it is all kinetic. The velocity at the midpoint has its maximum (absolute) value v_{max}:

$$\tfrac{1}{2}mv_{max}^2 = E,$$
$$v_{max} = \pm\sqrt{2E/m}, \qquad (11\text{-}3)$$

the sign of v being positive or negative depending on the direction of motion.

At the ends of the path the coordinate has its maximum (absolute) value x_{max}, equal to the amplitude A:

$$\tfrac{1}{2}kx_{max}^2 = E, \qquad x_{max} = \pm\sqrt{2E/k},$$
$$A = |x_{max}| = \sqrt{2E/k}. \qquad (11\text{-}4)$$

The velocity v at any coordinate x, from Eq. (11-2), is

$$v = \pm\sqrt{\frac{2E - kx^2}{m}}. \qquad (11\text{-}5)$$

Making use of Eq. (11-4), this can be written

$$v = \sqrt{k/m}\sqrt{A^2 - x^2}. \qquad (11\text{-}6)$$

The expression for the coordinate x as a function of time can now be obtained by replacing v with dx/dt, in Eq. (11-6), and integrating. This gives

$$\int \frac{dx}{\sqrt{A^2 - x^2}} = \sqrt{\frac{k}{m}}\int dt,$$
$$\sin^{-1}\frac{x}{A} = \sqrt{\frac{k}{m}}\,t + C_2. \qquad (11\text{-}7)$$

Let x_0 be the value of x when $t = 0$. The integration constant C_2 is then

$$C_2 = \sin^{-1}\frac{x_0}{A}.$$

That is, C_2 is the *angle* (in radians) whose sine equals x_0/A. Let us represent this angle by θ_0:

$$\sin\theta_0 = \frac{x_0}{A}, \qquad \theta_0 = \sin^{-1}\frac{x_0}{A}. \qquad (11\text{-}8)$$

Equation (11-7) can now be written

$$\sin^{-1}\frac{x}{A} = \sqrt{\frac{k}{m}}\,t + \theta_0,$$

$$x = A\sin\left(\sqrt{\frac{k}{m}}\,t + \theta_0\right). \qquad (11\text{-}9)$$

The coordinate x is therefore a *sinusoidal* function of the time t. The term in parentheses is an *angle*, in radians. It is called the *phase angle*, or simply the *phase* of the motion. The angle θ_0 is the *initial phase angle*, and is also referred to as the *epoch angle*.

The period T is the time required for one complete oscillation. That is, the coordinate x has the same value at the times t and $t + T$. In other words, the phase angle $(\sqrt{k/m}\,t + \theta_0)$ increases by 2π radians in the time T:

$$\sqrt{k/m}\,(t + T) + \theta_0 = (\sqrt{k/m}\,t + \theta_0) + 2\pi,$$

$$\boxed{T = 2\pi\sqrt{\frac{m}{k}}}. \qquad (11\text{-}10)$$

The period T therefore depends only on the mass m and the force constant k. It does not depend on the amplitude (or on the total energy). For given values of m and k, the time of one complete oscillation is the same whether the amplitude is large or small. A motion that has this property is said to be *isochronous* (equal time).

The frequency f, or the number of complete oscillations per unit time, is the reciprocal of the period T:

$$f = \frac{1}{T} = \frac{1}{2\pi}\sqrt{\frac{k}{m}}.$$

The *angular frequency* ω is defined as $\omega = 2\pi f$ and is expressed in radians per second. It follows from the two preceding equations that

$$\omega = \sqrt{k/m},$$

and Eqs. (11-9) and (11-6) can be written more compactly as

$$\boxed{x = A\sin(\omega t + \theta_0)}, \qquad (11\text{-}11)$$

$$v = \omega\sqrt{A^2 - x^2}. \qquad (11\text{-}12)$$

Expressions for the velocity and acceleration as functions of time can now be obtained by differentiation:

$$v = \frac{dx}{dt} = \omega A\cos(\omega t + \theta_0), \qquad (11\text{-}13)$$

$$a = \frac{dv}{dt} = -\omega^2 A\sin(\omega t + \theta_0). \qquad (11\text{-}14)$$

Since $A\sin(\omega t + \theta_0) = x$, the expression for the acceleration as a function of x is

$$a = -\omega^2 x. \qquad (11\text{-}15)$$

It follows from Eq. (11-13) that if v_0 is the velocity when $t = 0$, then

$$\cos\theta_0 = \frac{v_0}{\omega A}. \qquad (11\text{-}16)$$

This equation, together with Eq. (11-8), $\sin\theta_0 = x_0/A$, completely determines the initial phase angle θ_0. That is, the angle θ_0 depends both on the initial position x_0 and the initial velocity v_0.

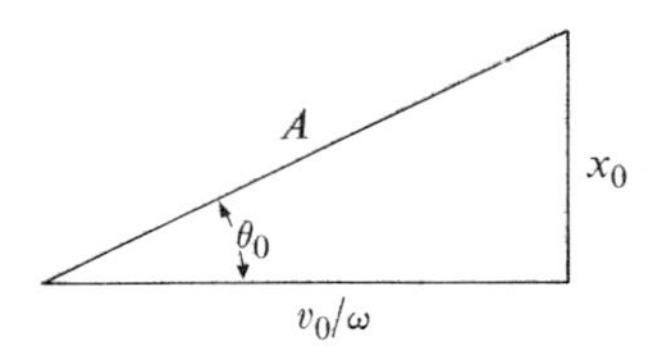

11-4 Relation between the initial phase angle θ_0, initial position x_0, and initial velocity v_0.

Figure 11-4 shows how the initial phase angle depends on the initial position and initial velocity. It follows from this triangle that

$$A = \sqrt{x_0^2 + (v_0/\omega)^2}, \qquad (11\text{-}17)$$

a relation that can also be obtained by setting the initial total energy, $\frac{1}{2}kx_0^2 + \frac{1}{2}mv_0^2$, equal to the potential energy at maximum displacement. The motion is therefore completely determined, for given values of m and k, when the initial position and initial velocity are known.

The equations of simple harmonic motion may be summarized by comparing them with similar equations

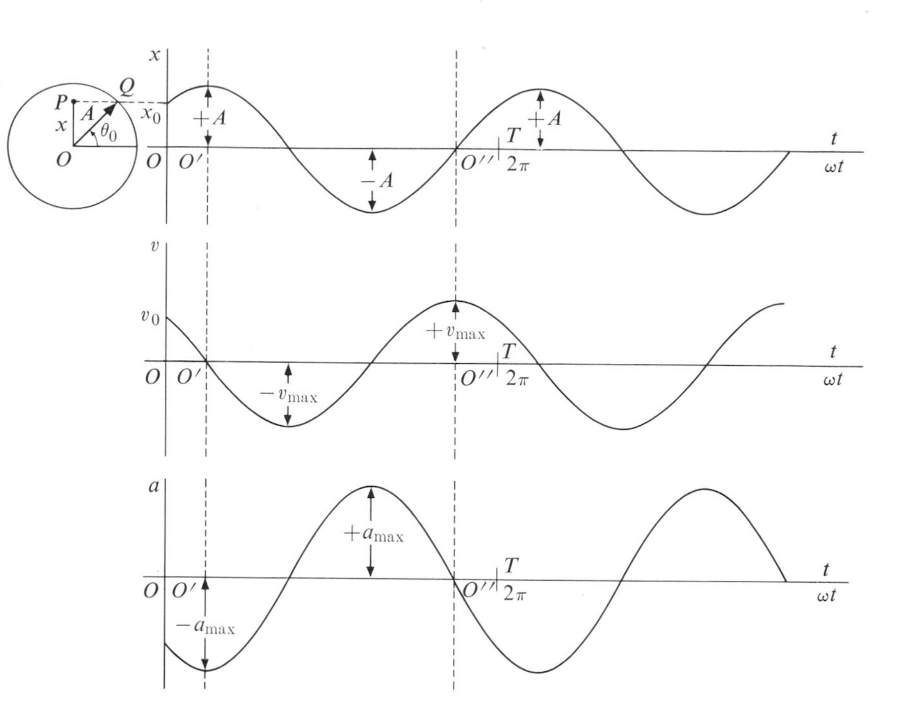

11-5 Graphs of coordinate x, velocity v, and acceleration a of a body oscillating with simple harmonic motion.

for rectilinear motion with constant acceleration (Table 11-1).

Figure 11-5 shows corresponding graphs of the coordinate x, velocity v, and acceleration a of a body oscillating with simple harmonic motion, plotted as functions of the time t (or of the angle ωt). The initial coordinate is x_0 and the initial velocity is v_0. The angle θ_0 has been taken as $\pi/4$ rad. Each curve repeats itself in a time interval equal to the period T, during which the angle ωt increases by 2π rad. Note that when the body is at either end of its path, i.e., when the coordinate x has its maximum positive or negative value ($\pm A$), the velocity is zero and the acceleration has its maximum negative or positive value ($\mp a_{max}$). Note also that when the body passes through its equilibrium position ($x = 0$), the velocity has its maximum positive or negative value ($\pm v_{max}$) and the acceleration is zero.

TABLE 11-1

Rectilinear motion with constant acceleration	Simple harmonic motion (in terms of ω and θ_0)
$a = \text{constant}$	$a = -\omega^2 x$ $a = -\omega^2 A \sin(\omega t + \theta_0)$
$v^2 = v_0^2 + 2a(x - x_0)$ $v = v_0 + at$	$v = \pm\omega\sqrt{A^2 - x^2}$ $v = \omega A \cos(\omega t + \theta_0)$
$x = x_0 + v_0 t + \frac{1}{2}at^2$	$x = A \sin(\omega t + \theta_0)$

$$\omega = 2\pi/T = 2\pi f = \sqrt{k/m}, \qquad A = \sqrt{x_0^2 + (v_0/\omega)^2},$$
$$\sin\theta_0 = x_0/A, \qquad \cos\theta_0 = v_0/\omega A.$$

The equations of motion take a simpler form if we set $t = 0$ when the body is at the midpoint, or is at one end of its path. For example, suppose we let $t = 0$ when the body has its maximum positive displacement. Then $x_0 = +A$, $\sin\theta_0 = 1$, $\theta_0 = \pi/2$, and

$$x = A \sin(\omega t + \pi/2) = A \cos \omega t,$$
$$v = -\omega A \sin \omega t, \tag{11-18}$$
$$a = -\omega^2 A \cos \omega t.$$

This corresponds to moving the origin from the point O, in Fig. 11-5, to the point O'. The graph of x versus t becomes a cosine curve, that of v versus t a negative sine curve, and that of a a negative cosine curve.

If we set $t = 0$ when the body is at the midpoint and moving toward the right, then

$$x_0 = 0, \qquad \sin\theta_0 = 0, \qquad \theta_0 = 0$$

and

$$\begin{aligned} x &= A \sin \omega t, \\ v &= \omega A \cos \omega t, \qquad (11\text{–}19) \\ a &= -\omega^2 A \sin \omega t. \end{aligned}$$

This corresponds to displacing the origin to the point O'' in Fig. 11–5.

Example. Let the mass of the body in Fig. 11–2 be 25 g, the force constant k be 400 dyn cm^{-1}, and let the motion be started by displacing the body 10 cm to the right of its equilibrium position and imparting to it a velocity toward the right of 40 cm s^{-1}. Compute (a) the period T, (b) the frequency f, (c) the angular frequency ω, (d) the total energy E, (e) the amplitude A, (f) the angle θ_0, (g) the maximum velocity v_{max}, (h) the maximum acceleration a_{max}, (i) the coordinate, velocity, and acceleration at a time $\pi/8$ s after the start of the motion.

a) $T = 2\pi\sqrt{\dfrac{m}{k}} = 2\pi\sqrt{\dfrac{25 \text{ g}}{400 \text{ dyn cm}^{-1}}} = \dfrac{\pi}{2} \text{ s} = 1.57 \text{ s}.$

b) $f = \dfrac{1}{T} = \dfrac{2}{\pi} \text{ Hz} = 0.638 \text{ Hz}.$

c) $\omega = 2\pi f = 4 \text{ rad s}^{-1}.$

d) $E = \frac{1}{2}mv_0^2 + \frac{1}{2}kx_0^2 = 40{,}000 \text{ ergs}.$

e) $A = \sqrt{2E/k} = 10\sqrt{2} \text{ cm}.$

f) $\sin\theta_0 = x_0/A = 1/\sqrt{2}, \qquad \theta_0 = \pi/4 \text{ rad}.$

g) $|v_{max}| = \sqrt{2E/m} = 40\sqrt{2} \text{ cm s}^{-1} = 56.6 \text{ cm s}^{-1}.$

The maximum velocity occurs at the midpoint, where $x = 0$. Hence, from Eq. (11–12),

$$|v_{max}| = \omega A = 40\sqrt{2} \text{ cm s}^{-1}.$$

h) The maximum acceleration occurs at the ends of the path where the force is a maximum. From Eq. (11–15),

$$|a_{max}| = \omega^2 x_{max} = 160\sqrt{2} \text{ cm s}^{-2}.$$

i) The equations of motion are:

$$\begin{aligned} x &= 10\sqrt{2} \sin\left(4t + \frac{\pi}{4}\right), \\ v &= 40\sqrt{2} \cos\left(4t + \frac{\pi}{4}\right), \\ a &= -160\sqrt{2} \sin\left(4t + \frac{\pi}{4}\right) \end{aligned}$$

When $t = \pi/8$ s, the phase angle is

$$\left(4t + \frac{\pi}{4}\right) = \frac{3\pi}{4} \text{ rad},$$

$$\begin{aligned} x &= 10\sqrt{2} \sin(3\pi/4) = 10 \text{ cm}, \\ v &= 40\sqrt{2} \cos(3\pi/4) = -40 \text{ cm s}^{-1}, \\ a &= -160\sqrt{2} \sin(3\pi/4) = -160 \text{ cm s}^{-2}. \end{aligned}$$

The curves of Fig. 11–5 represent the motion of the body in this example, if the scales of x, v, a, and t are such that

$$A = 10\sqrt{2} \text{ cm}, \qquad v_{max} = 40\sqrt{2} \text{ cm s}^{-1},$$
$$a_{max} = 160\sqrt{2} \text{ cm s}^{-2}, \qquad \text{and} \qquad T = \pi/2 \text{ s}.$$

The equations of simple harmonic motion can be given a geometrical interpretation as follows. Let the line segment OQ in Fig. 11–6(a), of length equal to the amplitude A, rotate with an angular velocity ω about the fixed point O. The rotating line segment is often referred

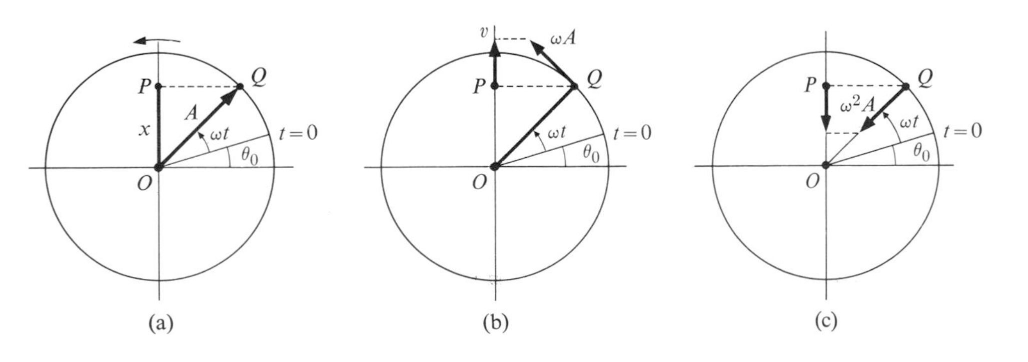

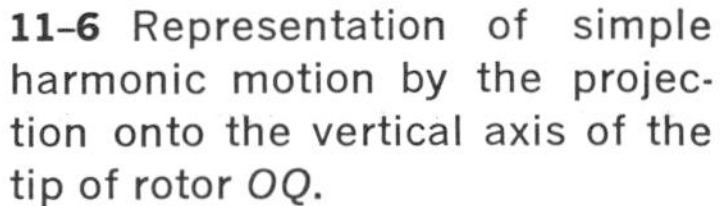

11-6 Representation of simple harmonic motion by the projection onto the vertical axis of the tip of rotor OQ.

to as a *rotating vector*, but strictly speaking it is not a vector quantity. That is, while it has a specified direction *in a diagram*, it has no specified direction *in space.* It is better described as a *rotor.* (The German term is "Zeiger," meaning a pointer in the sense of a clock hand or the pointer on a pressure gauge.) Let the rotor OQ make an angle with the horizontal axis, at time $t = 0$, equal to the initial phase angle θ_0. Point P is the projection of point Q onto the vertical axis, and as OQ rotates, point P oscillates along this axis.

We now show that the equations of motion of P are the same as those of a body oscillating with simple harmonic motion of amplitude A, angular frequency ω, and initial phase angle θ_0. Let x represent the length of OP. At any time t, the angle between the radius OQ and the horizontal axis equals the phase angle $\omega t + \theta_0$ and

$$x = A \sin(\omega t + \theta_0).$$

The velocity of point Q (see Fig. 11-6b) is ωA, and its vertical component, equal to the velocity of P, is

$$v = \omega A \cos(\omega t + \theta_0).$$

The acceleration of Q is its radial acceleration $\omega^2 A$ (see Fig. 11-6c) and its vertical component, equal to the acceleration of P, is

$$a = -\omega^2 A \sin(\omega t + \theta_0).$$

The negative sign must be included because the acceleration is negative whenever the sine of the phase angle is positive, and vice versa. These equations are, of course, just the general equations of harmonic motion. In the special case corresponding to Eqs. (11-18), the initial phase angle is 90° and the reference point Q is at the top of the circle when $t = 0$. If the reference point is at the right-hand end of the horizontal diameter when $t = 0$, then $\theta_0 = 0$ and the motion is described by Eqs. (11-19).

11-5 MOTION OF A BODY SUSPENDED FROM A COIL SPRING

Figure 11-7(a) shows a coil spring of force constant k and no-load length ℓ. When a body of mass m is attached to the spring as in part (b), it hangs in equilibrium with the spring extended by an amount $\Delta\ell$ such that the upward force P exerted by the spring is equal to the weight of the body, mg. But $P = k\,\Delta\ell$, so

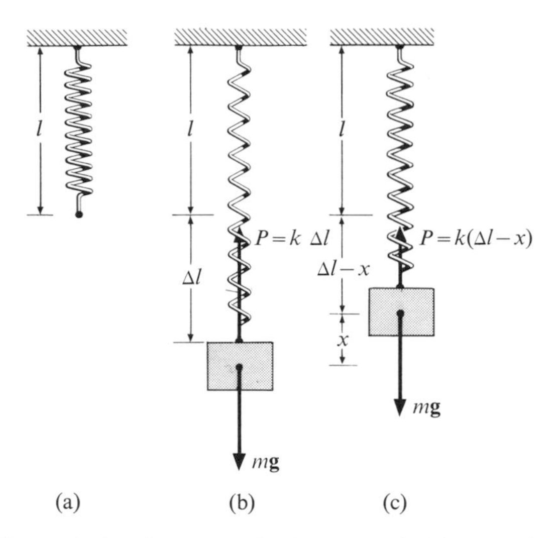

11-7 The restoring force on a body suspended by a spring is proportional to the coordinate measured from the equilibrium position.

$$k\,\Delta\ell = mg.$$

Now suppose the body is at a distance x above its equilibrium position, as in part (c). The extension of the spring is now $\Delta\ell - x$, the upward force it exerts on the body is $k(\Delta\ell - x)$, and the resultant force F on the body is

$$F = k(\Delta\ell - x) - mg = -kx.$$

The resultant force is therefore proportional to the displacement of the body *from its equilibrium position*, and if set in vertical motion the body oscillates with an angular frequency $\omega = \sqrt{k/m}$.

Except in the idealized case of a spring of zero mass, allowance must be made for the fact that the spring also oscillates. However, we cannot simply add the mass of the spring to that of the suspended body because not all portions of the spring oscillate with the same amplitude; the amplitude at the lower end equals that of the suspended body, while that at the upper end is zero. The correction term can be computed as follows.

Let L represent the length of the spring when the body is in its equilibrium position, and let m_s be the mass of the spring. Let us calculate the kinetic energy of the spring at an instant when the velocity of the lower end is v. Consider an element of the spring of length dy, at a distance y below the fixed upper end. The mass of the element, dm_s, is

$$dm_s = \frac{m_s}{L}\, dy.$$

All portions of the spring will be assumed to oscillate in phase and the velocity of the element, v_s, to be proportional to its distance from the fixed end: $v_s = (y/L)v$.

The kinetic energy of the element is

$$dE_k = \frac{1}{2} \cdot dm_s \cdot v_s^2 = \frac{1}{2} \cdot \frac{m_s}{L_s}\, dy \cdot \left(\frac{y}{L} v\right)^2,$$

and the total kinetic energy of the spring is

$$E_k = \frac{1}{2} \cdot \frac{m_s v^2}{L_3} \cdot \int_0^L y^2\, dy = \frac{1}{2}\left(\frac{1}{3} m_s\right) v^2.$$

This equals the kinetic energy of a body of mass one-third the mass of the spring, moving with the same velocity as that of the suspended body. In other words, the equivalent mass of the *vibrating system* equals that of the suspended body plus one-third the mass of the spring.

Example. A body of mass 1 kg is suspended from a coil spring whose mass is 0.09 kg and whose force constant is 66 N m^{-1}. Find the frequency and amplitude of the ensuing motion if the body is displaced 0.03 m below its equilibrium position and given a downward velocity of 0.4 m s^{-1}. The angular frequency is

$$\omega = \sqrt{\frac{k}{m + m_s/3}} = \sqrt{\frac{66 \text{ N m}^{-1}}{1.03 \text{ kg}}} = 8.00 \text{ rad s}^{-1}.$$

The amplitude is expressed in terms of the initial coordinate and velocity by means of Eq. (11-17). Thus,

$$A = \sqrt{x_0^2 + (v_0/\omega)^2} = \sqrt{(0.03 \text{ m})^2 + (0.4/8 \text{ m})^2} = 0.0582 \text{ m}.$$

11-6 THE SIMPLE PENDULUM

A *simple pendulum* (also called a *mathematical pendulum*) is defined as a particle suspended from a fixed point by a weightless, inextensible string. When the pendulum is displaced from the vertical by an angle θ, as in Fig. 11-8, the restoring force is $mg \sin\theta$, and the displacement s from the equilibrium position equals $L\theta$, where L is the length of the string and θ is in radians. The motion is therefore *not* harmonic, since the restoring force is proportional to $\sin\theta$ while the displacement is propor-

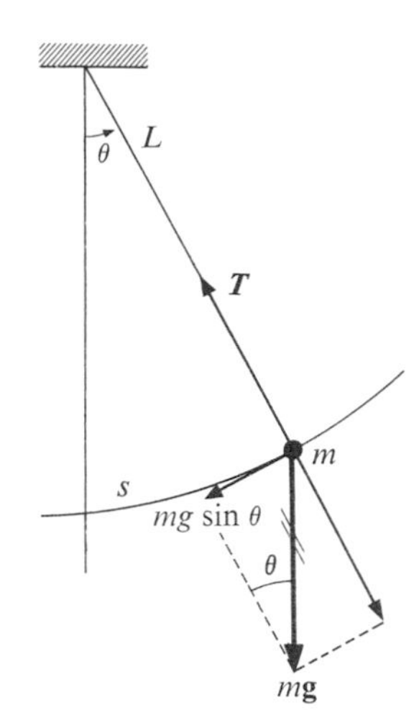

11-8 Forces on the bob of a simple pendulum.

tional to θ. However, if the angle θ is small we can approximate $\sin\theta$ by θ, and the restoring force is then

$$F \approx -mg\theta \approx -\left(\frac{mg}{L}\right) s.$$

The effective force constant of the pendulum is therefore $k = mg/L$, and the period is

$$T \approx 2\pi\sqrt{m/k} \approx 2\pi\sqrt{L/g}. \qquad (11\text{-}20)$$

It can be shown that the exact equation for the period, when the maximum angular displacement is ϕ, is given by the infinite series

$$T = 2\pi\sqrt{\frac{L}{g}}\left(1 + \frac{1^2}{2^2}\sin^2\frac{\phi}{2} + \frac{1^2 \cdot 3^2}{2^2 \cdot 4^2}\sin^4\frac{\phi}{2} + \cdots\right). \qquad (11\text{-}21)$$

The period can be computed to any desired degree of precision by taking enough terms in the series. When $\phi = 15°$, the true period differs from that given by the approximate equation (11-20) by less than 0.5%.

The utility of the pendulum as a timekeeper is based on the fact that the period is practically independent of the amplitude. Thus, as a pendulum clock runs down and as the amplitude of the swings becomes slightly smaller, the clock will still keep very nearly correct time.

11-9 A single swing of a simple pendulum.

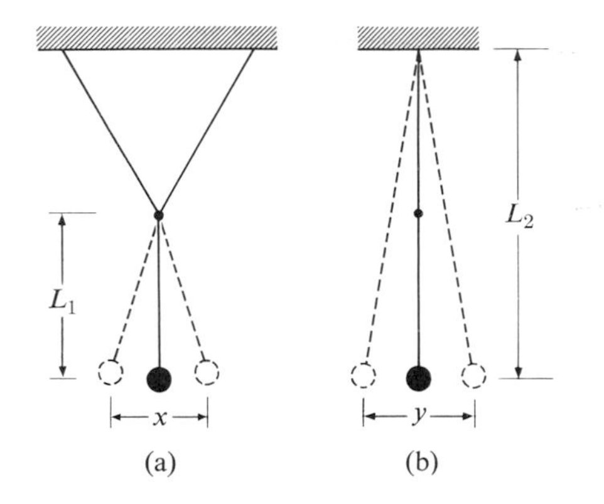

11-10 Double pendulum for producing Lissajous figures.

The simple pendulum is also a precise and convenient method of measuring the acceleration of gravity, g, without actually resorting to free fall, since L and T may readily be measured. More complicated pendulums find considerable application in the field of geophysics. Local deposits of ore or oil, if their density differs from that of their surroundings, affect the local value of g, and precise measurements of this quantity over an area which is being prospected often furnish valuable information regarding the nature of underlying deposits.

Figure 11-9 is a multiflash photograph of a single swing of a simple pendulum.

11-7 LISSAJOUS FIGURES

The curves known as *Lissajous figures* are the paths traced out by a particle which is oscillating simultaneously in two mutually perpendicular directions. In general, the amplitude and frequency may be different in the two directions, and the two oscillations may have an arbitrary initial phase difference.

The pendulum bob in Fig. 11-10, supported by three cords forming a Y, illustrates one means of producing this type of motion. When vibrating in the x-direction, as in (a), the frequency is that of a simple pendulum of length L_1. In the y-direction the frequency is that of a pendulum of length L_2. If displaced in both the x- and y-directions, and released, the bob executes vibrations of both frequencies simultaneously.

The spot on the screen of a cathode-ray tube, produced by the impact of a rapidly moving stream of electrons, will also move in a Lissajous figure if sinusoidal alternating voltages are applied simultaneously to the horizontal and vertical deflecting plates.

The general expressions for the x- and y-coordinates of the oscillating particle are

$$x = A_x \sin(\omega_x t + \theta_1), \qquad y = A_y \sin(\omega_y t + \theta_2),$$

where A_x and A_y are the respective amplitudes, ω_x and ω_y the corresponding angular frequencies, and θ_1 and θ_2 the initial phase angles. These are the equations of the path in parametric form.

The equations above are represented graphically by the rotor diagrams of Fig. 11-11. The x-coordinate of the tip of the rotor in the lower diagram gives the x-coordinate of the oscillating particle, and the y-coordinate of the tip of the rotor in the upper diagram gives its y-coordinate. Hence by projecting up, and across, from the tips of these rotors, the position of the particle at any time can be determined. The diagram shows its position at time $t = 0$ and at some arbitrary later time t.

The curves in Fig. 11-12 are a few Lissajous figures for various frequency ratios and initial phase differences $(\theta_2 - \theta_1)$. The amplitudes A_x and A_y are equal in each

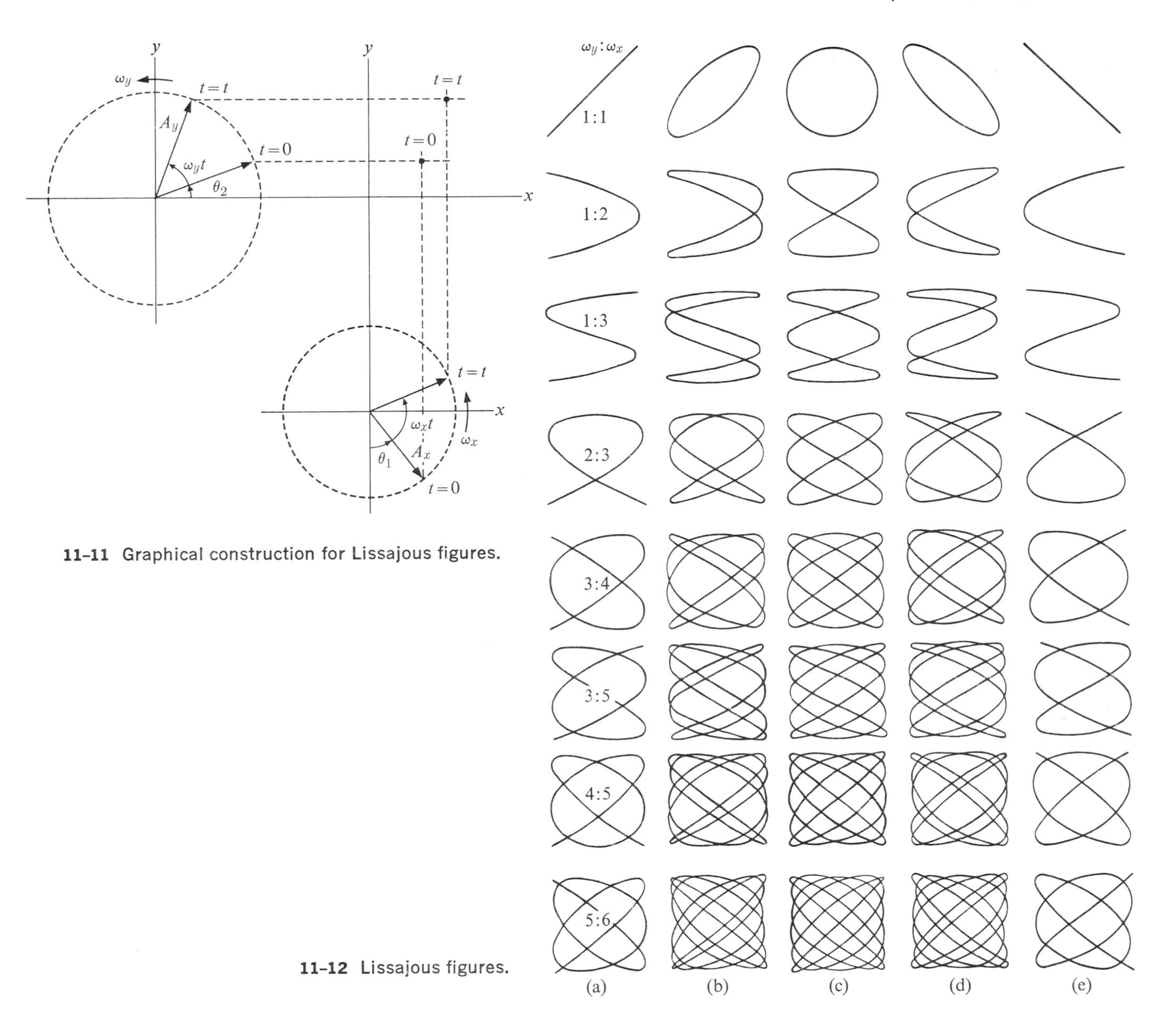

11-11 Graphical construction for Lissajous figures.

11-12 Lissajous figures.

case. If the frequencies are commensurable, as in the figures shown, the particle retraces a closed path over and over. If they are not, the path does not close on itself, and the pattern may be extremely complicated. If the frequencies are very nearly commensurable, the path changes very slowly, and if the motion itself is rapid, as it often is when the figures are formed on an oscilloscope, the impression is that of a closed curve which gradually alters its form. Thus, if the frequencies are very nearly equal ($\omega_x/\omega_y \approx 1$), the path changes slowly from a straight line at 45°, as in Fig. 11-12(a), to an ellipse as in (b), then to a circle as in (c), then to an ellipse as in (d), with its major axis at right angles to that in (b), then to a straight line as in (e), etc.

11-8 ANGULAR HARMONIC MOTION

Angular harmonic motion is an exact mathematical analog of linear harmonic motion. Let a body be pivoted about a fixed axis and acted on by a restoring torque Γ proportional to the angular displacement ϕ from some reference position. Then

$$\Gamma = -k'\phi,$$

where the factor k', or the restoring torque per unit of angular displacement, is called the *torque constant*. If frictional torques are negligible, the differential equation of motion is

$$\Gamma = -k'\phi = I\alpha = I\omega \frac{d\omega}{d\phi},$$

or

$$I\omega \frac{d\omega}{d\phi} + k'\phi = 0, \tag{11-22}$$

where I is the moment of inertia about the fixed axis. This equation has exactly the same form as Eq. (11-1): angular displacement ϕ corresponds to linear displacement x, moment of inertia I corresponds to mass m, and the torque constant k' corresponds to the force constant k. By analogy, the equation of motion is

$$\phi = \phi_m \sin(\omega t + \theta_0),$$

where $\omega = \sqrt{k'/I}$, and ϕ_m is the maximum angular displacement or the *angular amplitude*.

The balance wheel of a watch is a common example of a body vibrating with angular harmonic motion. In the ideal case assumed, the motion in isochronous and the watch "keeps time" even though the amplitude decreases as the mainspring unwinds.

11-9 THE PHYSICAL PENDULUM

In Fig. 11-13, a body of arbitrary shape is pivoted about a fixed axis through O, and the line joining O and the center of gravity is displaced by an angle ϕ from the vertical. Let h represent the distance from the pivot to the center of gravity. The weight mg gives rise to a restoring torque

$$\Gamma = -mgh \sin\phi.$$

When released, the body will oscillate about its equilibrium position but, as in the case of the simple

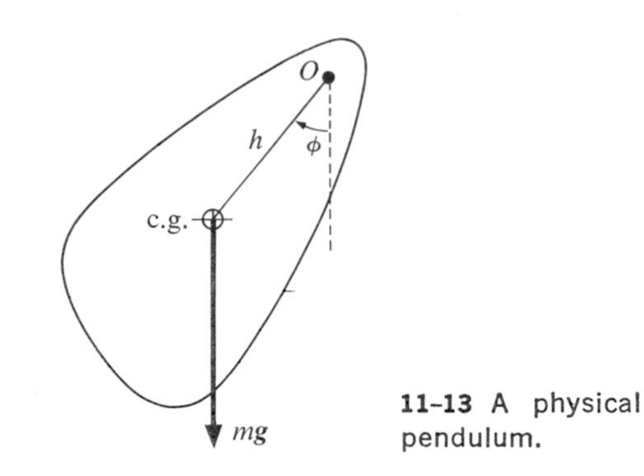

11-13 A physical pendulum.

pendulum, the motion is not (angular) harmonic, since the torque Γ is not proportional to ϕ but to $\sin\phi$. However, if ϕ is small, we can approximate $\sin\phi$ by ϕ, and the motion is approximately harmonic. Making this approximation, we get

$$\Gamma \approx -(mgh)\phi,$$

and the effective torque constant is

$$k' = -\Gamma/\phi = mgh.$$

The angular frequency is

$$\omega \approx \sqrt{k'/I} \approx \sqrt{mgh/I},$$

and the period T is

$$T = \frac{1}{f} = \frac{2\pi}{\omega} \approx 2\pi\sqrt{\frac{I}{mgh}}. \tag{11-23}$$

Such a pivoted body is called a *physical pendulum*, as contrasted with the ideal simple pendulum, which is a point mass attached to a weightless cord. Of course, every real pendulum is a physical pendulum.

Example. Equation (11-23) may be solved for the moment of inertia I, giving

$$I = \frac{T^2 mgh}{4\pi^2}.$$

The quantities on the right of the equation are all directly measurable. Hence the moment of inertia of a body of any complex shape may be found by suspending the body as a physical pendulum and measuring its period of vibration. The location of the center of gravity can be found by balancing. Since

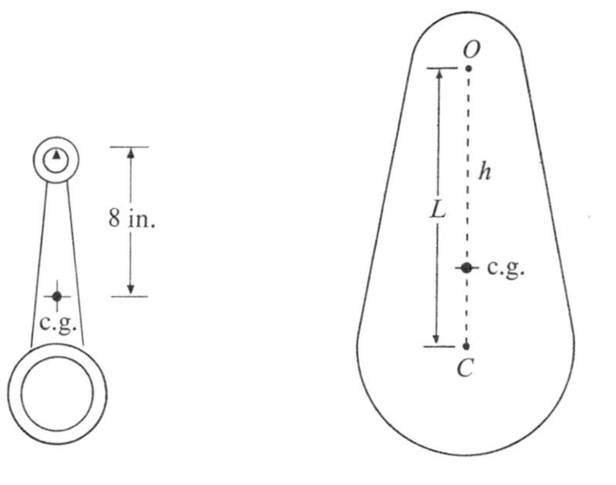

Fig. 11-14

11-15 Center of oscillation. The length L equals that of the equivalent simple pendulum.

T, m, g, and h are known, I can be computed. For example, Fig. 11-14 illustrates a connecting rod pivoted about a horizontal knife edge. The connecting rod weighs 4 lb and its center of gravity has been found by balancing to be 8 in. below the knife edge. When set into oscillation, it is found to make 100 complete vibrations in 120 s, so that $T = 120/100 = 1.2$ s. Therefore

$$I = \frac{(1.2)^2 \text{ s}^2 \times 4 \text{ lb} \times \frac{2}{3} \text{ ft}}{4\pi^2} = 0.097 \text{ slug ft}^2.$$

11-10 CENTER OF OSCILLATION

It is always possible to find an *equivalent* simple pendulum whose period is equal to that of a given physical pendulum. If L is the length of the equivalent simple pendulum,

$$T = 2\pi\sqrt{\frac{L}{g}} = 2\pi\sqrt{\frac{I}{mgh}}$$

or

$$L = \frac{I}{mh}. \qquad (11\text{-}24)$$

Thus, so far as its period of vibration is concerned, the mass of a physical pendulum may be considered to be concentrated at a point whose distance from the pivot is $L = I/mh$. This point is called the *center of oscillation* of the pendulum.

Example. A slender uniform rod of length a is pivoted at one end and swings as a physical pendulum. Find the center of oscillation of the pendulum.

The moment of inertia of the rod about an axis through one end is

$$I = \tfrac{1}{3}ma^2.$$

The distance from the pivot to the center of gravity is $h = a/2$. The length of the equivalent simple pendulum is

$$L = \frac{I}{mh} = \frac{\frac{1}{3}ma^2}{m(a/2)} = \frac{2}{3}a,$$

and the center of oscillation is at a distance $2a/3$ from the pivot.

Figure 11-15 shows a body pivoted about an axis through O and whose center of oscillation is at point C. The center of oscillation and the point of support have the following interesting property, namely, if the pendulum is pivoted about a new axis through point C, its period is unchanged and point O becomes the new center of oscillation. The point of support and the center of oscillation are said to be *conjugate* to each other.

The center of oscillation has another important property. Figure 11-16 shows a baseball bat pivoted at O. If a ball strikes the bat at its center of oscillation, no impulsive force is exerted on the pivot and hence no "sting" is felt if the bat is held at that point. Because of this property, the center of oscillation is called the *center of percussion*.

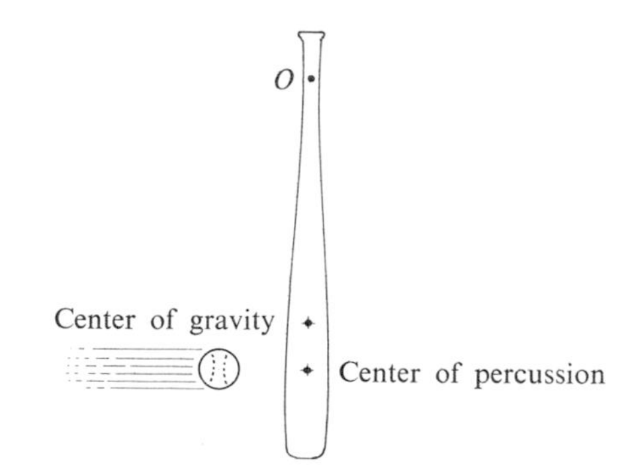

11-16 The center of percussion coincides with the center of oscillation.

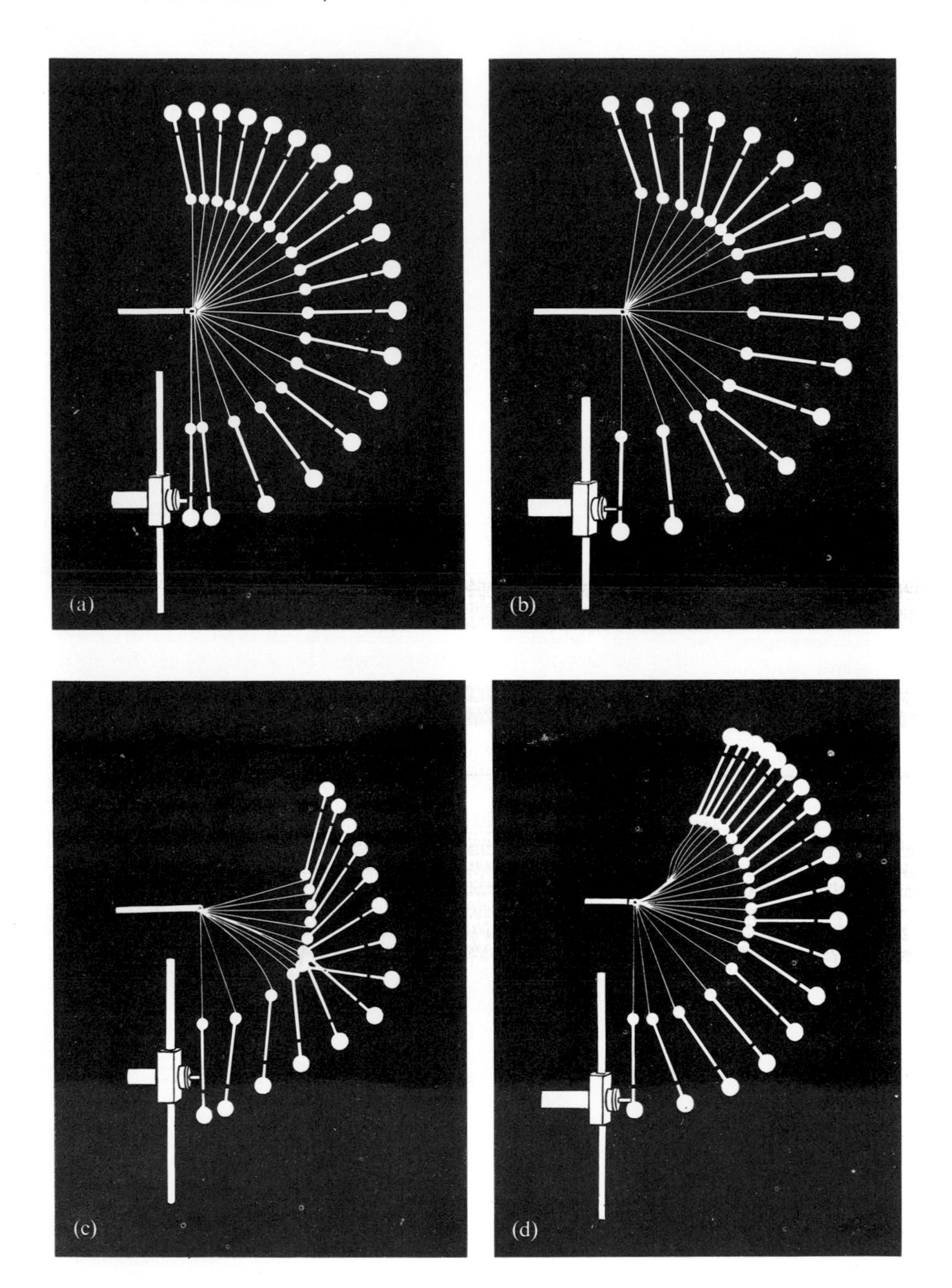

11-17 The motion of a body suspended by a cord when the body is struck a horizontal blow.

In Fig. 11-17, a series of multiflash photographs, the *center of gravity* is marked by a black band. In (a), the body is struck at its center of percussion relative to a pivot at the upper end of the cord, and it starts to swing smoothly about this pivot. In (b), the body is struck at its center of gravity. Note that it does not start to rotate about the pivot, but that its initial motion is one of pure translation. That is, the center of percussion does not coincide with the center of gravity. In (c), the body is struck above, and in (d) below its center of percussion.

Problems

11-1 The general equation of simple harmonic motion,

$$y = A \sin (\omega t + \theta_0),$$

can be written in the equivalent form

$$y = B \sin \omega t + C \cos \omega t.$$

(a) Find the expressions for the amplitudes B and C in terms of the amplitude A and the initial phase angle θ_0. (b) Interpret these expressions in terms of a rotating vector diagram.

11-2 A body of mass 0.25 kg is acted on by an elastic restoring force of force constant $k = 25$ N m^{-1}. (a) Construct the graph of elastic potential energy E_p as a function of displacement x, over a range of x from -0.3 m to $+0.3$ m. Let 1 in. = 0.25 J vertically, and 1 in. = 0.1 m horizontally.

The body is set into oscillation with an initial potential energy of 0.6 J and an initial kinetic energy of 0.2 J. Answer the following questions by reference to the graph: (b) What is the amplitude of oscillation? (c) What is the potential energy when the displacement is one-half the amplitude? (d) At what displacement are the kinetic and potential energies equal? (e) What is the speed of the body at the midpoint of its path? Find (f) the period T, (g) the frequency f, and (h) the angular frequency ω. (i) What is the initial phase angle θ_0 if the amplitude $A = 15$ cm, the initial displacement $x_0 = 7.5$ cm, and the initial velocity v_0 is negative?

11-3 A body is vibrating with simple harmonic motion of amplitude 15 cm and frequency 4 Hz. Compute (a) the maximum values of the acceleration and velocity, (b) the acceleration and velocity when the coordinate is 9 cm, (c) the time required to move from the equilibrium position to a point 12 cm distant from it.

11-4 A body of mass 10 g moves with simple harmonic motion of amplitude 24 cm and period 4 s. The coordinate is +24 cm when $t = 0$. Compute (a) the position of the body when $t = 0.5$ s, (b) the magnitude and direction of the force acting on the body when $t = 0.5$ s, (c) the minimum time required for the body to move from its initial position to the point where $x = -12$ cm, (d) the velocity of the body when $x = -12$ cm.

11-5 The motion of the piston of an automobile engine is approximately simple harmonic. (a) If the stroke of an engine (twice the amplitude) is 4 in. and the angular velocity is 3600 rev min^{-1}, compute the acceleration of the piston at the end of its stroke. (b) If the piston weighs 1 lb, what resultant force must be exerted on it at this point? (c) What is the velocity of the piston, in miles per hour, at the midpoint of its stroke?

11-6 A body whose weight is 4 lb is suspended from a spring of negligible mass, and is found to stretch the spring 8 in. (a) What is the force constant of the spring? (b) What is the period of oscillation of the body, if pulled down and released? (c) What would be the period of a body weighing 8 lb, hanging from the same spring?

11-7 The scale of a spring balance reading from zero to 32 lb is 6 in. long. A body suspended from the balance is observed to oscillate vertically at 1.5 Hz. What is the weight of the body? Neglect the mass of the spring.

11-8 A body weighing 8 lb is attached to a coil spring and oscillates vertically in simple harmonic motion. The amplitude is 2 ft, and at the highest point of the motion the spring has its natural unstretched length. Calculate the elastic potential energy of the spring, the kinetic energy of the body, its gravitational potential energy relative to the lowest point of the motion, and the sum of these three energies, when the body is (a) at its lowest point, (b) at its equilibrium position, and (c) at its highest point.

11-9 A load of 320 lb suspended from a wire whose unstretched length ℓ_0 is 10 ft is found to stretch the wire by 0.12 in. The cross-sectional area of the wire, which can be assumed constant, is 0.016 in^2. (a) If the load is pulled down a small additional distance and released, find the frequency at which it will vibrate. (b) Compute Young's modulus for the wire.

11-10 A small block is executing simple harmonic motion in a horizontal plane with an amplitude of 10 cm. At a point 6 cm away from equilibrium the velocity is ± 24 cm s^{-1}. (a) What is the period? (b) What is the displacement when the velocity is ± 12 cm s^{-1}? (c) If a small object placed on the oscillating block is just on the verge of slipping at the endpoint of the path, what is the coefficient of friction?

11-11 A force of 6 lb stretches a vertical spring 9 in. (a) What weight must be suspended from the spring so that the system will oscillate with a period of $\pi/4$ s? (b) If the amplitude of the motion is 3 in., where is the body and in what direction is it moving $\pi/12$ s after it has passed the equilibrium position, moving downward? (c) What force does the spring exert on the body when it is 1.8 in. below the equilibrium position, moving upward?

11-12 A body of mass m is suspended from a coil spring and the time for 100 complete oscillations is measured for the following values of m:

m (g)	100	200	400	1000
Time of 100 oscillations (s)	23.4	30.6	41.8	64.7

Plot graphs of the measured values of (a) T vs. m, and (b) T^2 vs. m. (c) Are the experimental results in agreement with theory? (d) Is either graph a straight line? (e) Does the straight line pass through the origin? (f) What is the force constant of the spring? (g) What is the mass of the spring?

11-13 A body of mass 100 g hangs from a long spiral spring. When pulled down 10 cm below its equilibrium position and released, it vibrates with a period of 2 s. (a) What is its velocity as it passes through the equilibrium position? (b) What is its acceleration when it is 5 cm above the equilibrium position? (c) When it is moving upward, how long a time is required for it to move from a point 5 cm below its equilibrium position to a point 5 cm above it? (d) How much will the spring shorten if the body is removed?

11-14 A body whose mass is 4.9 kg hangs from a spring and oscillates with a period of 0.5 s. How much will the spring shorten when the body is removed?

11-15 Four passengers whose combined weight is 600 lb are observed to compress the springs of an automobile by 2 in. when they enter the automobile. If the total load supported by the springs is 1800 lb, find the period of vibration of the loaded automobile.

11-16 (a) A block suspended from a spring vibrates with simple harmonic motion. At an instant when the displacement of the block is equal to one-half the amplitude, what fraction of the total energy of the system is kinetic and what fraction is potential? (b) When the block is in equilibrium, the length of the spring is an amount s greater than in the unstretched state. Prove that $T = 2\pi\sqrt{s/g}$.

11-17 (a) With what additional force must a vertical spring carrying an 8-lb body in equilibrium be stretched so that, when released, it will perform 48 complete oscillations in 32 s with an amplitude of 3 in.? (b) What force is exerted by the spring on the body when it is at the lowest point, the middle, and the highest point of the path? (c) What is the kinetic energy of the system when the body is 1 in. below the middle of the path? its potential energy?

11-18 A force of 12 lb stretches a certain spring 9 in. A body weighing 8 lb is hung from this spring and allowed to come to rest. It is then pulled down 4 in. and released. (a) What is the period of the motion? (b) What is the magnitude and direction of the acceleration of the body when it is 2 in. above the equilibrium position, moving upward? (c) What is the tension in the spring when the body is 2 in. above the equilibrium position? (d) What is the shortest time required to go from the equilibrium position to the point 2 in. above? (e) If a small object were placed on the oscillating body would it remain or leave? (f) If a small object were placed on the oscillating body and its amplitude doubled, where would the object and the oscillating body begin to separate?

11-19 Two springs with different force constants k_1 and k_2 are attached to a block of mass m on a level frictionless surface. Calculate the effective force constant in each of the three cases (a), (b), and (c) depicted in Fig. 11-18. (d) A body of mass m, suspended from a spring with a force constant k, vibrates with a frequency f_1. When the spring is cut in half and the same body is suspended from one of the halves, the frequency is f_2. What is the ratio of f_2/f_1?

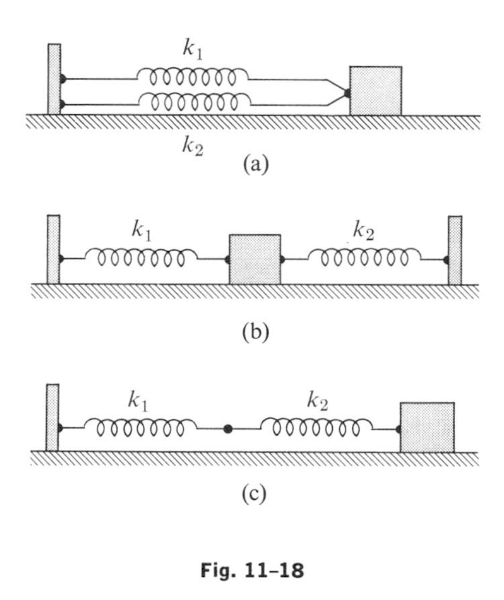

Fig. 11-18

11-20 Two springs, each of unstretched length 20 cm but having different force constants k_1 and k_2, are attached to opposite ends of a block of mass m on a level frictionless surface. The outer ends of the springs are now attached to two pins P_1 and P_2, 10 cm from the original positions of the springs. Let $k_1 = 1000$ dyn cm^{-1}, $k_2 = 3000$ dyn cm^{-1}, $m = 100$ g. (See Fig. 11-19.) (a) Find the length of each spring when the block is in its new equilibrium position, after the springs have been attached to the pins. (b) Find the period of vibration of the block if it is slightly displaced from its new equilibrium position and released.

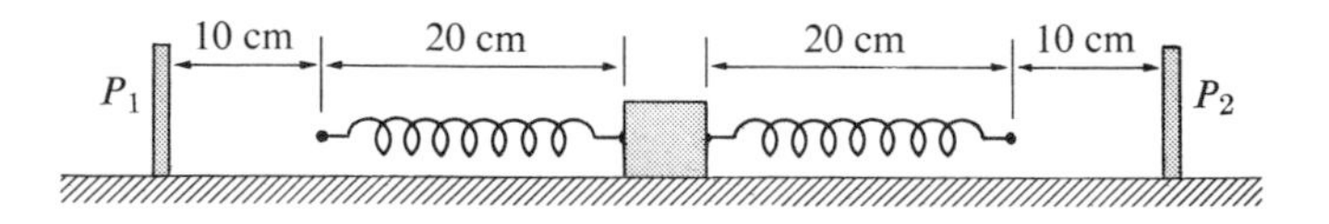

Fig. 11-19

11-21 The block in Problem 11-20 and Fig. 11-19 is oscillating with an amplitude of 5 cm. At the instant it passes through its equilibrium position, a lump of putty of mass 100 g is dropped vertically onto the block and sticks to it. (a) Find the new period and amplitude. (b) Was there a loss of energy and, if so, where did it go? (c) Would the answers be the same if the putty had been dropped on the block when it was at one end of its path?

11-22 A simple pendulum 8 ft long swings with an amplitude of 1 ft. (a) Compute the velocity of the pendulum at its lowest point. (b) Compute its acceleration at the ends of its path.

11-23 Find the length of a simple pendulum whose period is exactly 1 s at a point where $g = 32.2$ ft s^{-2}.

11-24 (a) What is the change dT in the period of a simple pendulum when the acceleration due to gravity changes by dg? (b) What is the fractional change in period, dT/T, in terms of the fractional change dg/g? (c) A pendulum clock which keeps correct time at a point where $g = 980.0$ cm sec^{-2} is found to lose 10 s day^{-1} at a higher altitude. Use the result of part (b) above to find the value of g at the new location, approximating the differentials dT and dg by the small finite changes in T and g.

(This problem illustrates a useful technique when it is desired to find the new value of a quantity, given a small change in another quantity on which the first depends. It enables one to compute the *change* in the quantity instead of computing the new value directly. To compute the new value of g directly to within 0.1 cm s^{-2}, computations must be carried out to four significant figures, which is beyond the precision of an ordinary slide rule and requires the use of four-place logarithms, or longhand multiplication and division. The *change* in g, on the other hand, is readily computed to within 0.1 cm s^{-2} on a small slide rule.)

11-25 A simple pendulum with a supporting steel wire of cross-sectional area 0.01 cm^2 is observed to have a period of 2 s when a 10-kg lead bob is used. The lead bob is replaced by an aluminum bob of the same dimensions having a mass of 2 kg, and the period is remeasured. (a) What was the length of the pendulum with the load bob? (b) By what fraction is the period changed when the aluminum bob is used? Is it an increase or decrease? [*Hint.* Use differentials.]

11-26 The balance wheel of a watch vibrates with an angular amplitude of π radians and with a period of 0.5 s. (a) Find its maximum angular velocity. (b) Find its angular velocity when its displacement is one-half its amplitude. (c) Find its angular acceleration when its displacement is 45°.

11-27 A monkey wrench is pivoted at one end and allowed to swing as a physical pendulum. The period is 0.9 s and the pivot is 6 in. from the center of gravity. (a) What is the radius of gyration of the wrench about an axis through the pivot? (b) If the wrench was initially displaced 0.1 rad from its equilibrium position, what is the angular velocity of the wrench as it passes through the equilibrium position?

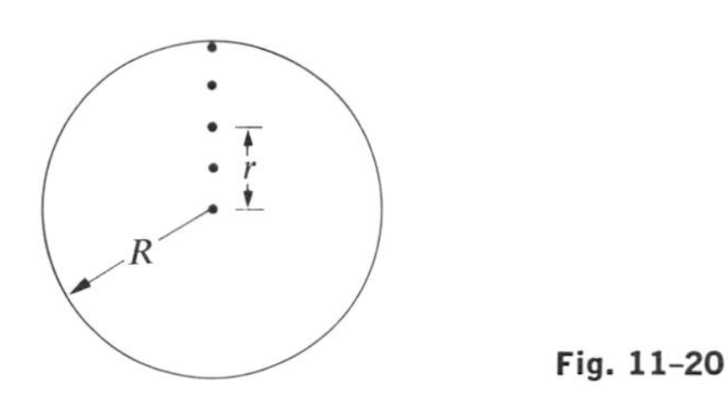

Fig. 11-20

11-28 It is shown in textbooks on mechanics that the moment of inertia I of a body about *any* axis through any point is given by

$$I = I_G + mh^2,$$

where I_G is the moment of inertia about a *parallel axis* through the center of gravity, m is the mass, and h is the perpendicular distance between the two parallel axes. A solid disk of radius $R = 12$ cm oscillates as a physical pendulum about an axis perpendicular to the plane of the disk at a distance r from its center. (See Fig. 11-20.) (a) Calculate the period of oscillation (for small amplitudes) for the following values of r: 0, $R/4$, $R/2$, $3R/4$, R. (b) Let T_0 represent the period when $r = R$, and T the period at any other value of r. Construct a graph of the dimensionless ratio T/T_0 as a function of the dimensionless ratio r/R. (Note that the graph then describes the behavior of *any* solid disk, whatever its radius.) (c) Prove by the methods of calculus that the period is a minimum when $r = R/\sqrt{2}$. Does this result agree with your graph?

11-29 It is desired to construct a pendulum of period 10 s. (a) What is the length of a *simple* pendulum having this period? (b) Suppose the pendulum must be mounted in a case not over 2 ft high. Can you devise a pendulum, having a period of 10 s, that will satisfy this requirement?

11-30 A meter stick hangs from a horizontal axis at one end and oscillates as a physical pendulum. A body of small dimensions, and of mass equal to that of the meter stick, can be clamped to the stick at a distance h below the axis. Let T represent the period of the system with the body attached, and T_0 the period of the meter stick alone. Find the ratio T/T_0, (a) when $h = 50$ cm, (b) when $h = 100$ cm. (c) Is there any value of h for which $T = T_0$? If so, find it and explain why the period is unchanged when h has this value.

11-31 A meter stick is pivoted at one end. At what distance below the pivot should it be struck in order that it start swinging smoothly about the pivot?

12 Hydrostatics

12-1 INTRODUCTION

The term "hydrostatics" is applied to the study of fluids at rest, and "hydrodynamics" to fluids in motion. The special branch of hydrodynamics relating to the flow of gases and of air in particular is called "aerodynamics."

A fluid is a substance which can flow. Hence the term includes both liquids and gases. Liquids and gases differ markedly in their compressibilities; a gas is easily compressed, while a liquid is practically incompressible. The small volume changes of a liquid under pressure can usually be neglected in this part of the subject.

The density of a homogeneous material is defined as its mass per unit volume. Units of density in the three systems are one kilogram per cubic meter (1 kg m^{-3}), one gram per cubic centimeter (1 g cm^{-3}), and one slug per cubic foot (1 slug ft^{-3}). We shall represent density by the greek letter ρ (rho):

$$\rho = \frac{m}{V}, \qquad m = \rho V. \tag{12-1}$$

For example, the weight of 1 ft^3 of water is 62.5 lb; its density is $62.5/32.2 = 1.94$ slugs ft^{-3}. Typical values of density at room temperature are given in Table 12-1.

The *specific gravity* of a material is the ratio of its density to that of water and is therefore a pure number. "Specific gravity" is an exceedingly poor term, since it has nothing to do with gravity. "Relative density" would describe the concept more precisely.

TABLE 12-1 DENSITIES

Material	Density, g cm^{-3}	Material	Density, g cm^{-3}
Aluminum	2.7	Silver	10.5
Brass	8.6	Steel	7.8
Copper	8.9	Mercury	13.6
Gold	19.3	Ethyl alcohol	0.81
Ice	0.92	Benzene	0.90
Iron	7.8	Glycerine	1.26
Lead	11.3	Water	1.00
Platinum	21.4		

12-2 PRESSURE IN A FLUID

When the concept of hydrostatic pressure was introduced in Section 10-1, the weight of the fluid was neglected and the pressure was assumed the same at all points. It is a familiar fact, however, that atmospheric pressure decreases with increasing altitude and that the pressure in a lake or in the ocean decreases with increasing distance from the bottom. We therefore generalize the definition of pressure and define the pressure *at any point* as the ratio of the normal force dF

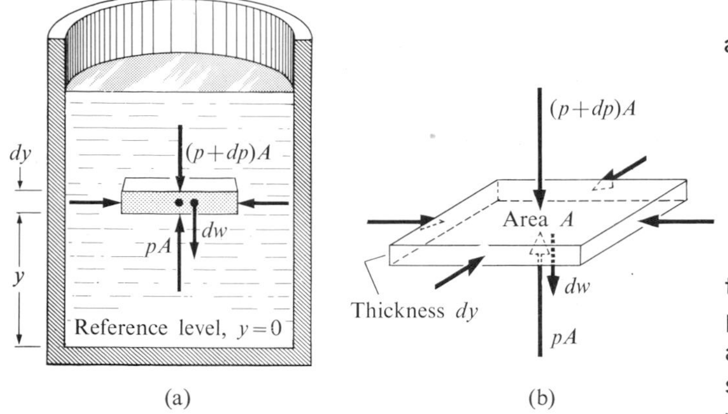

12–1 Forces on an element of fluid in equilibrium.

exerted on a small area dA including the point, to the area dA:

$$p = \frac{dF}{dA}, \qquad dF = p\,dA. \tag{12–2}$$

If the pressure is the same at all points of a finite plane surface of area A, these equations reduce to Eq. (10–3):

$$p = \frac{F}{A}, \qquad F = pA.$$

Let us find the general relation between the pressure p at any point in a fluid and the elevation of the point y. If the fluid is in equilibrium, every volume element is in equilibrium. Consider an element in the form of a thin slab, shown in Fig. 12–1, whose thickness is dy and whose faces have an area A. If ρ is the density of the fluid, the mass of the element is $\rho A\,dy$ and its weight dw is $\rho gA\,dy$. The force exerted on the element by the surrounding fluid is everywhere normal to its surface. By symmetry, the resultant horizontal force on its rim is zero. The upward force on its lower face is pA, and the downward force on its upper face is $(p + dp)A$. Since it is in equilibrium,

$$\sum F_y = 0,$$
$$pA - (p + dp)A - \rho gA\,dy = 0,$$

and therefore

$$\boxed{\frac{dp}{dy} = -\rho g.} \tag{12–3}$$

Since ρ and g are both positive quantities, it follows that a positive dy (an increase of elevation) is accompanied by a negative dp (decrease of pressure). If p_1 and p_2 are the pressures at elevations y_1 and y_2 above some reference level, then integration of Eq. (12–3), when ρ and g are constant, gives

$$p_2 - p_1 = -\rho g(y_2 - y_1).$$

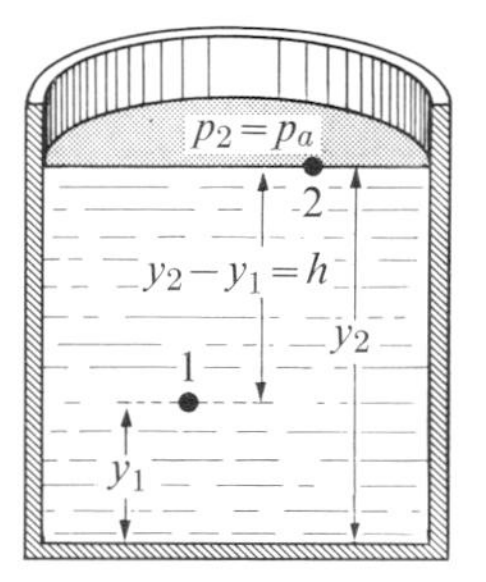

Fig. 12–2

Let us apply this equation to a liquid in an open vessel, such as that shown in Fig. 12–2. Take point 1 at any level and let p represent the pressure at this point. Take point 2 at the top where the pressure is atmospheric pressure, p_a. Then

$$p_a - p = -\rho g(y_2 - y_1),$$
$$p = p_a + \rho gh. \tag{12–4}$$

Note that the shape of the containing vessel does not affect the pressure, and that the pressure is the same at all points at the same depth. It also follows from Eq. (12–4) that if the pressure p_a is increased in any way, say by inserting a piston on the top surface and pressing down on it, the pressure p at any depth must increase by exactly the same amount. This fact was stated by the

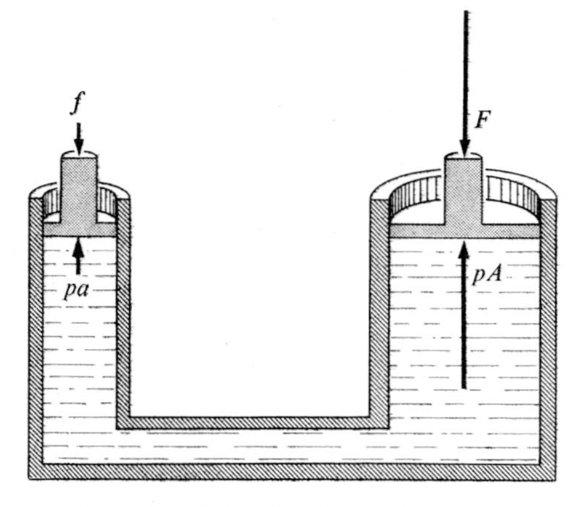

12-3 Principle of the hydraulic press.

French scientist Blaise Pascal (1623–1662) in 1653 and is called "Pascal's law." It is often stated: "*Pressure applied to an enclosed fluid is transmitted undiminished to every portion of the fluid and the walls of the containing vessel.*" We can see now that it is not an independent principle but a necessary consequence of the laws of mechanics.

Pascal's law is illustrated by the operation of a hydraulic press, shown in Fig. 12-3. A piston of small cross-sectional area a is used to exert a small force f directly on a liquid such as oil. The pressure $p = f/a$ is transmitted through the connecting pipe to a larger cylinder equipped with a larger piston of area A. Since the pressure is the same in both cylinders,

$$p = \frac{f}{a} = \frac{F}{A} \quad \text{and} \quad F = \frac{A}{a} \times f.$$

It follows that the hydraulic press is a force-multiplying device with a multiplication factor equal to the ratio of the areas of the two pistons. Barber chairs, dentist chairs, car lifts, and hydraulic brakes are all devices that make use of the principle of the hydraulic press.

12-3 THE HYDROSTATIC PARADOX

If a number of vessels of different shapes are interconnected as in Fig. 12-4(a), it will be found that a liquid poured into them will stand at the same level in each. Before the principles of hydrostatics were completely understood, this seemed a very puzzling phenomenon and was called the "hydrostatic paradox." It would appear at first sight, for example, that vessel C should develop a greater pressure at its base than should B, and hence that liquid would be forced from C into B.

Equation (12-4), however, states that the pressure depends only on the depth below the liquid surface and not at all on the shape of the containing vessel. Since the depth of the liquid is the same in each vessel, the pressure at the base of each is the same and hence the system is in equilibrium.

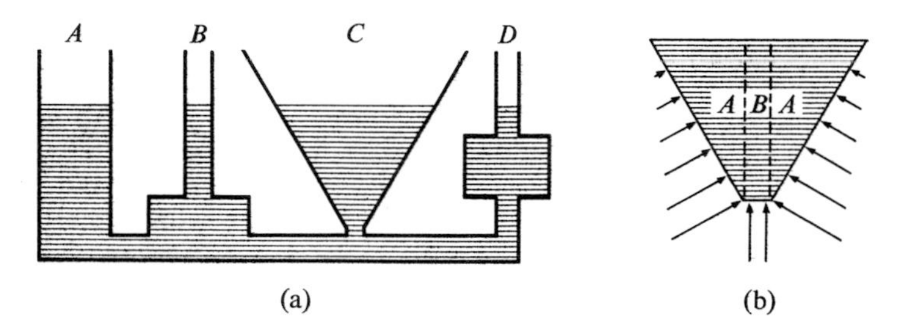

12-4 (a) The hydrostatic paradox. The top of the liquid stands at the same level in each vessel. (b) Forces on the liquid in vessel C.

A more detailed explanation may be helpful in understanding the situation. Consider vessel C in Fig. 12-4(b). The forces exerted against the liquid by the walls are shown by arrows, the force being everywhere perpendicular to the walls of the vessel. The inclined forces at the sloping walls may be resolved into horizontal and vertical components. The weight of the liquid in the sections lettered A is supported by the vertical components of these forces. Hence the pressure at the base of the vessel is due only to the weight of the liquid in the cylindrical column B. Any vessel, regardless of its shape, may be treated the same way.

12-4 PRESSURE GAUGES

The simplest type of pressure gauge is the open-tube manometer, illustrated in Fig. 12-5(a). It consists of a U-shaped tube containing a liquid, one end of the tube being at the pressure p which it is desired to measure, while the other end is open to the atmosphere.

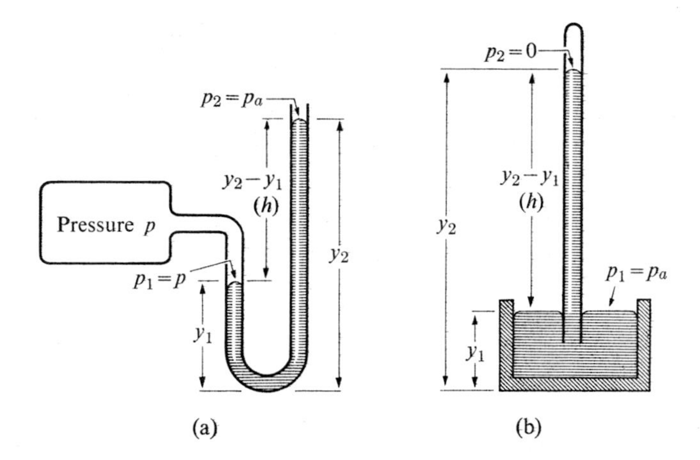

12–5 (a) The open-tube manometer. (b) The barometer.

The pressure at the bottom of the left column is $p + \rho g y_1$, while that at the bottom of the right column is $p_a + \rho g y_2$, where ρ is the density of the manometric liquid. Since these pressures both refer to the same point, it follows that

$$p + \rho g y_1 = p_a + \rho g y_2,$$

and

$$p - p_a = \rho g(y_2 - y_1) = \rho g h.$$

The pressure p is called the *absolute pressure*, whereas the difference $p - p_a$ between this and the atmospheric pressure is called the *gauge pressure*. It is seen that the gauge pressure is proportional to the difference in height of the liquid columns.

The mercury barometer is a long glass tube that has been filled with mercury and then inverted in a dish of mercury, as shown in Fig. 12–5(b). The space above the mercury column contains only mercury vapor, whose pressure, at room temperature, is so small that it may be neglected. It is easily seen that

$$p_a = \rho g(y_2 - y_1) = \rho g h.$$

Because mercury manometers and barometers are used so frequently in laboratories, it is customary to express atmospheric pressure and other pressures as so many "inches of mercury," "centimeters of mercury," or "millimeters of mercury." Although these are not real units of pressure, they are so descriptive that they are widely used. The pressure exerted by a column of mercury one millimeter high is commonly called *one Torr*, after the Italian physicist Torricelli who first investigated the mercury barometric column.

Example. Compute the atmospheric pressure on a day when the height of the barometer is 76.0 cm.

The height of the mercury column depends on ρ and g as well as on the atmospheric pressure. Hence both the density of mercury and the local acceleration of gravity must be known. The density varies with the temperature, and g with the latitude and elevation above sea level. All accurate barometers are provided with a thermometer and with a table or chart from which corrections for temperature and elevation can be found. If we assume $g = 980 \text{ cm s}^{-2}$ and $\rho = 13.6 \text{ g cm}^{-3}$,

$$p_a = \rho g h = 13.6 \text{ g cm}^{-3} \times 980 \text{ cm s}^{-2} \times 76 \text{ cm}$$
$$= 1{,}013{,}000 \text{ dyn cm}^{-2}$$

(about a million dynes per square centimeter).

In English units,

$$76 \text{ cm} = 30 \text{ in.} = 2.5 \text{ ft},$$
$$\rho g = 850 \text{ lb ft}^{-3},$$
$$p_a = 2120 \text{ lb ft}^{-2} = 14.7 \text{ lb in}^{-2}.$$

A pressure of $1.013 \times 10^6 \text{ dyn cm}^{-2} = 1.013 \times 10^5 \text{ N m}^{-2} = 14.7 \text{ lb in}^{-2}$, is called *one atmosphere* (1 atm). A pressure of exactly one million dynes per square centimeter is called one *bar*, and a pressure one one-thousandth as great is one *millibar*. Atmospheric pressures are of the order of 1000 millibars, and are now stated in terms of this unit by the United States Weather Bureau.

The Bourdon type pressure gauge is more convenient for most purposes than a liquid manometer. It consists of a flattened brass tube closed at one end and bent into a circular form. The closed end of the tube is connected by a gear and pinion to a pointer which moves over a scale. The open end of the tube is connected to the apparatus, the pressure within which is to be measured. When pressure is exerted within the flattened tube, it straightens slightly, just as a bent rubber hose straightens when water is admitted. The resulting motion of the closed end of the tube is transmitted to the pointer.

12–5 VACUUM PUMPS

Many of the devices of modern life consist of glass or metal tubes from which practically all the air has been exhausted. Among these are, for example, electric light bulbs, radio tubes, cathode-ray oscilloscopes, photo-

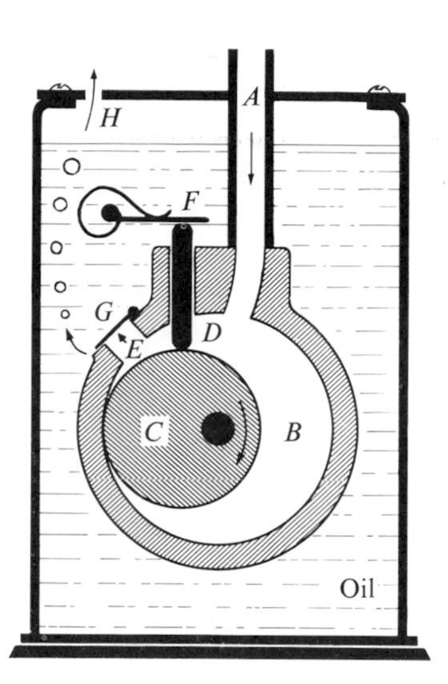

12-6 The rotary oil pump.

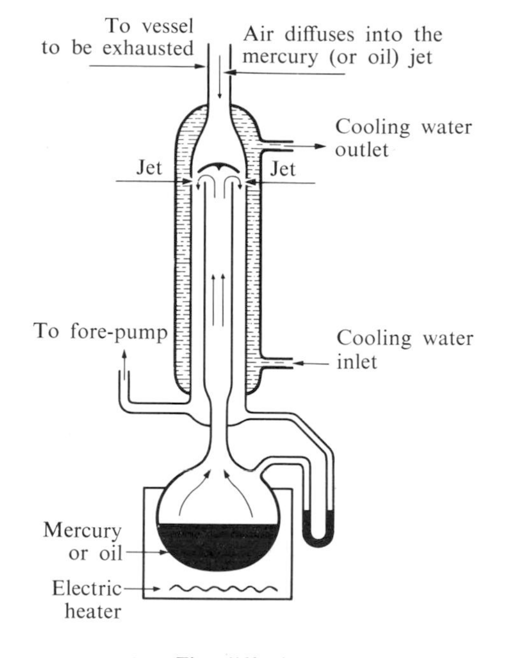

12-7 The diffusion pump.

electric cells, x-ray tubes, and many others. These devices could not have been developed if means had not been available for producing high vacua. It is therefore important to learn how pressures as low as 10^{-8} mm of mercury are obtained. Of all the pumps that have been developed, two are particularly important in physics laboratories: the *rotary oil pump* for pressures as low as 10^{-4} mm of mercury, and the *mercury or oil diffusion pump* for pressures as low as 10^{-8} mm Hg.

Figure 12-6 depicts schematically a type of rotary oil pump in common use in the United States. The vessel to be exhausted is connected to the tube *A*, which communicates directly with the space marked *B*. As an eccentric cylinder *C* rotates in the direction shown, the point of contact between it and the inner walls of the stationary cylinder moves around in a clockwise direction, thereby trapping some air in the space marked *E*. The sliding vane *D* is kept in contact with the rotating cylinder by the pressure of the rod *F*. When the air in *E* is compressed enough to increase the pressure slightly above atmospheric, the valve G opens and the air bubbles through the oil and leaves through an opening *H* in the upper plate. The cylinder is caused to rotate by means of a small electric motor.

To reduce the pressure below about 10^{-3} or 10^{-4} mm of mercury, a diffusion pump is usually employed. In this pump a rapidly moving jet of mercury or of a special oil of extremely low vapor pressure (octoil, butyl phthallate, etc.) sweeps or pushes the air molecules away from the vessel to be exhausted. Air molecules from the vessel keep diffusing into the jet. There are many types and sizes on the market, some of glass and some of metal; some vertical, some horizontal; some with water cooling and some with air cooling; some with one jet and some with multiple jets. A common type is depicted in Fig. 12-7.

A rotary oil pump is used to reduce the pressure within the diffusion pump to the low value necessary to ensure a well-defined jet in which there are fewer air molecules per unit of volume than in the vessel to be exhausted. Otherwise, air molecules might diffuse from the jet to the vessel. The air molecules that diffuse into the jet are eventually removed by the oil pump (called in this case the fore-pump) and the mercury or oil is

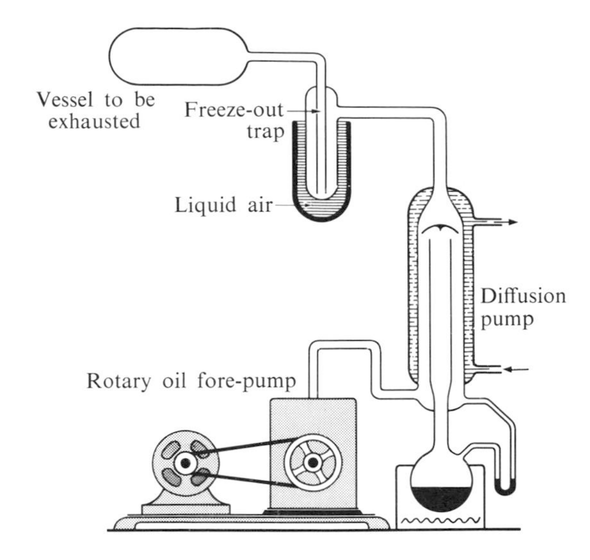

12-8 Typical pumping assembly for the production of a high vacuum.

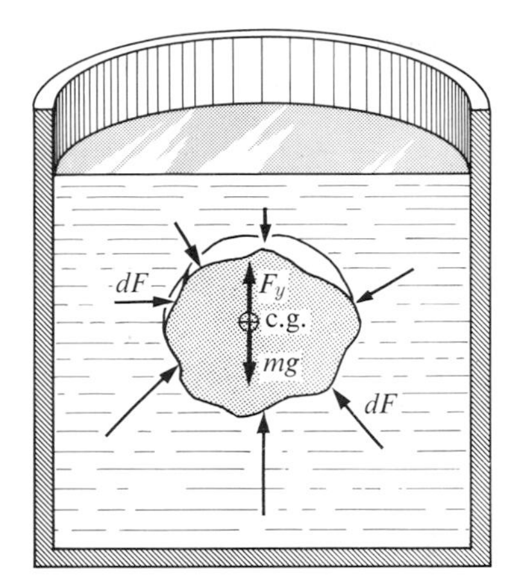

12-9 Archimedes' principle. The buoyant force F_z equals the weight of the displaced fluid.

condensed on the cool walls of the pump and returns to the well at the bottom.

If a liquid air trap is placed between the diffusion pump and the vessel to be exhausted to condense water vapor and other undesirable vapors, and the walls of the vessel are heated to drive out adsorbed gas (outgassing operation), the pressure in the vessel may be reduced to a value as low as 10^{-8} mm of mercury. A typical pumping assembly is depicted in Fig. 12-8.

12-6 ARCHIMEDES' PRINCIPLE

The irregular outline in Fig. 12-9 represents an imaginary surface bounding an arbitrary portion of a fluid at rest. The short arrows represent the forces exerted by the surrounding fluid against small elements of the boundary surface of equal area dA. The force dF against each element is normal to that element and equal to $p\,dA$, where p depends only on the vertical depth below the free surface and not on the shape or orientation of the boundary surface.

Since the entire fluid is at rest, the x-component of the resultant of these surface forces is zero. The y-component of the resultant, F_y, must equal the weight of the fluid inside the arbitrary surface, mg, and its line of action must pass through the center of gravity of this fluid.

Now suppose that the fluid inside the surface is removed and replaced by a solid body having exactly the same shape. The pressure at every point will be exactly the same as before, so the force exerted on the body by the surrounding fluid will be unaltered. That is, *the fluid exerts on the body an upward force F_y which is equal to the weight mg of the fluid originally occupying the boundary surface, and whose line of action passes through the original center of gravity.*

The submerged body, in general, will *not* be in equilibrium. Its weight may be greater or less than F_y, and if it is not homogeneous, its center of gravity may not lie on the line of F_y. Therefore, in general, it will be acted on by a resultant force through its own center of gravity and by a couple, and will rise or fall and also rotate.

The fact that a body immersed in a fluid should be "buoyed up" with a force equal to the weight of the displaced fluid was deduced by Archimedes (287–212 B.C.) from reasoning along the same lines as above. It is called *Archimedes' principle* and is, of course, a consequence of Newton's laws and the properties of a fluid. The position of the line of action of the upward force, usually omitted from a statement of the principle, is equal in importance to the magnitude of the force.

The weight of a dirigible floating in air, or of a submarine floating at some depth below the surface of the water, is just equal to the weight of a volume of air, or water, that is equal to the volume of the dirigible or submarine. That is, the average density of the dirigible equals that of air, and the average density of the submarine equals the density of water.

A body whose average density is less than that of a liquid can float *partially* submerged at the free upper surface of the liquid. However, we not only want a ship to float, but to float upright in stable equilibrium without capsizing. This requires that normally the line of action of the buoyant force should pass through the center of gravity of the ship and also, when the ship heels, the couple set up by its weight and the buoyant force should be in such a direction as to right it.

When making "weighings" with a sensitive analytical balance, correction must be made for the buoyant force of the air if the density of the body being "weighed" is very different from that of the standard "weights," which are usually of brass. For example, suppose a block of wood of density 0.4 g cm^{-3} is balanced on an equal-arm balance by brass "weights" of 20 g, density 8.0 g cm^{-3}. The apparent weight of each body is the difference between its true weight and the buoyant force of the air. If ρ_w, ρ_b, and ρ_a are the densities of the wood, brass, and air, and V_w and V_b are the volumes of the wood and brass, the apparent weights, which are equal, are

$$\rho_w V_w g - \rho_a V_w g = \rho_b V_b g - \rho_a V_b g.$$

The true mass of the wood is $\rho_w V_w$, and the true mass of the standard is $\rho_b V_b$. Hence,

$$\begin{aligned}\text{True mass} &= \rho_w V_w = \rho_b V_b + \rho_a(V_w - V_b)\\ &= \text{mass of standard} + \rho_a(V_w - V_b).\end{aligned}$$

In the specific example cited

$$V_w = \frac{20}{0.4} = 50 \text{ cm}^3 \text{ (very nearly)},$$

$$V_b = \frac{20}{8} = 2.5 \text{ cm}^3,$$

$$\rho_a = 0.0013 \text{ g cm}^{-3}.$$

Hence

$$\begin{aligned}\rho_a(V_w - V_b) &= 0.0013 \times 47.5\\ &= 0.062 \text{ g}.\end{aligned}$$

Therefore,

$$\text{true mass} = 20.062 \text{ g}.$$

If measurements are being made to one one-thousandth of a gram (0.001 g), it is obvious that the correction of 62 thousandths is of the greatest importance.

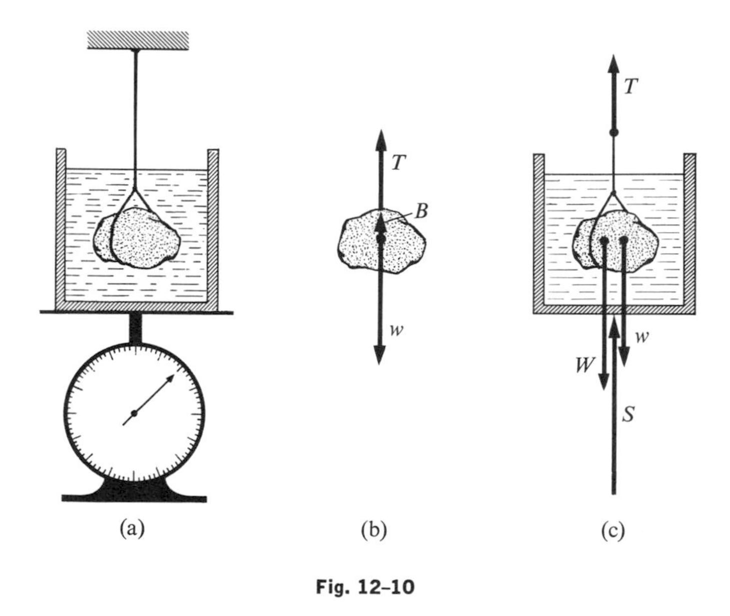

Fig. 12-10

Example. A tank containing water is placed on a spring scale, which registers a total weight W. A stone of weight w is hung from a string and lowered into the water without touching the sides or bottom of the tank (Fig. 12–10a). What will be the reading on the spring scale?

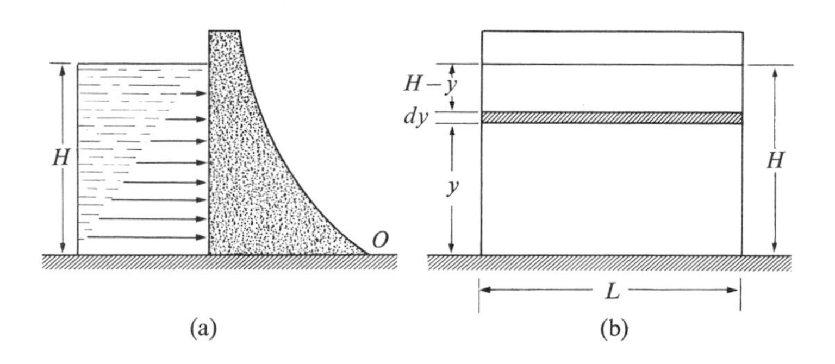

12-11 Forces on a dam.

First, for the stone alone, the forces are as shown in Fig. 12-10(b), where B is the buoyant force and T is the tension in the string. Since $\sum F_y = 0$,

$$T + B = w.$$

Next, for the tank with the water and stone in it, the forces are as shown in Fig. 12-10(c), where S is the force exerted by the spring scale on the isolated system and, by Newton's third law, is equal in magnitude and opposite in direction to the force exerted on the scale. The condition for equilibrium yields the equation

$$T + S = w + W.$$

Subtracting the first equation from the second, we get

$$S = W + B.$$

That is, the reading of the spring scale has been increased by an amount equal to the buoyant force.

12-7 FORCES AGAINST A DAM

Water stands at a depth H behind the vertical upstream face of a dam (Fig. 12-11). It exerts a certain resultant horizontal force on the dam, tending to slide it along its foundation, and a certain moment tending to overturn the dam about the point O. We wish to find the horizontal force and its moment.

Figure 12-11(b) is a view of the upstream face of the dam. The pressure at an elevation y is

$$p = \rho g(H - y).$$

(Atmospheric pressure can be omitted, since it also acts upstream against the other face of the dam.) The force against the shaded strip is

$$\begin{aligned} dF &= p\,dA \\ &= \rho g(H - y) \times L\,dy. \end{aligned}$$

The total force is

$$\begin{aligned} F = \int dF &= \int_0^H \rho g L(H - y)\,dy \\ &= \tfrac{1}{2}\rho g L H^2. \end{aligned}$$

The moment of the force dF about an axis through O is

$$d\Gamma = y\,dF = \rho g L y(H - y)\,dy.$$

The total torque about O is

$$\begin{aligned} \Gamma = \int d\Gamma &= \int_0^H \rho g L y(H - y)\,dy \\ &= \tfrac{1}{6}\rho g L H^3. \end{aligned}$$

If $\overline{H}$ is the height above O at which the total force F would have to act to produce this torque,

$$\begin{aligned} F\overline{H} &= \tfrac{1}{2}\rho g L H^2 \times \overline{H} = \tfrac{1}{6}\rho g L H^3, \\ \overline{H} &= \tfrac{1}{3}H. \end{aligned}$$

Hence the line of action of the resultant force is at 1/3 of the depth above O, or at 2/3 of the depth below the surface.

Problems

12–1 The piston of a hydraulic automobile lift is 12 in. in diameter. What pressure, in pounds per square inch, is required to lift a car weighing 2400 lb?

12–2 The expansion tank of a household hot-water heating system is open to the atmosphere and is 30 ft above a pressure gauge attached to the furnace. What is the gauge pressure at the furnace, in pounds per square inch?

12–3 Why can't a skin diver obtain an air supply at any desired depth by breathing through a "snorkel," a tube connected to his face mask and having its upper end above the water surface?

12–4 Suppose the door of a room makes an airtight but frictionless fit in its frame. Do you think you could open the door if the air pressure on one side were standard atmospheric pressure and that on the other side differed from standard by 1%?

12–5 The liquid in the open-tube manometer in Fig. 12–5(a) is mercury, and $y_1 = 3$ cm, $y_2 = 8$ cm. Atmospheric pressure is 970 millibars. (a) What is the absolute pressure at the bottom of the U-tube? (b) What is the absolute pressure in the open tube, at a depth of 5 cm below the free surface? (c) What is the absolute pressure of the gas in the tank? (d) What is the gauge pressure of the gas, in "cm of mercury"? (e) What is the gauge pressure in "cm of water"?

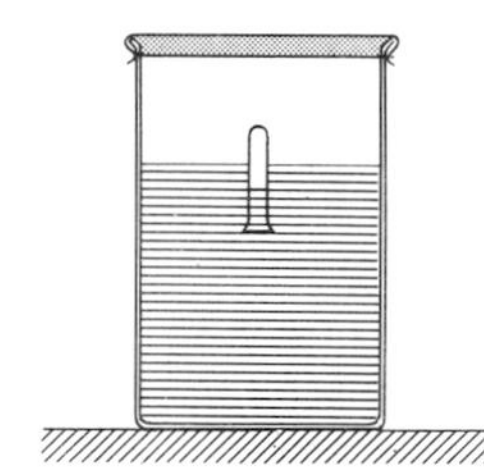

Fig. 12–12

12–6 (a) A small test tube, partially filled with water, is inverted in a large jar of water and floats as shown in Fig. 12–12. The lower end of the test tube is open and the top of the large jar is covered by a tightly fitting rubber membrane. When the membrane is pressed down the test tube sinks, and when the membrane is released it rises again. Explain. (A hollow glass figure in human form is often used instead of the test tube, and is called a "Cartesian diver.") (b) A torpedoed ship sinks below the surface of the ocean. If the depth is sufficiently great, is it possible for the ship to remain suspended in equilibrium at some point above the ocean bottom?

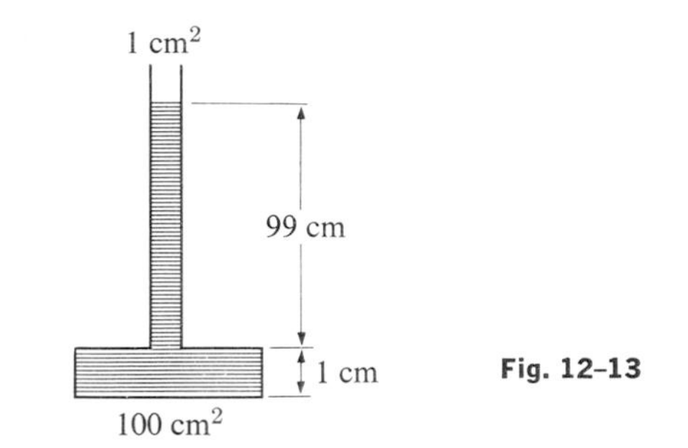

Fig. 12–13

12–7 A tube 1 cm^2 in cross section is attached to the top of a vessel 1 cm high and of cross section 100 cm^2. Water is poured into the system, filling it to a depth of 100 cm above the bottom of the vessel, as in Fig. 12–13. (a) What is the force exerted by the water against the bottom of the vessel? (b) What is the weight of the water in the system? (c) Explain why (a) and (b) are not equal.

12–8 A piece of gold-aluminum alloy weighs 10 lb. When suspended from a spring balance and submerged in water, the balance reads 8 lb. What is the weight of gold in the alloy if the specific gravity of gold is 19.3 and the specific gravity of aluminum is 2.5?

12–9 What is the area of the smallest block of ice 1 ft thick that will just support a man weighing 180 lb? The specific gravity of the ice is 0.917, and it is floating in fresh water.

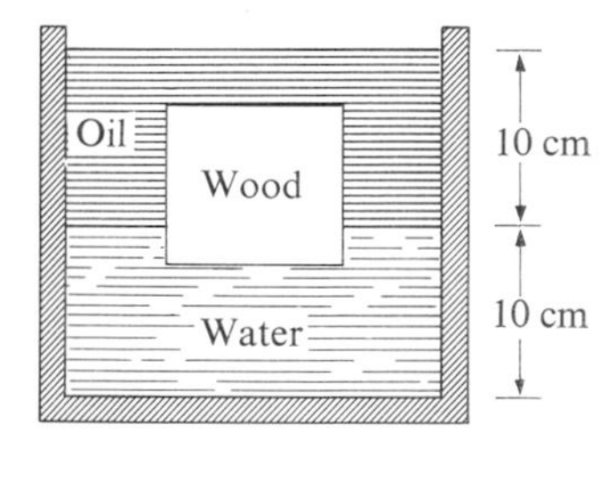

Fig. 12–14

12–10 A cubical block of wood 10 cm on a side floats at the interface between oil and water as in Fig. 12–14, with its lower surface 2 cm below the interface. The density of the oil is 0.6 g cm^{-3}. (a) What is the mass of the block? (b) What is the gauge pressure at the lower face of the block?

12–11 The densities of air, helium, and hydrogen (at standard conditions) are, respectively, $0.00129 \text{ g cm}^{-3}$, $0.000178 \text{ g cm}^{-3}$, and $0.0000899 \text{ g cm}^{-3}$. What is the volume in cubic feet displaced by a hydrogen-filled dirigible which has a total "lift" of 10 tons? What would be the "lift" if helium were used instead of hydrogen?

12–12 A piece of wood is 2 ft long, 1 ft wide, and 2 in. thick. Its specific gravity is 0.6. What volume of lead must be fastened underneath to sink the wood in calm water so that its top is just even with the water level?

12–13 A cubical block of wood 10 cm on a side and of density 0.5 g cm^{-3} floats in a jar of water. Oil of density 0.8 g cm^{-3} is poured on the water until the top of the oil layer is 4 cm below the top of the block. (a) How deep is the oil layer? (b) What is the gauge pressure at the lower face of the block?

12–14 A cubical block of steel (density $= 7.8 \text{ g cm}^{-3}$) floats on mercury (density $= 13.6 \text{ g cm}^{-3}$). (a) What fraction of the block is above the mercury surface? (b) If water is poured on the mercury surface, how deep must the water layer be so that the water surface just rises to the top of the steel block?

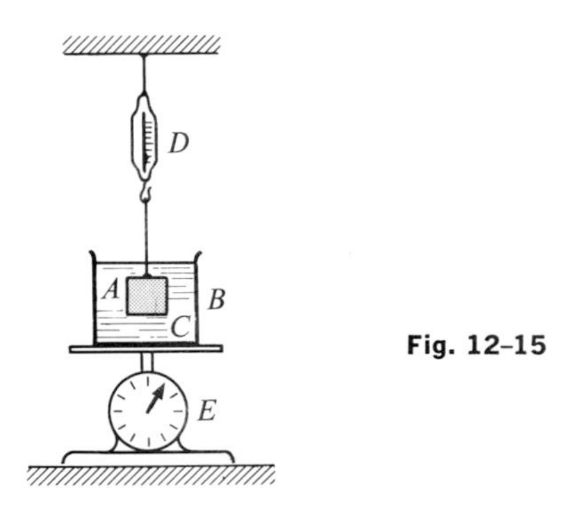

Fig. 12–15

12–15 Block *A* in Fig. 12–15 hangs by a cord from spring balance *D* and is submerged in a liquid *C* contained in beaker *B*. The weight of the beaker is 2 lb, the weight of the liquid is 3 lb. Balance *D* reads 5 lb and balance *E* reads 15 lb. The volume of block *A* is 0.1 ft^3. (a) What is the weight per unit volume of the liquid? (b) What will each balance read if block *A* is pulled up out of the liquid?

12–16 A hollow sphere of inner radius 9 cm and outer radius 10 cm floats half submerged in a liquid of specific gravity 0.8. (a) Calculate the density of the material of which the sphere is made. (b) What would be the density of a liquid in which the hollow sphere would just float completely submerged?

12–17 Two spherical bodies having the same diameter are released simultaneously from the same height. If the mass of one is 10 times that of the other and if the air resistance on each is the *same*, show that the heavier body will arrive at the ground first.

12–18 When a life preserver having a volume of 0.75 ft^3 is immersed in sea water (specific gravity 1.1) it will just support a 160-lb man (specific gravity 1.2) with 2/10 of his volume above water. What is the weight per unit volume of the material composing the life preserver?

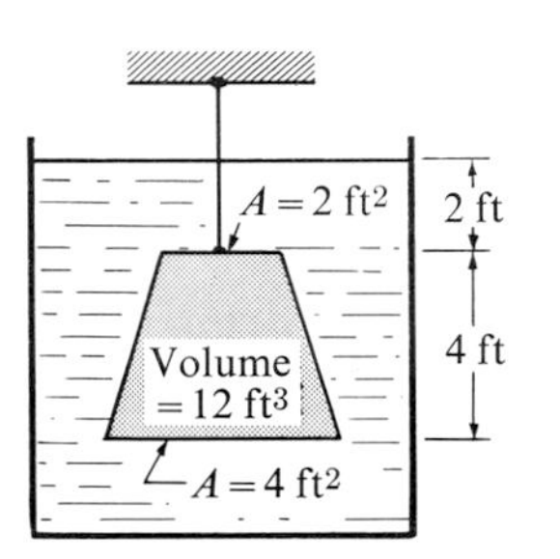

Fig. 12–16

12–19 An object in the shape of a truncated cone weighs 1000 lb in vacuum and is suspended by a rope in an open tank of liquid density 2 slugs ft^{-3}, as in Fig. 12–16. (a) Find the total downward force exerted by the liquid on the top of the object, of area 2 ft^2. (b) Find the total upward force exerted by the liquid on the bottom of the object, of area 4 ft^2. (c) Find the tension in the cord supporting the object.

12–20 A hollow cylindrical can 20 cm in diameter floats in water with 10 cm of its height above the water line when a 10-kg iron block hangs from its bottom. If the block is now placed inside the can, how much of the cylinder's height will be above the water line? The density of iron is 7.8 g cm^{-3}.

12–21 A block of balsa wood placed in one scale pan of an equal-arm balance is found to be exactly balanced by a 100-g brass "weight" in the other scale pan. Find the true mass of the balsa wood, if its specific gravity is 0.15.

12–22 A 3200-lb cylindrical can buoy floats vertically in salt water (specific gravity $= 1.03$). The diameter of the buoy is 3 ft. Calculate (a) the additional distance the buoy will sink when a 150-lb man stands on top, (b) the period of the resulting vertical simple harmonic motion when the man dives off.

12–23 A uniform rod *AB*, 12 ft long, weighing 24 lb, is supported at end *B* by a flexible cord and weighted at end *A* with a 12-lb lead weight. The rod floats as shown in Fig. 12–17 with one-half its length submerged. The buoyant force on the lead weight can be neglected. (a) Show in a diagram all the forces acting on the rod. (b) Find the tension in the cord. (c) Find the total volume of the rod.

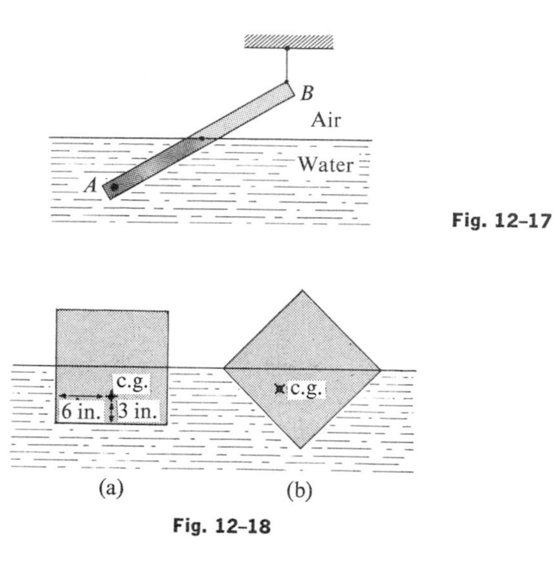

Fig. 12-17

Fig. 12-18

12-24 A cubical block of wood 1 ft on a side is weighted so that its center of gravity is at the point shown in Fig. 12-18(a), and it floats in water with one-half its volume submerged. Compute the restoring torque when the block is "heeled" at an angle of 45° as in Fig. 12-18(b).

12-25 A hydrometer consists of a spherical bulb and a cylindrical stem of cross section 0.4 cm^2. The total volume of bulb and stem is 13.2 cm^3. When immersed in water the hydrometer floats with 8 cm of the stem above the water surface. In alcohol, 1 cm of the stem is above the surface. Find the density of the alcohol.

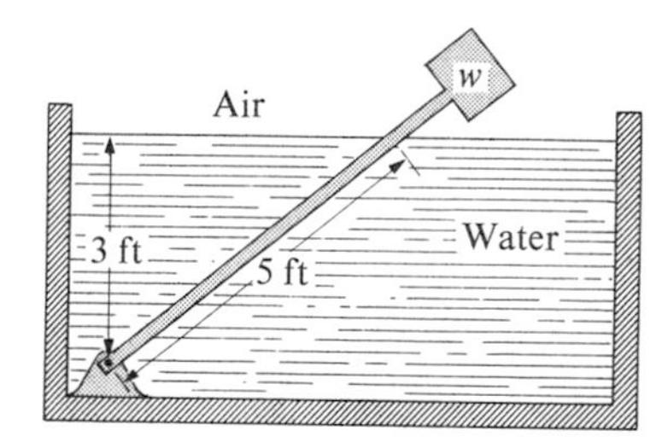

Fig. 12-19

12-26 A 12-lb uniform rod 6 ft long, whose specific gravity is 0.50, is hinged at one end 3 ft below a water surface, as in Fig. 12-19. (a) What weight w must be attached to the other end of the rod so that 5 ft of the rod are submerged? (b) Find the magnitude and direction of the force exerted by the hinge on the rod.

12-27 The following is quoted from a letter. How would you reply?

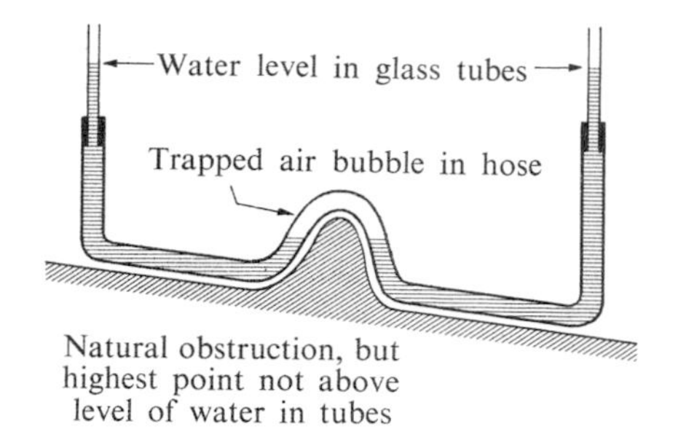

Fig. 12-20

"It is the practice of carpenters hereabouts, when laying out and leveling up the foundations of relatively long buildings, to use a garden hose filled with water, into the ends of the hose being thrust glass tubes 10 to 12 inches long.

"The theory is that the water, seeking a common level, will be of the same height in both the tubes and thus effect a level. Now the question rises as to what happens if a bubble of air is left in the hose. Our greybeards contend the air will not affect the reading from one end to the other. Others say that it will cause important inaccuracies.

"Can you give a relatively simple answer to this question, together with an explanation? I include a rough sketch (Fig. 12-20) *of the situation that caused the dispute."*

12-28 A swimming pool measures 75 × 25 × 8 ft deep. Compute the force exerted by the water against either end and against the bottom.

12-29 The upper edge of a vertical gate in a dam lies along the water surface. The gate is 6 ft wide and is hinged along the bottom edge, which is 10 ft below the water surface. What is the torque about the hinge?

12-30 The upper edge of a gate in a dam runs along the water surface. The gate is 6 ft high and 10 ft wide and is hinged along a horizontal line through its center. Calculate the torque about the hinge.

12-31 The cross section of a certain dam is a rectangle 10 ft wide and 20 ft high. The depth of water behind the dam is 20 ft and the dam is 500 ft long. (a) What is the torque tending to overturn the dam about the bottom edge of the downstream face? (b) If the material of the dam weighs 100 lb ft^{-3}, show whether or not the restoring torque due to the weight of the dam is greater than the torque due to water pressure.

12-32 Figure 12-21 is a cross-sectional view of a masonry dam whose length perpendicular to the diagram is 100 ft. The depth of water behind the dam is 30 ft. The masonry of which the dam is constructed weighs 150 lb ft^{-3}. Water weighs 62.5 lb ft^{-3}, but for an order-of-magnitude calculation this may be rounded

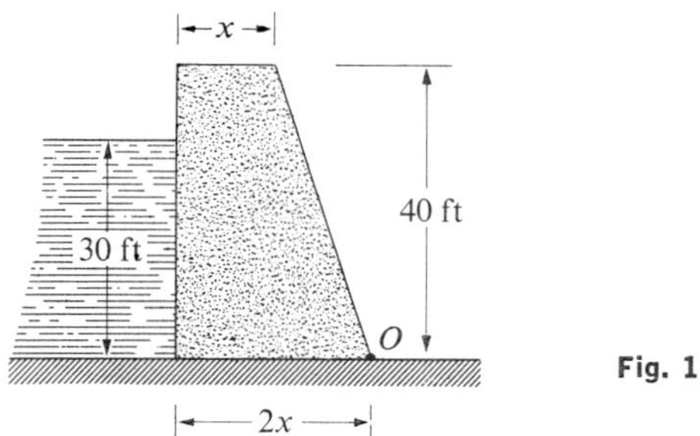

Fig. 12-21

off to 60 lb ft^{-3}. (a) Find the dimensions x and $2x$, if the weight of the dam is to be 10 times as great as the horizontal force exerted on it by the water. (b) Is the dam then stable with respect to its overturning about the edge through point O? (c) How does the size of the reservoir behind the dam affect the answers above?

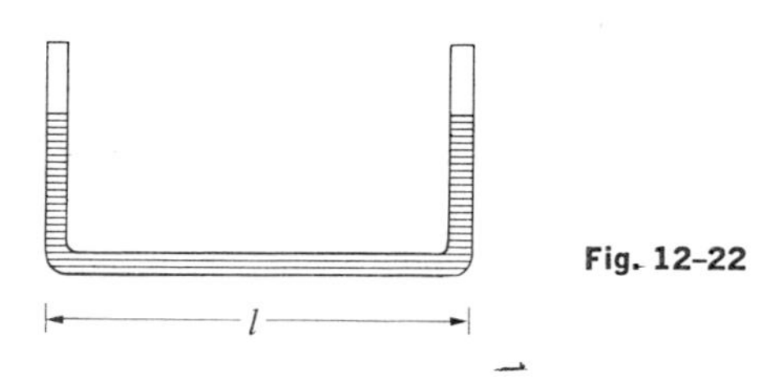

Fig. 12-22

12-33 A U-tube of length ℓ (see Fig. 12-22) contains a liquid. What is the difference in height between the liquid columns in the vertical arms (a) if the tube has an acceleration a toward the right, and (b) if the tube is mounted on a horizontal turntable rotating with an angular velocity ω, with one of the vertical arms on the axis of rotation? (c) Explain why the difference in height does not depend on the density of the liquid, or on the cross-sectional area of the tube. Would it be the same if the vertical tubes did not have equal cross sections? Would it be the same if the horizontal portion were tapered from one end to the other?

13 Surface Tension

13-1 SURFACE TENSION

A liquid flowing slowly from the tip of a medicine dropper emerges, not as a continuous stream, but as a succession of drops. A sewing needle, if placed carefully on a water surface, makes a small depression in the surface and rests there without sinking, even though its density may be as much as 10 times that of water. When a clean glass tube of small bore is dipped into water, the water rises in the tube, but if the tube is dipped in mercury, the mercury is depressed. All these phenomena, and many others of a similar nature, are associated with the existence of a *boundary surface* between a liquid and some other substance.

All surface phenomena indicate that the surface of a liquid can be considered to be in a state of stress such that if one considers any line lying in or bounding the surface, the material on either side of the line exerts a pull on the material on the other side. This pull lies in the plane of the surface and is perpendicular to the line. The effect can be demonstrated with the simple apparatus shown in Fig. 13–1. A wire ring a few inches in diameter has attached to it a loop of thread, as shown. When the ring and thread are dipped in a soap solution and removed, a thin film of liquid is formed in which the thread "floats" freely, as shown in part (a). If the film inside the loop of thread is punctured, the thread springs out into a circular shape as in part (b), as if the surfaces of the liquid were pulling radially outward on it, as shown by the arrows. Presumably, the same forces were acting before the film was punctured, but since there was film on *both* sides of the thread the net force exerted by the film on every portion of the thread was zero.

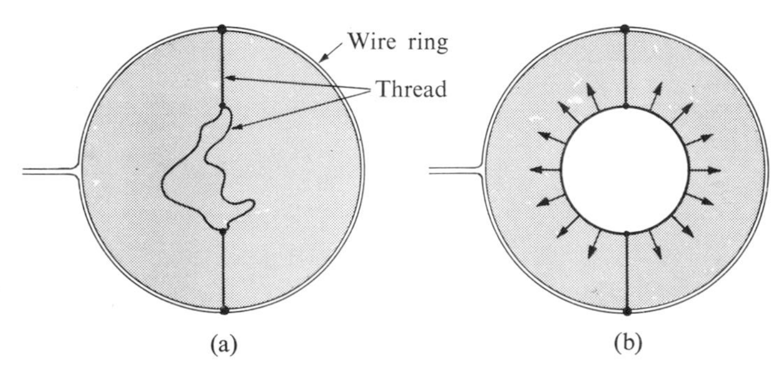

13–1 A wire ring with a flexible loop of thread, dipped in a soap solution, (a) before and (b) after puncturing the surface films inside the loop.

Another simple apparatus for demonstrating surface effects is shown in Fig. 13–2. A piece of wire is bent into the shape of a U and a second piece of wire is used as a slider. When the apparatus is dipped in a soap solution and removed, the slider (if its weight w_1 is not too great) is quickly pulled up to the top of the U. It may be held in equilibrium by adding a second weight w_2. Surprisingly, the same total force $F = w_1 + w_2$ will hold the slider at rest in *any* position, regardless of the area of the liquid film, provided the film remains at constant temperature. This is very different from the elastic behavior of a sheet of rubber, for which the force would be greater as the sheet was stretched.

Although a soap film like that in Fig. 13–2 is very thin, its thickness is still enormous compared with the size of a molecule. Hence it can be considered as made up chiefly of bulk liquid, bounded by two surface layers

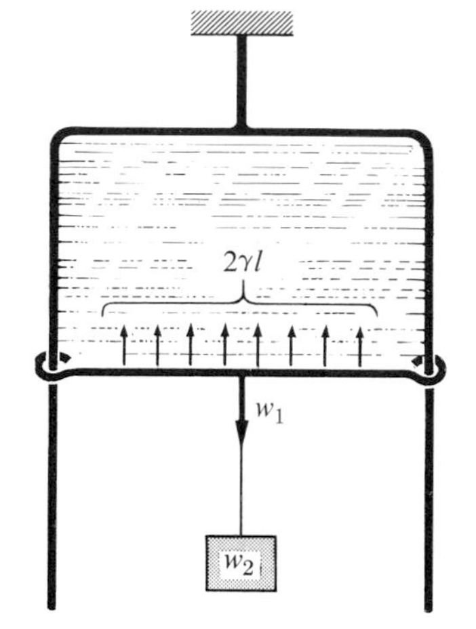

13-2 The horizontal slide wire is in equilibrium under the action of the upward surface force $2\gamma\ell$ and the downward pull $w_1 + w_2$.

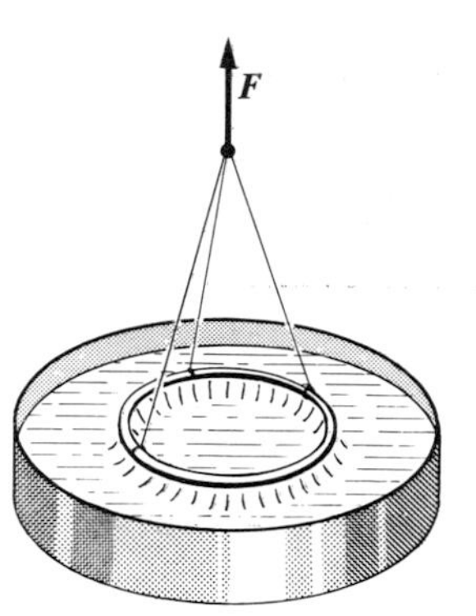

13-3 Lifting a circular wire of length ℓ out of a liquid requires an additional force F to balance the surface forces $2\gamma\ell$. This method is commonly used to measure surface tension.

a few molecules thick. When the crossbar in Fig. 13-2 is pulled down and the area of the film is increased, molecules formerly in the main body of the liquid move into the surface layers. That is, these layers are not "stretched" as a rubber sheet would be, but more surface is created by molecules moving from the bulk liquid.

Let ℓ be the length of the wire slider. Since the film has two surfaces, the total length along which the surface force acts is 2ℓ. The *surface tension* in the film, γ, is defined as *the ratio of the surface force to the length* (perpendicular to the force) *along which the force acts.* Hence in this case,

$$\gamma = \frac{F}{2\ell}. \tag{13-1}$$

The unit of surface tension in the cgs system is one dyne per centimeter (1 dyn cm^{-1}).

Another less spectacular way of showing a surface force is embodied in the actual apparatus, shown in Fig. 13-3, that is often used to measure surface tension. A circular wire whose circumference is of length ℓ is lifted out from the body of a liquid. The additional force F needed to balance the surface forces $2\gamma\ell$ due to the two surface films on each side is measured either by the stretch of a delicate spring or by the twist of a torsion wire. The surface tension is then given by

$$\gamma = \frac{F}{2\ell}.$$

Other methods of measuring surface tension will be apparent in what is to follow. Some typical values are shown in Table 13-1.

TABLE 13-1 EXPERIMENTAL VALUES OF SURFACE TENSION

Liquid in contact with air	t, °C	Surface tension, dyn cm^{-1}
Benzene	20	28.9
Carbon tetrachloride	20	26.8
Ethyl alcohol	20	22.3
Glycerine	20	63.1
Mercury	20	465
Olive oil	20	32.0
Soap solution	20	25.0
Water	0	75.6
Water	20	72.8
Water	60	66.2
Water	100	58.9
Oxygen	−193	15.7
Neon	−247	5.15
Helium	−269	0.12

The surface tension of a liquid surface in contact with its own vapor or with air is found to depend only on the nature of the liquid and on the temperature. The values for water in Table 13-1 are typical of the general result that surface tension decreases as the temperature increases. Measurements of the surface tension of an extremely thin layer of oil on the surface of water indicate that, in this case, the surface tension depends on the area of the oil film as well as on the temperature.

13-2 SURFACE TENSION AND SURFACE ENERGY

Another useful viewpoint regarding surface effects is the following. Suppose the horizontal wire in Fig. 13-2 is moved down a distance y by applying a downward force $F = w_1 + w_2$. The force F will remain constant provided the temperature of the surface film is kept constant during the motion. The work done is Fy, and the total surface area of the film is increased by $2\ell y$. The work done per unit area, in increasing the area, is therefore

$$\frac{\text{Work}}{\text{Increase in area}} = \frac{Fy}{2\ell y} = \frac{F}{2\ell}. \tag{13-2}$$

But, from Eq. (13-1), this is equal to the surface tension γ, and hence γ can be considered either as the *force per unit length* at right angles to the force, or as the *work per unit area* to increase the area. Using the second definition, the cgs unit of γ would be 1 erg cm^{-2}, which is equivalent to 1 dyn cm^{-1}, since 1 erg = 1 dyn cm.

However, the work done when the area of a surface film is increased at constant temperature is not equal to the increase in energy of the film, because in order to keep the temperature constant some heat must be supplied to the film. (If heat is not supplied, the temperature of the film decreases. The phenomenon is similar to the drop in temperature when a liquid evaporates.) The increase in surface energy of the film is equal to the sum of the work done and the heat supplied. For further details consult a text on thermodynamics.

If we let U represent the energy of a surface film of area A, its surface energy per unit area, U/A, is related to its surface tension by the equation

$$\frac{U}{A} = \gamma - T\frac{d\gamma}{dT},$$

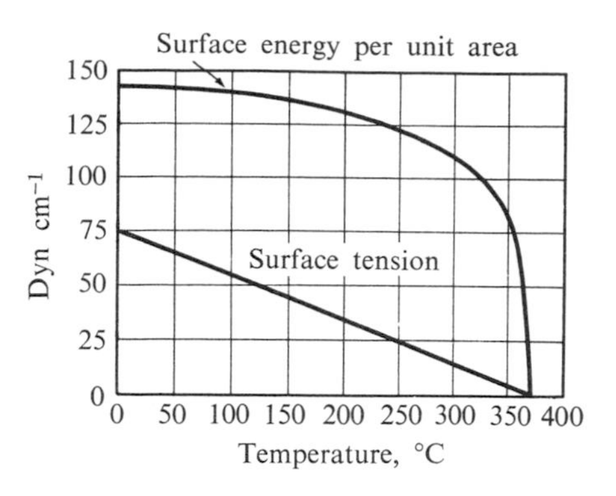

13-4 Surface energy and surface tension of water, as functions of temperature.

where T is the absolute temperature. Since surface tension always decreases with increasing temperature, $d\gamma/dT$ is negative and the surface energy per unit area is larger than the surface tension. Graphs of surface energy per unit area and of surface tension, as functions of temperature, are given in Fig. 13-4 for water. The temperature of 374°C, where both become zero, is the *critical temperature*.

13-3 PRESSURE DIFFERENCE ACROSS A SURFACE FILM

A soap bubble consists of two spherical surface films very close together, with liquid between. If we isolate one-half of the bubble and apply the principles of statics to the equilibrium of this half-bubble, we obtain a simple relation between the surface tension and the difference in pressure of the air inside and outside the bubble. Consider first a small element dA of a surface, shown in Fig. 13-5. Suppose the air pressure on the left of this element is p and that on the right is p_a. The force normal to the element is therefore $(p - p_a)\,dA$. The component of this force in the x-direction is

$$(p - p_a)\,dA \cos\theta.$$

But $dA \cos\theta$ is the area projected on a plane perpendicular to the x-axis. The force in the x-direction is therefore the difference of pressure multiplied by the projected area in the x-direction.

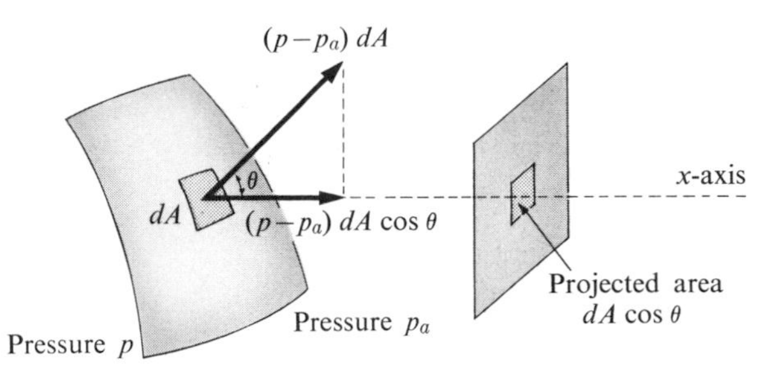

13-5 The force in the x-direction is the difference of pressure multiplied by the projected area in the x-direction.

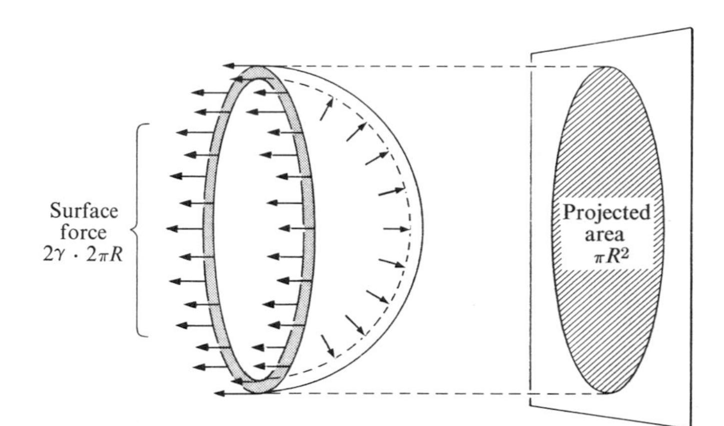

13-6 Equilibrium of half a soap bubble. The force exerted by the other half is $2 \cdot \gamma \cdot 2\pi R$, and the net force exerted by the air inside and outside the bubble is the pressure difference times projected area, or $(p - p_a)\pi R^2$.

Now consider the half-bubble shown in Fig. 13-6. The other half exerts a force to the left equal to twice the surface tension times the perimeter or

$$F \text{ (to the left)} = 2\gamma \times 2\pi R.$$

The force to the right is equal to the pressure difference $p - p_a$ multiplied by the area obtained by projecting the half-bubble on a plane perpendicular to the direction in question. Since this projected area is πR^2,

$$F \text{ (to the right)} = (p - p_a)\pi R^2.$$

Since the half-bubble is in equilibrium,

$$(p - p_a)\pi R^2 = 4\pi R\gamma,$$

or

$$p - p_a = \frac{4\gamma}{R} \quad \text{(soap bubble).} \qquad (13\text{-}3)$$

It follows from this result that if the surface tension remains constant (this means constant temperature), the pressure difference is larger the smaller the value of R. If two bubbles, therefore, are blown at opposite ends of a pipe, the smaller of the two will force air into the larger. In other words, the smaller one will get still smaller, and the larger will increase in size.

It may easily be verified that in the case of a liquid drop which has only one surface film, the difference between the pressure of the liquid and that of the outside air is given by

$$p - p_a = 2\gamma/R \quad \text{(liquid drop).} \qquad (13\text{-}4)$$

Example. Calculate the excess pressure inside a drop of mercury whose temperature is 20°C and whose diameter is 4 mm.

$$\begin{aligned} p - p_a &= \frac{2\gamma}{R} \\ &= \frac{2 \times 465 \text{ dyn cm}^{-1}}{0.4 \text{ cm}} \\ &= 2325 \text{ dyn cm}^{-2}. \end{aligned}$$

Any small portion of any curved surface can be fitted to a surface shaped like a blowout patch, that is, having two different radii of curvature in mutually perpendicular directions. If these radii, known as the *principal radii of curvature*, are called R_1 and R_2, the general expression for the pressure differential is

$$\Delta p = \gamma\left(\frac{1}{R_1} + \frac{1}{R_2}\right). \qquad (13\text{-}5)$$

The radii of curvature of a sphere are equal in any two mutually perpendicular directions and each is equal to the radius of the sphere, R. If $R_1 = R_2 = R$, then from Eq. (13-5) $\Delta p = 2\gamma/R$.

One of the radii of curvature of a cylinder is infinite and the other equals the radius of the cylinder. Hence for a cylinder

$$\Delta p = \frac{\gamma}{R}.$$

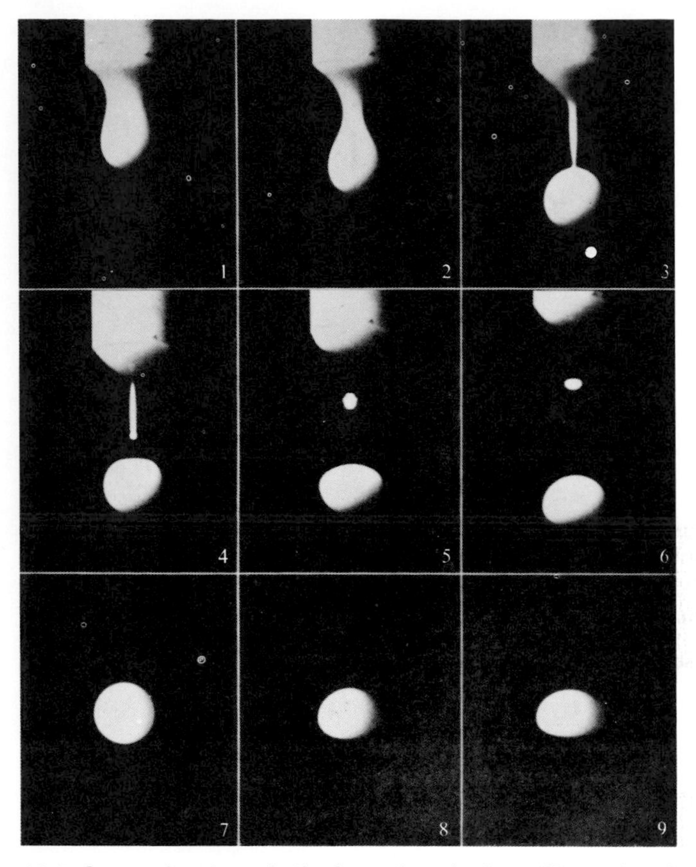

13-7 Successive stages in the formation of a drop. [Reproduced from *Flash*, courtesy of Ralph S. Hale & Co.]

13-4 MINIMAL SURFACES

Any surface under tension tends to contract until it occupies the minimum area consistent with the boundaries of the surface and with the difference of pressure on opposite sides of the surface. A small volume of heavy engine oil injected into the center of a mixture of alcohol and water whose density is the same as that of oil will therefore contract until it has the smallest surface area consistent with its volume. The shape of such a surface is spherical.

A more interesting situation is depicted in Fig. 13-7, which shows a series of high-speed photographs of successive stages in the formation of a drop of milk at the end of a vertical tube. The photographs were taken by Dr. Edgerton of M.I.T. It will be seen that the process,

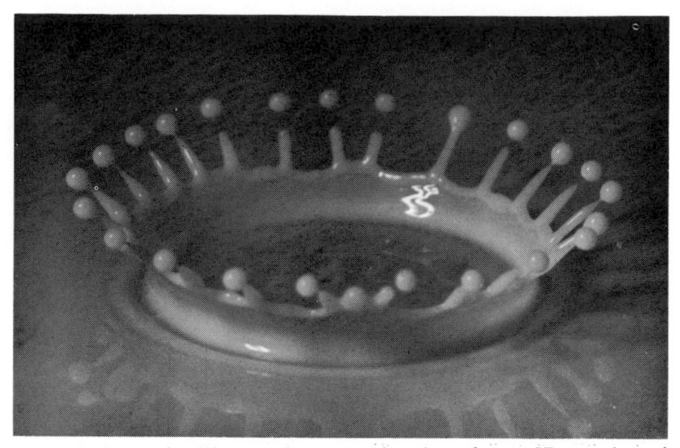

13-8 A drop of milk splashes on a hard surface. [Reproduced from *Flash*, courtesy of Ralph S. Hale & Co.]

if examined in detail, is exceedingly complex. An interesting feature is the small drop that follows the larger one. Both drops execute a few oscillations after their formation (4, 5, and 6) and eventually assume a spherical shape (7) which would be retained but for the effects of air resistance, as shown in (8) and (9). The drop in (9) has fallen 14 ft.

An approximate relation between the weight of the drop, w, and the radius of the tube from which it falls can be derived by assuming that the limiting size of the drop that can be supported by the tube is reached when the weight of the drop equals the surface tension force around a circle whose diameter equals that of the tube. This leads to the equation

$$w = 2\pi r\gamma.$$

Examination of Fig. 13-7 shows that the assumption that $w = 2\pi r\gamma$ is seriously in error. For one thing, the edge of the drop at the tip of the tube, at the instant the drop breaks away, is rarely vertical, and for another only a part of the drop falls. A very thorough study has been made of the relation between drop weight, radius of tube, and surface tension. When the proper corrections are made to the simple formula this method is one of the most satisfactory means of measuring surface tensions.

A beautiful photograph of the splash made by a drop of milk falling on a hard surface is reproduced in Fig. 13-8. It also was taken by Dr. Edgerton.

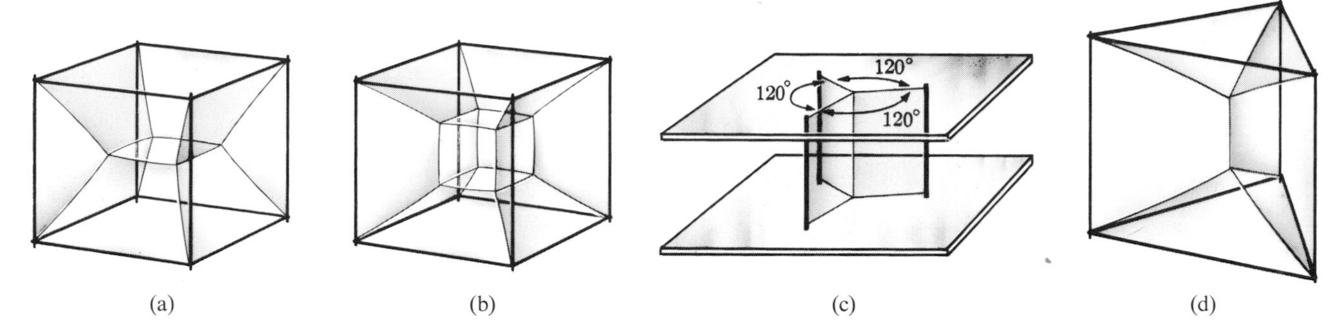

13-9 Solution of mathematical problems involving minimal surfaces by means of surface films. (a) Cubical wire framework dipped once. (b) Cubical wire framework dipped twice to entrap an air bubble in the center. (c) Two plastic plates connected by three wires form three plane surface films at angles of 120° to each other. (d) A wire framework in the form of a prism shows that at most three surface films can intersect in a line and at most four edges can intersect at a point.

Surface films may be used to solve problems in mathematics whose analytic solution presents great difficulties. If it is required to find the minimal surface (surface of minimum area) bounded by a wire framework bent into an arbitrary shape, the problem may be solved by dipping the wire framework into a soap solution and waiting a few seconds for the film to contract. The results for two frames in the form of a cube and one in the form of a prism are shown in Fig. 13-9. These results could hardly have been guessed, and their prediction by purely mathematical methods would have been attended by considerable difficulty.

For further details, the reader is referred to a fascinating book called *What is Mathematics* by Courant and Robbins.

13-5 CONTACT ANGLE

In the preceding sections we have limited the discussion of surface phenomena to surface films lying in the boundary between a liquid and a gas. There are other boundaries, however, in which surface films exist. One is the boundary between a solid wall and a liquid, and another is the boundary between a solid and a vapor. The three boundaries and their accompanying films are shown schematically in Fig. 13-10. The films are only a few molecules thick. Associated with each film is an appropriate surface tension. Thus

γ_{SL} = surface tension of the solid-liquid film,

γ_{SV} = surface tension of the solid-vapor film,

γ_{LV} = surface tension of the liquid-vapor film.

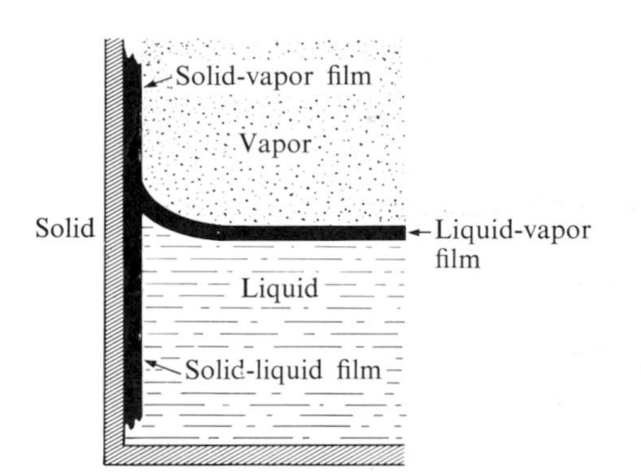

13-10 Surface films exist at the solid-vapor boundary and at the solid-liquid boundary as well as at the liquid-vapor boundary.

The symbol γ without subscripts, defined and used in the preceding sections, now appears as γ_{LV}.

The curvature of the surface of a liquid near a solid wall depends upon the difference between γ_{SV} and γ_{SL}. Consider a portion of a glass wall in contact with methylene iodide, as shown in Fig. 13-11(a). At the wall the

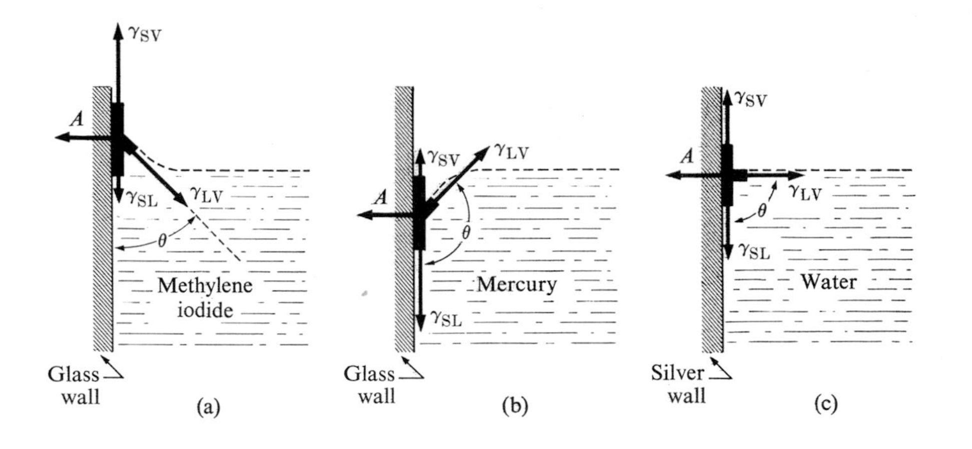

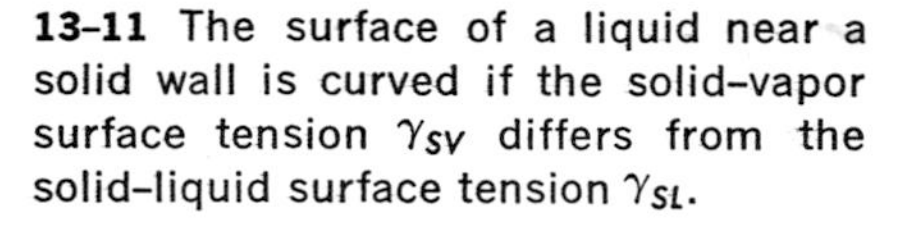
13-11 The surface of a liquid near a solid wall is curved if the solid–vapor surface tension γ_{SV} differs from the solid–liquid surface tension γ_{SL}.

three films meet. Let us isolate a small portion of all three films at their junction and imagine the films to extend unit distance in a direction perpendicular to the diagram. The isolated portion will be in equilibrium under the action of four forces, three of which are the surface tensions of the three films. The fourth force A is an attraction between the isolated portion and the wall, and is called the *adhesive force*. Applying the conditions for equilibrium, we get

$$\sum F_x = \gamma_{LV} \sin\theta - A = 0,$$
$$\sum F_y = \gamma_{SV} - \gamma_{SL} - \gamma_{LV}\cos\theta = 0,$$

from which

$$A = \gamma_{LV} \sin\theta, \tag{13-6}$$
$$\gamma_{SV} - \gamma_{SL} = \gamma_{LV}\cos\theta. \tag{13-7}$$

The first equation enables us to calculate the adhesive force from measurements of γ_{LV} and the angle θ, known as the *contact angle*. The second equation shows that the contact angle, which is a measure of the curvature of the liquid-vapor surface adjacent to the wall, depends on the difference between γ_{SV} and γ_{SL}. Thus, in Fig. 13-11(a), γ_{SV} is greater than γ_{SL}, cos θ is positive, and θ lies between 0° and 90°. The liquid is said to *wet* the glass.

In Fig. 13-11(b), a glass wall is in contact with mercury. The contact angle θ is about 140°, cos θ is negative, and hence γ_{SV} is less than γ_{SL}. When θ lies between 90° and 180°, as it does here, we say that the liquid does *not wet* the glass.

In Fig. 13-11(c) a situation closely approximated by silver in contact with water is shown. In this case, γ_{SV} is very nearly equal to γ_{SL}, cos θ is zero, and θ is 90°.

TABLE 13-2 CONTACT ANGLES

Liquid	Wall	Contact angle
α-Bromonaphthalene ($C_{10}H_7Br$)	Soda-lime glass	5°
	Lead glass	6°45′
	Pyrex	20°30′
	Fused quartz	21°
Methylene iodide (CH_2I_2)	Soda-lime glass	29°
	Lead glass	30°
	Pyrex	29°
	Fused quartz	33°
Water	Paraffin	107°
Mercury	Soda-lime glass	140°

There are a number of liquids whose contact angles are zero when the liquids are in contact with soda-lime glass, lead glass, pyrex, and fused quartz. This is the case for water, alcohol, ether, carbon tetrachloride, xylene, glycerine, and acetic acid. Some liquids, how-

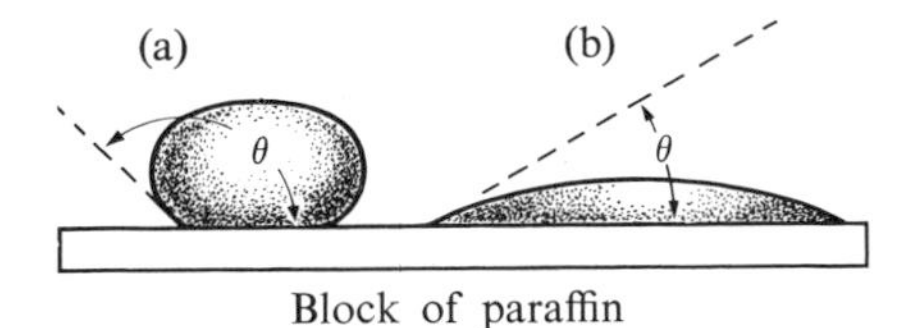

13-12 Effect of decreasing contact angle by a wetting agent.

ever, such as the first two listed in Table 13-2, have contact angles which depend upon the nature of the wall with which they are in contact.

Impurities and adulterants present in or added to a liquid may alter the contact angle considerably. In recent years a number of chemicals have been developed which are very potent as *wetting agents* or *detergents*. These compounds change the contact angle from a large value, greater than 90°, to a value much smaller than 90°. Conversely, waterproofing agents applied to a cloth cause the contact angle of water in contact with the cloth to be larger than 90°.

The effect of a detergent on a drop of water resting on a block of paraffin is shown in Fig. 13-12.

13-6 CAPILLARITY

The most familiar surface effect is the elevation of a liquid in an open tube of small cross section. The term *capillarity*, used to describe effects of this sort, originates from the description of such tubes as "capillary" or "hairlike." In the case of a liquid that wets the tube, the contact angle is less than 90° and the liquid rises until an equilibrium height y is reached, as shown in Fig. 13-13(a). The curved liquid surface in the tube is called a *meniscus*.

If the tube radius is r, the liquid makes contact with the tube along a line of length $2\pi r$. When we isolate the cylinder of liquid of height y and radius r, along with its liquid-vapor film, the total upward force is

$$F = 2\pi r \gamma_{LV} \cos\theta.$$

The downward force is the weight w of the cylinder, which is equal to the weight-density ρg times the volume $\pi r^2 y$, or

$$w = \rho g \pi r^2 y.$$

13-13 Surface tension forces on a liquid in a capillary tube. The liquid rises if $\theta < 90°$ and is depressed if $\theta > 90°$.

Since the cylinder is in equilibrium,

$$\rho g \pi r^2 y = 2\pi r \gamma_{LV} \cos\theta,$$

or

$$y = \frac{2\gamma_{LV}\cos\theta}{\rho g r}. \tag{13-8}$$

The same equation holds for capillary depression, shown in Fig. 13-13(b). Capillarity accounts for the rise of ink in blotting paper, the rise of lighter fluid in the wick of a cigarette lighter, and many other common phenomena.

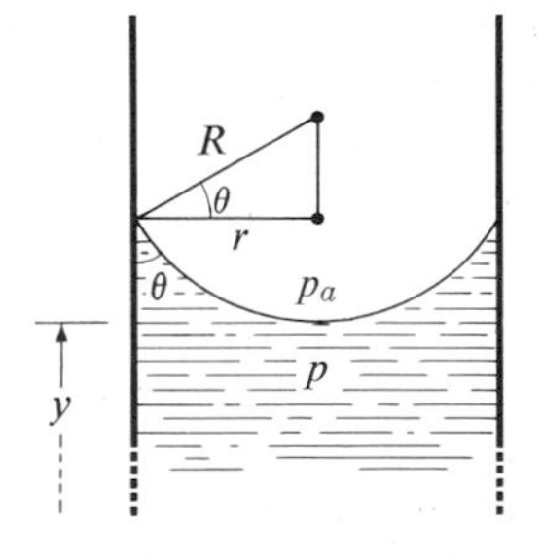

Fig. 13-14

The expression for the height of rise or fall of a liquid in a capillary tube can also be derived from a consideration of the pressure differential across a surface film. Figure 13-14 shows the top of the liquid column in the tube of Fig. 13–13(a). Let us assume the meniscus to be a portion of a spherical surface. The radius R of the surface is $R = r/\cos\theta$, where r is the tube radius. The pressure differential across the surface, from Eq. (13–4), is $2\gamma_{LV}/R$, so that if the pressure above the surface is atmospheric pressure p_a, the pressure p just below the surface is

$$p = p_a - \frac{2\gamma_{LV}\cos\theta}{r}.$$

The pressure in the liquid column, at the elevation of the flat liquid surface in Fig. 13–13(a), is also atmospheric. The pressure in the column at a height y is then

$$p = p_a - \rho g y.$$

Equating the expressions for p, we get

$$\rho g y = \frac{2\gamma_{LV}\cos\theta}{r},$$

which is the same as Eq. (13–8).

Problems

13–1 Compare the tension of a soap bubble with that of a rubber balloon in the following respects: (a) Has each a surface tension? (b) Does the surface tension depend on area? (c) Is Hooke's law applicable?

13–2 Water can rise to a height y in a certain capillary. Suppose that this tube is immersed in water so that only a length $y/2$ is above the surface. Will you have a fountain or not? Explain your reasoning.

13–3 A capillary tube is dipped in water with its lower end 10 cm below the water surface. Water rises in the tube to a height of 4 cm above that of the surrounding liquid, and the angle of contact is zero. What gauge pressure is required to blow a hemispherical bubble at the lower end of the tube?

13–4 A glass tube of inside diameter 1 mm is dipped vertically into a container of mercury, with its lower end 1 cm below the mercury surface. (a) What must be the gauge pressure of air in the tube to blow a hemispherical bubble at its lower end? (b) To what height will mercury rise in the tube if the air pressure in the tube is 3×10^4 dyn cm^{-2} below atmospheric? The angle of contact between mercury and glass is 140°.

13–5 On a day when the atmospheric pressure is 950 millibars, (a) what would be the height of the mercury column in a barometric tube of inside diameter 2 mm? (b) What would be the height in the absence of any surface tension effects? (c) What is the minimum diameter a barometric tube may have in order that the correction for capillary depression shall be less than 0.01 cm of mercury?

13–6 (a) Derive the expression for the height of capillary rise in the space between two parallel plates dipping in a liquid. (b) Two glass plates, parallel to each other and separated by 0.5 mm, are dipped in water. To what height will the water rise between them? Assume zero angle of contact.

13–7 A tube of circular cross section and outer radius 0.14 cm is closed at one end. This end is weighted and the tube floats vertically in water, heavy end down. The total mass of the tube and weights is 0.20 g. If the angle of contact is zero, how far below the water surface is the bottom of the tube?

13–8 Find the gauge pressure, in dynes per square centimeter, in a soap bubble 5 cm in diameter. The surface tension is 25 dyn cm^{-1}.

13–9 Two large glass plates are clamped together along one edge and separated by spacers a few millimeters thick along the opposite edge to form a wedge-shaped air film. These plates are then placed vertically in a dish of colored liquid. Show that the edge of the liquid forms an equilateral hyperbola.

13–10 When mercury is poured onto a flat glass surface which it does not wet, it spreads out into a pool of uniform thickness regardless of the size of the pool. Find the thickness of the pool.

14 Hydrodynamics and Viscosity

14–1 INTRODUCTION

The subject of hydrodynamics deals with the motion of fluids. To begin with, we shall consider only a so-called *ideal fluid*, that is, one which is incompressible and which has no internal friction or viscosity. The assumption of incompressibility is a good approximation when dealing with liquids. A gas can also be treated as incompressible provided the flow is such that pressure differences are not too great. Internal friction in a fluid gives rise to shearing stresses when two adjacent layers of fluid move relative to each other, or when the fluid flows in a tube or around an obstacle. In many actual flows, the shearing forces can be neglected in comparison with gravitational forces and forces arising from pressure differences.

The path followed by an element of a moving fluid is called a *line of flow*. In general, the velocity of the element changes in both magnitude and direction along its line of flow. If every element passing through a given point follows the same line of flow as that of preceding elements, the flow is said to be *steady* or *stationary*. When any given flow is first started, it passes through a nonsteady state, but in many instances the flow becomes steady after a certain period of time has elapsed. In steady flow, the velocity at every point of space remains constant in time, although the velocity of a particular particle of the fluid may change as it moves from one point to another.

A *streamline* is defined as a curve whose tangent, at any point, is in the direction of the fluid velocity at that point. In steady flow, the streamlines coincide with the lines of flow.

If we construct all of the streamlines passing through the periphery of an element of area, such as the area A in Fig. 14–1, these lines enclose a tube called a *flow tube* or *tube of flow*. From the definition of a streamline, no fluid can cross the side walls of a tube of flow; in steady flow there can be no mixing of the fluids in different flow tubes.

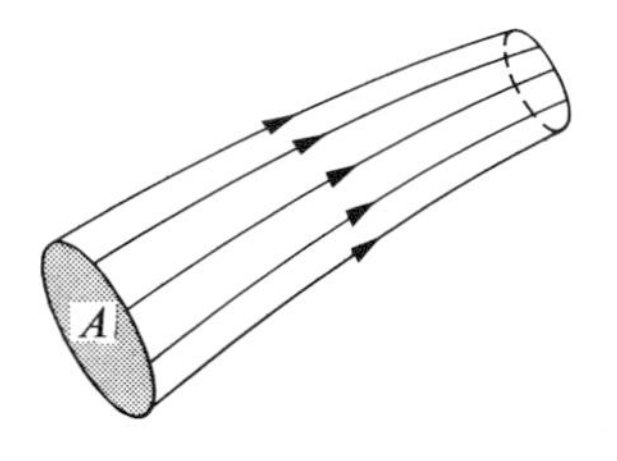

14–1 A flow tube bounded by streamlines.

The simplest type of fluid flow is a *homogeneous* flow, in which all flow tubes are straight and parallel, and the velocity is the same in each. Figure 14–2 illustrates the nature of the flow around a number of obstacles, and in a channel of varying cross section. The photographs were made using an apparatus designed by Pohl, in which alternate streams of clear and colored water flow between two closely spaced glass plates. The obstacles, and the channel walls, are opaque flat plates which fit between the glass plates. It will be noted that each obstacle is completely surrounded by a tube of flow. The tube splits into two portions at a so-called *stagnation point* on the upstream side of the obstacle. These portions rejoin at a second stagnation point on

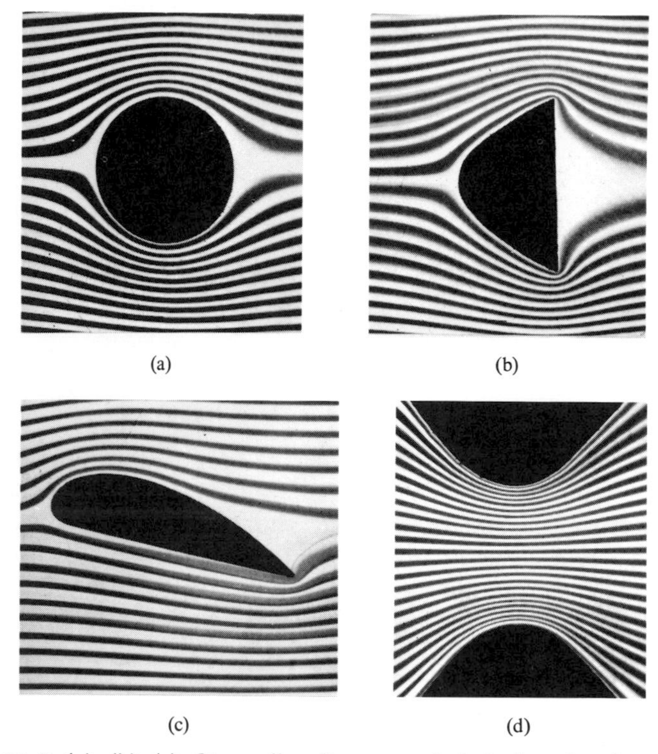

14-2 (a), (b), (c): Streamline flow around obstacles of various shapes. (d) Flow in a channel of varying cross-sectional area.

the downstream side. The velocity at the stagnation points is zero. It will also be noted that the cross sections of all flow tubes decrease at a constriction and increase again when the channel widens.

14-2 THE EQUATION OF CONTINUITY

If we consider any fixed, closed surface in a moving fluid, then, in general, fluid flows into the volume enclosed by the surface at some points and flows out at other points. The *equation of continuity* is a mathematical statement of the fact that the *net* rate of flow of mass *inward* across any closed surface is equal to the rate of increase of the mass within the surface.

For an incompressible fluid in steady flow, the equation takes the following form. Figure 14-3 represents a portion of a tube of flow, between two fixed cross

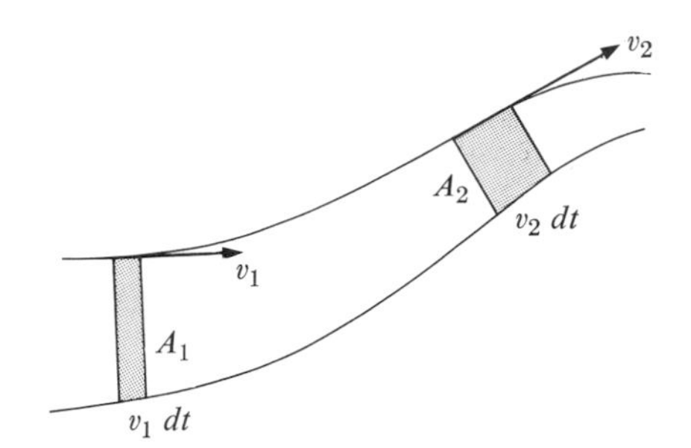

14-3 Flow into and out of a portion of a tube of flow.

sections of areas A_1 and A_2. Let v_1 and v_2 be the speeds at these sections. There is no flow across the side walls of the tube. The volume of fluid that will flow into the tube across A_1 in a time interval dt is that contained in the short cylindrical element of base A_1 and height $v_1\ dt$, or is $A_1 v_1\ dt$. If the density of the fluid is ρ, the mass flowing in is $\rho A_1 v_1\ dt$. Similarly, the mass that will flow out across A_2 in the same time is $\rho A_2 v_2\ dt$. The volume between A_1 and A_2 is constant, and since the flow is steady the mass flowing out equals that flowing in. Hence

$$\rho A_1 v_1\ dt = \rho A_2 v_2\ dt,$$

or

$$A_1 v_1 = A_2 v_2, \tag{14-1}$$

and the product Av is constant along any given tube of flow. It follows that when the cross section of a flow tube decreases, as in the constriction in Fig. 14-2(d), the velocity increases. This can readily be shown by introducing small particles in the fluid and observing their motion.

14-3 BERNOULLI'S EQUATION

When an incompressible fluid flows along a flow tube of varying cross section its velocity changes, that is, it accelerates or decelerates. It must therefore be acted on by a resultant force, and this means that the pressure must vary *along* the flow tube even though the elevation does not change. For two points at different elevations, the pressure difference depends not only on the difference in level but also on the difference between the velocities at the points. The general expression for the

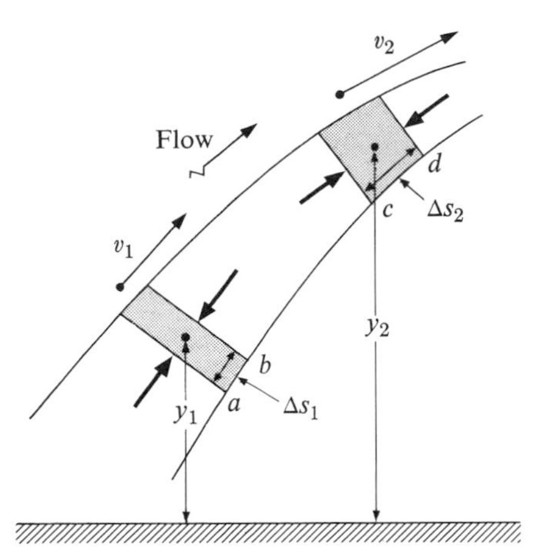

14–4 The net work done on the shaded element equals the increase in its kinetic and potential energy.

pressure difference can be obtained directly from Newton's second law, but it is simpler to make use of the work-energy theorem. The problem was first solved by Daniel Bernoulli in 1738.

Figure 14–4 represents a portion of a tube of flow. We are to follow a small element of the fluid, indicated by shading, as it moves from one point to another along the tube. Let y_1 be the elevation of the first point above some reference level, v_1 the speed at that point, A_1 the cross-sectional area of the tube, and p_1 the pressure. All these quantities may vary from point to point, and y_2, v_2, A_2, and p_2 are their values at the second point.

Since the fluid is under pressure at all points, inward forces, shown by the heavy arrows, are exerted against both faces of the element. As the element moves from the first point to the second, positive work is done by the force acting on its left face, and negative work by the force acting on its right face. The net work, or the difference between these quantities, equals the change in kinetic energy of the element plus the change in its potential energy.

If A represents the cross-sectional area of the tube at any point and p represents the corresponding pressure, the force against a face of the element at any point is pA. The work of the force acting on the left face of the element, in the motion in the diagram, is

$$\int_a^c F_s\,ds = \int_a^c pA\,ds,$$

where ds is a short distance measured along the tube of flow. The limits of integration are from a to c, since these are the initial and final positions of the left face. This integral may be written

$$\int_a^c pA\,ds = \int_a^b pA\,ds + \int_b^c pA\,ds.$$

Similarly, the work of the force acting on the right face of the element is

$$\int_b^d pA\,ds = \int_b^c pA\,ds + \int_c^d pA\,ds.$$

The *net* work is

$$\begin{aligned}\text{Net work} &= \int_a^b pA\,ds + \int_b^c pA\,ds - \int_b^c pA\,ds - \int_c^d pA\,ds\\ &= \int_a^b pA\,ds - \int_c^d pA\,ds.\end{aligned}$$

The distances from a to b and from c to d are sufficiently small so that the pressures and areas may be considered constant along their extent. Then

$$\int_a^b pA\,ds = p_1A_1\,\Delta s_1, \qquad \int_c^d pA\,ds = p_2A_2\,\Delta s_2.$$

But $A_1\,\Delta s_1 = A_2\,\Delta s_2 = V$, where V is the volume of the element. Hence

$$\text{Net work} = (p_1 - p_2)V. \tag{14–2}$$

Let ρ be the density of the fluid and m the mass of the element. Then $V = m/\rho$ and Eq. (14–2) becomes

$$\text{Net work} = (p_1 - p_2)\frac{m}{\rho}\cdot \tag{14–3}$$

We now equate the net work to the sum of the changes in potential and kinetic energy of the element:

$$(p_1 - p_2)\frac{m}{\rho} = (mgy_2 - mgy_1) + (\tfrac{1}{2}mv_2^2 - \tfrac{1}{2}mv_1^2).$$

After canceling m and multiplying through by ρ, we obtain

$$p_1 - p_2 = \rho g(y_2 - y_1) + \tfrac{1}{2}\rho(v_2^2 - v_1^2). \tag{14–4}$$

The first term on the right is the pressure difference arising from the weight of the fluid and the difference in elevation between points 1 and 2. The second is the additional pressure difference associated with the change in velocity, or the acceleration of the fluid.

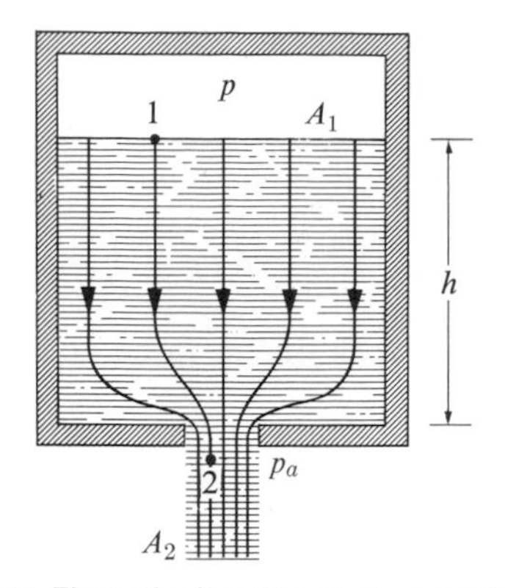

14-5 Flow of a liquid out of an orifice.

Equation (14-4) can be written

$$p_1 + \rho g y_1 + \tfrac{1}{2}\rho v_1^2 = p_2 + \rho g y_2 + \tfrac{1}{2}\rho v_2^2, \qquad (14\text{-}5)$$

and since the subscripts 1 and 2 refer to any two points along the tube of flow,

$$\boxed{p + \rho g y + \tfrac{1}{2}\rho v^2 = \text{constant.}} \qquad (14\text{-}6)$$

This is known as *Bernoulli's equation*. Note carefully that p is the *absolute* (not gauge) pressure, and must be expressed in pounds per square foot, newtons per square meter, or dynes per square centimeter. The density must be expressed in slugs per cubic foot, kilograms per cubic meter, or grams per cubic centimeter.

14-4 APPLICATIONS OF BERNOULLI'S EQUATION

1) The equations of hydrostatics are special cases of Bernoulli's equation, when the velocity is everywhere zero. Thus when v_1 and v_2 are zero, Eq. (14-4) reduces to

$$p_1 - p_2 = \rho g(y_2 - y_1),$$

which is the same as Eq. (12-4).

2) *Speed of efflux. Torricelli's theorem.* Figure 14-5 represents a tank of cross-sectional area A_1, filled to a depth h with a liquid of density ρ. The space above the top of the liquid contains air at pressure p, and the liquid flows out of an orifice of area A_2. Let us consider the entire volume of moving fluid as a single tube of flow, and let v_1 and v_2 be the speeds at points 1 and 2. The quantity v_2 is called the *speed of efflux*. The pressure at point 2 is atmospheric, p_a. Applying Bernoulli's equation to points 1 and 2, and taking the bottom of the tank as our reference level, we get

$$p + \tfrac{1}{2}\rho v_1^2 + \rho g h = p_a + \tfrac{1}{2}\rho v_2^2,$$

or

$$v_2^2 = v_1^2 + 2\,\frac{p - p_a}{\rho} + 2gh. \qquad (14\text{-}7)$$

From the equation of continuity,

$$v_2 = \frac{A_1}{A_2}\,v_1. \qquad (14\text{-}8)$$

Because of the converging of the streamlines as they approach the orifice, the cross section of the stream continues to diminish for a short distance outside the tank. It is the area of smallest cross section, known as the *vena contracta*, which should be used in Eq. (14-8). For a sharp-edged circular opening, the area of the *vena contracta* is about 65% as great as the area of the orifice.

Now let us consider some special cases. Suppose the tank is open to the atmosphere, so that

$$p = p_a \quad \text{and} \quad p - p_a = 0.$$

Suppose also that $A_1 \gg A_2$. Then v_1^2 is very much less than v_2^2 and can be neglected, and from Eq. (14-7),

$$v_2 = \sqrt{2gh}. \qquad (14\text{-}9)$$

That is, *the speed of efflux is the same as that acquired by any body in falling freely through a height h*. This is *Torricelli's theorem*. It is not restricted to an opening in the bottom of a vessel, but applies also to a hole in the side walls at a depth h below the surface.

Now suppose again that the ratio of areas is such that v_1^2 is negligible and that the pressure p (in a closed vessel) is so large that the term $2gh$ in Eq. (14-7) can be neglected, compared with $2(p - p_a)/\rho$. The speed of efflux is then

$$v_2 = \sqrt{2(p - p_a)/\rho}. \qquad (14\text{-}10)$$

The density ρ is that of the fluid escaping from the orifice. If the vessel is partly filled with a liquid, as in Fig. 14-5, ρ is the density of the liquid. On the other hand, if the vessel contains only a gas, ρ is the density of the gas. The efflux speed of a gas may be very great, even for small pressures, since its density is small. However, if the pressure is too great, it may no longer be permissible to treat a gas as incompressible, and if the speed is too great, the motion may become turbulent (see Section 14–10). Bernoulli's equation can no longer be applied to the motion under these conditions.

A flow of fluid out of an orifice in a vessel gives rise to a *thrust* or *reaction force* on the remainder of the system. The mechanics of the problem are the same as those involved in rocket propulsion. The thrust can be computed as follows, provided conditions are such that Bernoulli's equation is applicable. If A is the area of the orifice, ρ the density of the escaping fluid, and v the speed of efflux, the mass of fluid flowing out in time dt is $\rho Av\,dt$, and its momentum (mass × velocity) is $\rho Av^2\,dt$. Since we are neglecting the relatively small velocity of the fluid in the container, we can say that the escaping fluid started from rest and acquired the momentum above in time dt. Its *rate of change* of momentum was therefore ρAv^2, and from Newton's second law this equals the force acting on it. By Newton's third law, an equal and opposite reaction force acts on the remainder of the system. Taking the expression for v^2 from Eq. (14–10), the reaction force can be written

$$F = \rho Av^2 = \rho A\,\frac{2(p - p_a)}{\rho},$$

or

$$F = 2A(p - p_a). \qquad (14\text{–}11)$$

Thus while the *speed* of efflux is inversely proportional to the density, the *thrust* is independent of the density and depends only on the area of the orifice and the gauge pressure $p - p_a$.

Example. Figure 14–6 shows a toy "water rocket." Above the water is air at a pressure $p = 2$ atm. (a) If the rocket is held at rest, what is the speed of efflux out of an opening in the base of the rocket? (b) What is the upward thrust if the area of the opening is 0.5 cm^2?

(a) The gauge pressure $p - p_a = 1\text{ atm} \approx 10^6\text{ dyn cm}^{-2}$. The efflux speed is therefore

$$v \approx \sqrt{\frac{2\times 10^6\text{ dyn cm}^{-2}}{1\text{ gm cm}^{-3}}}$$
$$\approx 1.4\times 10^3\text{ cm s}^{-1}.$$

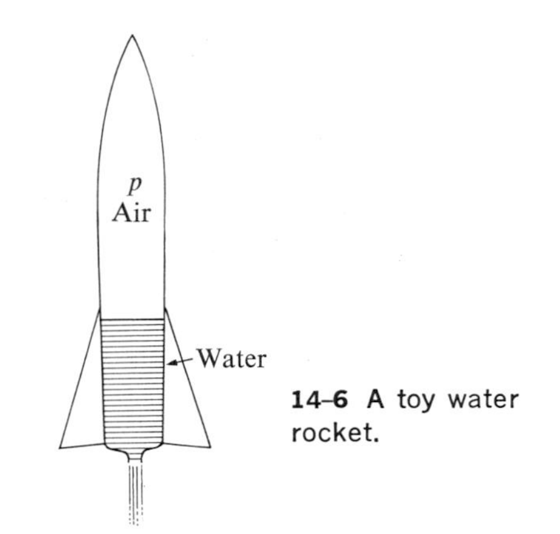

14–6 A toy water rocket.

The upward thrust, or reaction force, is

$$F \approx 2\times 0.5\text{ cm}^2\times 10^6\text{ dyn cm}^{-2}$$
$$\approx 10^6\text{ dyn}.$$

This force is much larger than the weight of the rocket and contents. Note that the reaction force, for the same gauge pressure, would be the same if the rocket initially contained air only. What is the reason for partially filling it with water?

It is interesting to treat this problem in the same way that the rocket was treated earlier. We have the rocket equation, Eq. (8–13), namely

$$m\,dv = v_r\mu\,dt - mg\,dt, \qquad (8\text{–}13)$$

where v_r is the speed of ejection and $\mu = dm/dt$. Dividing by dt, we get the relation

$$m\frac{dv}{dt} = v_r\frac{dm}{dt} - mg,$$

each term of which has a simple interpretation. The left-hand term is the resultant force which is seen to be the difference between the upward reaction force $v_r\mu$ and the weight mg downward. For the water rocket, $v_r \approx 1.4\times 10^3\text{ cm s}^{-1}$; $\mu = dm/dt = \rho v_r A \approx 0.7\times 10^3\text{ g s}^{-1}$. Hence the reaction force is

$$F = \mu v_r \approx 0.7\times 10^3\times 1.4\times 10^3 \approx 10^6\text{ dyn}.$$

3) The *Venturi tube*, illustrated in Fig. 14–7, consists of a constriction or throat inserted in a pipeline and having properly designed tapers at inlet and outlet to avoid turbulence. Bernoulli's equation, applied to the wide and to the constricted portions of the pipe, becomes

$$p_1 + \tfrac{1}{2}\rho v_1^2 = p_2 + \tfrac{1}{2}\rho v_2^2.$$

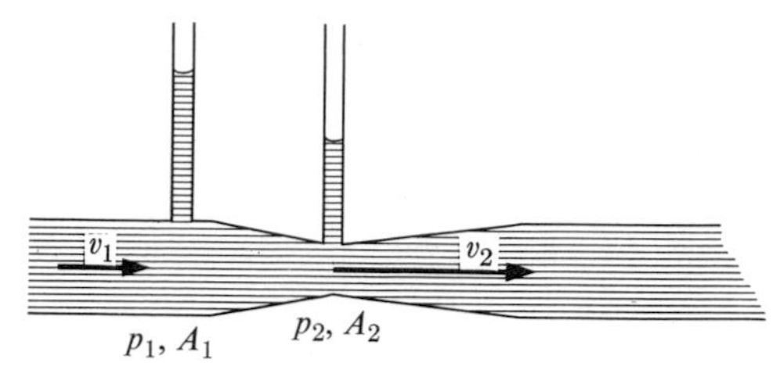

14-7 The Venturi tube.

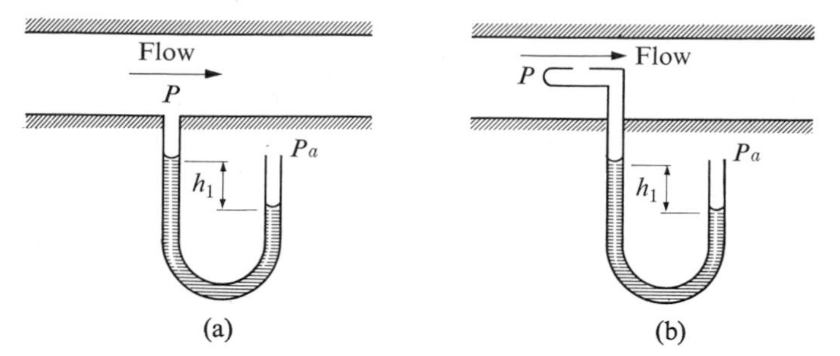

14-8 Pressure gauges for measuring the static pressure p in a fluid flowing in an enclosed channel.

From the equation of continuity, the speed v_2 is greater than the speed v_1 and hence the pressure p_2 in the throat is less than the pressure p_1. Thus a net force to the right acts to accelerate the fluid as it enters the throat, and a net force to the left decelerates it as it leaves. The pressures p_1 and p_2 can be measured by attaching vertical side tubes as shown in the diagram. From a knowledge of these pressures and of the cross-sectional areas A_1 and A_2, the velocities and the mass rate of flow can be computed. When used for this purpose, the device is called a *Venturi meter.*

The reduced pressure at a constriction finds a number of technical applications. Gasoline vapor is drawn into the manifold of an internal combustion engine by the low pressure produced in a Venturi throat to which the carburetor is connected. The aspirator pump is a Venturi throat through which water is forced. Air is drawn into the low-pressure water rushing through the constricted portion.

4) *Measurement of pressure in a moving fluid.* The so-called *static pressure* p in a fluid flowing in an enclosed channel can be measured with an open-tube manometer, as shown in Fig. 14-8. In (a), one arm of the manometer is connected to an opening in the channel wall. In (b), a *probe* is inserted in the stream. The probe should be small enough so that the flow is not appreciably disturbed and should be shaped so as to avoid turbulence. The difference h_1 in height of the liquid in the arms of the manometer is proportional to the difference between atmospheric pressure p_a and the static pressure p. That is,

$$p_a - p = \rho_m g h_1,$$
$$p = p_a - \rho_m g h_1, \qquad (14\text{-}12)$$

where ρ_m is the density of the manometric liquid.

The *Pitot tube*, shown in Fig. 14-9, is a probe with an opening at its upstream end. A stagnation point forms at the opening, where the pressure is p_2 and the speed is zero. Applying Bernoulli's equation to the stagnation point, and to a point at a large distance from the probe where the pressure is p and the speed is v, we get

$$p_2 = p + \tfrac{1}{2}\rho v^2$$

(ρ is the density of the flowing fluid). The pressure p_2 at the stagnation point is therefore the sum of the static pressure p and the quantity $\frac{1}{2}\rho v^2$, called the *dynamic pressure.*

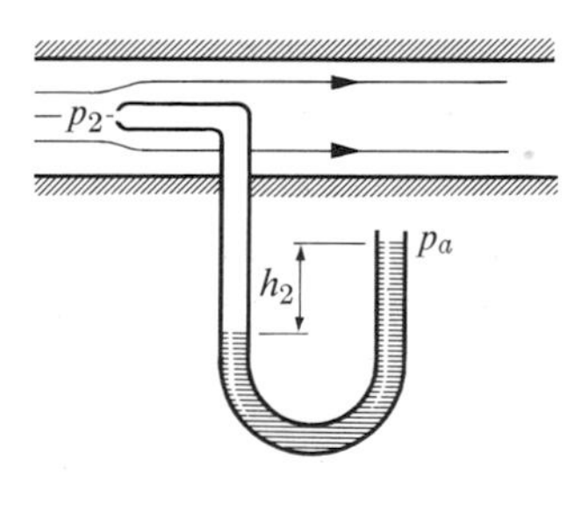

14-9 The Pitot tube.

The open-tube manometer reads the difference between atmospheric pressure p_a and the pressure p_2:

$$p_a - p_2 = \rho_m g h_2,$$
$$p_2 = p + \tfrac{1}{2}\rho v^2 = p_a - \rho_m g h_2. \qquad (14\text{-}13)$$

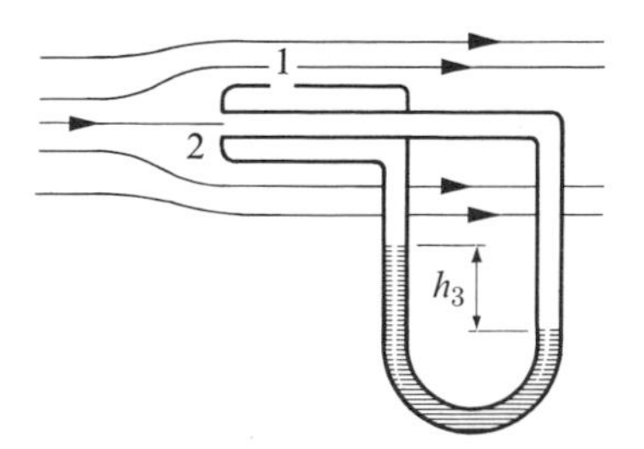

14–10 The Prandtl tube.

The instrument shown in Fig. 14–10 is known as a *Prandtl tube* (it is also sometimes called a Pitot tube). The pressure at opening 1 is the static pressure p and that at opening 2 is $p + \frac{1}{2}\rho v^2$. The manometric height h_3 is proportional to the difference between these, or to the dynamic pressure $\frac{1}{2}\rho v^2$. Hence

$$\tfrac{1}{2}\rho v^2 = \rho_m g h_3. \tag{14–14}$$

This instrument is self-contained and its reading does not depend on atmospheric pressure. If held at rest, it can be used to measure the velocity of a stream of fluid flowing past it. If mounted on an aircraft, it indicates the velocity of the aircraft relative to the surrounding air and is known as an *airspeed indicator.*

5) *The curved flight of a spinning ball.* Figure 14–11(a) represents a top view of a ball spinning about a vertical axis. Because of friction between the ball and the surrounding air, a thin layer of air is dragged around by the spinning ball.

Figure 14–11(b) represents a stationary ball in a blast of air moving from right to left. The motion of the air stream around and past the ball is the same as though the ball were moving through still air from left to right. If the ball is moving from left to right and spinning at the same time, the actual velocity of the air at any point is the resultant of the velocities at the same point in (a) and (b). At the top of the diagram the two velocities are in opposite directions, while the reverse is true at the bottom of the diagram. The top is a region of low velocity and high pressure, while the bottom is a region of high velocity and low pressure. There is therefore an excess pressure forcing the ball down in the diagram, so that if moving from left to right and spinning at the same time, it deviates from a straight line as shown in the top view in Fig. 14–11(c).

6) *Lift on an aircraft wing.* Figure 14–12 is a photograph of streamline flow around a section in the shape of an aircraft wing or an airfoil, at three different angles of attack. The apparatus consists of two parallel glass plates spaced about 1 mm apart. The wing section, whose thickness equals the separation of the plates, is inserted between them and alternate streams of clear water and ink flow by gravity between the plates and past the section. The photographs have been turned through 90° to give the effect of horizontal air flow past an aircraft wing. Because the fluid is water flowing relatively slowly, the nature of the flow pattern is not identical with that of air moving at high speed past an actual wing.

Consider the first photograph, which corresponds to a plane in level flight. It will be seen that there is relatively little disturbance of the flow below the wing, but because of the shape of the airfoil there is a marked crowding together of the streamlines above it, much as if they were being forced through the throat of a Venturi. Hence the region above the wing is one of increased velocity and reduced pressure, while below the wing the pressure is nearly atmospheric. It is this pressure

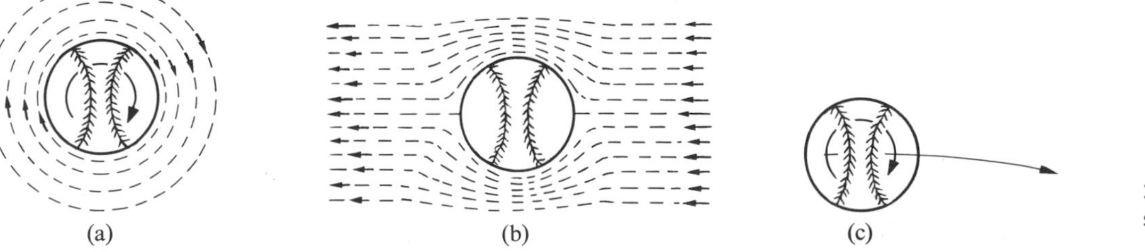

14–11 Curved flight of a spinning ball.

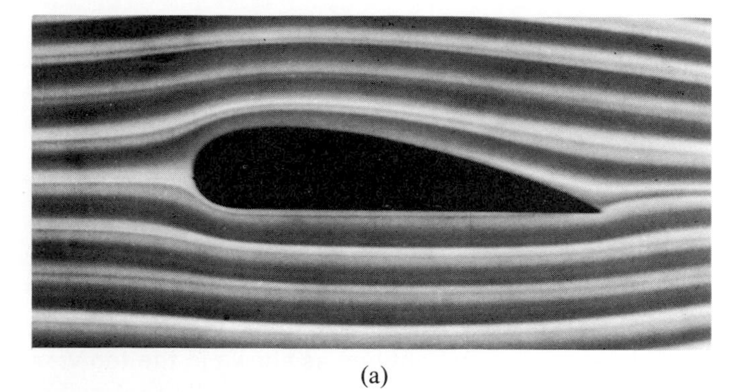

(a)

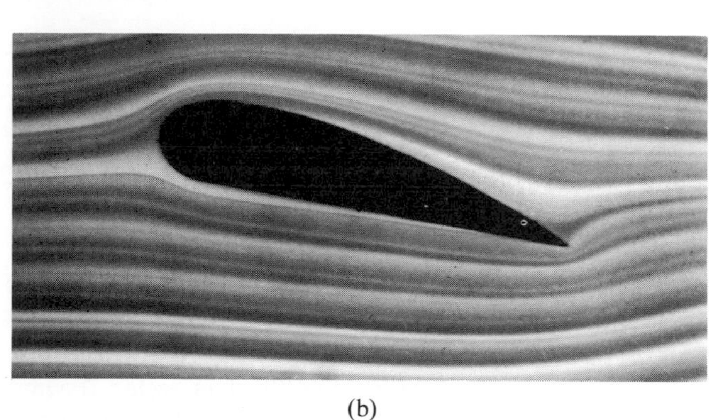

(b)

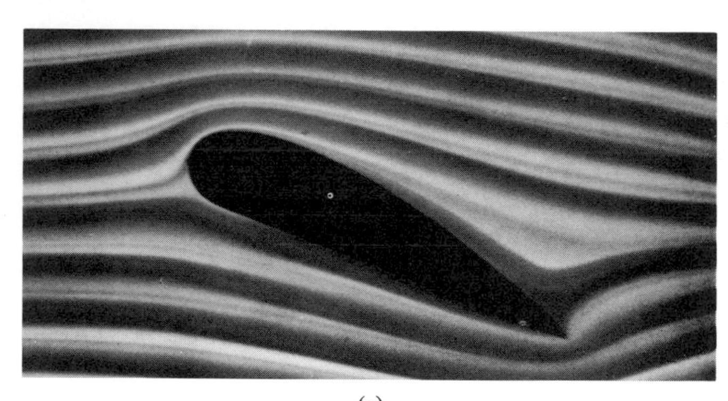

(c)

14-12 Lines of flow around an airfoil.

differential between upper and lower wing surfaces which gives rise to the lift on the wing. The wing is not simply forced up by air blowing against its lower surface.

There is a mistaken impression that the flow around an aircraft wing results in an upward "pull" on the upper surface of the wing. Of course this cannot happen. The air presses against all portions of the wing surface, but the reduction below atmospheric pressure, at the upper surface, usually exceeds the increase above atmospheric pressure at the lower surface.

The second and third photographs show how, as the angle of attack is increased, the streamlines above the wing have to change direction sharply to follow the contour of the wing surface and join smoothly with the streamline flow below the wing. While the slowly moving water in Fig. 14-12 does retain its streamline form even at the large angle of attack in the third photograph, it is much more difficult for the air moving rapidly past an aircraft wing to do so. As a consequence, if the angle of attack is too great, the streamline flow in the region above and behind the wing breaks down and a complicated system of whirls and eddies known as *turbulence* is set up. Bernoulli's equation no longer applies, the pressure above the wing rises, and the lift on the wing decreases and the plane stalls.

14-5 VISCOSITY

Viscosity may be thought of as the internal friction of a fluid. Because of viscosity, a force must be exerted to cause one layer of a fluid to slide past another, or to cause one surface to slide past another if there is a layer of fluid between the surfaces. Both liquids and gases exhibit viscosity, although liquids are much more viscous than gases. In developing the fundamental equations of viscous flow, it will be seen that the problem is very similar to that of the shearing stress and strain in a solid.

Figure 14-13 illustrates one type of apparatus for measuring the viscosity of a liquid. A cylinder is pivoted on nearly frictionless bearings so as to rotate coaxially within a cylindrical vessel. The liquid whose viscosity is to be measured is poured into the annular space between the cylinders. A torque can be applied to the inner cylinder by the weight-pulley system. When the weight is released, the inner cylinder accelerates momentarily but very quickly reaches a constant angular velocity and continues to rotate at that velocity so long as the torque acts. It is obvious that this velocity will be smaller with a liquid such as glycerine in the annular space than it will be if the liquid is water or kerosene. From a knowledge of the torque, the dimensions of the

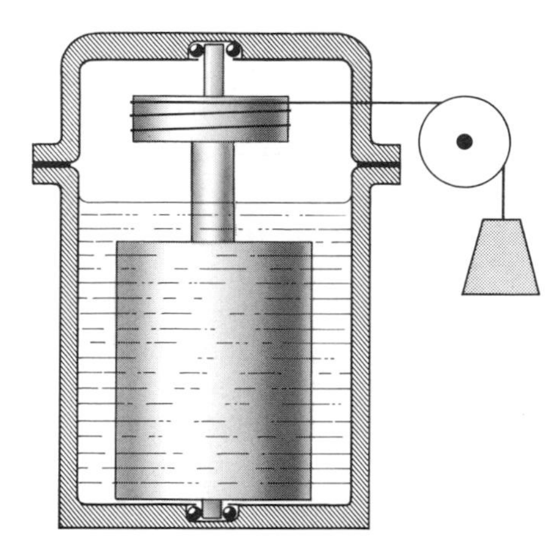

14–13 Schematic diagram of one type of viscosimeter.

apparatus, and the angular velocity, the viscosity of the liquid may be computed.

To reduce the problem to its essential terms, imagine that the cylinders are of nearly the same size, so that the liquid layer between them is thin. A short arc of this layer will then be approximately a straight line. Figure 14–14 shows a portion of the liquid layer between the moving inner wall and the stationary outer wall. The liquid in contact with the moving surface is found to have the same velocity as that surface; the liquid adjacent to the stationary inner wall is at rest. The velocities of intermediate layers of the liquid increase uniformly from one wall to the other, as shown by the arrows.

Flow of this type is called *laminar*. (A lamina is a thin sheet.) The layers of liquid slide over one another much as do the leaves of a book when it is placed flat on a table and a horizontal force applied to the top cover. As a consequence of this motion, a portion of the liquid which at some instant has the shape *abcd*, will a moment later take the shape *abc′d′*, and will become more and more distorted as the motion continues. That is, the liquid is in a state of continually increasing shearing strain.

In order to maintain the motion, it is necessary that a force be continually exerted to the right on the upper, moving plate, and hence indirectly on the upper liquid surface. This force tends to drag the liquid and the lower plate as well to the right. Therefore an equal

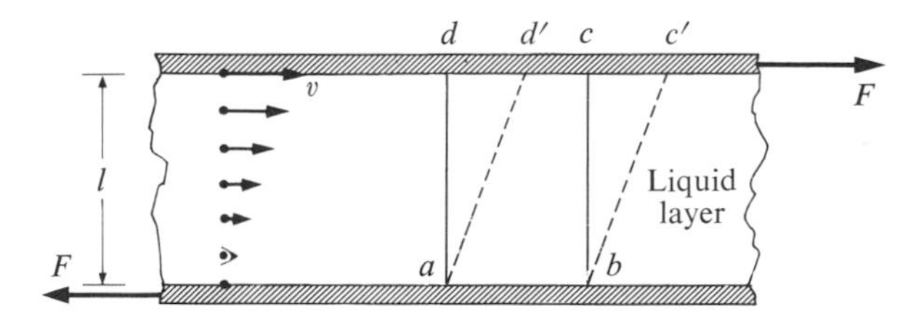

14–14 Laminar flow of a viscous fluid.

force must be exerted toward the left on the lower plate to hold it stationary. These forces are lettered F in Fig. 14–14. If A is the area of the liquid over which these forces are applied, the ratio F/A is the shearing stress exerted on the liquid.

When a shearing stress is applied to a solid, the effect of the stress is to produce a certain displacement of the solid, such as dd'. The shearing strain is defined as the ratio of this displacement to the transverse dimension ℓ, and within the elastic limit the shearing stress is proportional to the shearing strain. With a fluid, on the other hand, the shearing strain increases without limit so long as the stress is applied, and the stress is found by experiment to depend not on the shearing strain, but on its *rate of change*. The strain in Fig. 14–14, at the instant when the volume of fluid has the shape $abc'd'$, is dd'/ad, or dd'/ℓ. Since ℓ is constant, the rate of change of strain equals $1/\ell$ times the rate of change of dd'. But the rate of change of dd' is simply the velocity of point d', or the velocity v of the moving wall. Hence

$$\text{Rate of change of shearing strain} = \frac{v}{\ell}.$$

The *coefficient of viscosity* of the fluid, or simply its *viscosity* η, is defined as the ratio of the shearing stress, F/A, to the rate of change of shearing strain:

$$\eta = \frac{\text{shearing stress}}{\text{rate of change of shearing strain}} = \frac{F/A}{v/\ell},$$

or,

$$F = \eta A \frac{v}{\ell}. \tag{14–15}$$

For a liquid which flows readily, like water or kerosene, the shearing stress is relatively small for a given rate of change of shearing strain and the viscosity also is relatively small. For a liquid like molasses or glycerine, a greater shearing stress is necessary for the same rate of change of shearing strain and the viscosity is correspondingly greater. Viscosities of gases are very much less than those of liquids. The viscosities of all fluids are markedly dependent on temperature, increasing for gases and decreasing for liquids as the temperature is increased.

Equation (14–15) was derived for the special case in which the velocity increased at a uniform rate with increasing distance from the lower plate. The general term for the *space* rate of change of velocity, in a direction at right angles to the flow, is the *velocity gradient* in this direction. In this special case it is equal to v/ℓ. In the general case, the velocity gradient is not uniform and its value at any point can be written as dv/dy, where dv is the small difference in velocity between two points separated by a distance dy measured at right angles to the direction of flow. Hence the general form of Eq. (14–15) is

$$F = \eta A \frac{dv}{dy}. \qquad (14\text{–}16)$$

The unit of viscosity is that of force times distance divided by area times velocity. Thus in the cgs system the unit is 1 dyn cm $\text{cm}^{-2} \times (\text{cm s}^{-1})$, which reduces to 1 dyn s cm^{-2}. This unit is called 1 *poise*, in honor of the French scientist Poiseuille. Small viscosities are expressed in *centipoises* (1 cp $= 10^{-2}$ poise) or *micropoises* (1 μp $= 10^{-6}$ poise). Some typical values are given in Table 14–1.

TABLE 14-1 TYPICAL VALUES OF VISCOSITY

Temperature, °C	Viscosity of castor oil, poise	Viscosity of water, centipoise	Viscosity of air, micropoise
0	53	1.792	171
20	9.86	1.005	181
40	2.31	0.656	190
60	0.80	0.469	200
80	0.30	0.357	209
100	0.17	0.284	218

14-6 POISEUILLE'S LAW

It is evident from the general nature of viscous effects that the velocity of a viscous fluid flowing through a tube will not be the same at all points of a cross section. The outermost layer of fluid clings to the walls of the tube and its velocity is zero. The tube walls exert a backward drag on this layer which in turn drags backward on the next layer beyond it, and so on. Provided the velocity is not too great, the flow is laminar, with a velocity which is a maximum at the center of the tube and which decreases to zero at the walls. The flow is like that of a number of telescoping tubes sliding relative to one another, the central tube advancing most rapidly and the outer tube remaining at rest.

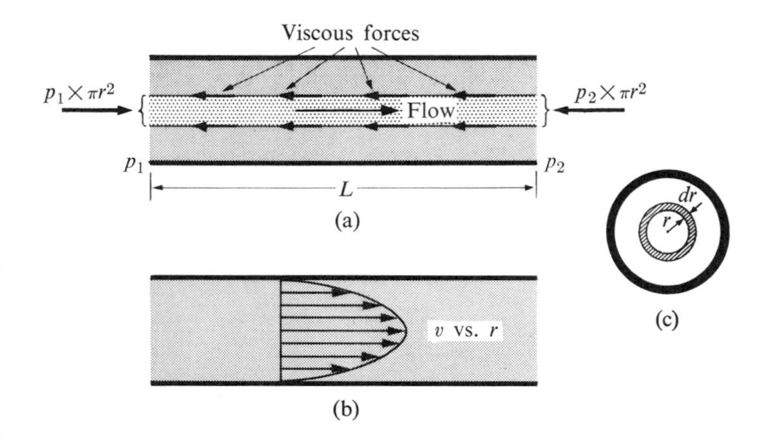

14-15 (a) Forces on a cylindrical element of a viscous fluid. (b) Velocity distribution for viscous flow. (c) End view.

Consider a portion of a tube of inner radius R and of length L, through which a fluid of viscosity η is flowing in laminar flow (Fig. 14–15). A small cylinder of radius r is in equilibrium (moving with constant velocity) under the driving force due to the pressure difference between its ends and the retarding viscous force at its outer surface. The driving force is

$$(p_1 - p_2)\pi r^2.$$

The viscous force, from Eq. (14–16), is

$$-\eta A \frac{dv}{dr} = -\eta \times 2\pi r L \times \frac{dv}{dr},$$

where dv/dr is the velocity gradient at a radial distance r from the axis. The negative sign must be introduced because v decreases as r increases. Equating the forces, and integrating, we get

$$-\int_v^0 dv = \frac{p_1 - p_2}{2\eta L} \int_r^R r\, dr,$$

and therefore

$$v = \frac{p_1 - p_2}{4\eta L}(R^2 - r^2), \tag{14–17}$$

which is the equation of a parabola. The curve in Fig. 14–15(b) is a graph of this equation. The lengths of the arrows are proportional to the velocities at their respective positions. The velocity gradient, dv/dr, at any radius, is the slope of this curve measured with respect to a vertical axis. We say that the flow has a *parabolic velocity profile.*

To find the discharge rate Q, or the volume of fluid crossing any section of the tube per unit time, consider the thin-walled element in Fig. 14–15(c). The volume of fluid dV crossing the ends of this element in a time dt is $v\,dA\,dt$, where v is the velocity at the radius r and dA is the shaded area, equal to $2\pi r\,dr$. Taking the expression for v from Eq. (14–17), we get

$$dV = \frac{p_1 - p_2}{4\eta L}(R^2 - r^2) \times 2\pi r\, dr \times dt.$$

The volume flowing across the entire cross section is obtained by integrating over all elements between $r = 0$ and $r = R$. Dividing by dt, for the volume rate of flow Q, we get

$$Q = \frac{\pi(p_1 - p_2)}{2\eta L}\int_0^R (R^2 - r^2) r\, dr = \frac{\pi}{8}\frac{R^4}{\eta}\frac{p_1 - p_2}{L}. \tag{14–18}$$

This relation was first derived by Poiseuille and is called *Poiseuille's law.*

The volume rate of flow is inversely proportional to the viscosity, as might be expected. It is proportional to the fourth power of the tube radius, so that if, for example, the radius is halved, the volume rate of flow is reduced by a factor of 16. The ratio $(p_1 - p_2)/L$ is the *pressure gradient* along the tube. The flow is directly proportional to the pressure gradient, and we see that for a viscous fluid there is a pressure drop even along a level tube of constant cross section. If the cross section varies from point to point, and if the tube is not horizontal, there will of course be further pressure differences resulting from the acceleration of the fluid or from gravitational effects, these differences being given by Bernoulli's equation.

The difference between the flow of an ideal nonviscous fluid and one having viscosity is illustrated in Fig. 14–16, where fluid is flowing along a horizontal tube of varying cross section. The height of the fluid in the small vertical tubes is proportional to the gauge pressure.

In part (a), the fluid is assumed to have no viscosity. The pressure at b is very nearly the static pressure ρgy, since the velocity is small in the large tank. The pressure at c is less than at b because the fluid must accelerate between these points. The pressures at c and d are equal, however, since the velocity and elevation at these points are the same. There is a further pressure drop between d and e, and between f and g. The pressure at g is atmospheric and the gauge pressure at this point is zero.

Part (b) of the diagram illustrates the effect of viscosity. Again, the pressure at b is nearly the static pressure ρgy. There is a pressure drop from b to c, due now in part to viscous effects, and also a further drop from c to d. The pressure gradient in this part of the tube is represented by the slope of the dotted line. The

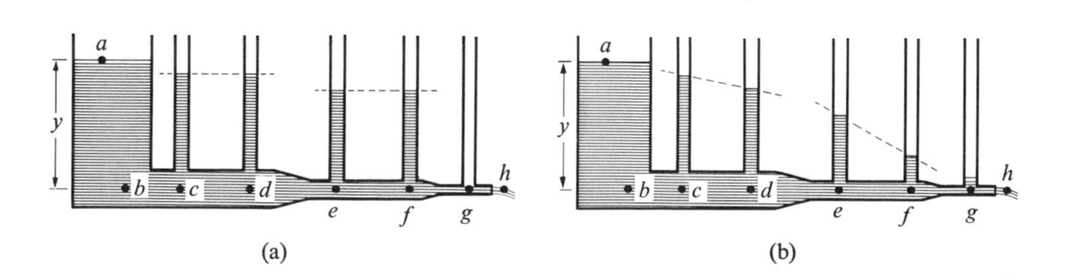

14–16 Pressures along a horizontal tube in which is flowing (a) an ideal fluid, (b) a viscous fluid.

drop from d to e results in part from acceleration and in part from viscosity. The pressure gradient between e and f is greater than between c and d because of the smaller radius in this portion. Finally, the pressure at g is somewhat above atmospheric, since there is now a pressure gradient between this point and the end of the tube.

14-7 STOKES' LAW

When an ideal fluid of zero viscosity flows past a sphere, or when a sphere moves through a stationary fluid, the streamlines form a perfectly symmetrical pattern around the sphere, as shown in Fig. 14-2(a). The pressure at any point on the upstream hemispherical surface is exactly the same as that at the corresponding point on the downstream face, and the resultant force on the sphere is zero. If the fluid has viscosity, however, there will be a viscous drag on the sphere. (A viscous drag will, of course, be experienced by a body of any shape, but only for a sphere is the drag readily calculable.)

We shall not attempt to derive the expression for the viscous force directly from the laws of flow of a viscous fluid. The only quantities on which the force can depend are the viscosity η of the fluid, the radius r of the sphere, and its velocity v relative to the fluid. A complete analysis shows that the force F is given by

$$F = 6\pi\eta r v. \tag{14-19}$$

This equation was first deduced by Sir George Stokes in 1845 and is called *Stokes' law*. We have already used it in Section 5-6 (Example 9) to study the motion of a sphere falling in a viscous fluid, although at that point it was necessary to know only that the viscous force on a given sphere in a given fluid is proportional to the relative velocity.

It will be recalled that a sphere falling in a viscous fluid reaches a *terminal velocity* v_T at which the viscous retarding force plus the buoyant force equals the weight of the sphere. Let ρ be the density of the sphere and ρ' the density of the fluid. The weight of the sphere is then $\frac{4}{3}\pi r^3 \rho g$, the buoyant force is $\frac{4}{3}\pi r^3 \rho' g$, and when the terminal velocity is reached,

$$\tfrac{4}{3}\pi r^3 \rho' g + 6\pi\eta r v_T = \tfrac{4}{3}\pi r^3 \rho g,$$

or

$$v_T = \frac{2}{9}\frac{r^2 g}{\eta}(\rho - \rho'). \tag{14-20}$$

By measuring the terminal velocity of a sphere of known radius and density, the viscosity of the fluid in which it is falling can be found from the equation above. This equation was also used by Millikan to calculate the radius of the tiny submicroscopic, electrically charged oil drops by means of which he determined the charge on an individual electron. In this case, the terminal velocity of the drops was measured as they fell in air of known viscosity.

14-8 REYNOLDS NUMBER

When the velocity of a fluid flowing in a tube exceeds a certain critical value (which depends on the properties of the fluid and the diameter of the tube) the nature of the flow becomes extremely complicated. Within an extremely thin layer adjacent to the tube walls, called the *boundary layer*, the flow is still laminar. The flow velocity in the boundary layer is zero at the tube walls and increases uniformly throughout the layer. The properties of the boundary layer are of the greatest importance in determining the resistance to flow, and the transfer of heat to or from the moving fluid.

Beyond the boundary layer, the motion is highly irregular. Random local circular currents called *vortices* develop within the fluid, with a large increase in the resistance to flow. Flow of this sort is called *turbulent*.

Experiment indicates that a combination of four factors determines whether the flow of a fluid through a tube or pipe is laminar or turbulent. This combination is known as the *Reynolds number*, N_R, and is defined as

$$N_R = \frac{\rho v D}{\eta},$$

where ρ is the density of the fluid, v the average forward velocity, η the viscosity, and D the diameter of the tube. (The average velocity is defined as the uniform velocity over the entire cross section of the tube, which would result in the same volume rate of flow.) The Reynolds number, $\rho v D/\eta$, is a *dimensionless* quantity and has the same numerical value in any consistent system of units.

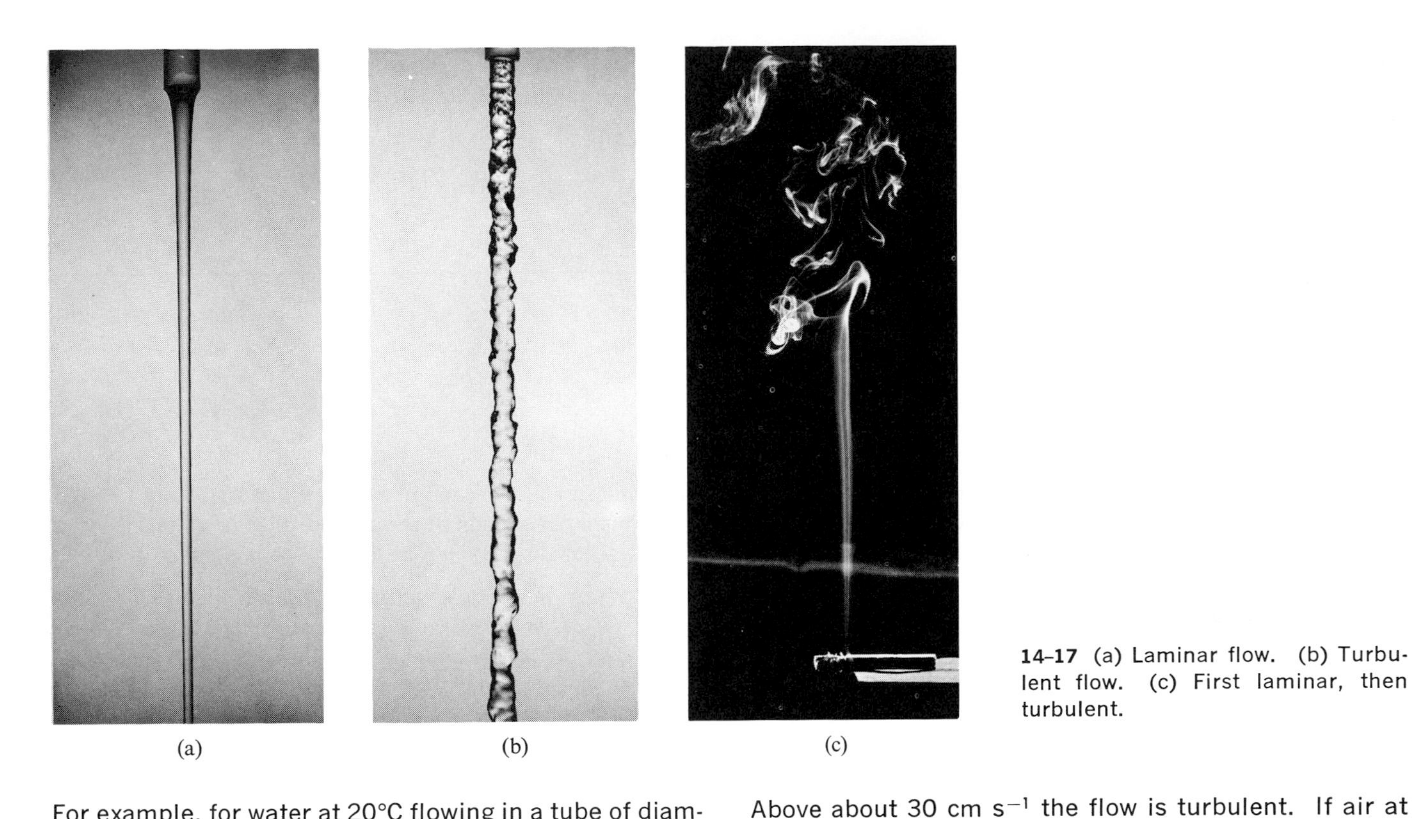

14-17 (a) Laminar flow. (b) Turbulent flow. (c) First laminar, then turbulent.

For example, for water at 20°C flowing in a tube of diameter 1 cm with an average velocity of 10 cm s^{-1}, the Reynolds number is

$$N_R = \frac{\rho v D}{\eta} = \frac{1 \text{ g cm}^{-3} \times 10 \text{ cm s}^{-1} \times 1 \text{ cm}}{0.01 \text{ dyn s cm}^{-2}} = 1000.$$

Had the four quantities been expressed originally in the engineering system of units, the same value of 1000 would have been obtained.

All experiments show that when the Reynolds number is less than about 2000 the flow is laminar, whereas above about 3000 the flow is turbulent. In the transition region between 2000 and 3000 the flow is unstable and may change from one type to the other. Thus for water at 20°C flowing in a tube 1 cm in diameter, the flow is laminar when

$$\frac{\rho v D}{\eta} \lesseqgtr 2000,$$

or when

$$v \lesseqgtr \frac{2000 \times 0.01}{1 \times 1} \text{ cm s}^{-1} = 20 \text{ cm s}^{-1}.$$

Above about 30 cm s^{-1} the flow is turbulent. If air at the same temperature were flowing at 30 cm s^{-1} in the same tube, the Reynolds number would be

$$N_R = \frac{0.0013 \times 30 \times 1}{181 \times 10^{-6}} = 216.$$

Since this is much less than 3000, the flow would be laminar and would not become turbulent unless the velocity were as great as 420 cm s^{-1}.

The distinction between laminar and turbulent flow is shown most effectively in the photographs of Fig. 14-17. In (a) and (b) the fluid is water and in (c) air and smoke particles.

The Reynolds number of a system forms the basis for the study of the behavior of real systems through the use of small scale models. A common example is the wind tunnel, in which one measures the aerodynamic forces on a scale model of an aircraft wing. The forces on a full-size wing are then deduced from these measurements.

Two systems are said to be *dynamically similar* if the Reynolds number, $\rho v D/\eta$, is the same for both. The letter D may refer, in general, to any dimension of a system, such as the span or chord of an aircraft wing. Thus the flow of a fluid of given density ρ and viscosity η, about a half-scale model, is dynamically similar to that around the full-size object if the velocity v is twice as great.

Problems

14-1 A circular hole 1 in. in diameter is cut in the side of a large standpipe, 20 ft below the water level in the standpipe. Find (a) the velocity of efflux, and (b) the volume discharged per unit time. Neglect the contraction of the streamlines after emerging from the hole.

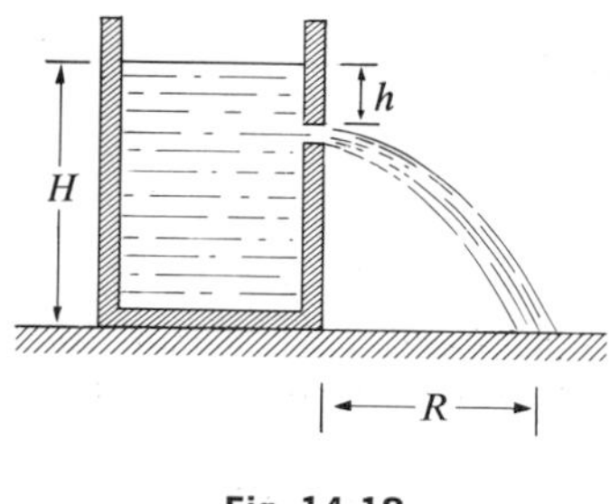

Fig. 14-18

14-2 Water stands at a depth H in a large open tank whose side walls are vertical (Fig. 14-18). A hole is made in one of the walls at a depth h below the water surface. (a) At what distance R from the foot of the wall does the emerging stream of water strike the floor? (b) At what height above the bottom of the tank could a second hole be cut so that the stream emerging from it would have the same range?

14-3 A cylindrical vessel, open at the top, is 20 cm high and 10 cm in diameter. A circular hole whose cross-sectional area is 1 cm^2 is cut in the center of the bottom of the vessel. Water flows into the vessel from a tube above it at the rate of 140 $cm^3\ s^{-1}$. (a) How high will the water in the vessel rise? (b) If the flow into the vessel is stopped after the above height has been reached, how long a time is required for the vessel to empty? Neglect the convergence of the streamlines.

14-4 At a certain point in a pipeline the velocity is 4 ft s^{-1} and the gauge pressure is 2 lb in^{-2} above atmospheric. Find the gauge pressure at a second point in the line 2 ft lower than the first, if the cross section at the second point is one-half that at the first. The liquid in the pipe is sea water weighing 64 lb ft^{-3}.

14-5 Water in an enclosed tank is subjected to a gauge pressure of 4 lb in^{-2} applied by compressed air introduced into the top of the tank. There is a small hole in the side of the tank 16 ft below the level of the water. Calculate the speed with which water escapes from this hole.

14-6 What gauge pressure is required in the city mains in order that a stream from a fire hose connected to the mains may reach a vertical height of 60 ft?

14-7 A tank of large area is filled with water to a depth of one foot. A hole of 1 in^2 cross section in the bottom allows water to drain out in a continuous stream. (a) What is the rate at which water flows out of the tank, in $ft^3\ s^{-1}$? (b) At what distance below the bottom of the tank is the cross-sectional area of the stream equal to one-half the area of the hole?

14-8 A sealed tank containing sea water to a height of 5 ft also contains air above the water at a gauge pressure of 580 lb ft^{-2}. Water flows out from a hole at the bottom. The cross-sectional area of the hole is 1.6 in^2. (a) Calculate the efflux velocity of the water. (b) Calculate the reaction force on the tank exerted by the water in the emergent stream.

14-9 A pipeline 6 in. in diameter, flowing full of water, has a constriction of diameter 3 in. If the velocity in the 6-in. portion is 4 ft s^{-1}, find (a) the velocity in the constriction, and (b) the discharge rate in cubic feet per second.

14-10 A horizontal pipe of 6 in^2 cross section tapers to a cross section of 2 in^2. If sea water of density 2 slugs ft^{-3} is flowing with a velocity of 180 ft min^{-1} in the large pipe where a pressure gauge reads 10.5 lb in^{-2}, what is the gauge pressure in the adjoining part of the small pipe? The barometer reads 30 in. of mercury.

14–11 At a certain point in a pipeline the velocity is 2 ft s^{-1} and the gauge pressure is 35 lb in^{-2}. Find the gauge pressure at a second point in the line 50 ft lower than the first, if the cross section at the second point is one-half that at the first. The liquid in the pipe is water.

14–12 Water stands at a depth of 4 ft in an enclosed tank whose side walls are vertical. The space above the water surface contains air at a gauge pressure of 120 lb in^{-2}. The tank rests on a platform 8 ft above the floor. A hole of cross-sectional area 0.5 in^2 is made in one of the side walls just above the bottom of the tank. (a) Where does the stream of water from the hole strike the floor? (b) What is the vertical force exerted on the floor by the stream? (c) What is the horizontal force exerted on the tank? Assume the water level and the pressure in the tank to remain constant, and neglect any effect of viscosity.

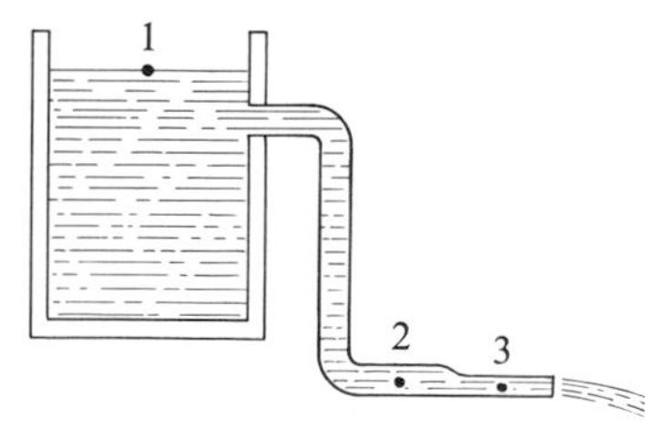

Fig. 14–19

14–13 Water flows steadily from a reservoir, as in Fig. 14–19. The elevation of point 1 is 40 ft; of points 2 and 3 it is 4 ft. The cross section at point 2 is 0.5 ft^2 and at point 3 it is 0.25 ft^2. The area of the reservoir is very large compared with the cross sections of the pipe. (a) Compute the gauge pressure at point 2. (b) Compute the discharge rate in cubic feet per second.

14–14 Sea water of density 2 slugs ft^{-3} flows steadily in a pipeline of constant cross section leading out of an elevated tank. At a point 4.5 ft below the water level in the tank the gauge pressure in the flowing stream is 1 lb in^{-2}. (a) What is the velocity of the water at this point? (b) If the pipe rises to a point 9 ft above the level of the water in the tank, what are the velocity and the pressure at the latter point?

14–15 Sea water weighing 64 lb ft^{-3} flows through a horizontal pipe of cross-sectional area 1.44 in^2. At one section the cross-sectional area is 0.72 in^2. The pressure difference between the two sections is 0.048 lb in^{-2}. How many cubic feet of water will flow out of the pipe in 1 min?

14–16 Two very large open tanks, *A* and *F* (Fig. 14–20), both contain the same liquid. A horizontal pipe *BCD* having a constriction at *C* leads out of the bottom of tank *A*, and a vertical pipe *E* opens into the constriction at *C* and dips into the liquid in

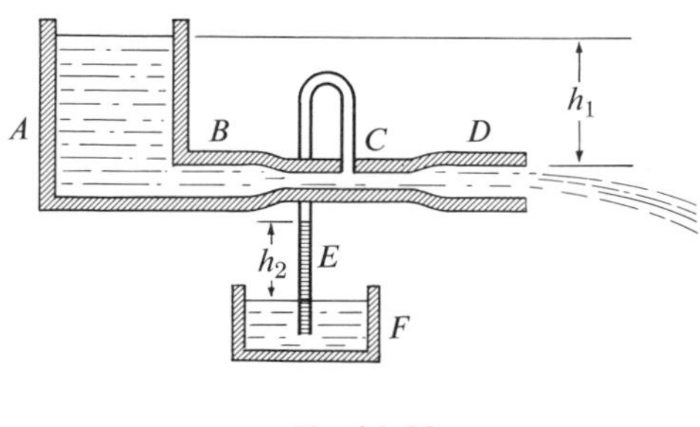

Fig. 14–20

tank *F*. Assume streamline flow and no viscosity. If the cross section at *C* is one-half that at *D*, and if *D* is at a distance h_1 below the level of the liquid in *A*, to what height h_2 will liquid rise in pipe *E*? Express your answer in terms of h_1. Neglect changes in atmospheric pressure with elevation.

14–17 At a certain point in a horizontal pipeline the gauge pressure is 6.24 lb in^{-2}. At another point the gauge pressure is 4.37 lb in^{-2}. If the areas of the pipe at these two points are 3 in^2 and 1.5 in^2, respectively, compute the number of cubic feet of water which flow across any cross section of the pipe per minute.

14–18 Water flowing in a horizontal pipe discharges at the rate of 0.12 ft^3 s^{-1}. At a point in the pipe where the cross section is 0.01 ft^2, the absolute pressure is 18 lb in^{-2}. What must be the cross section of a constriction in the pipe such that the pressure there is reduced to 15 lb in^{-2}?

14–19 The pressure difference between the main pipeline and the throat of a Venturi meter is 15 lb in^{-2}. The areas of the pipe and the constriction are 1 ft^2 and 0.5 ft^2. How many cubic feet per second are flowing through the pipe? The liquid in the pipe is water.

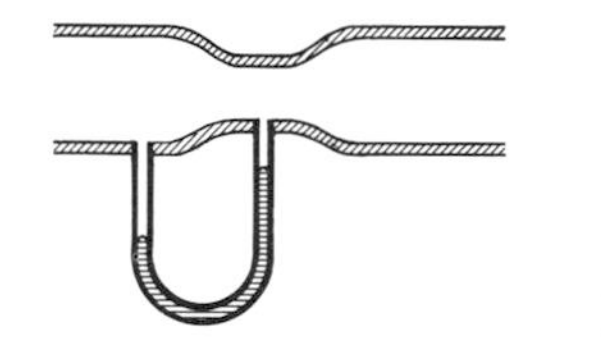

Fig. 14–21

14–20 The section of pipe shown in Fig. 14–21 has a cross section of 0.04 ft^2 at the wider portions and 0.01 ft^2 at the constriction. One cubic foot of water is discharged from the pipe in 5 s. (a) Find the velocities at the wide and the narrow portions. (b) Find the pressure difference between these portions. (c) Find the difference in height between the mercury columns in the U-tube.

14–21 Water is used as the manometric liquid in a Prandtl tube mounted in an aircraft to measure airspeed. If the maximum difference in height between the liquid columns is 10 cm, what is the maximum airspeed that can be measured? The density of air is 1.3×10^{-3} g cm^{-3}.

14–22 The inner, rotating cylinder of the viscosimeter in Fig. 14–13 is 5 cm in diameter. The inner diameter of the outer, fixed cylinder is 5.4 cm, and the diameter of the pulley attached to the inner cylinder is 4 cm. A liquid whose viscosity is 6 poise fills the space between inner and outer cylinders to a depth of 8 cm. A body of mass 30 g is supported by a thread wrapped around the pulley attached to the inner cylinder and hangs vertically, as shown in Fig. 14–13. Find the speed of descent of the body after it has reached its terminal velocity.

14–23 A viscous liquid flows through a tube with laminar flow, as in Fig. 14–15(b). Prove that the volume rate of flow is the same as if the velocity were uniform at all points of a cross section and equal to half the velocity at the axis.

14–24 (a) With what terminal velocity will an air bubble 1 mm in diameter rise in a liquid of viscosity 150 cp and density 0.90 g cm^{-3}? (b) What is the terminal velocity of the same bubble in water?

14–25 (a) With what velocity is a steel ball 1 mm in radius falling in a tank of glycerine at an instant when its acceleration is one-half that of a freely falling body? (b) What is the terminal velocity of the ball? The densities of steel and of glycerine are 8.5 g cm^{-3} and 1.32 g cm^{-3}, respectively.

14–26 Assume that air is streaming horizontally past an aircraft wing such that the velocity is 100 ft s^{-1} over the top surface and 80 ft s^{-1} past the bottom surface. If the wing weighs 600 lb and has an area of 40 ft^2, what is the net force on the wing? The density of air is 0.0013 g cm^{-3}.

14–27 Modern airplane design calls for a "lift" of about 20 lb ft^{-2} of wing area. Assume that air flows past the wing of an aircraft with streamline flow. If the velocity of flow past the lower wing surface is 300 ft s^{-1}, what is the required velocity over the upper surface to give a "lift" of 20 lb ft^{-2}? The density of air is 0.0013 g cm^{-3}.

14–28 Water at 20°C flows with a speed of 50 cm s^{-1} through a pipe of diameter 3 mm. (a) What is the Reynolds number? (b) What is the nature of the flow?

14–29 Water at 20°C is pumped through a horizontal smooth pipe 15 cm in diameter and discharges into the air. If the pump maintains a flow velocity of 30 cm s^{-1}, (a) what is the nature of the flow? (b) What is the discharge rate in liters per second?

14–30 The tank at the left of Fig. 14–16(a) has a very large cross section and is open to the atmosphere. The depth $y = 40$ cm. The cross sections of the horizontal tubes leading out of the tank are respectively 1 cm^2, 0.5 cm^2, and 0.2 cm^2. The liquid is ideal, having zero viscosity. (a) What is the volume rate of flow out of the tank? (b) What is the velocity in each portion of the horizontal tube? (c) What are the heights of the liquid in the vertical side tubes?

Suppose that the liquid in Fig. 14–16(b) has a viscosity of 0.5 poise, a density of 0.8 g cm^{-3}, and that the depth of liquid in the large tank is such that the volume rate of flow is the same as in part (a) above. The distance between the side tubes at c and d, and between those at e and f, is 20 cm. The cross sections of the horizontal tubes are the same in both diagrams. (d) What is the difference in level between the tops of the liquid columns in tubes c and d? (e) In tubes e and f? (f) What is the flow velocity on the axis of each part of the horizontal tube?

14–31 (a) Is it reasonable to assume that the flow in the second part of Problem 14–30 is laminar? (b) Would the flow be laminar if the liquid were water?

15 Temperature-Expansion

15-1 CONCEPT OF TEMPERATURE

To describe the equilibrium states of mechanical systems, as well as to study and predict the motions of rigid bodies and fluids, only three fundamental indefinables were needed: length, mass, and time. All other physical quantities of importance in mechanics could be expressed in terms of these three indefinables. We come, now, however, to a series of phenomena, called *thermal effects* or *heat phenomena*, which involve aspects that are essentially nonmechanical and which require for their description a fourth fundamental indefinable, the *temperature*.

Ever since early childhood we have experienced the sensations of hotness and coldness and described these sensations with the aid of adjectives such as cold, cool, tepid, warm, hot, etc. When we touch an object, we use our *temperature sense* to ascribe *to the object* a property called *temperature*, which determines whether it will feel hot or cold to the touch. The hotter it feels, the higher the temperature. This procedure plays the same role in "qualitative science" that hefting a body does in determining its weight or that kicking an object does in estimating its mass. To determine the mass of an object quantitatively, we must first arrive at the concept of mass by means of *quantitative* operations such as measuring the acceleration imparted to the object by a measured force, and then taking the ratio of F to a. This set of operations is performed without appeal to the sense perceptions associated with flexed muscles, or the discomfort connected with kicking. Similarly, the quantitative determination of temperature requires a set of operations that are independent of our sense perceptions of hotness or coldness, and which involve measurable quantities. How this is done will be explained in the following paragraphs.

There are certain simple systems the state of any one of which may be specified by measuring the value of one physical quantity. Consider for example a liquid such as mercury or alcohol contained in a very thin-walled bulb which communicates with a very narrow tube or capillary, as shown in Fig. 15-1(a). The state of this system is specified by noting the length of the liquid column, starting at an arbitrarily chosen point. The length L is called a *state coordinate*. Another simple system is shown in Fig. 15-1(b) which depicts a thin-walled vessel containing a gas whose volume remains constant and whose state coordinate, the pressure, is read on any convenient pressure gauge. In the next section, use will be made of a coil of fine wire (at constant tension) whose state coordinate is the value of its electrical resistance, and also a junction of two dis-

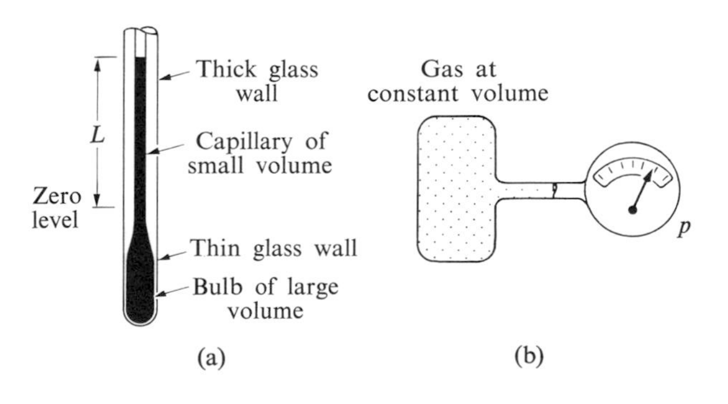

15-1 (a) A system whose state is specified by the value of L. (b) A system whose state is given by the value of p.

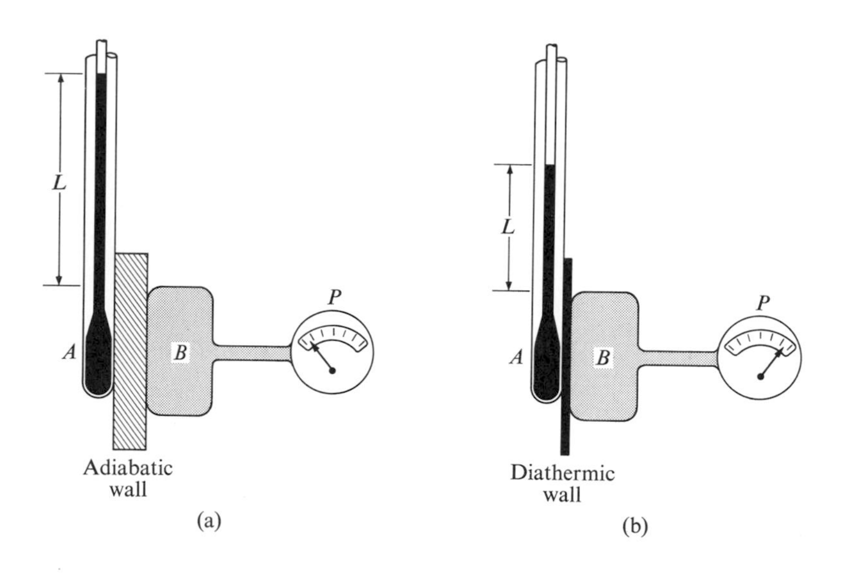

15-2 System A, a liquid column, and system B, a gas at constant volume, separated by (a) an adiabatic wall, p and L independent, and (b) a diathermic wall, p and L dependent.

similar metals whose state coordinate is the value of the electromotive force between its ends.

Let A stand for the liquid-in-capillary system with state coordinate L and let B stand for the gas at constant volume with state coordinate p. If A and B are brought into contact, their state coordinates, in general, are found to change. When A and B are separated, however, the change is slower, and when thick walls of various materials like wood, plaster, felt, asbestos, etc., are used to separate A and B, the values of the respective state coordinates L and p are almost independent of each other. Generalizing from these observations, we postulate the existence of an ideal partition, called an **adiabatic wall,** *which, when used to separate two systems, allows their state coordinates to vary over a large range of values independently.* An adiabatic wall is an idealization that cannot be realized perfectly but may be approximated closely. It is represented as a thick cross-shaded region, as shown in Fig. 15-2(a).

When systems A and B are first put into actual contact or are separated by a thin metallic partition, their state coordinates may or may not change. *A wall which enables a state coordinate of one system to influence that of another is called a* **diathermic wall.** A thin sheet of copper is the most practical diathermic wall. As shown in Fig. 15-2(b), a diathermic wall is depicted as a thin, darkly shaded region. Eventually, a time will be reached when no further change in the coordinates of A and B takes place. *The joint state of both systems that exists when all changes in the coordinates have ceased is called* **thermal equilibrium.**

Imagine two systems A and B separated from each other by an adiabatic wall but each in contact with a third system C through diathermic walls, the whole assembly being surrounded by an adiabatic wall as shown in Fig. 15-3(a). Experiment shows that the two systems will come to thermal equilibrium with the third and that no further change will occur if the adiabatic wall separating them is then replaced by a diathermic

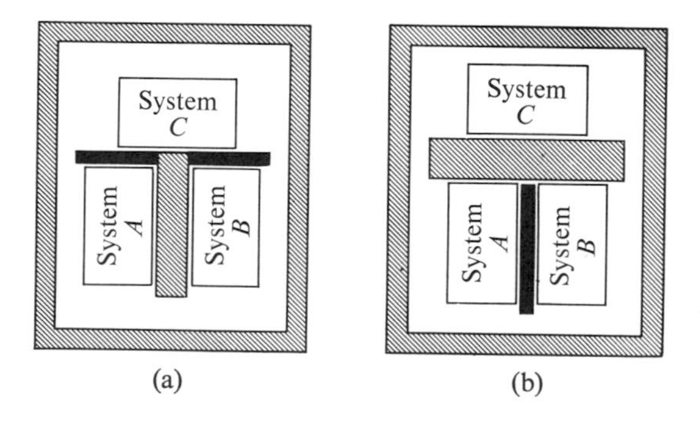

15-3 The zeroth law of thermodynamics. (a) If A and B are each in thermal equilibrium with C, then (b) A and B are in thermal equilibrium with each other.

wall (Fig. 15–3b). If, instead of allowing both systems *A* and *B* to come to equilibrium with *C* at the same time, we first have equilibrium between *A* and *C* and then equilibrium between *B* and *C* (the state of system *C* being the same in both cases), then, when *A* and *B* are brought into communication through a diathermic wall, they will be found to be in thermal equilibrium. We shall use the expression "two systems are in thermal equilibrium" to mean that the two systems are in states such that if the two *were* to be connected through a diathermic wall, the combined system *would be* in thermal equilibrium.

These experimental facts may then be stated concisely in the following form: *Two systems in thermal equilibrium with a third are in thermal equilibrium with each other.* Following R. H. Fowler, we shall call this postulate *the zeroth law of thermodynamics.* At first thought it might seem that the zeroth law is obvious, but this is not so. An amber rod *A* that has been rubbed with fur will attract a neutral pith ball *C*. So will another similarly rubbed amber rod *B*, but the two amber rods will not attract each other. (Woman *A* loves man *C*. Woman *B* loves man *C*, but does woman *A* love woman *B*?)

When two systems *A* and *B* are first put in contact through a diathermic wall, they may or may not be in thermal equilibrium. One is entitled to ask, "What is there about *A* and *B* that determines whether or not they are in thermal equilibrium?" Experiment shows that not the mass, density, elastic modulus, electric charge, or the magnetic state, in fact none of the quantities of importance in mechanics, electricity, or magnetism, is the determining factor. *We therefore infer the existence of a new property called the* **temperature.** *The temperature of a system is that property which determines whether or not it will be in thermal equilibrium with other systems. When two or more systems are in thermal equilibrium, they are said to have the same temperature.*

The temperature of all systems in thermal equilibrium may be represented by a number. The establishment of a temperature scale is merely the adoption of a set of rules for assigning numbers to temperatures. Once this is done, the condition for thermal equilibrium between two systems is that they have the same temperature. Also, when the temperatures are different, we may be sure that the systems are not in thermal equilibrium.

The preceding operational treatment of the concept of temperature merely expresses the fundamental idea that the temperature of a system is a property which eventually attains the same value as that of other systems when all these systems are put in contact or separated by thin metallic walls within an enclosure of thick asbestos walls. It will be recognized that this concept is identical with the everyday idea of temperature as a measure of the hotness or coldness of a system, since, so far as our senses may be relied upon, the hotness of all objects becomes the same after they have been together long enough. However, it was necessary to express this simple idea in technical language in order to be able to establish a rational set of rules for measuring temperature and also to provide a solid foundation for the study of advanced physics.

15–2 THERMOMETERS

If we want to determine the temperatures of a number of systems, the simplest procedure is to choose *one* of the systems as an indicator of thermal equilibrium between it and the other systems. The system so chosen is called a *thermometer.* The reading of the thermometer is the temperature of all systems in thermal equilibrium with it. The important characteristics of a thermometer are *sensitivity* (an appreciable change in the state coordinate produced by a small change in temperature), *accuracy* in the measurement of the state coordinate, and *reproducibility.* Another often desirable property is *speed* in coming to thermal equilibrium with other systems. The thermometers which satisfy these requirements best will be described in the following paragraphs.

The most useful thermometer in research and engineering laboratories is the *thermocouple,* which consists of a junction of two different metals or alloys, such as *A* and *B*, labeled "test junction" in Fig. 15–4. The test junction is usually embedded in the material whose temperature is to be measured. Since the test junction is small and has a small mass, it can follow temperature changes rapidly and come to equilibrium quickly. The reference junction consists of two junctions: one of *A* and copper and the other of *B* and copper. These two junctions are maintained at any desired constant temperature, called the reference temperature. The state coordinate of this thermometer is an electrical quantity

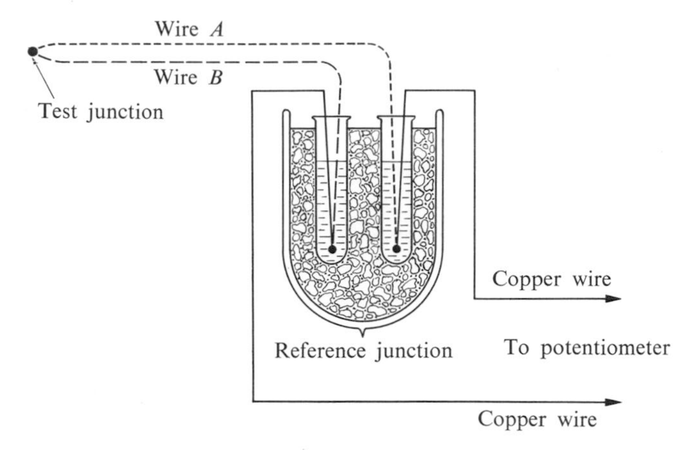

15-4 Thermocouple, showing the test junction and the reference junction.

called the emf (ee-em-eff, electromotive force) which is measured with an instrument known as a potentiometer. A thermocouple with one junction of pure platinum and the other of 90% platinum and 10% rhodium is often used. Copper and an alloy called constantan are also frequently used.

A *resistance thermometer* consists of a fine wire, often enclosed in a thin-walled silver tube for protection. Copper wires lead from the thermometer unit to a resistance-measuring device such as a Wheatstone bridge. Since resistance may be measured with great precision, the resistance thermometer is one of the most precise instruments for the measurement of temperature. In the region of extremely low temperatures a small carbon cylinder or a small piece of germanium crystal is used instead of a coil of platinum wire.

To measure temperatures above the range of thermocouples and resistance thermometers an *optical pyrometer* is used. As shown in Fig. 15-5, it consists essentially of a telescope *T*, in the tube of which is mounted a filter *F* of red glass and a small electric lamp bulb *L*. When the pyrometer is directed toward a furnace, an observer looking through the telescope sees the dark lamp filament against the bright background of the furnace. The lamp filament is connected to a battery *B* and a rheostat *R*. By turning the rheostat knob, the current in the filament, and hence the brightness, may be gradually increased until the brightness of the filament

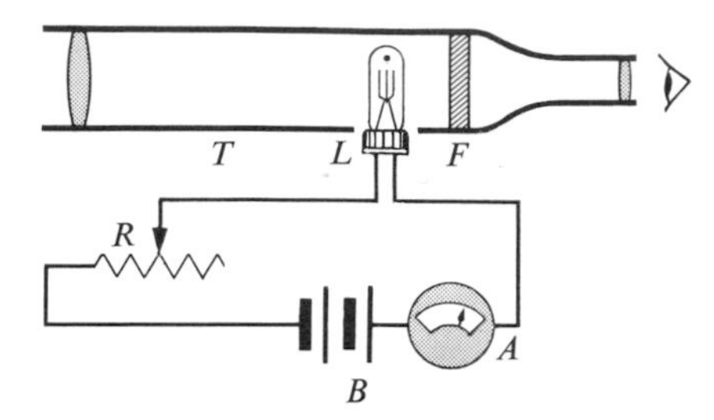

15-5 Principle of the optical pyrometer.

just matches the brightness of the background. From previous calibration of the instrument at known temperatures, the scale of the ammeter *A* in the circuit may be marked to read the unknown temperature directly. Since no part of the instrument needs to come into contact with the hot body, the optical pyrometer may be used at temperatures above the melting points of metals.

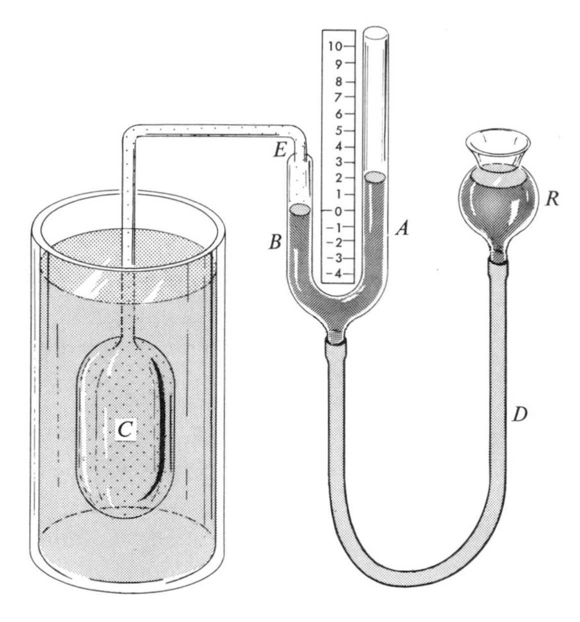

15-6 The constant-volume gas thermometer.

Of all the state coordinates or, as they are often called, *thermometric properties*, the pressure of a gas whose volume is maintained constant stands out for its sensitivity, accuracy of measurement, and reproducibility. The constant-volume gas thermometer is illustrated schematically in Fig. 15-6. The materials, construction, and dimensions differ in various laboratories throughout the world and depend on the nature of the gas and the temperature range to be covered.

The gas, usually helium, is contained in bulb *C* and the pressure exerted by it can be measured by the open-tube mercury manometer. As the temperature of the gas increases, the gas expands, forcing the mercury down in tube *B* and up in tube *A*. Tubes *A* and *B* communicate through a rubber tube *D* with a mercury reservoir *R*. By raising *R*, the mercury level in *B* may be brought back to a reference mark *E*. The gas is thus kept at constant volume.

Gas thermometers are used mainly in bureaus of standards and in some university research laboratories. They are usually large, bulky, and slow in coming to thermal equilibrium.

15-3 THE ESTABLISHMENT OF A TEMPERATURE SCALE

Any one of the thermometers described in the preceding section may be used to indicate the constancy of a temperature if its state coordinate or thermometric property remains constant. By this means, it has been found that a system composed of a solid and a liquid of the same material maintained at constant pressure will remain in *phase equilibrium* (that is, the liquid and solid exist together without the liquid changing into solid or the solid changing into liquid) only at one definite temperature. Similarly, a liquid will remain in phase equilibrium with its vapor at only one definite temperature when the pressure is maintained constant.

The temperature at which a solid and liquid of the same material coexist in phase equilibrium *at atmospheric pressure* is called the *normal melting point*, abbreviated NMP. The temperature at which a liquid and its vapor exist in phase equilibrium at atmospheric pressure is called the *normal boiling point*, abbreviated NBP.

Phase equilibrium between a solid and its vapor is sometimes possible at atmospheric pressure. The temperature at which this takes place is the *normal sublimation point*, NSP.

It is possible for all three phases—solid, liquid, and vapor—to coexist in equilibrium, but only at one definite pressure and temperature known as the *triple point*, abbreviated TP. The triple-point pressure of water is 4.58 mm of mercury.

The NMP, NBP, NSP, and TP of any material can be chosen as a standard for the purpose of setting up a temperature scale. Any temperature so chosen is called a *fixed point*. Before 1954 there were two standard fixed points, the NBP of water and the equilibrium temperature of pure ice and air-saturated water. Both of these have been given up. *There is only one standard fixed point in modern thermometry and that is the triple point of water* to which is given the arbitrary number

$$273.16°K,$$

read 273.16 degrees Kelvin.*

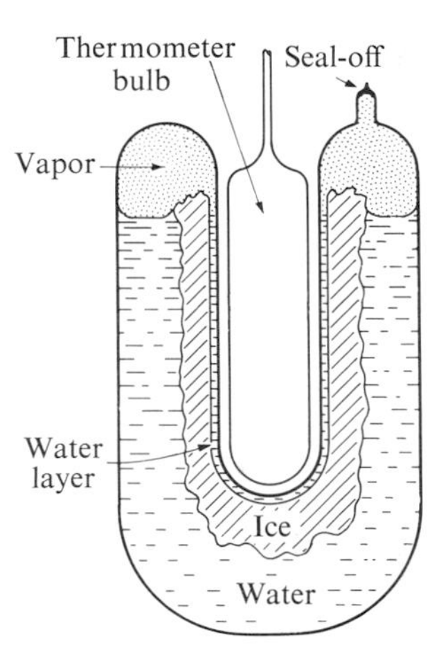

15-7 Triple-point cell with a thermometer in the well, which melts a thin layer of ice nearby.

To achieve the triple point, water of the highest purity is distilled into a vessel like that shown schematically in Fig. 15-7. When all air has been removed, the vessel is sealed off. With the aid of a freezing mixture in the inner well, a layer of ice is formed around the well. When the freezing mixture is replaced by a thermometer

* At a meeting of the Thirteenth General Conference on Weights and Measures on Oct. 13, 1967, the name of the unit of temperature was changed from *degree Kelvin* (symbol °K) to *kelvin* (symbol K). The kelvin, now the unit of temperature, is the fraction 1/273.16 of the thermodynamic temperature of the triple point of water.

TABLE 15-1 COMPARISON OF THERMOMETERS

Fixed point	(Cu-constantan) $\mathcal{E}$, mV	$T(\mathcal{E})$	(Pt) R, ohms	$T(R)$	(H_2, const. V) p, atm	$T(p)$	(H_2, const. V) p, atm	$T(p)$
N_2 (NBP)	0.73	32.0	1.96	54.5	1.82	73	0.29	79
O_2 (NBP)	0.95	41.5	2.50	69.5	2.13	86	0.33	90
CO_2 (NSP)	3.52	154	6.65	185	4.80	193	0.72	196
H_2O (TP)	$\mathcal{E}_3 = 6.26$	273	$R_3 = 9.83$	273	$p_3 = 6.80$	273	$p_3 = 1.00$	273
H_2O (NBP)	10.05	440	13.65	380	9.30	374	1.37	374
Sn (NMP)	17.50	762	18.56	516	12.70	510	1.85	505

bulb, a thin layer of ice is melted nearby. So long as the solid, liquid, and vapor phases coexist in equilibrium, the system is at the triple point.

We start our program of setting up a temperature scale by denoting with the letter X *any one* of the thermometric properties mentioned previously:

the emf of a thermocouple, $\mathcal{E}$,

the resistance of a wire, R,

the pressure of a gas at constant volume, p, etc.

We define the ratio of two temperatures to be the same as the ratio of the two corresponding values of X. Thus, if a thermometer with thermometric property X is put in thermal equilibrium with a system and registers a value X, and is then put in thermal equilibrium with another system and registers a value X_3, the ratio of the temperatures of those two systems is given by

$$\frac{T(X)}{T(X_3)} = \frac{X}{X_3}. \qquad (15\text{-}1)$$

If now, we let the subscript 3 stand for the standard fixed point, the triple point of water, then $T(X_3) = 273.16°\text{K}$. Hence

$$T(X) = 273.16°\text{K}\,\frac{X}{X_3}. \qquad (15\text{-}2)$$

It is important to understand that the relation represented by Eq. (15–1) is an *arbitrary choice*. The ratio of two temperatures could have been chosen to be the ratio of the squares of the X's, or the logarithm of the X's, or the ratio of the negative reciprocals of the X's.

The next step is to see what results are obtained when different thermometers are used to measure the same temperature, following the rules laid down in Eq. (15–2). The results of such a test in which four different thermometers [a copper-constantan thermocouple, a platinum resistance thermometer, and two hydrogen gas thermometers (one at high pressures and one at low pressures)] were used to measure the temperatures of six different fixed points, are listed in Table 15–1. It is clear from Eq. (15–2) that the temperature of the triple point of water must come out to be 273.16°K. At any other fixed point, however, the two gas thermometers agree quite well, but differ markedly from the readings of the other two thermometers. Further experiments show that the greatest agreement is found among gas thermometers and that, *regardless of the nature of the gas, all gas thermometers at the same temperature approach the same reading as the pressure of the gas approaches zero*. The results of an experiment of this kind are shown in Fig. 15–8, where four gas thermometers

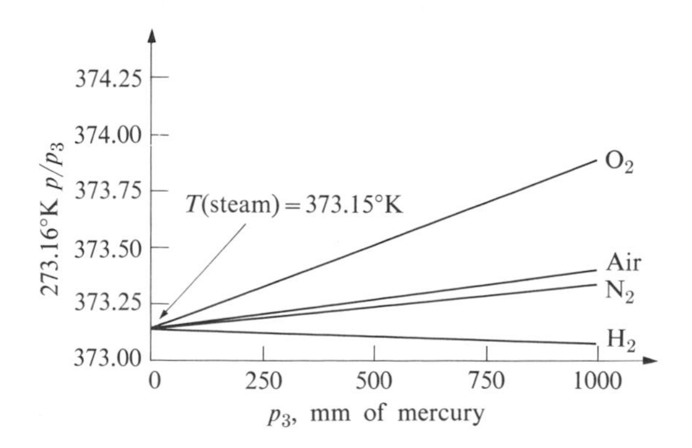

15-8 Readings of a constant-volume gas thermometer for the temperature of condensing steam, when different gases are used at various values of p_3.

give readings for the NBP of water which approach the same value, 373.15°K, as the pressure approaches zero.

We therefore define the gas temperature T by the equation

$$T = [273.16°\text{K}] \lim_{p_3 \to 0} \left(\frac{p}{p_3}\right)_{\text{const. vol.}} \tag{15-3}$$

It should be emphasized that this equation, or the set of rules for measuring temperature which it embodies, does not rest on the assumption that "the pressure of a gas at constant volume is directly proportional to the temperature." To make such a statement, before having a temperature scale set up, is meaningless.

Although the temperature scale of Eq. (15-3) is independent of the properties of any one particular gas, it still depends on the properties of gases in general. To measure a low temperature, a gas that does not liquefy at the low temperature must be used. The lowest temperature that can be measured with a gas thermometer is about 1°K, provided low-pressure helium is used as the gas. *The temperature $T = 0$ remains as yet undefined.*

The Kelvin temperature scale, which is independent of the properties of any particular substance, will be described in Chapter 19. It can be shown that in the temperature region in which a gas thermometer can be used, the gas scale and the Kelvin scale are identical. In anticipation of this result, we write °K after a gas temperature. It will also be shown in Chapter 19 how the absolute zero of temperature is defined on the Kelvin scale. Until then, the term "absolute zero" will have no meaning. The statement made so often that all molecular activity ceases at the temperature $T = 0$ is entirely erroneous. When it is necessary in statistical mechanics to correlate temperature with molecular activity, it is found that classical statistical mechanics must be modified with the aid of quantum mechanics. When this modification is carried out, the molecules of a substance at absolute zero have a *finite* amount of kinetic energy known as the *zero-point* energy.

15-4 THE CELSIUS, RANKINE, AND FAHRENHEIT SCALES

The Celsius* temperature scale (formerly called the centigrade scale in the United States and Great Britain) employs a degree of the same magnitude as that of the kelvin scale, but its zero point is shifted so that *the Celsius temperature of the triple point of water is* 0.01 *degree Celsius*, abbreviated 0.01°C. Thus, if t denotes the Celsius temperature,

$$t = T - 273.15°\text{K}. \tag{15-4}$$

The Celsius temperature t_S at which steam condenses at 1 atm pressure is

$$t_S = T_S - 273.15°\text{K},$$

and reading T_S from Fig. 15-8,

$$t_S = 373.15° - 273.15° \quad \text{or} \quad t_S = 100.00°\text{C}.$$

There are two other scales in common use in engineering and in everyday life in the United States and in Great Britain. The *Rankine*† temperature T_R (written °R) is proportional to the Kelvin temperature according to the relation

$$T_R = \tfrac{9}{5}T. \tag{15-5}$$

A degree of the same size is used in the *Fahrenheit*‡ scale t_F (written °F), but with the zero point shifted according to the relation

$$t_F = T_R - 459.67°\text{R}. \tag{15-6}$$

Substituting Eqs. (15-4) and (15-5) into Eq. (15-6), we get

$$t_F = \tfrac{9}{5}t + 32°\text{F}, \tag{15-7}$$

from which it follows that the Fahrenheit temperature of the ice point ($t = 0$°C) is 32°F and of the steam point ($t = 100$°C) is 212°F. The 100 Celsius or Kelvin degrees between the ice point and the steam point correspond to 180 Fahrenheit or Rankine degrees, as shown in Fig. 15-9, where the four scales are compared.

---

* Named after Anders Celsius (1701-1744).
† William John MacQuorn Rankine (1820-1872).
‡ Gabriel Fahrenheit (1686-1736).

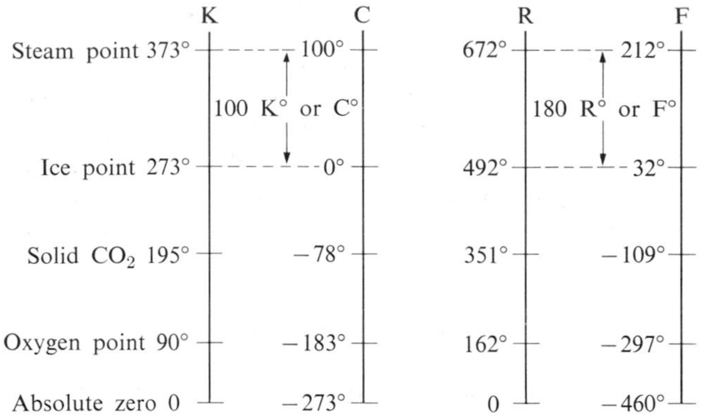

K C R F
Steam point 373° 100° 672° 212°
100 K° or C° 180 R° or F°
Ice point 273° 0° 492° 32°
Solid CO_2 195° −78° 351° −109°
Oxygen point 90° −183° 162° −297°
Absolute zero 0 −273° 0 −460°

15-9 Relations among Kelvin, Celsius, Rankine, and Fahrenheit temperature scales. Temperatures have been rounded off to the nearest degree.

The accurate measurement of a boiling point or melting point with the aid of a gas thermometer requires months of painstaking laboratory work and mathematical computation. Fortunately, this has been done for a large number of substances which are obtainable with high purity. Some of these results are shown in Table 15-2. With the aid of these basic fixed points other thermometers may be calibrated.

Suppose the temperature of a beaker of water is raised from 20°C to 30°C, through a temperature interval of 10 Celsius degrees. It is desirable to distinguish between such a temperature interval and the actual temperature of 10 degrees above the Celsius zero. Hence we shall use the phrase "10 degrees Celsius," or "10°C," when referring to an *actual temperature*, and "10 Celsius degrees," or "10 C°" to mean a temperature *interval*. Thus there is an interval of 10 Celsius degrees between 20 degrees Celsius and 30 degrees Celsius.

15-5 EXPANSION OF SOLIDS AND LIQUIDS

With a few exceptions, the volumes of all bodies increase with increasing temperature if the external pressure on the body remains constant. Suppose that a solid or a liquid undergoes a change of volume dV when the temperature is changed an amount dT (or dt, since the Kelvin degree and the Celsius degree are temperature intervals of equal magnitude). The *coefficient of volume expansion* β is defined as *the fractional change in volume* dV/V *divided by the change of temperature* dT, or

$$\beta = \frac{1}{V}\frac{dV}{dT} = \frac{1}{V}\frac{dV}{dt} \quad \text{(at constant pressure).} \quad (15\text{-}8)$$

The unit of β is 1 *reciprocal degree*, or 1 deg^{-1}. The numerical value depends, of course, on the size of the degree. Since the Kelvin and Celsius degrees are $\frac{9}{5}$ as large as the Rankine and Fahrenheit degrees, the fractional volume change per Kelvin or Celsius degree is $\frac{9}{5}$ as great as that per Rankine or Fahrenheit degree.

The coefficient of volume expansion is often calculated from an empirical equation between the density ρ and the temperature at constant pressure. When this method is not possible, optical methods involving the interference of light are used.

The coefficient of volume expansion is insensitive to a change of pressure, but varies markedly with the temperature. Many experiments indicate that β de-

TABLE 15-2 TEMPERATURES OF FIXED POINTS

Basic fixed points	T, °K	t, °C	T_R, °R	t_F, °F
Standard: Triple point of water	**273.16**	**0.01**	**491.688**	**32.018**
NBP of oxygen	90.18	−182.97	162.32	−297.35
Equilibrium of ice and air-saturated water (ice point)	273.15	0.00	491.67	32.00
NBP of water (steam point)	373.15	100.00	671.67	212.00
NMP of zinc	692.66	419.51	1246.78	787.11
NMP of antimony	903.65	630.50	1626.57	1166.90
NMP of silver	1233.95	960.80	2221.11	1761.44
NMP of gold	1336.15	1063.00	2405.07	1945.40

creases as the temperature is lowered, approaching zero as the Kelvin temperature approaches zero. It is also a peculiar circumstance that the higher the melting point of a metal, the lower the coefficient of volume expansion.

Calling $\bar{\beta}$ the average value of β within a moderate temperature interval $\Delta T = \Delta t$, we may approximate Eq. (15-8) by writing

$$\Delta V = \bar{\beta} V_0 \, \Delta T = \bar{\beta} V_0 \, \Delta t, \tag{15-9}$$

where V_0 is the original volume.

Some values of $\bar{\beta}$ in the neighborhood of room temperature are listed in Table 15-3. Note that the values for liquids are much larger than those for solids.

TABLE 15-3 COEFFICIENT OF VOLUME EXPANSION (APPROXIMATE)

Solids	$\bar{\beta}$, $(C^\circ)^{-1}$	Liquids	$\bar{\beta}$, $(C^\circ)^{-1}$
Aluminum	7.2×10^{-5}	Alcohol, ethyl	75×10^{-5}
Brass	6.0	Carbon disulfide	115
Copper	4.2	Glycerine	49
Glass	1.2-2.7	Mercury	18
Steel	3.6		
Invar	0.27		
Quartz (fused)	0.12		

If there is a hole in a solid body, the volume of the hole increases when the body expands, just as if the hole were a solid of the same material as the body. This remains true even if the hole becomes so large that the surrounding body is reduced to a thin shell. Thus the volume enclosed by a thin-walled glass flask or thermometer bulb increases just as would a solid body of glass of the same size.

Example. A glass flask of volume 200 cm^3 is just filled with mercury at 20°C. How much mercury will overflow when the temperature of the system is raised to 100°C? The coefficient of volume expansion of the glass is 1.2×10^{-5} $(C^\circ)^{-1}$.

The increase in the volume of the flask is

$$\Delta V = 1.2 \times 10^{-5} \times 200 \times (100^\circ - 20^\circ) = 0.192 \text{ cm}^3.$$

The increase in the volume of the mercury is

$$\Delta V = 18 \times 10^{-5} \times 200 \times (100^\circ - 20^\circ) = 2.88 \text{ cm}^3.$$

The volume of mercury overflowing is therefore

$$2.88 - 0.19 = 2.69 \text{ cm}^3.$$

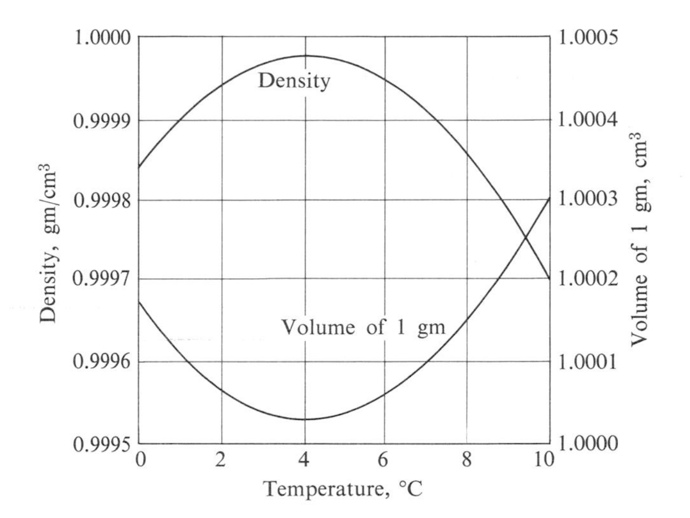

15-10 Density of water, and volume of 1 gram, in the temperature range from 0°C to 10°C.

Water, in the temperature range from 0°C to 4°C, *decreases* in volume with increasing temperature, contrary to the behavior of most substances. That is, between 0°C and 4°C the coefficient of expansion of water is *negative*. Above 4°C, water expands when heated. Since the volume of a given mass of water is smaller at 4°C than at any other temperature, the density of water is a maximum at 4°C. This behavior of water is the reason why lakes and ponds freeze first at their upper surface. Figure 15-10 illustrates this anomalous expansion of water in the temperature range from 0°C to 10°C. Table 15-4 covers a much wider range of temperatures.

TABLE 15-4 DENSITY AND VOLUME OF WATER

t°C	Density, g cm^{-3}	Volume of 1 g, cm^3
0	0.9998	1.0002
4	1.0000	1.0000
10	0.9997	1.0003
20	0.9982	1.0018
50	0.9881	1.0121
75	0.9749	1.0258
100	0.9584	1.0434

For a body in the form of a rod or cable, we are often interested only in the change of *length* with temperature, and we define a *coefficient of linear expansion* α. If L is the length,

$$\alpha = \frac{1}{L}\frac{dL}{dT} = \frac{1}{L}\frac{dL}{dt}, \tag{15-10}$$

and, over a moderate temperature change,

$$\Delta L = L_0\bar{\alpha}\,\Delta T = L_0\bar{\alpha}\,\Delta t, \tag{15-11}$$

where $\bar{\alpha}$ is the average coefficient within the temperature interval, and L_0 is the original length.

The volume coefficient may be calculated in terms of the linear coefficient as follows. Suppose that a solid body is in the form of a rectangular parallelepiped with dimensions L_1, L_2, and L_3. Then the volume is

$$V = L_1L_2L_3,$$

and

$$\frac{dV}{dT} = L_2L_3\frac{dL_1}{dT} + L_1L_3\frac{dL_2}{dT} + L_1L_2\frac{dL_3}{dT}.$$

Dividing by $L_1L_2L_3$, we obtain

$$\frac{1}{V}\frac{dV}{dT} = \frac{1}{L_1}\frac{dL_1}{dT} + \frac{1}{L_2}\frac{dL_2}{dT} + \frac{1}{L_3}\frac{dL_3}{dT}.$$

If the solid has the same properties in each of the three directions, each of the three expressions on the right is the coefficient of linear expansion α, and hence

$$\beta = 3\alpha. \tag{15-12}$$

The coefficient of linear expansion is usually measured with the aid of an optical interferometer.

15-6 THERMAL STRESSES

If the ends of a rod are rigidly fixed so as to prevent expansion or contraction and the temperature of the rod is changed, tensile or compressive stresses, called *thermal stresses*, will be set up in the rod. These stresses may become very large, sufficiently so to stress the rod beyond its elastic limit or even beyond its breaking strength. Hence in the design of any structure which is subject to changes in temperature, some provision must, in general, be made for expansion. In a long steam pipe this is accomplished by the insertion of expansion joints or a section of pipe in the form of a U. In bridges, one end may be rigidly fastened to its abutment while the other rests on rollers.

It is a simple matter to compute the thermal stress set up in a rod which is not free to expand or contract. Suppose that a rod at a temperature t has its ends rigidly fastened, and that while they are thus held the temperature is reduced to a lower value, t_0.

The fractional change in length if the rod were free to contract would be

$$\frac{\Delta L}{L_0} = \bar{\alpha}(t - t_0) = \bar{\alpha}\,\Delta t. \tag{15-13}$$

Since the rod is not free to contract, the tension must increase by a sufficient amount to produce the same fractional change in length. But from the definition of Young's modulus,

$$Y = \frac{F/A}{\Delta L/L_0}, \qquad F = AY\frac{\Delta L}{L_0}.$$

Introducing the expression for $\Delta L/L_0$ from Eq. (15-13), we have

$$F = AY\bar{\alpha}\,\Delta t, \tag{15-14}$$

which gives the tension F in the rod. The *stress* in the rod is

$$\frac{F}{A} = Y\bar{\alpha}\,\Delta t. \tag{15-15}$$

Problems

15–1 The limiting value of the ratio of the pressures of a gas at the melting point of lead and at the triple point of water, when the gas is kept at constant volume, is found to be 2.19816. What is the Kelvin temperature of the melting point of lead?

15–2 (a) If you feel sick in France and are told you have a fever of 40°C, should you be concerned? (b) What is normal body temperature on the Celsius scale? (c) The normal boiling point of liquid oxygen is −182.97°C. What is this temperature on the Kelvin and Rankine scales? (d) At what temperature do the Fahrenheit and Celsius scales coincide?

15–3 The pressure p, volume V, number of moles n, and Kelvin temperature T of an ideal gas are related by the equation $pV = nRT$. Prove that the coefficient of volume expansion is equal to the reciprocal of the Kelvin temperature.

15–4 (a) The relation among the density ρ, the mass m, and the volume V is $\rho = m/V$. Prove that

$$\beta = -\frac{1}{\rho}\frac{\partial \rho}{\partial T}.$$

(b) The density of rock salt between −193°C and −13°C is given by the empirical formula

$$\rho = 2.1680 \times (1 - 11.2 \times 10^{-5}t - 0.5 \times 10^{-7}t^2).$$

Calculate β at −100°C.

15–5 The total change of volume ΔV of a metal, when the temperature changes from room temperature ($T_r = 300$°K) to the melting point T_m, is still only a small fraction of the original volume, so that

$$\frac{\Delta V}{V} = \int_{T_r}^{T_m} \beta \, dT.$$

Given the data in Table 15–5, calculate the average value of $\Delta V/V$ for these three metals.

TABLE 15–5

	T_m, °K	β, $10^{-6}(\text{K}°)^{-1}$
Copper	1360	$43 + 0.022T$
Palladium	1830	$30 + 0.015T$
Platinum	2050	$24 + 0.0086T$

15–6 A glass flask whose volume is exactly 1000 cm^3 at 0°C is filled level full of mercury at this temperature. When flask and mercury are heated to 100°C, 15.2 cm^3 of mercury overflow. If the coefficient of volume expansion of mercury is 0.000182 per centigrade degree, compute the coefficient of linear expansion of the glass.

15–7 At a temperature of 20°C, the volume of a certain glass flask, up to a reference mark on the stem of the flask, is exactly 100 cm^3. The flask is filled to this point with a liquid whose cubical coefficient of expansion is 120×10^{-5} (C°)$^{-1}$, with both flask and liquid at 20°C. The linear coefficient of expansion of the glass is 8×10^{-6} (C°)$^{-1}$. The cross section of the stem is 1 mm^2 and can be considered constant. How far will the liquid rise or fall in the stem when the temperature is raised to 40°C?

15–8 To ensure a tight fit, the aluminum rivets used in airplane construction are made slightly larger than the rivet holes and cooled by "dry ice" (solid CO_2) before being driven. If the diameter of a hole is 0.2500 in., what should be the diameter of a rivet at 20°C if its diameter is to equal that of the hole when the rivet is cooled to −78°C, the temperature of dry ice? Assume the expansion coefficient to remain constant at the value given in Table 15–3.

15–9 A metal rod 30.0 cm long expands by 0.075 cm when its temperature is raised from 0°C to 100°C. A rod of a different metal of the same length expands by 0.045 cm for the same rise in temperature. A third rod, also 30.0 cm long, is made up of pieces of each of the above metals placed end-to-end and expands 0.065 cm between 0°C and 100°C. Find the length of each portion of the composite bar.

15–10 A hole 1.000 in. in diameter is bored in a brass plate at a temperature of 20°C. What is the diameter of the hole when the temperature of the plate is increased to 200°C? Assume the expansion coefficient to remain constant.

15–11 Suppose that a steel hoop could be constructed around the earth's equator, just fitting it at a temperature of 20°C. What would be the space between the hoop and the earth if the temperature of the hoop were increased by 1 C°?

15–12 A clock whose pendulum makes one vibration in 2 s is correct at 25°C. The pendulum shaft is of steel and its mass may be neglected compared with that of the bob. (a) What is the fractional change in length of the shaft when it is cooled to 15°C? (b) How many seconds per day will the clock gain or lose at 15°C? [*Hint:* Use differentials.]

15–13 A clock with a brass pendulum shaft keeps correct time at a certain temperature. (a) How closely must the temperature be controlled if the clock is not to gain or lose more than 1 second a day? Does the answer depend on the period of the pendulum? (b) Will an increase of temperature cause the clock to gain or lose?

15–14 The length of a bridge is 2000 ft. (a) If it were a continuous span, fixed at one end and free to move at the other, about what would be the range of motion of the free end between a cold winter day (—20°F) and a hot summer day (100°F)? (b) If both ends were rigidly fixed on the summer day, what would be the stress on the winter day?

15–15 The cross section of a steel rod is 1.5 in^2. What is the least force that will prevent it from contracting while cooling from 520°C to 20°C?

15–16 A steel wire which is 10 ft long at 20°C is found to increase in length by $\frac{3}{4}$ in. when heated to 520°C. Compute its average coefficient of linear expansion. (b) Find the stress in the wire if it is stretched taut at 520°C and cooled to 20°C without being allowed to contract.

15–17 A steel rod of length 40 cm and a copper rod of length 36 cm, both of the same diameter, are placed end-to-end between two rigid supports, with no initial stress in the rods. The temperature of the rods is now raised by 50C°. What is the stress in each rod?

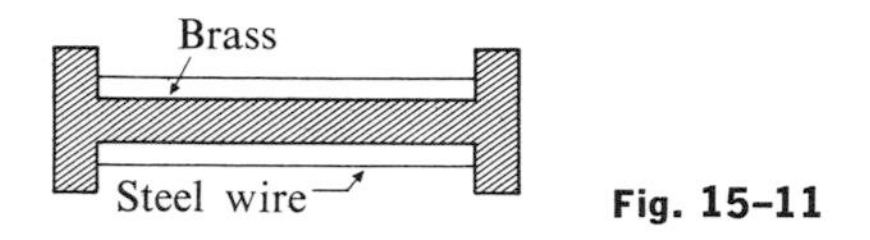

Fig. 15–11

15–18 A heavy brass bar has projections at its ends, as in Fig. 15–11. Two fine steel wires fastened between the projections are just taut (zero tension) when the whole system is at 0°C. What is the tensile stress in the steel wires when the temperature of the system is raised to 300°C? Make any simplifying assumptions you think are justified, but state what they are.

15–19 The steel rails of the missile testing track at Tularosa Basin, New Mexico, are welded into 10,000-ft lengths and each length is prestressed by stretching it 3 ft before it is fastened to the concrete base slab. (a) What is the strain in the rails? (b) If Young's modulus is 30×10^6 lb in^{-2}, what is the stress? (c) What stretching force is required if the cross-sectional area of a rail is 50 in^2? (d) If the stretched rails are fastened to the base slab at a temperature of 60°F, what is the stress when the temperature decreases to 0°F? (e) At what temperature will the stress change from tension to compression?

15–20 Prove that if a body under hydrostatic pressure is raised in temperature but not allowed to expand, the increase in pressure is

$$\Delta p = B\bar{\beta}\,\Delta t,$$

where the bulk modulus B and the average coefficient of volume expansion $\bar{\beta}$ are both assumed positive and constant.

15–21 (a) A block of metal at a pressure of 1 atm and a temperature of 20°C is kept at constant volume. If the temperature is raised to 32°C, what will be the final pressure? (b) If the block is maintained at constant volume by rigid walls that can withstand a maximum pressure of 1200 atm, what is the highest temperature to which the system may be raised? Assume B and $\bar{\beta}$ to remain practically constant at the values 1.5×10^{12} dyn cm^{-2} and 5.0×10^{-5}(C°)$^{-1}$, respectively.

15–22 What hydrostatic pressure is necessary to prevent a copper block from expanding when its temperature is increased from 20°C to 30°C?

15–23 Table 15–4 lists the density of water, and the volume of 1 g, at atmospheric pressure. A steel bomb is filled with water at 10°C and atmospheric pressure, and the system is heated to 75°C. What is then the pressure in the bomb? Assume the bomb to be sufficiently rigid so that its volume is not affected by the increased pressure.

15–24 A liquid is enclosed in a metal cylinder provided with a piston of the same metal. The system is originally at atmospheric pressure and at a temperature of 80°C. The piston is forced down until the pressure on the liquid is increased by 100 atm, and it is then clamped in this position. Find the new temperature at which the pressure of the liquid is again 1 atm. Assume that the cylinder is sufficiently strong so that its volume is not altered by changes in pressure, but only by changes in temperature.

Compressibility of liquid	$k = 50 \times 10^{-6}$ atm^{-1}.
Cubical coefficient of expansion of liquid	$\beta = 5.3 \times 10^{-4}$ (C°)$^{-1}$.
Linear coefficient of expansion of metal	$\alpha = 10 \times 10^{-6}$ (C°)$^{-1}$.

16 Heat and Heat Measurements

16–1 HEAT TRANSFER

Suppose that system *A*, at a higher temperature than system *B*, is put in contact with *B*. When thermal equilibrium has been reached, *A* will be found to have undergone a temperature decrease and *B* a temperature increase. It was therefore quite natural for the early investigators in this field to assume that *A* lost something and that this "something" flowed into *B*. While the temperature changes are taking place, it is customary to refer to a *heat flow* or a *heat transfer* from *A* to *B*. The process of heat transfer was formerly thought to be a flow of an invisible weightless fluid called *caloric*, which was produced when a substance burned and which moved from a region rich in caloric (where the temperature was high) to a region where there was less caloric (and a lower temperature). The abandonment of the caloric theory was a part of the general development of physics during the eighteenth and nineteenth centuries. Due to the experimental skill and physical insight of Count Rumford (1753–1814), a native of Woburn, Massachusetts, and Sir James Prescott Joule (1818–1889), the idea slowly emerged that a heat flow is an *energy transfer*.

When an energy transfer takes place by virtue of a temperature difference exclusively, it is called a heat flow. Thus, in Fig. 16–1(a), the hot flame of a Bunsen burner is in contact with a system consisting of water and water vapor at a lower temperature. Proceeding from left to right, water is converted to steam at a higher temperature and pressure. Under these conditions the steam is capable of doing more work (by impinging against a turbine blade, for example) than before, and therefore has received energy in a process involving a heat flow.

An energy transfer to a system may take place, however, without a flow of heat. Consider the three diagrams (from left to right) of Fig. 16–1(b). As an object hanging from a string attached to a cam descends, a piston is caused to descend and a gas is compressed. According to the definition given in Chapter 7, this compression is the result of the performance of work, since the point of application of a force has moved in the direction of the force. In its compressed state the gas is capable of doing more work than before and hence the gas has received energy in a process involving the performance of work.

The three diagrams of Fig. 16–1(c) show how a process may take place involving the simultaneous flow of heat and performance of work.

16–2 QUANTITY OF HEAT

A system consisting of a quantity of water and a small piece of resistance wire is depicted in Fig. 16–2. In part (a), the system is caused to undergo a temperature rise ΔT in a process involving a heat flow. In part (b), the same ΔT is produced in the same system by maintaining an electric current in the resistance wire. The electric generator is rotated by a string which is pulled by a descending attached object. If we include the resistance wire as part of the system, there is no temperature difference between this system and its surroundings. There is, however, a torque applied to the generator

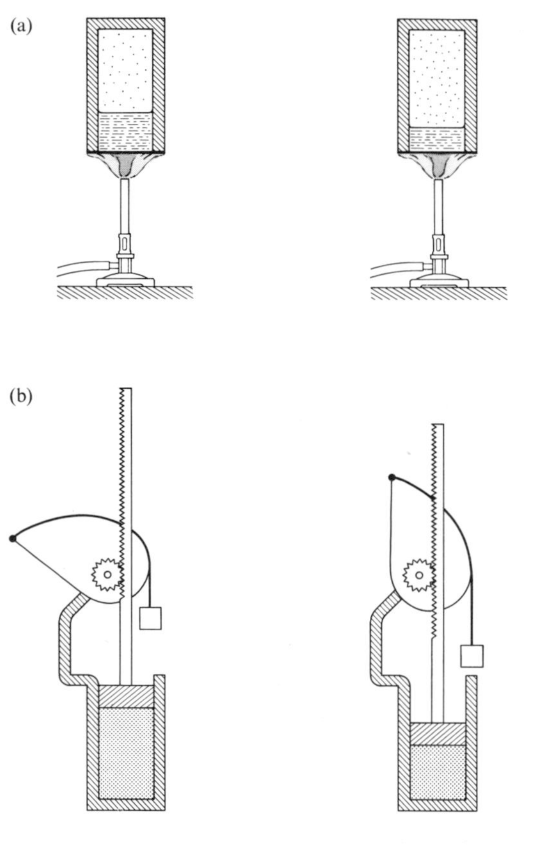

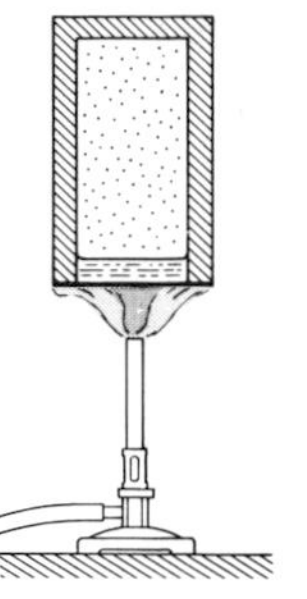

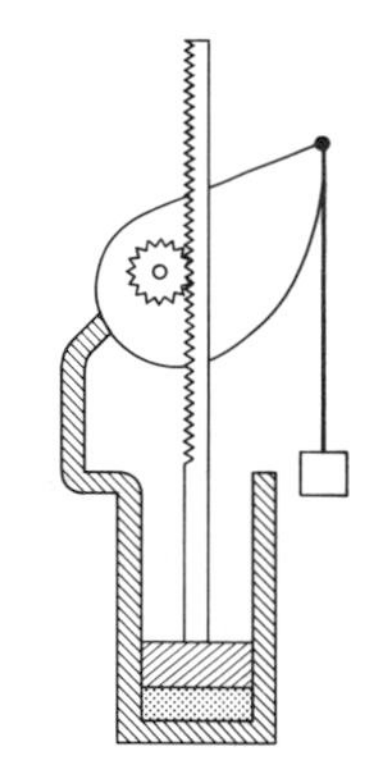

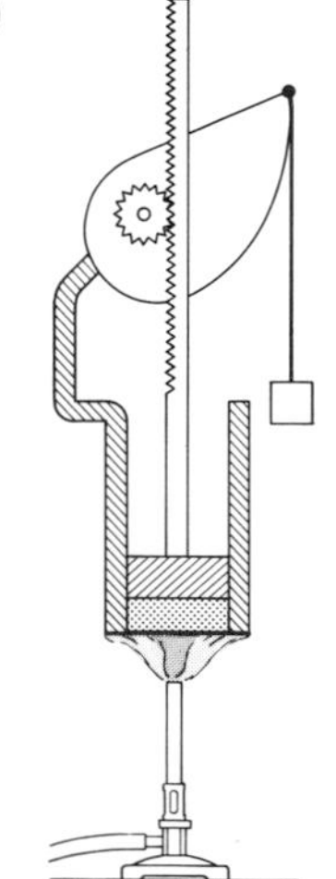

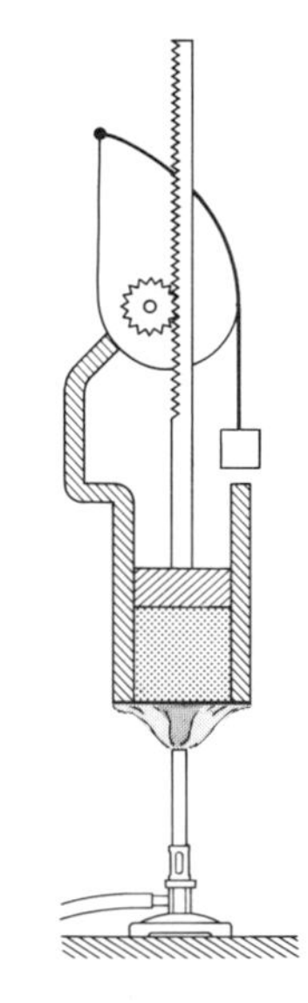

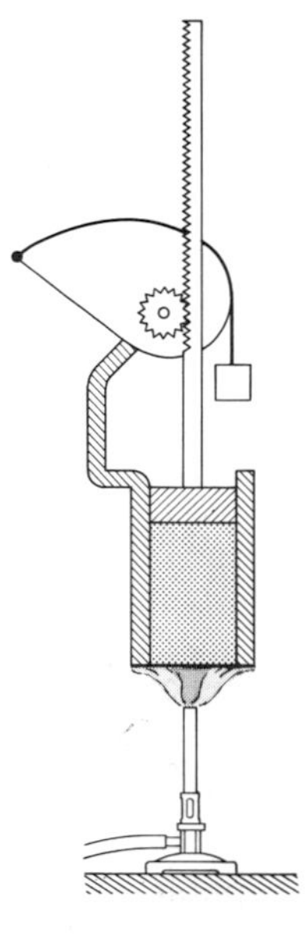

16–1 Distinction between flow of heat and performance of work.

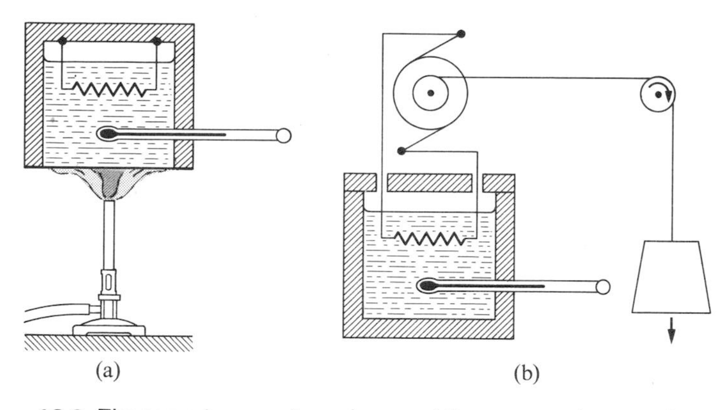

16–2 The same temperature change of the same system may be accomplished by either (a) a heat flow or (b) the performance of work.

which rotates with a definite angular velocity and therefore the method of energy transfer depicted in part (b) is the performance of work.

Suppose, however, that we regard the water only as our system in part (b). Then the flow of energy into the water is due to the presence of a resistance wire whose temperature is maintained a trifle higher than that of the water by means of the electric current in the wire. This process involves a heat flow from the wire to the water. Two conclusions may be drawn from these considerations:

1) Whether a process is regarded as involving a flow of heat or a performance of work depends on the *choice of the system*.
2) Whether the energy transfer depicted in Fig. 16–2(b) is regarded as a flow of heat or as a performance of work has no bearing on the quantity of energy delivered to the water.

We may describe the results of the two processes depicted in Fig. 16–2 by saying that a heat flow and a performance of work *are equivalent*. Once either process is completed, the energy of the system is greater than before, and no experiment can tell whether this energy increase had been caused by a flow of heat or a performance of work. They are both methods of energy transfer; both are expressed in power units, such as watts (joules per second) or foot-pounds per second.

To calculate the amount of energy transferred during a given heat flow or performance of work, the simplest procedure is to make use of the apparatus symbolized in Fig. 16–2(b). It will be shown in a later chapter that if V is the potential difference across the resistor and I is the current in it, then regardless of whether the energy transfer is labeled a flow of heat or a performance of work, the rate of increase of energy of the system is VI.

Up to this point, the word "heat" has been used only in conjunction with "flow" or "transfer." Thus, a heat flow is an energy transfer brought about by a temperature difference only. If a piece of resistance wire is immersed in a liquid or is wrapped around a solid, and is included as part of the system, then the establishment of a constant potential difference V and a constant current I in the resistance wire gives rise to an energy flow called a *performance of work*. If this goes on for a time τ (τ is the symbol for time), the total amount of work done, W, is

$$W = VI\tau,$$

and this is the total amount of energy added to the system. If the resistance is *not* part of the system, the energy transfer is labeled a flow of heat, and in time τ, the total energy transferred may be designated as a *quantity of heat* Q, where

$$Q = VI\tau.$$

It would be much better to use the word "heat" only when referring to a *method* of energy transfer, and when the transfer is completed to refer to the total amount of energy so transferred. The expression "quantity of heat," however, has played such a huge role in the development of the subject and is to be found in so many books and tables, that it is almost impossible to avoid.

In the eighteenth century, a unit of quantity of heat, the *calorie*, was defined as that amount of heat required to raise the temperature of one gram of water one celsius or kelvin degree. It was found later that more heat was required to raise the temperature of 1 g of water from, say, 90°C to 91°C than from 30°C to 31°C. The definition was then refined and the chosen calorie became known as the "15° calorie," that is, the quantity of heat required to change the temperature of 1 g of water from 14.5°C to 15.5°C.

16-3 HEAT CAPACITY

Suppose that a small quantity of heat dQ is transferred between a system and its surroundings. If the system undergoes a temperature change dt, the *specific heat capacity c of the system is defined as the ratio of the heat dQ to the product of the mass m and temperature change dt;* thus

$$c = \frac{dQ}{m\,dt}. \tag{16-1}$$

The specific heat capacity of water can be taken to be 1 cal g^{-1} $(C°)^{-1}$ or 1 Btu lb^{-1} $(F°)^{-1}$ for most practical purposes.

It is often convenient to use the *gram-mole* as a unit of mass. One gram-mole is a number of grams equal to the molecular weight M. To calculate the number of moles n, we divide the mass in grams by the molecular weight; thus $n = m/M$. Replacing the mass m in Eq. (16-1) by the product nM, we get

$$Mc = \frac{dQ}{n\,dt}.$$

The product Mc is called the *molar heat capacity* and is represented by the symbol C. Hence, by definition

$$C = Mc = \frac{dQ}{n\,dt}. \tag{16-2}$$

The molar heat capacity of water is practically 18 cal $mole^{-1}$ $(C°)^{-1}$.

From Eq. (16-1) the total quantity of heat Q that must be supplied to a body of mass m to change its temperature from t_1 to t_2 is

$$Q = m\int_{t_1}^{t_2} c\,dt. \tag{16-3}$$

The specific heat capacities of all materials vary with temperature, and of course c must be expressed as a function of t in order to carry out the integration. Over a temperature range in which c can be considered constant, Eq. (16-3) becomes

$$Q = mc(t_2 - t_1). \tag{16-4}$$

The *mean* specific heat capacity $\bar{c}$ over any range of temperature is defined as the constant value of c which would result in the same heat transfer. Hence, for the temperature range from t_1 to t_2,

$$Q = m\bar{c}(t_2 - t_1) = m\int_{t_1}^{t_2} c\,dt. \tag{16-5}$$

Equations similar to (16-3), (16-4), and (16-5) may be written in terms of n, C, and $\bar{C}$.

The specific or molar heat capacity of a substance is not the only physical property whose experimental determination requires the measurement of a quantity of heat. Heat conductivity, heat of fusion, heat of vaporization, heat of combustion, heat of solution, and heat of reaction are examples of other such properties which are called *thermal properties* of matter. The field of physics and physical chemistry concerned with the measurement of thermal properties is called *calorimetry*, and it is one of the most difficult branches of experimental science.

16-4 THE MEASUREMENT OF HEAT CAPACITY

Almost all modern calorimetry is electrical. The temperature difference necessary for a flow of heat is provided by maintaining an electric current I in a coil of resistance wire (the heater) usually wound around the material under investigation. Electrical resistance plays the same role as friction in mechanics. As soon as the temperature of the heater coil rises a trifle above that of the sample, heat flows in, and if proper precautions are taken, only a little of this heat is lost to the surroundings. The thermometer is usually a small resistance thermometer or thermocouple embedded in the sample and is chosen for its rapidity and its sensitivity. Readings are taken before the switch in the heater circuit is closed; these are shown on the left of Fig. 16-3. The fact that the temperature rises before energy is supplied to the heater indicates that the surroundings are at a temperature higher than that of the sample. The current is maintained in the heater for a time interval $\Delta\tau$ during which we do not measure the temperature. After the switch in the heater circuit is opened, temperature measurements are resumed, giving rise to the right-hand part of Fig. 16-3, where the positive slope indicates that the surroundings are still higher in temperature than the sample.

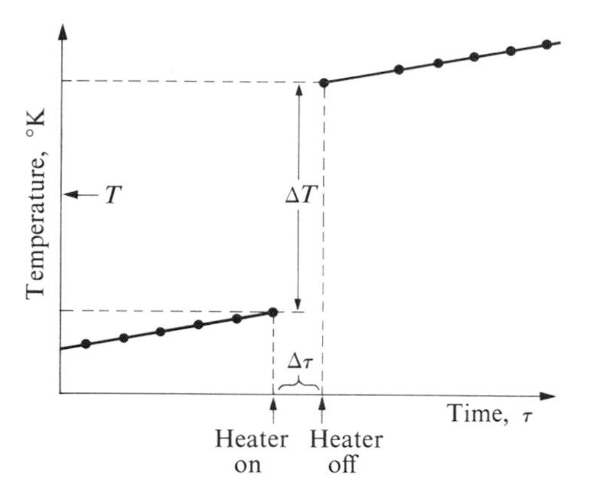

16-3 Temperature–time graph of the data taken during a heat-capacity measurement.

If the potential difference across the heater is V, and an electric current I is maintained for a time $\Delta\tau$, then the heat transferred to the sample is $VI\,\Delta\tau$. The molar heat capacity is therefore

$$C = VI\,\Delta\tau/n\,\Delta t = VI\,\Delta\tau/n\,\Delta T, \tag{16-6}$$

where C will be expressed in joules per mole·degree when I is expressed in amperes, V in volts, τ in seconds, and ΔT is read from the graph as in Fig. 16-3. The resulting molar heat capacity C is the value at the temperature T which is in the middle of the range ΔT. In careful experiments, ΔT may be as small as 0.01K°.

The shape, size, and construction of the calorimeter, heating coils, thermometers, etc., depend on the nature of the material to be studied and the temperature range desired. It is impossible to describe one calorimeter that is sufficient for all purposes. In general, the measurement of any heat capacity is a research problem requiring all the ability of a trained physicist or physical chemist, the facilities of a good shop, and the skill of a good glass blower.

The temperature variation of the specific or molar heat capacity provides the closest and most direct approach to the understanding of the energy of the particles constituting matter. The measurement of the heat capacities of chemical compounds, normal metals, metallic alloys, superconducting metals, alloys, compounds, etc., particularly at low temperatures, constitutes one of the liveliest and most interesting topics of modern physics.

Figure 16-4 shows the variation of the specific heat capacity of water with temperature. It may be seen that the quantity of heat necessary to raise the temperature of 1 g of water from 14.5°C to 15.5°C is

$$1\ 15°\ \text{cal} = 4.186\ \text{J}.$$

Two other calories are frequently used. The *international table calorie* (IT cal) is *defined* to be

$$1\ \text{IT cal} = \frac{1\ \text{W hr}}{860} = \frac{3600\ \text{J}}{860} = 4.186\ \text{J},$$

and is almost identical with the 15° cal. The thermochemical calorie, however, is equal to 4.1840 J and may be seen from Fig. 16-4 to correspond to about a 17° calorie. In the absence of any additional information, the word *calorie* should be taken to mean 4.186 J.

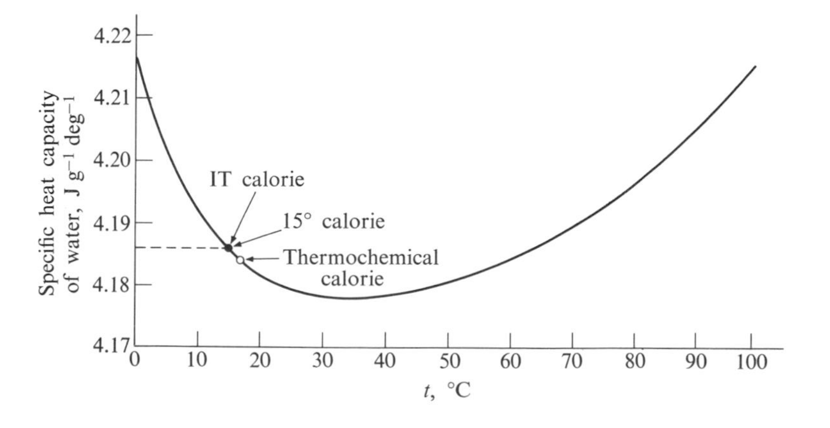

16-4 Specific heat of water as a function of temperature.

Mechanical engineers frequently use the British thermal unit (Btu), defined as the quantity of heat required to raise the temperature of 1 lb (mass) of water from 63°F to 64°F. The following relations hold:

$$1 \text{ Btu} = 778.3 \text{ ft lb} = 252.0 \text{ cal} = 1055 \text{ J}.$$

16–5 EXPERIMENTAL VALUES OF HEAT CAPACITIES

The amount of heat transferred to or from a system depends on the manner in which the system is controlled or constrained during the transfer, that is, whether the system is kept at *constant pressure* or at *constant volume*. The two corresponding specific or molar heat capacities are denoted by the respective symbols c_P (or C_P) and c_V (or C_V). In the electrical method of measuring heat capacities, the sample under investigation is maintained at constant pressure. As a matter of fact, it would be almost impossible to make a precise determination of c_V because there is no effective way of keeping the volume of a system constant and of allowing for the heat transferred to the containing walls.

TABLE 16–1 MEAN SPECIFIC AND MOLAR HEAT CAPACITIES OF METALS

Metal	(Specific) $\bar{c}_P$, cal $g^{-1}(C°)^{-1}$	Temperature range, °C	M, g mole^{-1}	(Molar) $\bar{C}_P = M\bar{c}_P$, cal mole$^{-1}(C°)^{-1}$
Beryllium	0.470	20–100	9.01	4.24
Aluminum	0.217	17–100	27.0	5.86
Iron	0.113	18–100	55.9	6.31
Copper	0.093	15–100	63.5	5.90
Silver	0.056	15–100	108	6.05
Mercury	0.033	0–100	201	6.64
Lead	0.031	20–100	207	6.42

In the design of practical devices for carrying currents of hot liquids, for raising the temperature of water, for generating steam, etc., it is sufficient to know, for the materials, merely the *average* specific or molar heat capacity at constant pressure. A few such values are listed in Table 16–1. The average values of the specific heat capacity are seen to be less than 1 cal g^{-1} $(C°)^{-1}$ and are smaller, the larger the molecular weight. In the last column, an interesting regularity, first noted in 1819 by two French physicists, Dulong and Petit, appears.

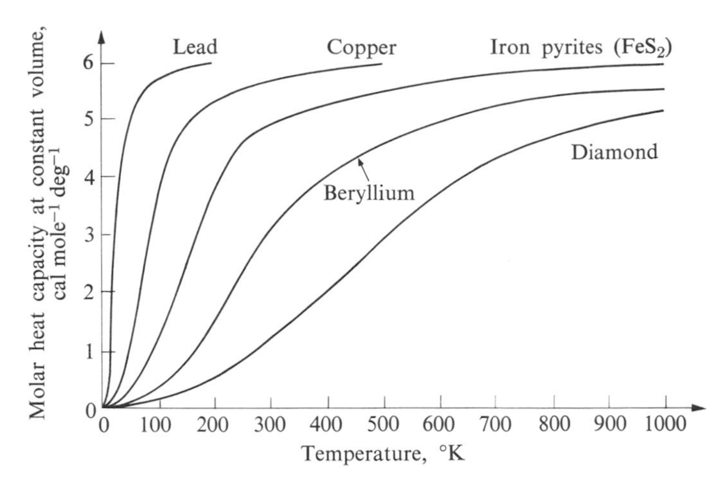

16–5 Temperature variation of C_v of solids.

The average molar heat capacity at constant pressure for all metals except the very lightest is approximately the same and equal to about 6 cal mole^{-1} $(C°)^{-1}$. This is known as the *Dulong and Petit law* and, although rough and applying to the average molar heat capacity at constant pressure, $\bar{C}_P$, averaged over a temperature range that is of no particular scientific importance, it contains the germ of a very important idea. We know from chemistry that the number of molecules in 1 g-mole is the same for all substances. It follows that (very nearly) the same amount of heat is required *per molecule* to raise the temperature of each of these metals by a given amount although, for example, the mass of a molecule of lead (At. wt. = 207) is nearly 10 times as great as that of a molecule of aluminum (At. wt. = 27). To put it in a different way, the heat required to raise the temperature of a sample of metal depends only on *how many* molecules the sample contains, and not on the mass of an individual molecule. This is the first time in our study of physics that we have met a property of matter so directly related to its molecular structure.

The really important theoretical interpretation of heat capacity measurements can be made only when the theoretician has at his disposal the complete temperature dependence, from the lowest to the highest possible temperatures, of the *molar heat capacity at constant volume*, C_V, because it is this quantity that is directly connected with the energy which, in turn, may be cal-

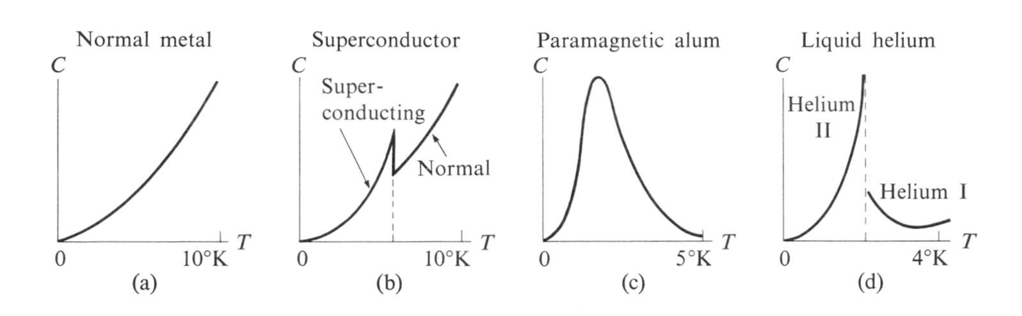

16–6 Different ways in which the molar heat capacity varies with the temperature at low temperatures, indicating widely different atomic processes.

culated by statistical methods. Fortunately, it is possible to convert experimental values of C_P to values of C_V with the aid of a partly theoretical, partly empirical equation due to Nernst and Lindemann, namely,

$$C_V = C_P\left(1 - 0.0214 \text{ mole K}^\circ \text{ cal}^{-1}\, C_P \frac{T}{T_m}\right), \quad (16\text{–}7)$$

where T_m is the melting temperature. When this equation was used in conjunction with measurements of C_P from about 4°K to 1000°K, the curves of Fig. 16–5 were obtained. These curves represent the extremes of behavior. All other metals and nonmetals lie within the boundaries of lead and diamond. Except for certain anomalies that are too complicated to dwell upon, all such curves approach about 6 cal mole^{-1} (K°)$^{-1}$ as the temperature approaches infinity. The Dulong and Petit value, therefore, is approached by all solids, lead arriving at this value at about 200°K (below room temperature) and diamond requiring well over 2000°K.

Each curve in Fig. 16–5 has exactly the same shape: it starts at zero, rises rapidly at first, then bends over and approaches the value 6 cal mole^{-1} (K°)$^{-1}$. It was shown by Debye that the behavior of the molar heat capacity of *nonmetals* over the entire temperature range was accounted for by the *vibrational motion of the molecules occupying regular positions in the crystal lattice*. There are, however, many substances whose temperature variation of heat capacity cannot be explained entirely on the basis of lattice vibrations. The behavior of four "abnormal" types of substance is depicted in Fig. 16–6. There are many more different types of heat capacity curves and all of them indicate some special molecular or atomic or ionic property of the particles that occupy the regular positions within the crystal lattice. The theoretical explanations of heat capacity curves are among the most interesting and complicated theories of modern solid state physics.

16–6 CHANGE OF PHASE

The term *phase* as used here relates to the fact that matter exists either as a solid, liquid, or gas. Thus the chemical substance H_2O exists in the *solid phase* as ice, in the *liquid phase* as water, and in the *gaseous phase* as steam. Provided they do not decompose at high temperatures, all substances can exist in any of the three phases under the proper conditions of temperature and pressure. Transitions from one phase to another are accompanied by the absorption or liberation of heat and usually by a change in volume.

As an illustration, suppose that ice is taken from a refrigerator where its temperature was, say, −25°C. Let the ice be crushed quickly, placed in a container, and a thermometer inserted in the mass. Imagine the container to be surrounded by a heating coil which supplies heat to the ice at a uniform rate, and suppose that no other heat reaches the ice. The temperature of the ice would be observed to increase steadily, as shown by the portion of the graph (Fig. 16–7) from *a* to *b*, or until the temperature has risen to 0°C. In this temperature range the specific heat capacity of ice is approximately 0.55 cal gm^{-1} (C°)$^{-1}$. As soon as this temperature is reached, some liquid water will be observed in the container. In other words, the ice begins to melt. The melting process is a *change of phase*, from the solid phase to the liquid phase. The thermometer, however, will show no *increase in temperature*, and even though heat is being supplied at the same rate as before, the temperature will remain at 0°C until all the ice is melted (point *c*, Fig. 16–7), if the pressure is maintained constant at one atmosphere.

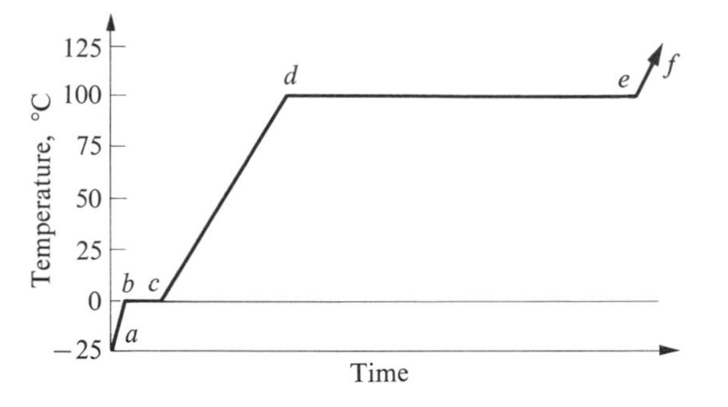

16-7 The temperature remains constant during each change of phase, provided the pressure remains constant.

As soon as the last of the ice has melted, the temperature begins to rise again at a uniform rate (from c to *d*, Fig. 16-7) although this rate will be slower than that from *a* to *b* because the specific heat of water is greater than that of ice. When a temperature of 100°C is reached (point *d*), bubbles of steam (gaseous water or water vapor) start to escape from the liquid surface, or the water begins to boil. The temperature remains constant at 100°C (at constant atmospheric pressure) until all the water has boiled away. Another change of phase has therefore taken place, from the liquid phase to the gaseous phase.

If all the water vapor had been trapped and not allowed to diffuse away (a very large container would be needed), the heating process could be continued as from *e* to *f*. The gas would now be called "superheated steam."

Although water was chosen as an example in the process just described, the same type of curve as in Fig. 16-7 is obtained for many other substances. Some, of course, decompose before reaching a melting or boiling point, and others, such as glass or pitch, do not change phase at a definite temperature but become gradually softer as their temperature is raised. Crystalline substances, such as ice, or a metal, melt at a definite temperature. Glass and pitch behave like supercooled liquids of very high viscosity.

The quantity of heat per unit mass that must be supplied to a material at its melting point to convert it completely to a liquid at the same temperature is called the *heat of fusion* of the material. The quantity of heat per unit mass that must be supplied to a material at its boiling point to convert it completely to a gas at the same temperature is called the *heat of vaporization* of the material. Heats of fusion and vaporization are expressed in calories per gram, or Btu per pound. Thus the heat of fusion of ice is about 80 cal g^{-1} or 144 Btu lb^{-1}. The heat of vaporization of water (at 100°C) is 539 cal g^{-1} or 970 Btu lb^{-1}. Some heats of fusion and vaporization are listed in Table 16-2.

TABLE 16-2 HEATS OF FUSION AND VAPORIZATION

Substance	Normal melting point		Heat of fusion, cal g^{-1}	Normal boiling point		Heat of vaporization, cal g^{-1}
	°K	°C		°K	°C	
Helium	3.5	−269.65	1.25	4.216	−268.93	5
Hydrogen	13.84	−259.31	14	20.26	−252.89	108
Nitrogen	63.18	−209.97	6.09	77.34	−195.81	48
Oxygen	54.36	−218.79	3.30	90.18	−182.97	51
Ethyl alcohol	159	−114	24.9	351	78	204
Mercury	234	−39	2.82	630	357	65
Water	273.15	0.00	79.7	373.15	100.00	539
Sulfur	392	119	9.1	717.75	444.60	78
Lead	600.5	327.3	5.86	2023	1750	208
Antimony	903.65	630.50	39.4	1713	1440	134
Silver	1233.95	960.80	21.1	2466	2193	558
Gold	1336.15	1063.00	15.4	2933	2660	377
Copper	1356	1083	32	1460	1187	1211

When heat is removed from a gas, its temperature falls and at the same temperature at which it boiled, it returns to the liquid phase, or *condenses*. In so doing it gives up to its surroundings the same quantity of heat which was required to vaporize it. The heat so given up, per unit mass, is called the *heat of condensation* and is equal to the heat of vaporization. Similarly, a liquid returns to the solid phase, or freezes, when cooled to the temperature at which it melted, and gives up heat called *heat of solidification* exactly equal to the heat of fusion. Thus the melting point and the freezing point are at the same temperature, and the boiling point and condensation point are at the same temperature.

Whether a substance, at its melting point, is freezing or melting depends on whether heat is being supplied or removed. That is, if heat is supplied to a beaker containing both ice and water at 0°C, some of the ice will melt; if heat is removed, some of the water will freeze; the temperature in either case remains at 0°C so long as both ice and water are present. If heat is *neither supplied nor removed*, no change at all takes place and the relative amounts of ice and water, and the temperature, all remain constant.

This furnishes, then, another point of view which may be taken regarding the melting point. That is, the melting (or freezing) point of a substance is *that temperature at which both the liquid and solid phases can exist together*. At any higher temperature, the substance can only be a liquid; at any lower temperature, it can only be a solid.

The general term *heat of transformation* is applied both to heats of fusion and heats of vaporization, and both are designated by the letter L. Since L represents the heat absorbed or liberated in the change of phase of unit mass, the heat Q absorbed or liberated in the change of phase of a mass m is

$$Q = mL. \qquad (16\text{–}8)$$

The household steam-heating system makes use of a boiling-condensing process to transfer heat from the furnace to the radiators. Each pound of water which is turned to steam in the furnace absorbs 970 Btu (the heat of vaporization of water) from the furnace, and gives up 970 Btu when it condenses in the radiators. (This figure is correct if the steam pressure is 1 atm. It will be slightly smaller at higher pressures.) Thus the steam-heating system does not need to circulate as much water as a hot-water heating system. If water leaves a hot-water furnace at 140°F and returns at 100°F, dropping 40 F°, about 24 lb of water must circulate to carry the same heat as is carried in the form of heat of vaporization by 1 lb of steam.

Under the proper conditions of temperature and pressure, a substance can change directly from the solid to the gaseous phase without passing through the liquid phase. The transfer from solid to vapor is called *sublimation*, and the solid is said to *sublime*. "Dry ice" (solid carbon dioxide) sublimes at atmospheric pressure. Liquid carbon dioxide cannot exist at a pressure lower than about 73 lb in^{-2}.

Heat is absorbed in the process of sublimation, and liberated in the reverse process. The quantity of heat per unit mass is called the *heat of sublimation*.

Problems

[*Note:* When no information is given about the temperature variation of specific or molar heat capacity, assume it to be constant and use the average value given in Table 16–1.]

16–1 A combustion experiment is performed by burning a mixture of fuel and oxygen in a constant-volume "bomb" surrounded by a water bath. During the experiment the temperature of the water is observed to rise. Regarding the mixture of fuel and oxygen as the system, (a) has heat been transferred? (b) has work been done?

16–2 A liquid is irregularly stirred in a well-insulated container and thereby undergoes a rise in temperature. Regarding the liquid as the system, (a) has heat been transferred? (b) has work been done?

16–3 An automobile weighing 2000 lb is traveling at 10 ft s^{-1}. How many Btu are transferred in the brake mechanism when it is brought to rest?

16–4 A copper vessel of mass 200 g contains 400 g of water. The water is heated by a friction device which dissipates mechanical energy, and it is observed that the temperature of the system rises at the rate of 3 C° min^{-1}. Neglect heat losses to surroundings. What power in watts is dissipated in the water?

16–5 How long could a 2000-hp motor be operated on the heat energy liberated by 1 mi^3 of ocean water when the temperature of the water is lowered by 1 C° if all this heat were converted to mechanical energy? Why do we not utilize this tremendous reservoir of energy?

16–6 (a) A certain house burns 10 tons of coal in a heating season. The heat of combustion of the coal is 11,000 Btu lb^{-1}. If stack losses are 15%, how many Btu were actually used to heat the house? (b) In some localities large tanks of water are heated by solar radiation during the summer and the stored energy is used for heating during the winter. Find the required dimensions of the storage tank, assuming it to be a cube, to store a quantity of energy equal to that computed in part (a). Assume that the water is raised to 120°F in the summer and cooled to 80°F in the winter.

16–7 An artificial satellite, constructed of aluminum, encircles the earth at a speed of 18,000 mi hr^{-1}. (a) Find the ratio of its kinetic energy to the energy required to raise its temperature by 600 C°. (The melting point of aluminum is 660°C.) Assume a constant specific heat capacity of 0.20 Btu $lbm^{-1}(F°)^{-1}$. (b) Discuss the bearing of your answer on the problem of the re-entry of a satellite into the earth's atmosphere.

16–8 A calorimeter contains 100 g of water at 0°C. A 1000-g copper cylinder and a 1000-g lead cylinder, both at 100°C, are placed in the calorimeter. Find the final temperature if there is no loss of heat to the surroundings.

16–9 (a) Compare the heat capacities (heat capacity is the heat per unit temperature change) of equal *masses* of water, copper, and lead. (b) Compare the heat capacities of equal *volumes* of water, copper, and lead.

16–10 An aluminum can of mass 500 g contains 117.5 g of water at a temperature of 20°C. A 200-g block of iron at 75°C is dropped into the can. Find the final temperature, assuming no heat loss to the surroundings.

16–11 A casting weighing 100 lb is taken from an annealing furnace where its temperature was 900°F and plunged into a tank containing 800 lb of oil at a temperature of 80°F. The final temperature is 100°F, and the specific heat capacity of the oil is 0.5 Btu lb^{-1} $(F°)^{-1}$. What was the specific heat capacity of the casting? Neglect the heat capacity of the tank itself and any heat losses.

16–12 A lead bullet, traveling at 350 m s^{-1}, strikes a target and is brought to rest. What would be the rise in temperature of the bullet if there were no heat loss to the surroundings?

16–13 A copper calorimeter (mass 300 g) contains 500 g of water at a temperature of 15°C. A 560-g block of copper, at a temperature of 100°C, is dropped into the calorimeter and the temperature is observed to increase to 22.5°C. Neglect heat losses to the surroundings. Find the specific heat capacity of copper.

16–14 A 50-g sample of a material, at a temperature of 100°C, is dropped into a calorimeter containing 200 g of water initially at 20°C. The calorimeter is of copper and its mass is 100 g. The final temperature of the calorimeter is 22°C. Compute the specific heat capacity of the sample.

16–15 In Fig. 16–8 an electric heater is shown whose purpose is to provide a continuous supply of hot water. Water is flowing at the rate of 300 g min^{-1}, the inlet thermometer registers 15°C, the voltmeter reads 120 V and the ammeter 10 amp. (a) When a steady state is finally reached, what is the reading of the outlet thermometer? (b) Why is it unnecessary to take into account the heat capacity (mc or nC) of the apparatus itself?

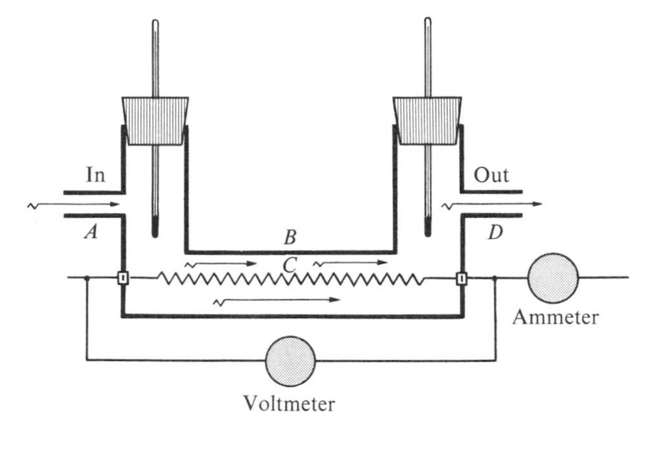

Fig. 16–8

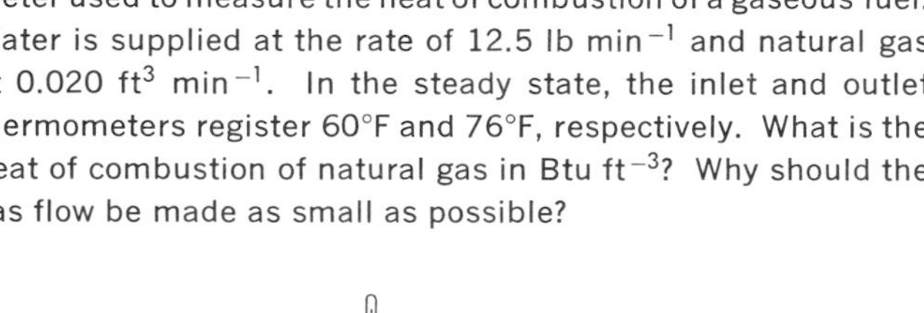

16–16 Figure 16–9 shows a sketch of a continuous-flow calorimeter used to measure the heat of combustion of a gaseous fuel. Water is supplied at the rate of 12.5 lb min^{-1} and natural gas at 0.020 ft^3 min^{-1}. In the steady state, the inlet and outlet thermometers register 60°F and 76°F, respectively. What is the heat of combustion of natural gas in Btu ft^{-3}? Why should the gas flow be made as small as possible?

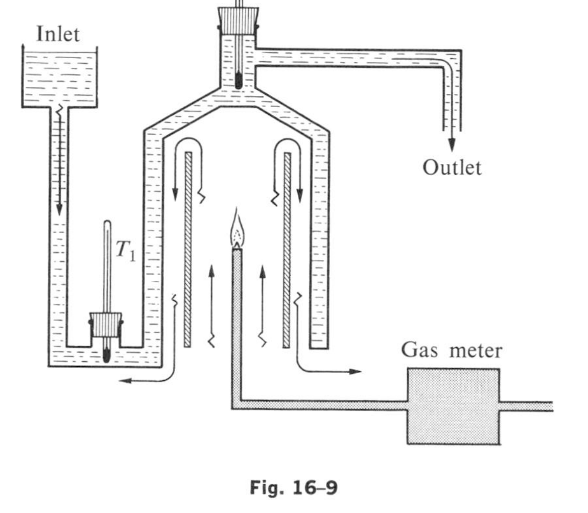

Fig. 16–9

16–17 (a) Make a rough sketch of the type of temperature-time graph (such as that in Fig. 16–3) that would result if the surroundings were much cooler than the sample. (b) An electrical resistor is immersed in a liquid and electrical energy is dissipated for 100 s at a constant rate of 50 W. The mass of the liquid is 530 g, and its temperature increases from 17.64°C to 20.77°C. Find the mean specific heat capacity of the liquid in this temperature range.

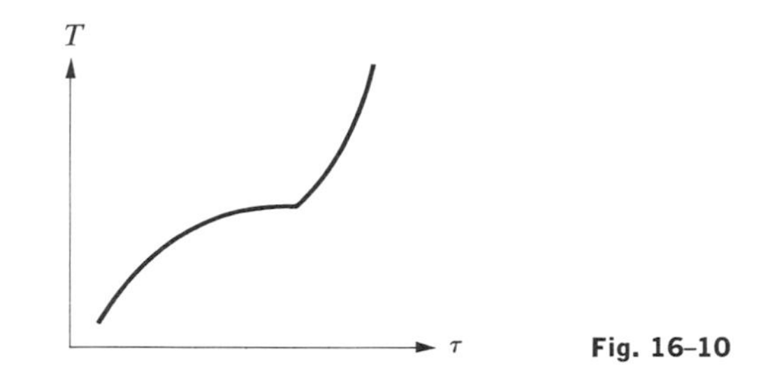

Fig. 16–10

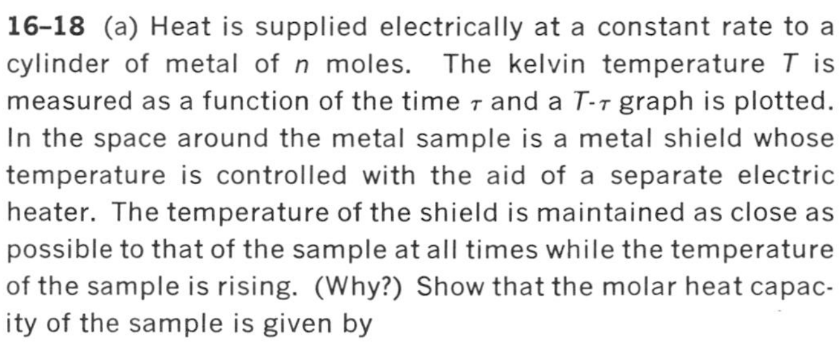

16–18 (a) Heat is supplied electrically at a constant rate to a cylinder of metal of n moles. The kelvin temperature T is measured as a function of the time τ and a T-τ graph is plotted. In the space around the metal sample is a metal shield whose temperature is controlled with the aid of a separate electric heater. The temperature of the shield is maintained as close as possible to that of the sample at all times while the temperature of the sample is rising. (Why?) Show that the molar heat capacity of the sample is given by

$$C = \frac{VI}{n(dT/d\tau)}.$$

(b) If the T-τ graph has the appearance of Fig. 16–10, draw a rough graph showing the dependence of molar heat capacity on temperature. (c) If the T-τ graph has the appearance of Fig. 16–11, draw a rough graph showing the dependence of molar heat capacity on temperature.

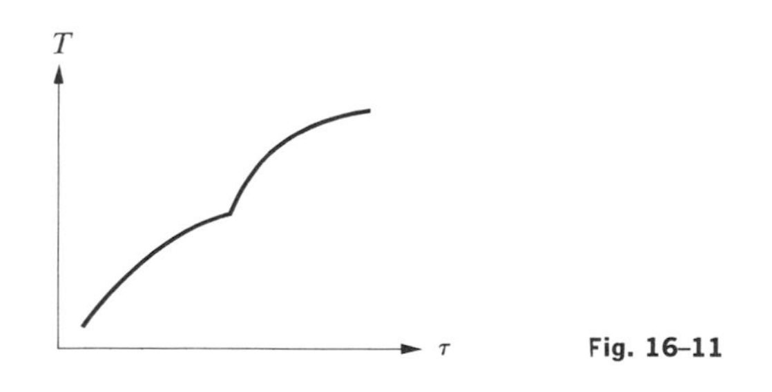

Fig. 16–11

16–19 The molar heat capacity at constant pressure of a substance varies with the temperature according to the empirical equation

$$c_P = 6.50 \text{ cal mole}^{-1}\,(\text{K}°)^{-1} + 10^{-3} \text{ cal mole}^{-1}\,(\text{K}°)^{-2} \cdot T.$$

How many calories of heat are necessary to change the temperature of 10 moles from 27°C to 527°C?

16–20 At very low temperatures, the molar heat capacity of rock salt varies with the temperature according to "Debye's T^3 law"; thus

$$C = k \frac{T^3}{\Theta^3},$$

where $k = 464$ cal mole^{-1} (K°)$^{-1}$ and $\Theta = 281$°K. (a) How much heat is required to raise the temperature of 2 moles of rock salt from 10°K to 50°K? (b) What is the mean molar heat capacity in this range? (c) What is the true molar heat capacity at 50°K?

16–21 (a) With the aid of the Nernst-Lindemann relation, Eq. (16–7), prove that as T approaches zero, C_P approaches C_V. (b) The true value of C_P of copper at 100°K is 3.88 cal mole^{-1} (K°)$^{-1}$ and at 800°K it is 6.70 cal mole^{-1} (K°)$^{-1}$. Calculate the difference $C_P - C_V$ at these two temperatures.

16–22 How much heat is required to convert 1 g of ice at −10°C to steam at 100°C?

16–23 A beaker of very small mass contains 500 g of water at a temperature of 80°C. How many grams of ice at a temperature of −20°C must be dropped in the water so that the final temperature of the system will be 50°C?

16–24 An open vessel contains 500 g of ice at −20°C. The mass of the container can be neglected. Heat is supplied to the vessel at the constant rate of 1000 cal min^{-1} for 100 min. Plot a curve showing the elapsed time as abscissa and the temperature as ordinate.

16–25 A copper calorimeter of mass 100 g contains 150 g of water and 8 g of ice in thermal equilibrium at atmospheric pressure. 100 g of lead at a temperature of 200°C are dropped into the calorimeter. Find the final temperature if no heat is lost to the surroundings.

16–26 500 g of ice at −16°C are dropped into a calorimeter containing 1000 g of water at 20°C. The calorimeter can is of copper and has a mass of 278 g. Compute the final temperature of the system, assuming no heat losses.

16–27 A tube leads from a flask in which water is boiling under atmospheric pressure to a calorimeter. The mass of the calorimeter is 150 g, its heat capacity is 15 cal (C°)$^{-1}$ and it contains originally 340 g of water at 15°C. Steam is allowed to condense in the calorimeter until its temperature increases to 71°C, after which the total mass of calorimeter and contents is found to be 525 g. Compute the heat of condensation of steam from these data.

16–28 An aluminum canteen whose mass is 500 g contains 750 g of water and 100 g of ice. The canteen is dropped from an aircraft to the ground. After landing, the temperature of the canteen is found to be 25°C. Assuming that no energy is given to the ground in the impact, what was the velocity of the canteen just before it landed?

16–29 A calorimeter contains 500 g of water and 300 g of ice, all at a temperature of 0°C. A block of metal of mass 1000 g is taken from a furnace where its temperature was 240°C and is dropped quickly into the calorimeter. As a result, all the ice is just melted. What would the final temperature of the system have been if the mass of the block had been twice as great? Neglect heat loss from and the heat capacity of the calorimeter.

16–30 An ice cube whose mass is 50 g is taken from a refrigerator where its temperature was −10°C and is dropped into a glass of water at 0°C. If no heat is gained or lost from outside, how much water will freeze onto the cube?

16–31 A copper calorimeter can ($mc = 30$ cal deg^{-1}) contains 50 g of ice. The system is initially at 0°C. 12 g of steam at 100°C and 1 atm pressure are run into the calorimeter. What is the final temperature of the calorimeter and its contents?

16–32 A vessel whose walls are thermally insulated contains 2100 g of water and 200 g of ice, all at a temperature of 0°C. The outlet of a tube leading from a boiler, in which water is boiling at atmospheric pressure, is inserted in the water. How many grams of steam must condense to raise the temperature of the system to 20°C? Neglect the heat capacity of the container.

16–33 A 2-kg iron block is taken from a furnace where its temperature was 650°C and placed on a large block of ice at 0°C. Assuming that all the heat given up by the iron is used to melt the ice, how much ice is melted?

16–34 In a household hot-water heating system, water is delivered to the radiators at 140°F and leaves at 100°F. The system is to be replaced by a steam system in which steam at atmospheric pressure condenses in the radiators, the condensed steam leaving the radiators at 180°F. How many pounds of steam will supply the same heat as was supplied by 1 lb of hot water in the first installation?

16–35 A "solar house" has storage facilities for 4 million Btu. Compare the space requirements for this storage on the assumption (a) that the heat is stored in water heated from a minimum temperature of 80°F to a maximum at 120°F, and (b) that the heat is stored in Glauber salt ($Na_2SO_4 \cdot 10H_2O$) heated in the same temperature range.

Properties of Glauber salt:

Specific heat capacity (solid)	0.46 Btu lb^{-1} (F°)$^{-1}$
Specific heat capacity (liquid)	0.68 Btu lb^{-1} (F°)$^{-1}$
Specific gravity	1.6
Melting point	90°F
Heat of fusion	104 Btu lb^{-1}

17 Transfer of Heat

17-1 CONDUCTION

If one end of a metal rod is placed in a flame while the other is held in the hand, that part of the rod one is holding will be felt to become hotter and hotter, although it was not itself in direct contact with the flame. Heat is said to reach the cooler end of the rod by *conduction* along or through the material of the rod. Conduction of heat can take place in a body only when different parts of the body are at different temperatures, and the direction of heat flow is always from points of higher to points of lower temperature.

To fix our ideas, let us consider the following idealized case. A rod of length L and cross-sectional area A is originally at a uniform temperature t_1. At some instant, the right end of the rod is brought into contact with a body which is maintained at the constant temperature t_1, and the left end of the rod is brought into contact with a body kept at a higher temperature t_2. The remainder of the rod is surrounded by a material which is a nonconductor of heat. (This condition cannot be fulfilled exactly, since all materials conduct heat to some extent.) Now, perhaps by means of small thermocouples inserted in holes in the rod, we measure the temperature at a number of points along the rod at times τ_1, τ_2, etc., after the experiment is started. Figure 17-1(a) is a diagram of the experimental setup, and Fig. 17-1(b) shows the graphs of the temperature t vs. distance x along the rod, at times τ_1, τ_2, etc.

Initially, at time $\tau = 0$, the graph is a horizontal straight line at a height t_1. At all later times τ_1, τ_2, etc., the temperature at the left end is t_2 and the temperature decreases from left to right as shown by the corresponding curves. After a sufficiently long time has elapsed the temperature at every point becomes constant (in time) and the rod is said to be in the *steady state*. The steady-state graph is labeled $\tau = \infty$.

We define the *temperature gradient* at any point and at any time as the rate of change of temperature t with distance x along the rod:

$$\text{Temperature gradient} = \frac{dt}{dx}.$$

Graphically, the temperature gradient is represented by the slope of a graph of t vs. x, at any coordinate x and any time τ.

At every instant, both in the transient and in the steady state, there is a flow of heat along the rod from left to right. Let dQ represent the heat flowing across a

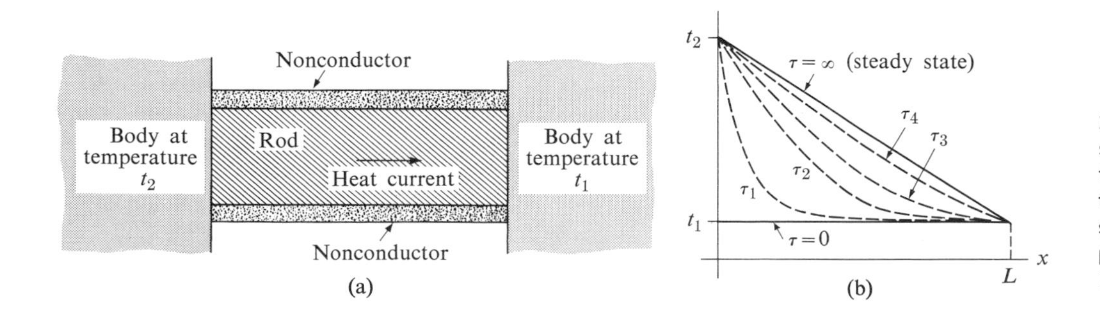

17-1 Transient and steady state temperature distribution along a rod initially at temperature t_1. The transient distributions were computed for $\tau_2 = 5\tau_1$, $\tau_3 = 10\tau_1$, $\tau_4 = 20\tau_1$.

section of the rod at the coordinate x, during the time interval $d\tau$ between τ and $\tau + d\tau$. The ratio $dQ/d\tau$, the heat flow per unit time, is called the *heat current* H:

$$H = \frac{dQ}{d\tau}.$$

The *thermal conductivity* k of the material of the rod is defined as the (negative of the) heat current per unit area perpendicular to the flow, and per unit temperature gradient:

$$k = -\frac{H}{A(dt/dx)}.$$

The negative sign is introduced in the definition because H is positive (heat flows from left to right) when the temperature gradient, as in Fig. 17–1, is negative. Thus k is a positive quantity. The preceding equation is more commonly written as

$$H = -kA\frac{dt}{dx}. \tag{17–1}$$

The thermal conductivity of most materials is a function of temperature, increasing slightly with increasing temperature, but the variation is small and can often be neglected. Some numerical values of k, at temperatures near room temperature, are given in Table 17–1. In the cgs system the unit of heat current is one calorie per second (1 cal s^{-1}), that of area is one square centimeter, and that of temperature gradient is one celsius degree per centimeter (1 C° cm^{-1}). The properties of materials used commercially as heat insulators are expressed in a system in which the unit of heat current is 1 Btu hr^{-1}, the unit of area is 1 ft^2, and the unit of temperature gradient is one fahrenheit degree per inch (1 F° in^{-1}).

It is evident from Eq. (17–1) that the larger the thermal conductivity k, the larger the heat current, other factors being equal. A material for which k is large is therefore a good heat conductor, while if k is small, the material is a poor conductor or a good insulator. A "perfect heat conductor" ($k = \infty$) or a "perfect heat insulator" ($k = 0$) does not exist. However, it will be seen from Table 17–1 that the metals as a group have much greater thermal conductivities than the nonmetals, and that those of gases are extremely small.

TABLE 17–1 THERMAL CONDUCTIVITIES

	k, cal s^{-1} $cm^{-1}$$(C°)^{-1}$
Metals	
Aluminum	0.49
Brass	0.26
Copper	0.92
Lead	0.083
Mercury	0.020
Silver	0.97
Steel	0.12
Various solids (Representative values)	
Insulating brick	0.00035
Red brick	0.0015
Concrete	0.002
Cork	0.0001
Felt	0.0001
Glass	0.002
Ice	0.004
Rock wool	0.0001
Wood	0.0003–0.0001
Gases	
Air	0.000057
Argon	0.000039
Helium	0.00034
Hydrogen	0.00033
Oxygen	0.000056

In the steady state, the heat current in the rod in Fig. 17–1 must be the same at all cross sections. If this were not so, the quantity of heat flowing into an element of the rod between two cross sections would not be the same as that flowing out. Heat would then accumulate in the element and its temperature would change, which contradicts the assumption of a steady state. (In this respect the steady-state flow of heat is like the flow of an incompressible fluid.) It follows that in the steady state the quantity $kA(dt/dx)$ is the same at all cross sections. Then *if A is constant and if k is independent of temperature*, the temperature gradient dt/dx is the same at all cross sections. In other words, the temperature decreases *linearly* along the rod and the steady-state temperature gradient is $(t_2 - t_1)/L$. Hence for a specimen of constant cross section and constant k, in the steady state, the magnitude of the heat current (omitting the negative sign) is

$$H = kA\frac{t_2 - t_1}{L}. \qquad (17\text{-}2)$$

Since there are no restrictions on the relative dimensions of the rod in Fig. 17-1, Eq. (17-2) applies also when the length L becomes relatively small and the area A relatively large, as in the case of a sheet or slab of material such as a portion of the wall of a house, a refrigerator, or a furnace. (The dimension L would then ordinarily be described as the "thickness" rather than the "length.")

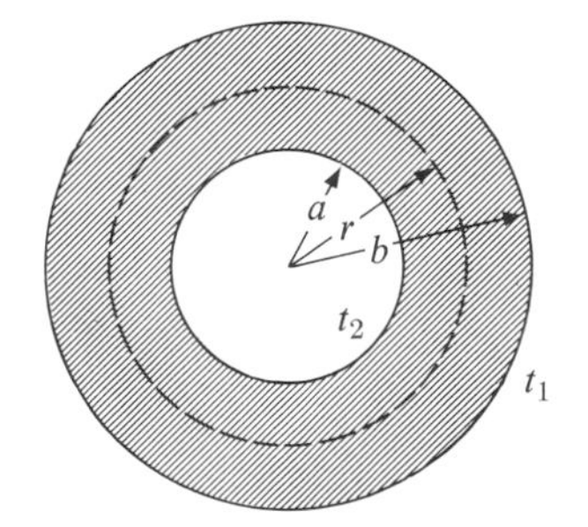

17-2 Radial heat flow in a cylinder or sphere.

17-2 RADIAL HEAT FLOW IN A SPHERE OR CYLINDER

We next consider two examples of heat flow in which the temperature gradient is not uniform along the direction of flow, even in the steady state. Figure 17-2 represents a steam pipe surrounded by a layer of insulating material, or a sphere surrounded by a spherical shell of insulation. Let t_2 and t_1 be the temperatures of the inner and outer surfaces of the insulation, and a and b be the inner and outer radii. If $t_2 > t_1$, heat flows outward, and in the steady state the heat current H is the same across all surfaces within the insulation, such as that of radius r, shown by a dashed circle. If A is the area of this surface and dt/dr the corresponding temperature gradient,

$$H = -kA\frac{dt}{dr} = \text{constant}.$$

The heat current can be expressed in terms of t_2, t_1, a, and b, by integrating this equation and inserting the appropriate boundary conditions. The calculations for the cylinder and the sphere in parallel have been carried out in Table 17-2.

TABLE 17-2

Cylinder (length L)	Sphere
$A = 2\pi rL$	$A = 4\pi r^2$
$H = -k \cdot 2\pi rL \cdot \dfrac{dt}{dr}$	$H = -k \cdot 4\pi r^2 \dfrac{dt}{dr}$
$H\dfrac{dr}{r} = -2\pi kL\,dt$	$H\dfrac{dr}{r^2} = -4\pi k\,dt$
$H \ln r = -2\pi kLt + C$	$-\dfrac{H}{r} = -4\pi kt + C$
$r = a, \quad t = t_2;$	$r = b, \quad t = t_1$
$H = \dfrac{2\pi k(t_2 - t_1)}{[\ln(b/a)]/L}$	$H = \dfrac{4\pi k(t_2 - t_1)}{(b - a)/ab}$

17-3 CONVECTION

The term *convection* is applied to the transfer of heat from one place to another by the actual motion of material. The hot-air furnace and the hot-water heating system are examples. If the heated material is forced to move by a blower or pump, the process is called *forced* convection; if the material flows because of differences in density, the process is called *natural* or *free* convection. To understand the latter, consider a U-tube as illustrated in Fig. 17-3.

In (a), the water is at the same temperature in both arms of the U and hence stands at the same level in each. In (b), the right side of the U has been heated. The water in this side expands and therefore, being of smaller density, a longer column is needed to balance the pressure produced by the cold water in the left column. The stopcock may now be opened and water will flow from the top of the warmer column into the colder column. This increases the pressure at the bottom of the U produced by the cold column, and decreases the pressure at this point due to the hot column. Hence at the bottom of the U, water is forced from the cold to the hot side. If heat is continually applied to the hot side and removed from the cold side,

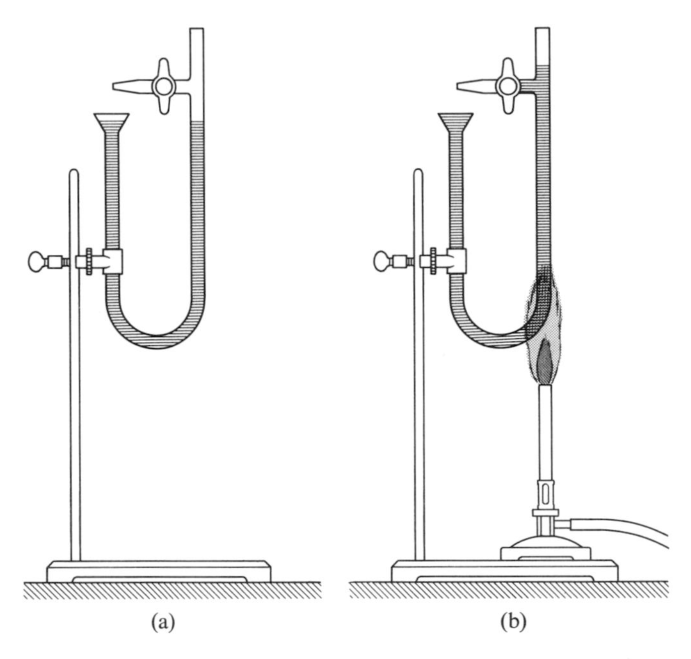

17-3 Convection is brought about by differences in density.

the circulation continues of itself. The net result is a continual transfer of heat from the hot to the cold side of the column. In the common household hot-water heating system, the "cold" side corresponds to the radiators and the "hot" side to the furnace.

The anomalous expansion of water which was mentioned in Chapter 15 has an important effect on the way in which lakes and ponds freeze in winter. Consider a pond at a temperature of, say, 20°C throughout, and suppose the air temperature at its surface falls to −10°C. The water at the surface becomes cooled to, say, 19°C. It therefore contracts, becomes more dense than the warmer water below it, and sinks in this less dense water, its place being taken by water at 20°C. The sinking of the cooled water causes a mixing process, which continues until all of the water has been cooled to 4°C. Now, however, when the surface water cools to 3°C, it expands, is less dense than the water below it, and hence floats on the surface. Convection and mixing then cease, and the remainder of the water can lose heat only by *conduction*. Since water is an extremely poor heat conductor, cooling takes place very slowly after 4°C is reached, with the result that the pond freezes first at its surface. Then, since the density of ice is even smaller than that of water at 0°C, the ice floats on the water below it, and further freezing can result only from heat flow upward by conduction.

The mathematical theory of heat convection is quite involved. There is no simple equation for convection, as there is for conduction. This arises from the fact that the heat lost or gained by a surface at one temperature in contact with a fluid at another temperature depends on many circumstances, such as (1) whether the surface is flat or curved, (2) whether the surface is horizontal or vertical, (3) whether the fluid in contact with the surface is a gas or a liquid, (4) the density, viscosity, specific heat, and thermal conductivity of the fluid, (5) whether the velocity of the fluid is small enough to give rise to laminar flow or large enough to cause turbulent flow, (6) whether evaporation, condensation, or formation of scale takes place.

The procedure adopted in practical calculations is first to define a *convection coefficient h* by means of the equation

$$H = hA\,\Delta t, \tag{17-3}$$

where H is the heat convection current (the heat gained or lost by convection by a surface per unit of time), A is the area of the surface, and Δt is the temperature difference between the surface and the main body of fluid. The next step is the determination of numerical values of h that are appropriate to a given piece of equipment. Such a determination is accomplished partly by dimensional analysis and partly by an elaborate series of experiments. An enormous amount of research in this field has been done in recent years so that, by now, there are in existence fairly complete tables and graphs from which the physicist or engineer may obtain the convection coefficient appropriate to certain standard types of apparatus.

A case of common occurrence is that of natural convection from a wall or a pipe which is at a constant temperature and is surrounded by air at atmospheric pressure whose temperature differs from that of the wall or pipe by an amount Δt. The convection coefficients applicable in this situation are given in Table 17-3.

Example. The air in a room is at a temperature of 25°C, and the outside air is at −15°C. How much heat is transferred per unit

TABLE 17-3 COEFFICIENTS OF NATURAL CONVECTION IN AIR AT ATMOSPHERIC PRESSURE

Equipment	Convection coefficient h, cal s^{-1} cm^{-2} $(C°)^{-1}$
Horizontal plate, facing upward	$0.595 \times 10^{-4}(\Delta t)^{1/4}$
Horizontal plate, facing downward	$0.314 \times 10^{-4}(\Delta t)^{1/4}$
Vertical plate	$0.424 \times 10^{-4}(\Delta t)^{1/4}$
Horizontal or vertical pipe $\left(\frac{\text{diameter}}{D}\right)$	$1.00 \times 10^{-4}\left(\frac{\Delta t}{D}\right)^{1/4}$

area of a glass windowpane of thermal conductivity 2.5×10^{-3} cgs units and of thickness 2 mm?

To assume that the inner surface of the glass is at 25°C and the outer surface is at −15°C is entirely erroneous, as anyone can verify by touching the inner surface of a glass windowpane on a cold day. One must expect a much smaller temperature difference across the windowpane, so that in the steady state the rates of transfer of heat (1) by convection in the room, (2) by conduction through the glass, and (3) by convection in the outside air, are all equal.

As a first approximation in the solution of this problem, let us assume that the window is at a uniform temperature t. If $t = 5$°C, then the temperature difference between the inside air and the glass is the same as that between the glass and the outside air, or 20 C°. Hence the convection coefficient in both cases is

$$h = 0.424 \times 10^{-4}(20)^{1/4} \text{ cal s}^{-1} \text{ cm}^{-2} (\text{C}°)^{-1}$$
$$= 0.895 \times 10^{-4} \text{ cal s}^{-1} \text{ cm}^{-2} (\text{C}°)^{-1},$$

and, from Eq. (17-3), the heat transferred per unit area is

$$\frac{H}{A} = 0.895 \times 10^{-4} \times 20 = 17.9 \times 10^{-4} \text{ cal s}^{-1} \text{ cm}^{-2}.$$

The glass, however, is not at a uniform temperature; there must be a temperature difference Δt across the glass sufficient to provide heat conduction at the rate of 17.9×10^{-4} cal s^{-1} cm^{-2}. Using the conduction equation, Eq. (17-2), we obtain

$$\Delta t = \frac{L}{k} \times \frac{H}{A} = \frac{0.2}{2.5 \times 10^{-3}} \times 17.9 \times 10^{-4} \text{ C}°$$
$$= 0.14 \text{ C}°.$$

With sufficient accuracy we may therefore say that the inner surface is at 5.07°C and the outer surface is at 4.93°C.

17-4 RADIATION

When one's hand is placed in direct contact with the surface of a hot-water or steam radiator, heat reaches the hand by *conduction* through the radiator walls. If the hand is held above the radiator but not in contact with it, heat reaches the hand by way of the upward-moving *convection* currents of warm air. If the hand is held at one side of the radiator it still becomes warm, even though conduction through the air is negligible and the hand is not in the path of the convection currents. Energy now reaches the hand by *radiation*.

The term radiation refers to the continual emission of energy from the surface of all bodies. This energy is called *radiant energy* and is in the form of electromagnetic waves. These waves travel with the velocity of light and are transmitted through a vacuum as well as through air. (Better, in fact, since they are absorbed by air to some extent.) When they fall on a body which is not transparent to them, such as the surface of one's hand or the walls of the room, they are absorbed.

The radiant energy emitted by a surface, per unit time and per unit area, depends on the nature of the surface and on its temperature. At low temperatures the rate of radiation is small and the radiant energy is chiefly of relatively long wavelength. As the temperature is increased, the rate of radiation increases very rapidly, in proportion to the fourth power of the absolute temperature. For example, a copper block at a temperature of 100°C (373°K) radiates about 300,000 ergs s^{-1} or 0.03 W from each square centimeter of its surface. At a temperature of 500°C (773°K), it radiates about 0.54 W from each square centimeter, and at 1000°C (1273°K), it radiates about 4 W per square centimeter. This rate is 130 times as great as that at a temperature of 100°C.

At each of these temperatures the radiant energy emitted is a mixture of waves of different wavelengths. At a temperature of 300°C the most intense of these waves has a wavelength of about 5×10^{-4} cm; for wavelengths either greater or less than this value the intensity decreases as shown by the curve in Fig. 17-4. The corresponding distribution of energy at higher temperatures is also shown in the figure. The area between each curve and the horizontal axis represents the total rate of radiation at that temperature. It is evident that this rate increases rapidly with increasing temperature,

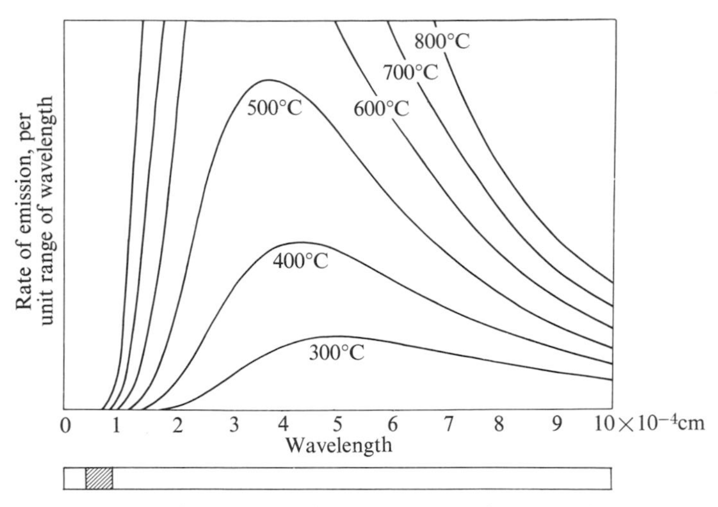

17-4 Rate of emission of radiant energy by a blackbody, per unit range of wavelength, as a function of wavelength. Shaded area indicates visible spectrum.

and also that the wavelength of the most intense wave shifts toward the left, or toward shorter wavelengths, with increasing temperature.

At a temperature of 300°C, practically all of the radiant energy emitted by a body is carried by waves longer than those corresponding to red light. Such waves are called *infrared*, meaning "beyond the red." At a temperature of 800°C a body emits enough visible radiant energy to be self-luminous and appears "red hot." By far the larger part of the energy emitted, however, is still carried by infrared waves. At 3000°C, which is about the temperature of an incandescent lamp filament, the radiant energy contains enough of the shorter wavelengths so that the body appears nearly "white hot."

17-5 STEFAN'S LAW

Experimental measurements of the rate of emission of radiant energy from the surface of a body were made by John Tyndall (1820–1893), and on the basis of these Josef Stefan (1835–1893), in 1879, concluded that the rate of emission could be expressed by the relation

$$\boxed{R = e\sigma T^4,} \tag{17-4}$$

which is *Stefan's law*. The quantity R is known as the *radiant emittance*, and is equal to the rate of emission of radiant energy per unit area and is expressed in ergs per second per square centimeter in the cgs system, and in watts per square meter in the mks system. The constant σ has a numerical value of 5.6699×10^{-5} in cgs units and 5.6699×10^{-8} in mks units. The quantity T is the Kelvin temperature of the surface, and e is a quantity called the emissivity of the surface. The emissivity lies between zero and unity, depending on the nature of the surface. The emissivity of copper, for example, is about 0.3. (Strictly speaking, the emissivity varies somewhat with temperature even for the same surface.) In general, the emissivity is larger for rough and smaller for smooth, polished surfaces.

It may be wondered why it is, if the surfaces of all bodies are continually emitting radiant energy, that all bodies do not eventually radiate away all of their internal energy and cool down to a temperature of absolute zero [where $R = 0$ by Eq. (17-4)]. The answer is that they would do so if energy were not supplied to them in some way. In the case of a Sunbowl heater element or the filament of an electric lamp, energy is supplied electrically to make up for the energy radiated. As soon as this energy supply is cut off, these bodies do, in fact, cool down very quickly to room temperature. The reason that they do not cool further is that their surroundings (the walls, and other objects in the room) are also radiating, and some of this radiant energy is intercepted, absorbed, and converted into internal energy. The same thing is true of all other objects in the room—each is both emitting and absorbing radiant energy simultaneously. If any object is hotter than its surroundings, its rate of emission will exceed its rate of absorption. There will thus be a net loss of energy and the body will cool down unless heated by some other method. If a body is at a lower temperature than its surroundings, its rate of absorption will be larger than its rate of emission and its temperature will rise. When the body is at the same temperature as its surroundings, the two rates become equal, there is no net gain or loss of energy, and no change in temperature.

If a small body of emissivity e is completely surrounded by walls at a temperature T, the rate of *absorption* of radiant energy per unit area by the body is

$$R = e\sigma T^4.$$

Hence for such a body at a temperature T_1, surrounded by walls at a temperature T_2, the *net* rate of loss (or gain) of energy per unit area by radiation is

$$R_{\text{net}} = e\sigma T_1^4 - e\sigma T_2^4$$
$$= e\sigma(T_1^4 - T_2^4). \qquad (17\text{-}5)$$

17-6 THE IDEAL RADIATOR

Imagine that the walls of the enclosure in Fig. 17-5 are kept at the temperature T_2 and a number of different bodies having different emissivities are suspended one after another within the enclosure. Regardless of their temperature when they were inserted, it will be found that eventually each comes to the same temperature, T_2, as that of the walls even if the enclosure is evacuated.

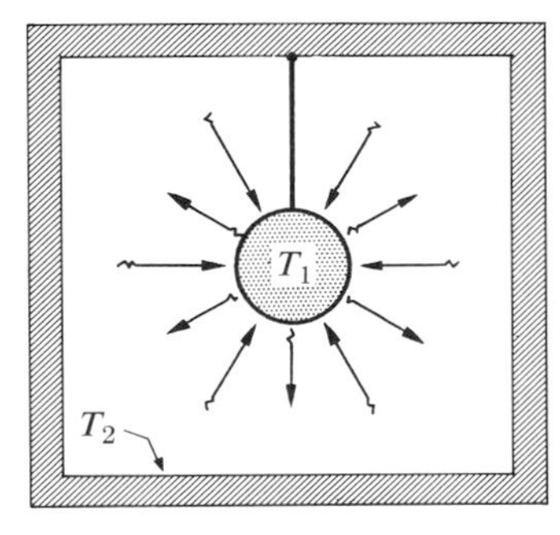

17-5 In thermal equilibrium, the rate of emission of radiant energy equals the rate of absorption. Hence a good absorber is a good emitter.

If the bodies are small compared with the size of the enclosure, radiant energy from the walls strikes the surface of each body at the same rate. Of this energy, a part is reflected and the remainder absorbed. In the absence of any other process, the energy absorbed will raise the temperature of the absorbing body, but since the temperature is observed *not* to change, each body must *emit* radiant energy at the same rate as it *absorbs* it. Hence a good absorber is a good emitter, and a poor absorber is a poor emitter. But since each body must either absorb or reflect the radiant energy reaching it, a poor absorber must also be a good reflector. Hence a *good reflector* is a *poor emitter*.

This is the reason for silvering the walls of vacuum ("thermos") bottles. A vacuum bottle is constructed with double glass walls, the space between the walls being evacuated so that heat flow by conduction and convection is practically eliminated. To reduce the radiant emission to as low a value as possible, the walls are covered with a coating of silver, which is highly reflecting and hence is a very poor emitter.

Since a good absorber is a good emitter, the *best* emitter will be that surface which is the best absorber. But no surface can absorb more than all of the radiant energy which strikes it. Any surface which does absorb all of the incident energy will be the best emitting surface possible. Such a surface would reflect no radiant energy, and hence would appear black in color (provided its temperature is not so high that it is self-luminous). It is called an *ideally black surface*, and a body having such a surface is called an ideal blackbody, an ideal radiator, or simply a *blackbody*.

No actual surface is ideally black, the closest approach being lampblack, which reflects only about 1%. Blackbody conditions can be closely realized, however, by a small opening in the walls of a closed container. Radiant energy entering the opening is in part absorbed by the interior walls. Of the part reflected, only a very little escapes through the opening, the remainder being eventually absorbed by the walls. Hence the *opening* behaves like an ideal absorber.

Conversely, the radiant energy emitted by the walls or by any body within the enclosure, and escaping through the opening, will, if the walls are of uniform temperature, be of the same nature as that emitted by an ideal radiator. This fact is of importance when using an optical pyrometer, described in Section 15-2. The readings of such an instrument are correct only when it is sighted on a blackbody. If used to measure the temperature of a red hot ingot of iron in the open, its readings will be too low, since iron is a poorer emitter than a blackbody. If, however, the pyrometer is sighted on the iron while still in the furnace, where it is surrounded by walls at the same temperature, "blackbody conditions" are fulfilled and the reading will be correct. The failure of the iron to emit as effectively as a blackbody is just compensated by the radiant energy which it reflects.

The emissivity e of an ideally black surface is equal to unity. For all real surfaces it is a fraction less than one.

Problems

17–1 Suppose that both ends of the rod in Fig. 17–1 are kept at a temperature of 0°C, and that the initial temperature distribution along the rod is given by $t = 100 \sin \pi x/L$, where t is in °C. Let the rod be of copper, of length $L = 10$ cm and of cross section 1 cm^2. (a) Show the initial temperature distribution in a diagram. (b) What is the final temperature distribution after a very long time has elapsed? (c) Sketch curves which you think would represent the temperature distribution at intermediate times. (d) What is the initial temperature gradient at the ends of the rod? (e) What is the initial heat current from the ends of the rod into the bodies making contact with its ends? (f) What is the initial heat current at the center of the rod? Explain. What is the heat current at this point at any later time? (g) What is the value of $k/\rho C$ for copper, and in what unit is it expressed? (This quantity is called the *diffusivity*.) (h) What is the initial rate of change of temperature at the center of the rod? (i) How long a time would be required for the rod to reach its final temperature, if the temperature continued to decrease at this rate? (This time can be described as the *relaxation time* of the rod.) (j) From the graphs in part (c), would you expect the rate of change of temperature at the midpoint to remain constant, increase, or decrease? (k) What is the initial rate of change of temperature at a point in the rod, 2.5 cm from its left end?

17–2 Suppose that the rod in Fig. 17–1 is of copper, of length 10 cm and of cross-sectional area 1 cm^2. Let $t_2 = 100$°C and $t_1 = 0$°C. (a) What is the final steady-state temperature gradient along the rod? (b) What is the heat current in the rod, in the final steady state? (c) What is the final steady-state temperature at a point in the rod, 2 cm from its left end?

17–3 A long rod, insulated to prevent heat losses, has one end immersed in boiling water (at atmospheric pressure) and the other end in a water-ice mixture. The rod consists of 100 cm of copper (one end in steam) and a length, L_2, of steel (one end in ice). Both rods are of cross-sectional area 5 cm^2. The temperature of the copper-iron junction is 60°C, after a steady state has been set up. (a) How many calories per second flow from the steam bath to the ice-water mixture? (b) How long is L_2?

17–4 A rod is initially at a uniform temperature of 0°C throughout. One end is kept at 0°C and the other is brought into contact with a steam bath at 100°C. The surface of the rod is insulated so that heat can flow only lengthwise along the rod. The cross-sectional area of the rod is 2 cm^2, its length is 100 cm, its thermal conductivity is 0.8 cgs units, its density is 10 g cm^{-3}, and its specific heat capacity is 0.10 cal gm^{-1}(C°)$^{-1}$. Consider a short cylindrical element of the rod 1 cm in length. (a) If the temperature gradient at one end of this element is 200 C° cm^{-1}, how many calories flow across this end per second? (b) If the average temperature of the element is increasing at the rate of 5 C° s^{-1}, what is the temperature gradient at the other end of the element?

17–5 One experimental method of measuring the thermal conductivity of an insulating material is to construct a box of the material and measure the power input to an electric heater, inside the box, which maintains the interior at a measured temperature above the outside surface. Suppose that in such an apparatus a power input of 120 W is required to keep the interior surface of the box 120 F° above the temperature of the outer surface. The total area of the box is 25 ft^2 and the wall thickness is 1.5 in. Find the thermal conductivity of the material in the commercial system of units.

17–6 A container of wall area 5000 cm^2 and thickness 2 cm is filled with water in which there is a stirrer. The outer surface of the walls is kept at a constant temperature of 0°C. The thermal conductivity of the walls is 0.000478 cgs units, and the effect of edges and corners can be neglected. The power required to run the stirrer at an angular velocity of 1800 rpm is found to be 100 W. What will be the final steady-state temperature of the water in the container? Assume that the stirrer keeps the entire mass of water at a uniform temperature.

17–7 A boiler with a steel bottom 1.5 cm thick rests on a hot stove. The area of the bottom of the boiler is 1500 cm^2. The water inside the boiler is at 100°C, and 750 g are evaporated every 5 min. Find the temperature of the lower surface of the boiler, which is in contact with the stove.

17–8 An icebox, having wall area of 2 m^2 and thickness 5 cm, is constructed of insulating material having a thermal conductivity of 10^{-4} cal s^{-1} cm^{-1} (C°)$^{-1}$. The outside temperature is 20°C, and the inside of the box is to be maintained at 5°C by ice. The melted ice leaves the box at a temperature of 15°C. If ice costs 1 cent per kilogram, what will it cost to run the icebox for 1 hr?

17–9 Heat flows radially outward through a cylindrical insulator of outside radius R_2 surrounding a steam pipe of outside radius R_1. The temperature of the inner surface of the insulator is t_1, that of the outer surface is t_2. (a) At what radial distance from the center of the pipe is the steady-state temperature exactly halfway between temperatures t_1 and t_2? (b) Sketch a graph of t versus r.

17–10 A steam pipe 2 cm in radius is surrounded by a cylindrical jacket of insulating material 2 cm thick. The temperature of the steam pipe is 100°C, and that of the outer surface of the jacket is 20°C. The thermal conductivity of the insulating material is 0.0002 cal s^{-1} cm^{-1} $(C°)^{-1}$. Compute the temperature gradient, dt/dr, at the inner and outer surfaces of the jacket, and sketch the graph of t versus r.

17–11 An electric transformer is in a cylindrical tank 60 cm in diameter and 1 m high, with flat top and bottom. If the tank transfers heat to the air only by natural convection, and the electrical losses are to be dissipated at the rate of 1 kwh, how many degrees will the tank surface rise above room temperature?

17–12 (a) What would be the difference in height between the columns in the U-tube in Fig. 17–3 if the liquid is water and the left arm is 1 m high at 4°C while the other is at 75°C? (b) What is the difference between the pressures at the foot of two columns of water each 10 m high if the temperature of one is 4°C and that of the other is 75°C?

17–13 A flat wall is maintained at a constant temperature of 100°C, and the air on both sides is at atmospheric pressure and at 20°C. How much heat is lost by natural convection from 1 m^2 of wall (both sides) in 1 hr if (a) the wall is vertical and (b) the wall is horizontal?

17–14 A vertical steam pipe of outside diameter 7.5 cm and height 4 m has its outer surface at the constant temperature of 95°C. The surrounding air is at atmospheric pressure and at 20°C. How much heat is delivered to the air by natural convection in 1 hr?

17–15 What is the radiant emittance of a blackbody at a temperature of (a) 300°K, (b) 3000°K?

17–16 The radiant emittance of tungsten is approximately 0.35 that of a blackbody at the same temperature. A tungsten sphere 1 cm in radius is suspended within a large evacuated enclosure whose walls are at 300°K. What power input is required to maintain the sphere at a temperature of 3000°K if heat conduction along the supports is neglected?

17–17 A small blackened solid copper sphere of radius 2 cm is placed in an evacuated enclosure whose walls are kept at 100°C. At what rate must energy be supplied to the sphere to keep its temperature constant at 127°C?

17–18 A cylindrical metal can 10 cm high and 5 cm in diameter contains liquid helium at 4°K, at which temperature its heat of vaporization is 5 cal g^{-1}. Completely surrounding the helium can are walls maintained at the temperature of liquid nitrogen, 80°K, the intervening space being evacuated. How much helium is lost per hour? Assume the radiant emittance of the helium can to be 0.2 that of a blackbody at 4°K.

17–19 A solid cylindrical copper rod 10 cm long has one end maintained at a temperature of 20.00°K. The other end is blackened and exposed to thermal radiation from a body at 300°K, no energy being lost or gained elsewhere. When equilibrium is reached, what is the temperature of the blackened end? [*Hint:* Since copper is a very good conductor of heat at low temperature, $K = 4$ cal s^{-1} cm^{-1} $(C°)^{-1}$, the temperature of the blackened end is only slightly greater than 20°K.]

18 Thermal Properties of Matter

18-1 EQUATIONS OF STATE

The volume V occupied by a definite mass m of any substance depends on the pressure p to which the substance is subjected, and on its temperature T. For every pure substance there is a definite relation between these quantities, called the *equation of state* of the substance. In formal mathematical language, the equation can be written

$$f(m, V, p, T) = 0. \quad (18\text{–}1)$$

The exact form of the function is usually very complicated. It often suffices to know only how some one of the quantities changes when some other is varied, the rest being kept constant. Thus the compressibility k describes the change in volume when the pressure is changed, for a constant mass at a constant temperature,

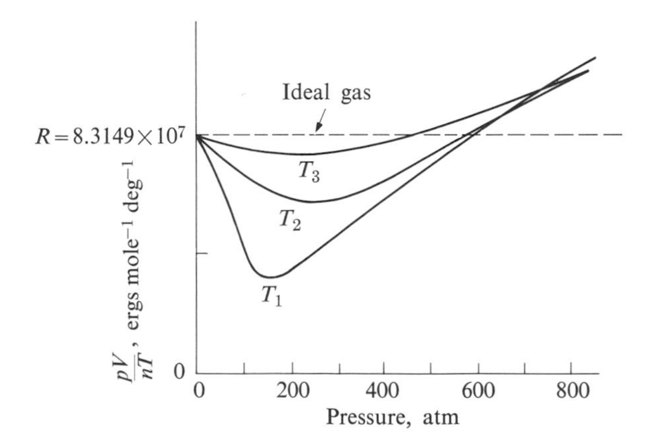

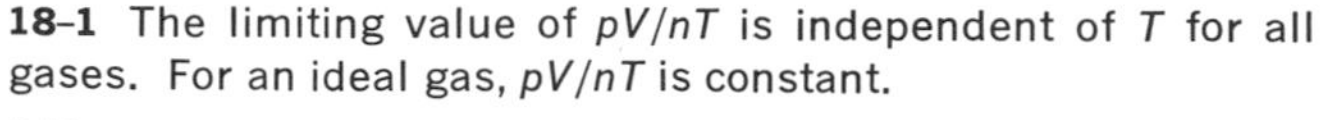

18-1 The limiting value of pV/nT is independent of T for all gases. For an ideal gas, pV/nT is constant.

and the coefficient of volume expansion gives the change in volume when the temperature is changed, for a constant mass at a constant pressure.

The term "state" as used here implies an *equilibrium state*. This means that the temperature and pressure are the same at all points. Hence if heat is added at some point to a system in an equilibrium state, we must wait until the processes of heat transfer within the system have brought about a new uniform temperature before the system is again in an equilibrium state.

18-2 THE IDEAL GAS

The simplest equation of state is that of a gas at low pressure. Consider a container whose volume can be varied, such as a cylinder provided with a movable piston. Provision is made for pumping any desired mass of any kind of gas into or out of the cylinder, and the cylinder is provided with a pressure gauge and with a thermometer for determining the Kelvin temperature T. Then corresponding values of m, p, V, and T can all be measured. Instead of the mass m, let us express the results in terms of the number of moles, n.

Let us first collect all the measurements made on one particular gas at a given temperature, say T_1. The pressure, volume, and the number of moles present will vary over a wide range. For each individual measurement we compute the quantity pV/nT_1, and construct a graph of the results, plotting pV/nT_1 vertically and p horizontally. The plotted points will all be found to lie on a smooth curve called an *isotherm*, such as the one labeled T_1 in Fig. 18–1. When we collect all the results at some other temperature T_2, the plotted points lie on another isotherm. Similar sets of curves can be constructed for other gases.

We now discover two very remarkable facts: (1) for any one gas, the curves at different temperatures, when extrapolated to zero pressure, all intersect the vertical axis at the same point, and (2) this point of intersection is the same for all gases. In other words, the quantity pV/nT, at low pressures, is *the same for all gases*. This quantity is called the *universal gas constant* and is represented by R. The numerical value of R depends, of course, on the units in which p, V, n, and T are expressed. The adjective "universal" means that in any one system of units R has the same value for *all* gases. In the cgs system, where the unit of p is 1 dyn cm^{-2}, and the unit of V is 1 cm^3,

$$\begin{aligned} R &= 8.3149 \times 10^7 \text{ (dyn cm}^{-2}\text{) cm}^3 \text{ mole}^{-1} \text{ (°K)}^{-1} \\ &= 8.3149 \times 10^7 \text{ ergs mole}^{-1} \text{ (°K)}^{-1} \\ &= 8.3149 \text{ J mole}^{-1} \text{ (°K)}^{-1} = 1.99 \text{ cal mole}^{-1} \text{ (°K)}^{-1}. \end{aligned}$$

In chemistry, volumes are commonly expressed in liters (ℓ), pressures in atmospheres, and temperatures in degrees kelvin. In this system of units,

$$R = 0.08207 \ \ell \text{ atm mole}^{-1} \text{ (°K)}^{-1}.$$

Then for all real gases at sufficiently low pressures,

$$\frac{pV}{nT} = R \qquad \text{or} \qquad pV = nRT.$$

We now define an *ideal gas* as one for which, by definition, the ratio pV/nT is equal to R at *all* pressures. In other words, an ideal gas behaves at *all* pressures in the same way that real gases behave at *low* pressures. The equation of state of an ideal gas is therefore

$$\boxed{pV = nRT,} \tag{18–2}$$

and for an ideal gas the curves in Fig. 18–1 all coalesce into a single horizontal straight line at a height R. Real gases, of course, do not obey this equation except at extremely low pressures. However, the deviations are not very great at moderate pressures, and at temperatures not too near those at which the gas condenses to a liquid. More complicated equations have been developed which describe the behavior of real gases more accurately, but in this chapter we shall assume that Eq. (18–2) is obeyed at all pressures and temperatures.

It follows from the equation above, and the value of the universal gas constant R, that one gram-mole of an ideal gas occupies a volume of 22,400 cm^3 or 22.4 ℓ at "standard conditions," or at "normal temperature and pressure" (NTP), that is, at a temperature of 0°C = 273°K, and a pressure of 1 atm = 1.01×10^6 dyn cm^{-2}. Thus, from Eq. (18–2),

$$\begin{aligned} V &= \frac{nRT}{p} \\ &= \frac{1 \text{ mole} \times 8.31 \times 10^7 \text{ ergs mole}^{-1} \text{ (°K)}^{-1} \times 273\text{°K}}{1.01 \times 10^6 \text{ dyn cm}^{-2}} \\ &= 22{,}400 \text{ cm}^3. \end{aligned}$$

For a *fixed mass* (or fixed number of moles) of an ideal gas, the product nR is constant and hence pV/T is constant also. Thus if the subscripts 1 and 2 refer to two states of the same mass of a gas, but at different pressures, volumes, and temperatures,

$$\frac{p_1V_1}{T_1} = \frac{p_2V_2}{T_2} = \text{constant}. \tag{18–3}$$

If the temperatures T_1 and T_2 are the same, then

$$p_1V_1 = p_2V_2 = \text{constant}. \tag{18–4}$$

The fact that the product of the pressure and volume of a fixed mass of gas is very nearly constant at constant temperature was discovered experimentally by Robert Boyle in 1660, and the equation above is called *Boyle's law*. Although exactly true, by definition, for an ideal gas, it is obeyed only approximately by real gases and is not a fundamental law like Newton's laws or the law of conservation of energy.

Example 1. The volume of an oxygen tank is 50 ℓ. As oxygen is withdrawn from the tank, the reading of a pressure gauge drops from 300 lb in^{-2} to 100 lb in^{-2}, and the temperature of the gas remaining in the tank drops from 30°C to 10°C. (a) How many kilograms of oxygen were there in the tank originally? (b) How many kilograms were withdrawn? (c) What volume would be occupied by the oxygen withdrawn from the tank at a pressure of 1 atm and a temperature of 20°C?

a) Let us express pressures in atmospheres, volumes in liters, and temperatures in degrees Kelvin. Then $R = 0.082$ ℓ atm mole^{-1} (°K)$^{-1}$. The initial and final gauge pressures are

$$300 \text{ lb in}^{-2} = \frac{300}{14.7} = 20.5 \text{ atm};$$

$$100 \text{ lb in}^{-2} = 6.8 \text{ atm}.$$

The corresponding *absolute* pressures are 21.5 atm and 7.8 atm.
Initially,

$$n_1 = \frac{p_1 V}{RT_1} = \frac{21.5 \times 50}{0.082 \times 303} = 43.2 \text{ moles}.$$

The original mass was therefore

$$m_1 = 43.2 \text{ moles} \times 32 \text{ g mole}^{-1} = 1380 \text{ g} = 1.38 \text{ kg}.$$

b) The number of moles remaining in the tank is

$$n_2 = \frac{p_2 V}{RT_2} = \frac{7.8 \times 50}{0.082 \times 283} = 16.7 \text{ moles},$$

$$m_2 = 16.7 \text{ moles} \times 32 \text{ g mole}^{-1} = 536 \text{ g} = 0.54 \text{ kg}.$$

The number of kilograms withdrawn is

$$m_1 - m_2 = 1.38 - 0.54 = 0.84 \text{ kg}.$$

c) The volume occupied would be

$$V = \frac{nRT}{p} = \frac{16.7 \times 0.082 \times 293}{1} = 403 \ \ell.$$

Example 2. The *McLeod gauge* (pronounced "McLoud"), illustrated in Fig. 18–2, can be used to measure pressures as low as 5×10^{-6} mm of mercury. In part (a), the entire space above point A is occupied by the gas at the low pressure p which is to be measured. When the mercury container B is raised as in (b), the gas in bulb C, whose volume V might be 500 cm^3, is trapped and eventually compressed into a much smaller volume V' above a reference mark on the capillary tube D. Assuming the temperature constant, the pressure p' of the compressed gas is given by applying Boyle's law,

$$p' = \frac{pV}{V'}.$$

The pressure at the upper surface of the mercury in capillary E remains at the value p, so that if h is the difference in elevation between the tops of the mercury columns in E and D,

$$p' = p + \rho g h,$$

where ρ is the density of mercury. Elimination of p' between these equations gives

$$p = \frac{\rho g V'}{V - V'} h \approx \frac{V'}{V} \rho g h,$$

to a good approximation, since $V' \ll V$.

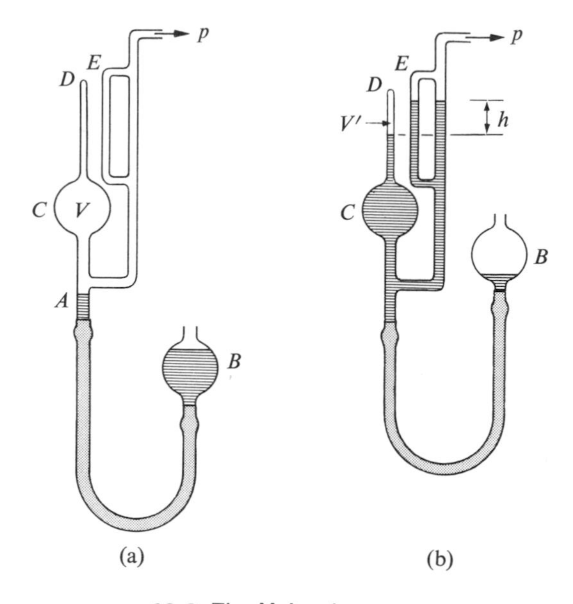

18–2 The McLeod gauge.

For example, if $V'/V = 10^{-4}$ and $h = 4$ mm, $p = 4 \times 10^{-4}$ torr.

Since ρ, g, V, and V' are constants, the pressure p is directly proportional to h and a uniform pressure scale can be mounted beside tube E. Corrections for capillary depression are eliminated if capillaries D and E have the same diameter.

Example 3. In the upper part of the atmosphere (the stratosphere) the temperature varies only slightly with changes in elevation. Find the law of variation of pressure with elevation.

The rate of change of pressure with elevation in a fluid is $dp/dy = -\rho g$.

Let us replace n by m/M in the ideal gas law, where m is the mass and M the molecular weight. Then

$$pV = \frac{m}{M} RT,$$

and the density ρ of an ideal gas is

$$\rho = \frac{m}{V} = \frac{pM}{RT}.$$

Hence

$$\frac{dp}{dy} = -\frac{pMg}{RT}, \qquad \frac{dp}{p} = -\frac{Mg}{RT} dy.$$

If g and T are constant, and if p_1 and p_2 are the pressures at two elevations y_1 and y_2, integration of the preceding equation gives

$$\ln \frac{p_2}{p_1} = -\frac{Mg}{RT}(y_2 - y_1). \tag{18–5}$$

This is known as the *barometric equation*.

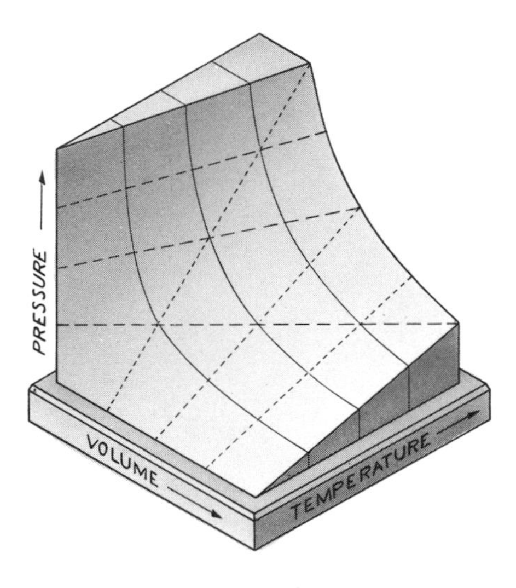

18–3 *pVT* surface for an ideal gas.

18–3 *pVT*-SURFACE FOR AN IDEAL GAS

Since the equation of state for a fixed mass of a substance is a relation among the three variables p, V, and T, it defines a *surface* in a rectangular coordinate system in which p, V, and T are plotted along the three axes. Figure 18–3 shows the *pVT*-surface of an ideal gas. The full lines on the surface show the relation between p and V when T is constant (Boyle's law), the dashed lines the relation between V and T when p is constant (Gay-Lussac's law), and the dotted lines the relation between p and T when V is constant. When viewed perpendicular to the *pV*-plane, the surface appears as in Fig. 18–4(a), and Fig. 18–4(b) is its appearance when viewed perpendicular to the *pT*-plane.

Every possible equilibrium state of the gas corresponds to a point on the surface, and every point of the surface corresponds to a possible equilibrium state. The gas cannot exist in a state that is not on the surface. For example, if the volume and temperature are given, thus locating a point in the base plane of Fig. 18–3, the pressure is then determined by the nature of the gas, and it can have only the value represented by the height of the surface above this point.

In any process in which the gas passes through a succession of equilibrium states, the point representing its state moves along a curve lying in the *pVT*-surface. Such a process must be carried out very slowly to give the temperature and pressure time to become uniform at all points of the gas.

18–4 *pVT*-SURFACE FOR A REAL SUBSTANCE

While all real substances approximate ideal gases at sufficiently low pressures, their behavior departs more and more from that of an ideal gas at high pressures and

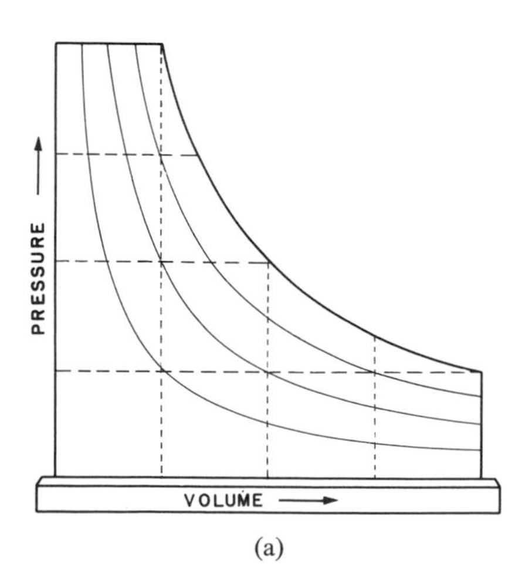

(a)

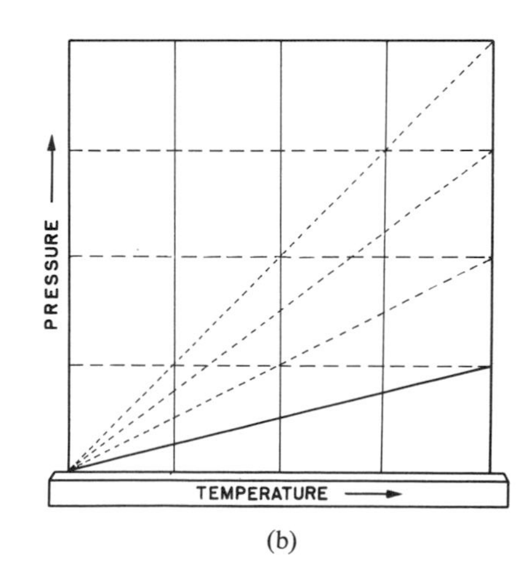

(b)

18–4 Projections of ideal gas *pVT* surface on (a) the *pV*-plane, (b) the *pT*-plane.

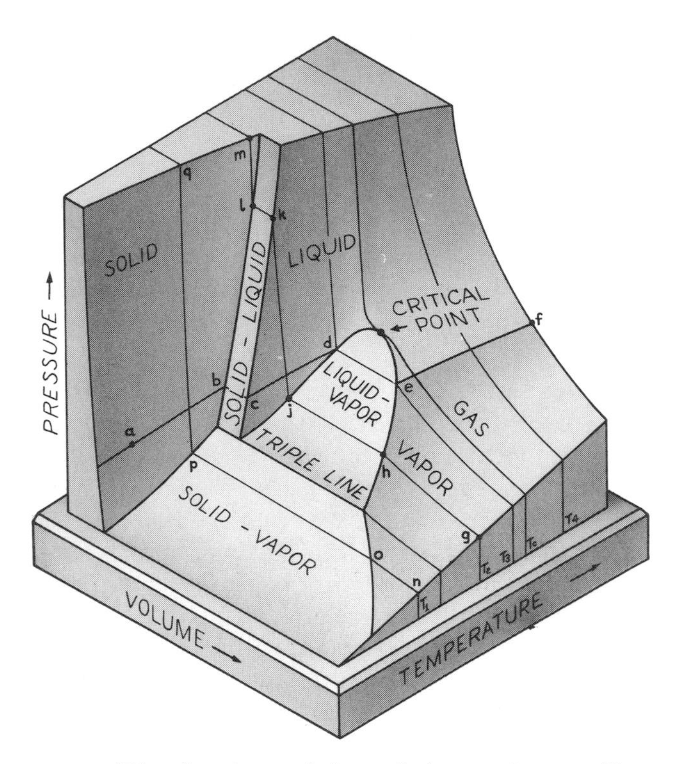

18–5 *pVT*-surface for a substance that expands on melting.

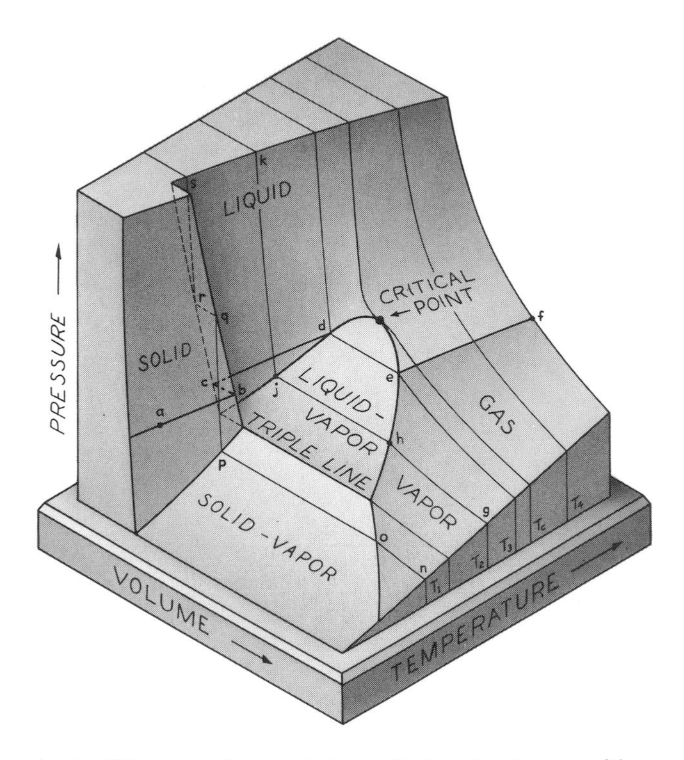

18–6 *pVT*-surface for a substance that contracts on melting.

low temperatures. As the temperature is lowered and the pressure increased, all substances change from the *gas phase* to the *liquid phase* or the *solid phase*. For a fixed mass of a substance, however, there is still a definite relation between the pressure, temperature, and total volume. In other words, the substance has an equation of state in any circumstances, and although the general form of the equation is much too complicated to express mathematically, we can represent it graphically by a *pVT*-surface. Figure 18–5 is a schematic diagram (to a greatly distorted scale) of the *pVT*-surface for a substance that expands on melting (the most common case) and Fig. 18–6 is the corresponding diagram for a substance which, like water, contracts on melting. We see that the substance can exist in either the solid, liquid, or gas phase, or in two phases simultaneously, or, along the triple line, in all three phases. (The distinction between the terms "gas" and "vapor" will be explained later. For the present they may be considered synonymous.)

In order that the diagram shall represent the properties of a particular *substance*, but not depend on *how much* of the substance is present, we plot along the volume axis not the actual volume V but the *specific volume* v, the volume per unit mass. Thus for a system of mass m,

$$v = \frac{V}{m}.$$

The actual volume of a unit mass of a substance is numerically equal to its specific volume. The specific volume is the reciprocal of the density ρ,

$$\rho = \frac{m}{V} = \frac{1}{v}.$$

To correlate these diagrams with our familiar experiences regarding the behavior of solids, liquids, and gases, let us start with a substance in the solid phase at point a in Fig. 18–5 or 18–6. Let the substance be contained in a cylinder as in Fig. 18–7, and let a constant force F be exerted on the piston so that the pressure remains constant as the substance expands or contracts.

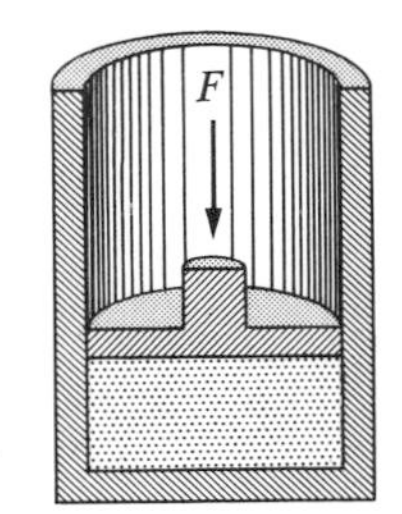

Fig. 18–7

Such a process is called *isobaric*. We now bring the cylinder in contact with some body such as an electric stove, whose temperature can be slowly increased and always kept slightly higher than that of the substance. There will then be a flow of heat into the substance.

At the start of the process, the temperature of the substance rises as heat is added, at a rate determined by the specific heat capacity of the solid, and the volume increases slightly at a rate determined by its coefficient of expansion. When point *b* is reached in either diagram the substance starts to melt, that is, to change from the solid to the liquid phase. The temperature ceases to rise even though heat is being continuously supplied. The volume increases in Fig. 18–5 and decreases in Fig. 18–6.

When point *c* is reached, the substance is wholly in the liquid phase. The temperature now starts to rise again, at a rate determined by the specific heat capacity of the liquid, and the volume increases at a rate determined by its coefficient of expansion. (For water at atmospheric pressure, there would be a small *decrease* in volume at first.)

When point *d* is reached, the temperature again ceases to rise, although heat is still supplied, and bubbles of vapor start to form in the cylinder and rise to its upper surface. Since the density of the vapor phase is very much smaller than that of the liquid, the volume increases markedly.

At point *e*, the substance is wholly in the vapor phase. (The relative volume increase from *d* to *e*, except at very high pressures, is enormously greater than the volume change between liquid and solid, so diagrams like Figs. 18–5 and 18–6 are not constructed on a uniform scale.) With further addition of heat the temperature rises again, the rate now being determined by the specific heat capacity of the vapor phase (at constant pressure). The volume also increases much more rapidly than did that of the solid or liquid.

If the cylinder is now brought in contact with a body whose temperature is kept slightly *lower* than that of the substance, heat flows out of the substance and all the changes in the original process take place in reversed order.

As a second illustration to show how the *pVT*-surface describes the behavior of a substance, suppose we start with a cylinder containing a gas, at a pressure, volume, and temperature corresponding to point *g* in Fig. 18–5 or 18–6. Let the external pressure be adjusted so that it is always slightly greater than the pressure exerted by the substance, and let the cylinder be kept in contact with a body at a constant temperature. The process is then *isothermal*.

The first effect of the increased external pressure will be to raise the temperature of the gas slightly. There will then be a flow of heat from the gas to the body with which it is in contact, and as the pressure increases the volume decreases along the line *gh* in a manner not very different from that of an ideal gas. When point *h* is reached, drops of liquid begin to form in the cylinder, and the volume continues to decrease without further increase of pressure.

At point *j*, all of the substance has condensed to the liquid phase. With further increase of pressure the volume decreases, but only slightly because of the small compressibility of liquids. For a substance like that in Fig. 18–6, no other change of phase occurs as the pressure is increased to point *k* and beyond (unless other forms of the solid state exist; see Fig. 18–8 and the accompanying discussion). In Fig. 18–5, however, another break in the curve takes place at point *k*. Crystals of the solid begin to appear, and the volume again decreases without an increase of external pressure. At point ℓ, in Fig. 18–5, the substance has been completely converted to the solid phase. Further increase of pressure reduces the volume of the solid only slightly and unless the substance can exist in more than one modification of the solid phase, no further changes in phase result.

As one more example, suppose we again start with a gas in the cylinder but at a lower temperature, corresponding to point *n* in Figs. 18–5 and 18–6. If the pressure is increased isothermally, crystals of the solid begin to appear at point *o* and the gas changes directly to a solid without passing through the liquid phase. The pressure remains constant along the line *op*, and at

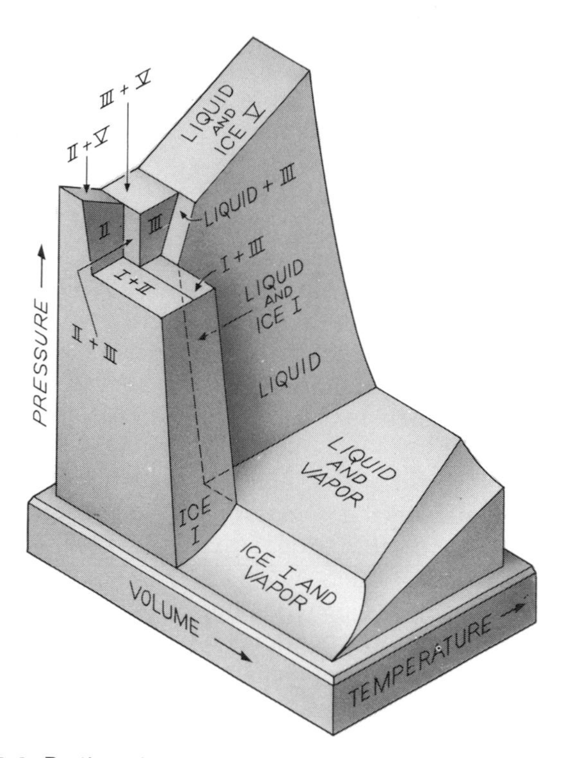

18-8 Portion of the pVT-surface of water at high pressure.

point p the substance is all in the solid phase. In Fig. 18-5 no further phase change takes place but in Fig. 18-6 the solid starts to melt at point q and has completely melted at point r.

Bear in mind that in all the processes described above, heat must be removed or added continuously in order to keep the pressure or temperature constant. If at any stage of a process the system is thermally insulated so that there can be no flow of heat in or out, and if the external pressure is kept constant, the system remains in equilibrium. Thus at any point on the surfaces lettered solid-liquid, solid-vapor, or liquid-vapor, *two* phases can coexist in equilibrium, and along the *triple line* all three phases can coexist. A vapor at the pressure and temperature at which it can exist in equilibrium with its liquid is called a *saturated vapor*, and the liquid is called a *saturated liquid*. Thus points e and h represent saturated vapor, and points j and d, saturated liquid. (The term "saturated" is poorly chosen. It does not have the same meaning as a "saturated solution" in chemistry. There is no question here of one substance being dissolved in another.)

18-5 CRITICAL POINT AND TRIPLE POINT

A study of Figs. 18-5 and 18-6 will show that the liquid and gas (or vapor) phases can exist together only if the temperature and pressure are less than those at the point lying at the top of the tongue-shaped surface lettered liquid-vapor. This point is called the *critical point*, and the corresponding values of T, p, and v are the *critical temperature, pressure*, and *specific volume*. A gas at a temperature above the critical temperature, such as T_4, does not separate into two phases when compressed isothermally but its properties change gradually and continuously from those we ordinarily associate with a gas (low density, large compressibility) to those of a liquid (high density, small compressibility). Table 18-1 lists the critical constants for a few substances. The very low critical temperatures of hydrogen and helium make it evident why these gases defied attempts to liquefy them for many years.

The term *vapor* is sometimes used to mean a gas at any temperature below its critical temperature, and sometimes is restricted to mean a gas in equilibrium with the liquid phase, that is, a saturated vapor. The term is really unnecessary; no sudden change takes

TABLE 18-1 CRITICAL CONSTANTS

Substance	Critical temperature, °K	Critical pressure, atm	Critical volume, $cm^3\ mole^{-1}$	Critical density, $g\ cm^{-3}$
Helium (4)	5.3	2.26	57.8	0.0693
Helium (3)	3.34	1.15	72.6	0.0413
Hydrogen (normal)	33.3	12.80	65.0	0.0310
Deuterium (normal)	38.4	16.4	60.3	0.0663
Nitrogen	126.2	33.5	90.1	0.311
Oxygen	154.8	50.1	78	0.41
Ammonia	405.5	111.3	72.5	0.235
Freon 12	384.7	39.6	218	0.555
Carbon dioxide	304.2	72.9	94.0	0.468
Sulfur dioxide	430.7	77.8	122	0.524
Water	647.4	218.3	56	0.32
Carbon disulfide	552	78	170	0.44

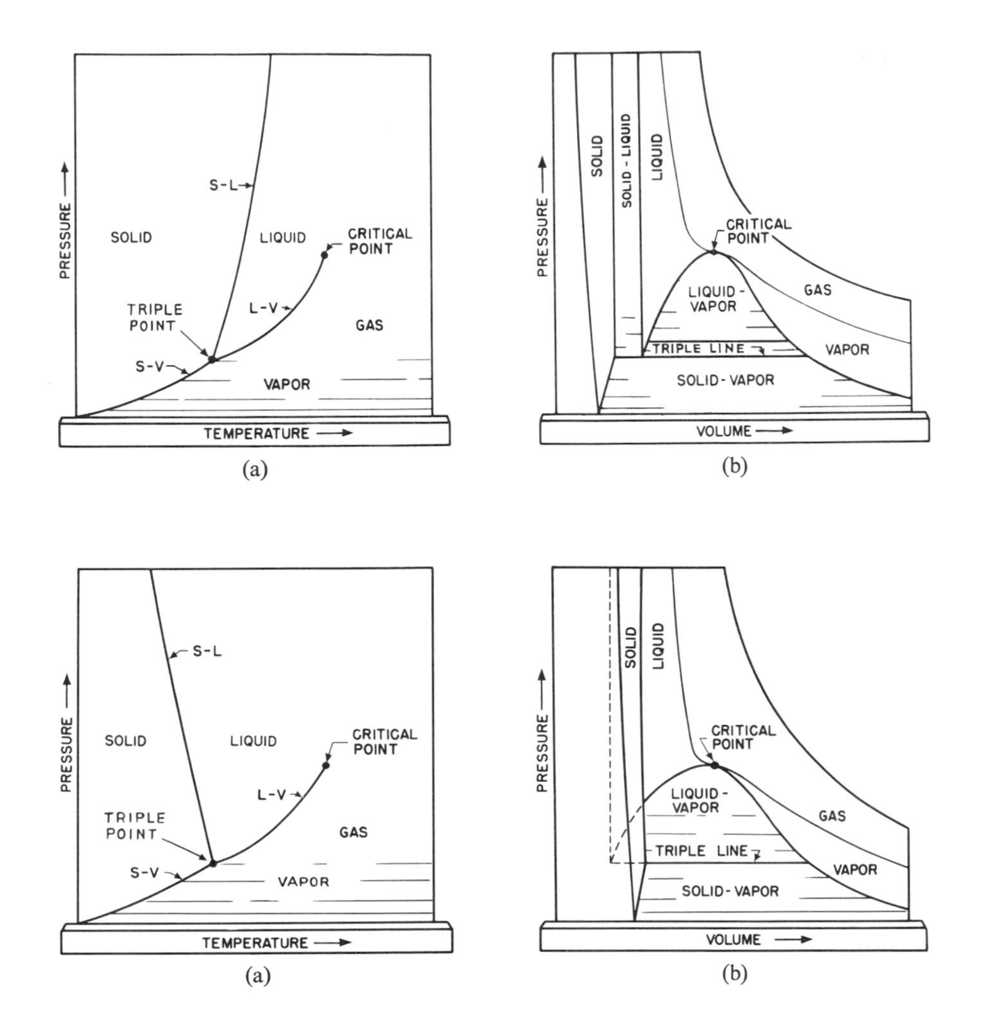

18-9 Projections of the surface in Fig. 18-5 on the pT- and pV-planes.

18-10 Projections of the surface in Fig. 18-6 on the pT- and pV-planes.

place in the properties of a substance when the critical isotherm is crossed either on the portion of the surface lettered gas and vapor, or on the portion lettered liquid.

It has been found that some substances can exist in more than one modification of the solid phase. Transitions from one modification to another occur at definite temperatures and pressures, like the phase changes from liquid to solid, etc. Water is one such substance, and at least eight types of ice have been observed at very high pressures. Figure 18-8 shows a portion of the pVT-surface of water at high pressure. Note that ordinary ice (ice I) is the only form whose specific volume is greater than that of the liquid phase.

Because of the difficulty of drawing three-dimensional diagrams, it is customary to represent the pVT-surface by its projections onto the pT- and pV-planes. Figure 18-9 shows the two projections of Fig. 18-5, and Fig. 18-10 shows those of Fig. 18-6. The reader should follow through on these diagrams the isobaric and isothermal processes indicated in Figs. 18-5 and 18-6.

The curves in Figs. 18-9(a) and 18-10(a) lettered S-L, S-V, and L-V, which are side views of the respective surfaces in 18-9(b) and 18-10(b), are loci of corresponding values of pressure and temperature at which the two phases can coexist if a substance is isolated, or at which one phase will transform to the other if heat is supplied or removed. Thus the S-L curve is also a graph of the melting-point temperature or freezing-point temperature of the substance as a function of pressure, the curve S-V is a graph of the *sublimation point* versus *pressure*, and the curve L-V is a graph of the *boiling point* versus *pressure*. The S-V and L-V curves always slope upward to the right. The S-L curve slopes upward to the right for a substance that expands on melting (Fig. 18-9) but upward to the left for a substance like water that contracts on melting (Fig. 18-10). Thus an increase of pressure always increases the temperature of the sublimation point or boiling point, but the temperature of the freezing point may be raised (Fig. 18-9) or lowered (Fig. 18-10) by an increase in pressure.

The pressure of a vapor in equilibrium with the liquid or solid at any temperature is called the *vapor pressure* of the substance at that temperature. Thus the curves S-V and L-V in Figs. 18-9(a) and 18-10(a) are graphs of vapor pressure versus temperature. The vapor pressure of a substance is a function of *temperature only*, not of volume. That is, in a vessel containing a liquid (or solid) and vapor in equilibrium *at a fixed temperature*, the pressure does not depend on the relative amounts of liquid and vapor present. If the volume is decreased some of the vapor condenses, and vice versa, but if the temperature is kept constant by removing or adding heat the pressure does not change.

The boiling-point temperature of a liquid is the temperature at which its vapor pressure equals the external pressure. Table 18-2 gives the vapor pressure of water as a function of temperature, and we see that the vapor pressure is 1 atm at a temperature of 100°C. If the external pressure is reduced to 17.5 mm of mercury, water will boil at room temperature (20°C), while under a pressure of 90 lb in^{-2} (about 6 atm) the boiling point is 160°C.

The point of intersection of the three equilibrium lines in Figs. 18-9(a) and 18-10(a), which is an end view of the triple line in Figs. 18-9(b) and 18-10(b), is called the *triple point*. There is only one pressure and temperature at which all three phases can coexist. Triple-point data for a few substances are given in Table 18-3.

As numerical examples, consider the *pT*-diagrams of water and carbon dioxide, in Fig. 18-11. In (a), a horizontal line at a pressure of 1 atm intersects the freezing-point curve at 0°C and the boiling-point curve at 100°C. The boiling point increases with increasing pressure up to the critical temperature of 374°C. Solid, liquid, and vapor can remain in equilibrium at the triple point, where the vapor pressure is 4.5 mm of mercury and the temperature is 0.01°C.

The freezing point of a substance like water, which expands on solidifying, is *lowered* by an increase in pressure. The reverse is true for substances which contract on solidifying. The change in the freezing-point temperature is much smaller than is that of the

TABLE 18-2 VAPOR PRESSURE OF WATER

T, °C	Vapor pressure, torr	Vapor pressure, lb in^{-2}	*T*, °F
0	4.58	0.0886	32
5	6.51	0.126	41
10	8.94	0.173	50
15	12.67	0.245	59
20	17.5	0.339	68
40	55.1	1.07	104
60	149	2.89	140
80	355	6.87	176
100	**760**	**14.7**	**212**
120	1490	28.8	248
140	2710	52.4	284
160	4630	89.6	320
180	7510	145	356
200	11650	225	392
220	17390	336	428

TABLE 18-3 TRIPLE-POINT DATA

Substance	Temperature, °K	Pressure, torr
Helium (4) (λ point)	2.186	38.3
Helium (3)	None	None
Hydrogen (normal)	13.84	52.8
Deuterium (normal)	18.63	128
Neon	24.57	324
Nitrogen	63.18	94
Oxygen	54.36	1.14
Ammonia	195.40	45.57
Carbon dioxide	216.55	3880
Sulfur dioxide	197.68	1.256
Water	**273.16**	**4.58**

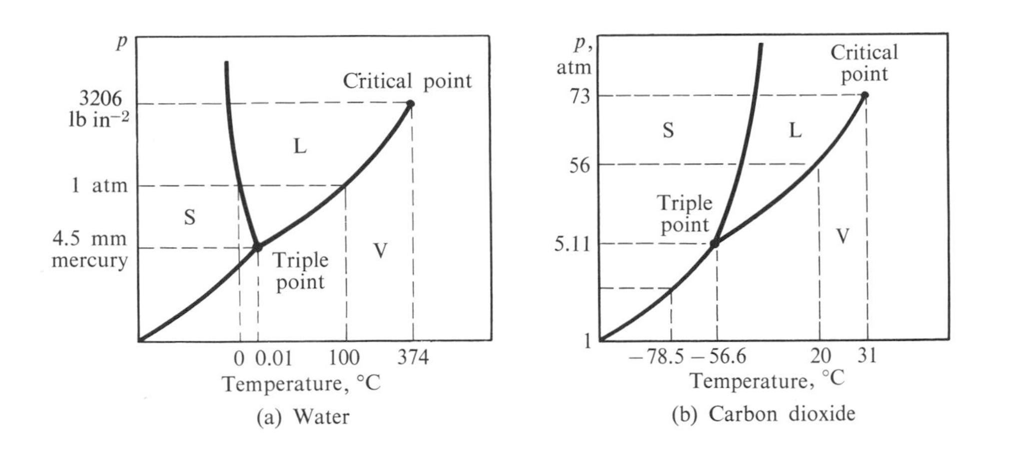

18-11 Pressure-temperature diagrams (not to a uniform scale).

boiling point; an increase of 1 atm lowers the freezing point of water by only about 0.007°C.

The lowering of the freezing point of water (or the melting point of ice) can be demonstrated by passing a loop of fine wire over a block of ice and hanging a weight of a few pounds from each end of the loop. Suppose that the main body of the ice is at 0°C and at atmospheric pressure. The temperature of the small amount of ice directly under the wire decreases until it achieves the melting point appropriate to the pressure under the wire. During this increase of pressure and decrease of temperature, a small amount of melting takes place. The water thus formed is squeezed out from under the wire and, coming to the top of the wire where the pressure is atmospheric, it refreezes and liberates heat which passes through the wire and serves to melt the next bit of ice below the wire.

The wire thus sinks farther and farther into the block, eventually cutting its way completely through, but leaving a solid block of ice behind it. This phenomenon is known as *regelation* (refreezing). Since heat is conducted from the top to the bottom of the wire while the wire is cutting through the ice, the greater the thermal conductivity of the wire, the faster will the wire cut through the ice. Even a perfectly conducting wire would not cut through the ice very rapidly, however, because of the very low thermal conductivity of the water film which is always present beneath the wire.

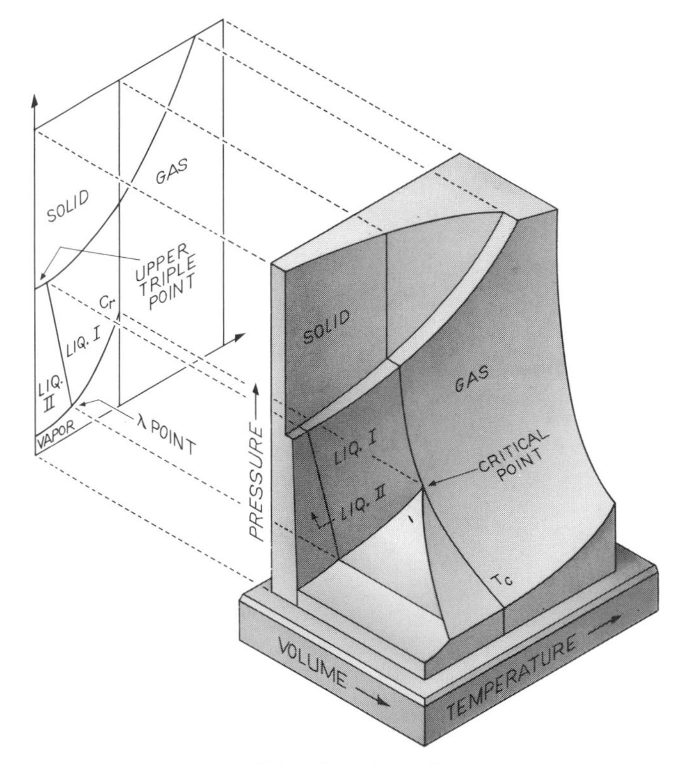

18-12 *pVT*-surface for helium.

For carbon dioxide, the triple-point temperature is −56.6°C and the corresponding pressure is 5.11 atm. Hence at atmospheric pressure CO_2 can exist only as a solid or a vapor. When heat is supplied to solid CO_2, open to the atmosphere, it changes directly to a vapor, without passing through the liquid phase, whence the name "dry ice." Liquid CO_2 can exist only at a pressure greater than 5.11 atm. The steel tanks in which CO_2 is commonly stored contain liquid and vapor (both saturated). The pressure in these tanks is the vapor pressure of CO_2 at the temperature of the tank. If the latter is 20°C the vapor pressure is 56 atm or about 830 lb in^{-2}.

The *pVT*-surface for ordinary helium (of mass number 4) is shown in Fig. 18-12. Two remarkable properties may be seen. (1) Helium has no triple point at which solid, liquid, and gas coexist in equilibrium. Instead, it has two triple points, the lower one (called the "lambda point") representing the temperature and pressure at which two liquids, I and II, coexist with vapor; and the upper one at which the two liquids coexist with solid. (2) As the temperature is lowered, helium does not solidify, but remains liquid all the way to absolute zero. To get solid helium, the pressure must be raised to a value above 25 atm, at which the helium atoms can attract one another and coalesce into a crystal lattice.

18-6 EFFECT OF DISSOLVED SUBSTANCES ON FREEZING AND BOILING POINTS

The freezing point of a liquid is lowered when some other substance is dissolved in the liquid. A common example is the use of an "antifreeze" to lower the freezing point of the water in the radiator of an automobile engine.

The freezing point of a saturated solution of common salt in water is about −20°C. To understand why a mixture of ice and salt may be used as a freezing mixture, let us make use of the definition of the freezing point as the only temperature at which the liquid and solid phases can exist in equilibrium. When a concentrated salt solution is cooled, it freezes at −20°C, and crystals of ice (pure H_2O) separate from the solution. In other words, ice crystals and a salt solution can exist in equilibrium only at −20°C, just as ice crystals and pure water can exist together only at 0°C.

When ice at 0°C is mixed with a salt solution at 20°C, some of the ice melts, abstracting its heat of fusion from the solution until the temperature falls to 0°C. But ice and salt solution cannot remain in equilibrium at 0°C, so that the ice continues to melt. Heat is now supplied by both the ice and the solution, and both cool down until the equilibrium temperature of −20°C is reached. If no heat is supplied from outside, the mixture remains unchanged at this temperature. If the mixture is brought into contact with a warmer body, say an ice-cream mixture at 20°C, heat flows from the ice-cream mixture to the cold salt solution, melting more of the ice but producing no rise in temperature as long as any ice remains. The flow of heat from the ice-cream mixture lowers its temperature to its freezing point (below 0°C, since it is itself a solution). Further loss of heat to the ice-salt mixture causes the ice cream to freeze.

The boiling point of a liquid is also affected by dissolved substances, but may be either increased or decreased. Thus the boiling point of a water-alcohol solution is *lower* than that of pure water, while the boiling point of a water-salt solution is *higher* than that of pure water.

18-7 HUMIDITY

Atmospheric air is a mixture of gases consisting of about 80% nitrogen, 18% oxygen, and small amounts of carbon dioxide, water vapor, and other gases. The mass of water vapor per unit volume is called the *absolute humidity*. The total pressure exerted by the atmosphere is the sum of the pressures exerted by its component gases. These pressures are called the *partial pressures* of the components. It is found that the partial pressure of each of the component gases of a gas mixture is very nearly the same as would be the actual pressure of that component alone if it occupied the same volume as does the mixture, a fact known as *Dalton's law*. That is, each of the gases of a gas mixture behaves independently of the others. The partial pressure of water vapor in the atmosphere is ordinarily equal to a few millimeters of mercury.

It should be evident that the partial pressure of water vapor at any given air temperature can never exceed the vapor pressure of water at that particular temperature. Thus at 10°C, from Table 18-2, the partial pressure cannot exceed 8.94 torr, or at 15°C it cannot exceed 12.67 torr. If the concentration of water vapor, or the absolute humidity, is such that the partial pressure equals the vapor pressure, the vapor is *saturated*. If the partial pressure is less than the vapor pressure, the vapor is *unsaturated*. The ratio of the partial pressure to the vapor pressure at the same temperature is called the *relative humidity*, and is usually expressed as a percentage:

Relative humidity (%)

$$= 100 \times \frac{\text{partial pressure of water vapor}}{\text{vapor pressure at same temperature}}.$$

The relative humidity is 100% if the vapor is saturated and zero if no water vapor at all is present.

Example. The partial pressure of water vapor in the atmosphere is 10 torr and the temperature is 20°C. Find the relative humidity.

From Table 18-2, the vapor pressure at 20°C is 17.5 torr. Hence,

$$\text{Relative humidity} = 10/17.5 \times 100 = 57\%.$$

Since the water vapor in the atmosphere is saturated when its partial pressure equals the vapor pressure at the air temperature, saturation can be brought about either by increasing the water vapor content or by lowering the temperature. For example, let the partial pressure of water vapor be 10 torr when the air temperature is 20°C, as in the preceding example. Saturation or 100% relative humidity could be attained either by introducing enough more water vapor (keeping the temperature constant) to increase the partial pressure to 17.5 torr, *or by lowering the temperature* to 11.4°C, at which, by interpolation from Table 18-2, the vapor pressure is 10 torr.

If the temperature were to be lowered *below* 11.4°C, the vapor pressure would be less than 10 torr. The partial pressure would then be higher than the vapor pressure and enough vapor would condense to reduce the partial pressure to the vapor pressure at the lower temperature. It is this process which brings about the formation of clouds, fog, and rain. The phenomenon is also of frequent occurrence at night when the earth's surface becomes cooled by radiation. The condensed moisture is called *dew*. If the partial pressure is so low that the temperature must fall below 0°C before saturation exists, the vapor condenses into ice crystals in the form of frost.

The temperature at which the water vapor in a given sample of air becomes saturated is called the *dew point*. Measuring the temperature of the dew point is the most accurate method of determining relative humidity. The usual method is to cool a metal container having a bright, polished surface, and to observe its temperature when the surface becomes clouded with condensed moisture. Suppose the dew point is observed in this way to be 10°C, when the air temperature is 20°C We then know that the water vapor in the air is saturated at 10°C; hence its partial pressure is 8.94 torr, equal to the vapor pressure at 10°C. The pressure necessary for saturation at 20°C is 17.5 torr. The relative humidity is therefore

$$\frac{8.94}{17.5} \times 100 = 51\%.$$

A simpler but less accurate method of determining relative humidity employs a *wet-* and *dry-bulb thermometer*. Two thermometers are placed side by side, the bulb of one being kept moist by a wick dipping in water. The lower the relative humidity, the more rapid will be the evaporation from the wet bulb, and the lower will be its temperature below that of the dry bulb. The relative humidity corresponding to any pair of wet- and dry-bulb temperatures is read from tables.

18-8 THE WILSON CLOUD CHAMBER AND THE BUBBLE CHAMBER

The Wilson cloud chamber is an apparatus for securing information about elementary particles such as electrons and α-particles. In principle (Fig. 18-13), it consists of a cylindrical enclosure having glass walls A and a glass top B, and provided with a movable piston C. The enclosed space contains air, water vapor, and sufficient excess water so that the vapor is saturated.

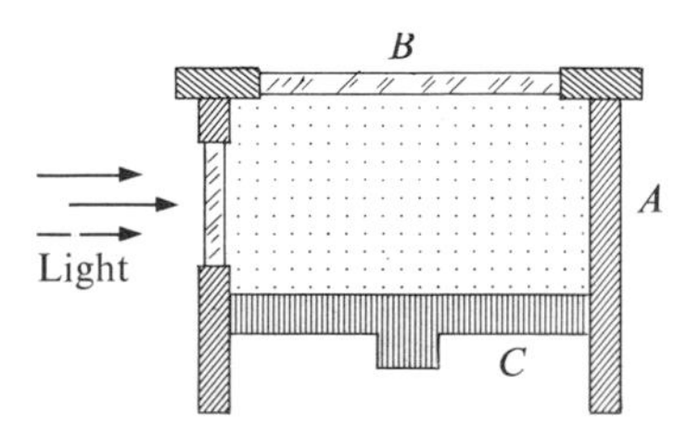

18-13 The Wilson cloud chamber.

(Other liquids such as alcohol are often used in place of water.) When the piston is suddenly pulled down a short distance, the mixture is lowered in temperature below the dew point. If the air is perfectly clean, the cooled vapor does not immediately condense, and it is said to be *supersaturated*. The point representing its state then lies *above* the *pVT*-surface, but this does not contradict a statement made earlier that all equilibrium states of a substance must lie on its *pVT*-surface, since the supersaturated state is not an equilibrium state. It is found that any ions which may be present serve as very efficient nuclei upon which the supersaturated vapor condenses to form liquid droplets. Hence if any ions were present just before the expansion, their presence is made evident by the appearance of tiny droplets immediately after the expansion.

Electrons, protons, and α-particles are all capable of traveling several centimeters through air, but as they collide with or pass near the air molecules they may knock off one or more of their electrons and hence leave behind them a trail of ions. Therefore if such a particle passed through the chamber just before the expansion, a trail of droplets appears after the expansion, indicating the path the particle followed. For photographing the tracks, an intense beam of light is projected transversely through the chamber and a camera mounted above it. Figure 8–11 is a photograph, made with a Wilson cloud chamber, of the paths of the particles resulting from the fission of a uranium nucleus.

The *bubble chamber*, a recently developed apparatus for studying ionizing particles, makes use of a *superheated liquid* instead of a supersaturated vapor. (A superheated liquid is a liquid at a temperature higher than that of its boiling point at the pressure to which it is subjected. It also is a nonequilibrium state.) When ions are produced in a superheated liquid it "boils" in the vicinity of the ion and forms a tiny bubble of vapor. The track of an ionizing particle through it is thus marked by a line of vapor bubbles rather than of liquid droplets. The advantage of the bubble chamber is that the molecules of a liquid are much closer together than those of a gas, so that there is a greater chance that a particle passing through the liquid will collide with a molecule and produce an ion.

Problems

(Assume all gases to be ideal.)

18–1 A tank contains 1.5 ft^3 of nitrogen at an absolute pressure of 20 lb in^{-2} and a temperature of 40°F. What will be the pressure if the volume is increased to 15 ft^3 and the temperature is raised to 440°F?

18–2 A tank having a capacity of 2 ft^3 is filled with oxygen which has a gauge pressure of 60 lb in^{-2} when the temperature is 47°C. At a later time it is found that because of a leak the gauge pressure has dropped to 50 lb in^{-2} and the temperature has decreased to 27°C. Find (a) the mass of the oxygen in the tank under the first set of conditions, (b) the amount of oxygen that has leaked out.

18–3 A flask of volume 2 ℓ, provided with a stopcock, contains oxygen at 300°K and atmospheric pressure. The system is heated to a temperature of 400°K, with the stopcock open to the atmosphere. The stopcock is then closed and the flask cooled to its original temperature. (a) What is the final pressure of the oxygen in the flask? (b) How many grams of oxygen remain in the flask?

18–4 A barrage balloon whose volume is 20,000 ft^3 is to be filled with hydrogen at atmospheric pressure. (a) If the hydrogen is stored in cylinders of volume 2 ft^3 at an absolute pressure of 200 lb in^{-2}, how many cylinders are required? (b) What is the total weight that can be supported by the balloon, in air at standard conditions? (c) What weight could be supported if the balloon were filled with helium instead of hydrogen?

18–5 Derive from the equation of state of an ideal gas an equation for the density of an ideal gas in terms of pressure, temperature, and appropriate constants.

18–6 At the beginning of the compression stroke, the cylinder of a diesel engine contains 48 in^3 of air at atmospheric pressure and a temperature of 27°C. At the end of the stroke, the air has been compressed to a volume of 3 in^3 and the gauge pressure has increased to 600 lb in^{-2}. Compute the temperature.

18–7 A bubble of air rises from the bottom of a lake, where the pressure is 3.03 atm, to the surface, where the pressure is 1 atm. The temperature at the bottom of the lake is 7°C and the temperature at the surface is 27°C. What is the ratio of the size (i.e., the volume) of the bubble as it reaches the surface to the size of the bubble at the bottom?

18–8 A liter of helium under a pressure of 2 atm and at a temperature of 27°C is heated until both pressure and volume are doubled. (a) What is the final temperature? (b) How many grams of helium are there?

18–9 A flask contains 1 g of oxygen at an absolute pressure of 10 atm and at a temperature of 47°C. At a later time it is found that because of a leak the pressure has dropped to $\frac{5}{8}$ of its original value and the temperature has decreased to 27°C. (a) What is the volume of the flask? (b) How many grams of oxygen leaked out between the two observations?

18–10 The submarine Squalus sank at a point where the depth of water was 240 ft. The temperature at the surface is 27°C and at the bottom it is 7°C. The density of sea water may be taken as 2 slugs ft^{-3}. (a) If a diving bell in the form of a circular cylinder 8 ft high, open at the bottom and closed at the top, is lowered to this depth, to what height will the water rise within it when it reaches the bottom? (b) At what gauge pressure must compressed air be supplied to the bell while on the bottom to expel all the water from it?

18–11 A bicycle pump is full of air at an absolute pressure of 15 lb in^{-2}. The length of stroke of the pump is 18 in. At what part of the stroke does air begin to enter a tire in which the gauge pressure is 40 lb in^{-2}? Assume the compression to be isothermal.

18–12 A vertical cylindrical tank 1 m high has its top end closed by a tightly fitting frictionless piston of negligible weight. The air inside the cylinder is at an absolute pressure of 1 atm. The piston is depressed by pouring mercury on it slowly. How far will the piston descend before mercury spills over the top of the cylinder? The temperature of the air is maintained constant.

18–13 A barometer is made of a tube 90 cm long and of cross section 1.5 cm^2. Mercury stands in this tube to a height of 75 cm. The room temperature is 27°C. A small amount of nitrogen is introduced into the evacuated space above the mercury and the column drops to a height of 70 cm. How many grams of nitrogen were introduced?

18–14 A large tank of water has a hose connected to it, as shown in Fig. 18–14. The tank is sealed at the top and has compressed air between the water surface and the top. When the water height h_2 is 10 ft, the gauge pressure p_1 is 15 lb in^{-2}. Assume that the air above the water surface expands isothermally and that water weighs 64 lb ft^{-3}. (a) What is the velocity of flow out of the hose when $h_2 = 10$ ft? (b) What is the velocity of flow of the hose when h_2 has decreased to 8 ft? Neglect friction.

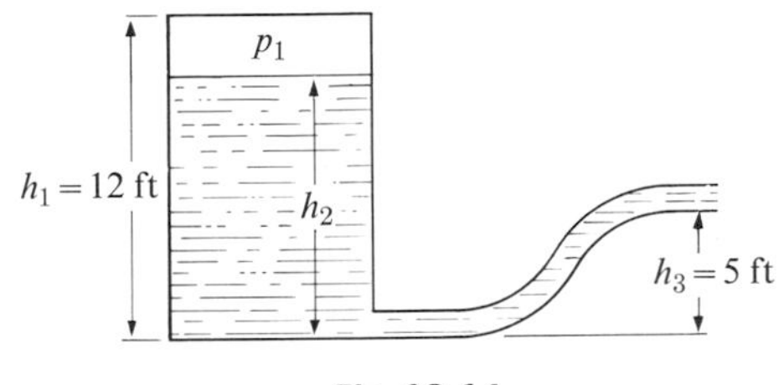

Fig. 18–14

18–15 The volume of an ideal gas is 4 ℓ, the pressure is 2 atm, and the temperature is 300°K. The gas first expands at constant pressure to twice its original volume; it is then compressed isothermally to its original volume, and finally cooled at constant volume to its original pressure. (a) Show the process in a pV-diagram. (b) Compute the temperature during the isothermal compression. (c) Compute the maximum pressure.

18–16 At the beginning of the compression stroke, the cylinder of a diesel engine contains 48 in^3 of air at atmospheric pressure and a temperature of 27°C. At the end of the stroke, the air has been compressed to a volume of 3 in^3, and the gauge pressure has increased to 600 lb in^{-2}. Compute the temperature, (a) in degrees Kelvin, (b) degrees Celsius, (c) degrees Fahrenheit.

18–17 In the lower part of the atmosphere (the troposphere) the temperature is not uniform but decreases with increasing elevation. Show that if the temperature variation is approximated by the linear relation

$$T = T_0 - \alpha y,$$

where T_0 is the temperature at the earth's surface and T is the temperature at a height y, the pressure is given by

$$\ln \frac{p_0}{p} = \frac{Mg}{R\alpha} \ln \frac{T_0}{T_0 - \alpha y}.$$

The coefficient α is called the "lapse rate of temperature." While it varies with atmospheric conditions, an average value is about 0.6 C°/100 m.

18–18 Construct two graphs for a real substance, one showing pressure as a function of volume, and the other showing pressure as a function of temperature. Show on each graph the region in which the substance exists as (a) a gas or vapor, (b) a liquid, (c) a solid. Show also the triple point and the critical point.

18–19 A small amount of liquid is introduced into a glass tube, all air is removed, and the tube is sealed off. Describe the behavior of the meniscus when the temperature of the system is raised: (a) If the volume of the tube is much greater than the critical volume. (b) If the volume of the tube is much less than the critical volume. (c) If the volume of the tube is only slightly different from the critical volume.

18–20 A piece of ice at 0°C is placed alongside a beaker of water at 0°C in a glass vessel, from which all air has been removed. If the ice, water, and vessel are all maintained at a temperature of 0°C by a suitable thermostat, describe the final equilibrium state inside the vessel.

18–21 A mass m of pure substance is placed in a tube of constant volume V. If there are two phases present, show that (a) the volume occupied by one phase will be

$$V_1 = \frac{m - \rho_2 V_2}{\rho_1 - \rho_2},$$

where ρ_1, V_1, and ρ_2, V_2 are the densities and volumes, respectively, of the two phases; (b) the condition that V_1 will not change as the temperature is increased is given by

$$\frac{V_1}{V_2} = -\frac{d\rho_2/dT}{d\rho_1/dT}.$$

18–22 (a) What is the relative humidity on a day when the temperature is 68°F and the dew point is 41°F? (b) What is the partial pressure of water vapor in the atmosphere? (c) What is the absolute humidity, in grams per cubic meter?

18–23 The temperature in a room is 40°C. A can is gradually cooled by adding cold water. At 10°C the surface of the can clouds over. What is the relative humidity in the room?

18–24 A pan of water is placed in a sealed room of volume 60 m^3 and at a temperature of 27°C. (a) What is the absolute humidity in g m^{-3} after equilibrium has been reached? (b) If the temperature of the room is then increased 1 C°, how many more grams of water will evaporate?

18–25 (a) What is the dew-point temperature on a day when the air temperature is 20°C and the relative humidity is 60%? (b) What is the absolute humidity, expressed in grams per cubic meter?

18–26 The volume of a closed room, kept at a constant temperature of 20°C, is 60 m^3. The relative humidity in the room is 10%. If a pan of water is brought into the room, how many grams will evaporate?

18–27 An air-conditioning system is required to increase the relative humidity of 10 ft^3 of air per second from 30% to 65%. The air temperature is 68°F. How many pounds of water are needed by the system per hour?

19 The Laws of Thermodynamics

19-1 WORK IN THERMODYNAMICS

In the study of the equilibrium or the acceleration of a mechanical system, it is necessary to calculate the resultant force F acting *on* the system, because it is this force which when used with Newton's second law provides the means for calculating the acceleration, or when set equal to zero provides one of the conditions of equilibrium. The product of the system's displacement ds and the component of F in the direction of the displacement, F_S, is the *work of the force*. The sign convention of mechanics provides that the work is positive when F_S and ds have the same sign. When this is the case, the energy of the system increases, and the expression "work is done *on* the system" is often used. In the reverse situation, where F_S and ds have opposite sign, the energy of the system decreases, and this is referred to in the words, "work is done *by* the system."

We shall be concerned in this chapter with the principles underlying the subject of *thermodynamics*, a branch of physics that owes its origin to the study of engines whose purpose is to do work. A gasoline engine, for example, is designed to propel an automobile; a steam turbine is often used to turn the rotor of an electric generator. As a result of the study of such devices, all of which have as their goal the performance of work *by* the system, it became the custom many years ago in thermodynamics to refer to such work as positive.

It is a simple matter to bring the mechanical sign convention for work into line with the thermodynamic point of view. All we have to do is to focus our attention on the *resultant force exerted by a system*. If *this* force has the same direction as the displacement, the product of force and displacement is positive and (1) work is done *by* the system, (2) the energy of the system decreases provided there is no other energy transfer.

To summarize:

a) Calculate the force F exerted *by* the system.
b) Calculate the displacement ds of the point of application of F.
c) If F and ds have the same sign, the work is positive.
d) Positive work is said to be done *by* the system.
e) When positive work is done, the energy of the system decreases, unless there is some other energy transfer.

19-2 WORK IN CHANGING THE VOLUME

Consider a solid or fluid contained in a cylinder equipped with a movable piston, as shown in Fig. 19-1(a). Suppose that the cylinder has a cross-sectional area A and that

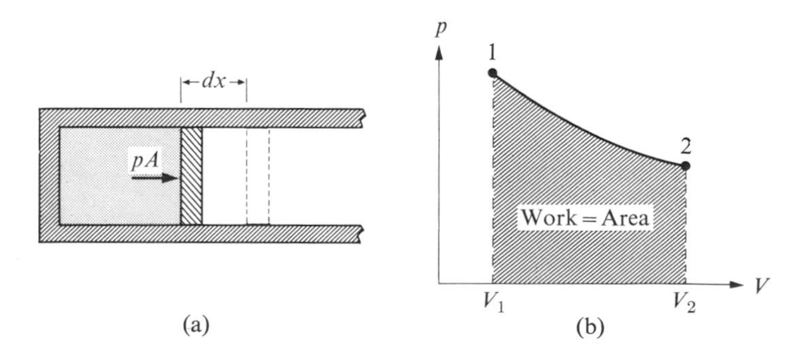

19-1 (a) Force exerted *by* a system during a small expansion. (b) Work is the area under the curve on a pV-diagram.

the pressure exerted by the system at the piston face is p. The force exerted *by* the system is therefore pA. If the piston moves out an infinitesimal distance dx, the work dW of this force is equal to

$$dW = pA\,dx.$$

But

$$A\,dx = dV,$$

where dV is the change of volume of the system. Therefore

$$dW = p\,dV, \tag{19-1}$$

and in a finite change of volume from V_1 to V_2,

$$\boxed{W = \int_{V_1}^{V_2} p\,dV.} \tag{19-2}$$

This integral may be interpreted graphically as the area under a curve on a pV-diagram, as shown in Fig. 19-1(b). If the substance *expands* from 1 to 2 in Fig. 19-1(b), the work and the area are positive. The *compression* from 2 to 1 gives rise to negative work and a negative area.

If the pressure remains constant while the volume changes, then the work is

$$W = p(V_2 - V_1) \qquad \text{(constant pressure only).}$$

Example. An ideal gas is kept in good thermal contact with a very large body of constant temperature and undergoes an *isothermal expansion* in which its volume changes from V_1 to V_2. How much work is done?

From Eq. (19-2)

$$W = \int_{V_1}^{V_2} p\,dV.$$

For an ideal gas

$$p = \frac{nRT}{V}.$$

Since n, R, and T are constant,

$$W = nRT \int_{V_1}^{V_2} \frac{dV}{V} = nRT \ln \frac{V_2}{V_1}.$$

In an expansion, $V_2 > V_1$ and W is positive. At constant T,

$$p_1V_1 = p_2V_2, \qquad \text{or} \qquad \frac{V_2}{V_1} = \frac{p_1}{p_2},$$

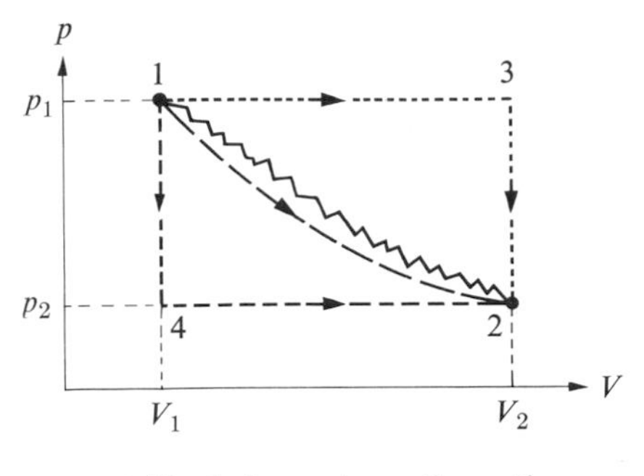

19-2 Work depends on the path.

and the isothermal work may be expressed also in the form

$$W = nRT \ln \frac{p_1}{p_2}.$$

On the pV-diagram in Fig. 19-2 an initial state 1 (characterized by pressure p_1 and volume V_1) and a final state 2 (characterized by pressure p_2 and volume V_2) are represented by the two points 1 and 2. There are many ways in which the system may be taken from 1 to 2. For example, the pressure may be kept constant from 1 to 3 (*isobaric* process) and then the volume kept constant from 3 to 2 (*isochoric* process), in which case the work is equal to the area under the line $1 \rightarrow 3$. Another possibility is the path $1 \rightarrow 4 \rightarrow 2$, in which case the work is the area under the line $4 \rightarrow 2$. The jagged line and the continuous curve from 1 to 2 represent other possibilities, in each of which the work is different. We can see, therefore, that *the work depends not only on the initial and final states but also on the intermediate states, i.e., on the path.*

19-3 HEAT DEPENDS ON THE PATH

As we have seen in Chapter 16, heat is the energy transferred to or from a system by virtue of a temperature difference between the system and its surroundings. Heat is regarded as positive when it enters a system and negative when it leaves. The performance of work and the transfer of heat are *methods of energy transfer*, that is, methods whereby the energy of a system may be increased or decreased.

Consider a small amount of gas as in Fig. 19-3(a), contained in a cylinder and confined by a piston to a

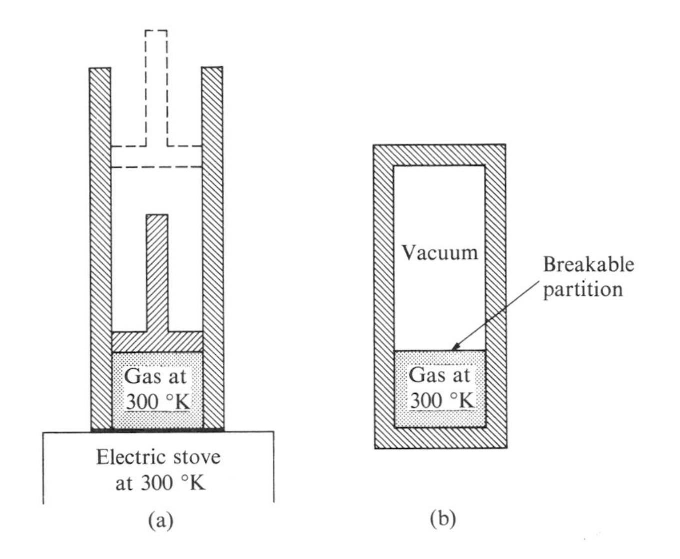

19-3 (a) Slow, controlled, isothermal expansion of a gas from an initial state 1 to a final state 2. (b) Rapid, uncontrolled expansion of the same gas starting at the same state 1 and ending at the same state 2.

volume of, say, two liters, and maintained by an electric stove at a temperature of 300°K. If the pressure exerted on the piston is only a trifle smaller than the gas pressure, the gas will expand slowly and heat will flow from the stove to the gas, thereby maintaining the temperature of the gas at the constant value 300°K. Suppose that the gas expands in this slow, controlled, isothermal manner until its volume becomes, say, seven liters. A finite amount of heat is absorbed by the gas during this process.

In Fig. 19-3(b), there is shown a vessel surrounded by adiabatic walls and divided into two compartments (the lower one of volume two liters and the upper one with a volume of five liters) by a thin, breakable partition. In the lower compartment the same gas has the same initial volume and temperature as in (a). Suppose the partition is broken and the gas in (b) undergoes a rapid, uncontrolled expansion into the vacuum above where it encounters no movable piston and therefore exerts no external force. Not only is no work done in this expansion but also, no heat passes through the adiabatic walls. The final volume is seven liters as in (a). The rapid, uncontrolled expansion of a gas into a vacuum is called a *free expansion* and will be discussed later in this chapter. Experiments have indicated that very little temperature change takes place, provided the initial gas pressure is less than a few atmospheres, and therefore the final state of the gas is also the same as the final state in (a). But the states traversed by the gas (the pressures and volumes) while proceeding from state 1 to state 2 are entirely different, so that (a) and (b) represent two *different paths* connecting the *same states* 1 and 2. During path (a) heat is transferred; during path (b) no heat is transferred. *Heat*, like work, *depends not only on the initial and final states but also on the path.*

It would be just as incorrect to refer to the "heat in a body" as it would be to speak about the "work in a body." For suppose we assigned an arbitrary value to "the heat in a body" in some standard reference state. The "heat in the body" in some other state would then equal the "heat" in the reference state plus the heat added when the body is carried to the second state. But the heat added depends entirely on the path by which we go from one state to the other, and since there are an infinite number of paths which might be followed, there are an infinite number of values which might equally well be assigned to the "heat in the body" in the second state. Since it is not possible to assign any one value to the "heat in the body" we conclude that this concept is meaningless.

19-4 THE FIRST LAW OF THERMODYNAMICS

The transfer of heat and the performance of work constitute two methods of adding energy to or subtracting energy from a system. Once the transfer of energy is over, the system is said to have undergone a change in *internal energy.*

Suppose a system is caused to change from state 1 to state 2 along a definite path and that the heat absorbed, Q, and the work W are measured. Expressing both Q and W either in thermal units or in mechanical units, we may then calculate the difference $Q - W$. If now we do the same thing over again for many different paths (between the same states 1 and 2), the important result is obtained that *$Q - W$ is the same for all paths connecting* 1 *and* 2. But Q is the energy that has been added to a system by the transfer of heat and W is equal to the energy that has been extracted from the system by the performance of work. The difference $Q - W$, therefore, must represent the internal energy change of the system. It follows that *the internal energy change of a system is independent of the path*, and is therefore equal to the energy of the system in state 2 minus the energy in state 1, or $U_2 - U_1$:

$$U_2 - U_1 = Q - W. \tag{19-3}$$

If some arbitrary value is assigned to the internal energy in some standard reference state, its value in any other state is uniquely defined, since $Q - W$ is the same for all processes connecting the states. Equation (19-3) is known as *the first law of thermodynamics*. In applying the law in this form it must be remembered that (1) all quantities must be expressed in the same units, (2) Q is positive when the heat goes *into* the system, (3) W is positive when the force exerted *by* the system and the displacement have the same sign.

From the thermodynamic standpoint, it is entirely unnecessary to interpret internal energy in terms of molecular energies. Equation (19-3) is the *definition* of the internal energy of a system or, more precisely, of the *change* in its internal energy in any process. As with other forms of energy, only *differences* in internal energy are defined, and not absolute values.

If a system is carried through a process which eventually returns it to its initial state (a *cyclic* process), then

$$U_2 = U_1 \quad \text{and} \quad Q = W.$$

Thus, although net work W may be done by the system in the process, energy has not been created, since an equal amount of energy must have flowed into the system as heat Q.

An *isolated* system is one which does no external work and into which there is no flow of heat. Then for any process taking place in such a system, $W = Q = 0$ and $U_2 - U_1 = 0$ or $\Delta U = 0$. That is, *the internal energy of an isolated system remains constant*. This is the most general statement of the *principle of conservation of energy*. The internal energy of an *isolated* system cannot be changed by any process (mechanical, electrical, chemical, nuclear, or biological) taking place *within* the system. The energy of a system can be changed only by a flow of heat across its boundary, or by the performance of work. (If either of these takes place, the system is no longer isolated.) The increase in energy of the system is then equal to the energy flowing in as heat, minus the energy flowing out as work.

19-5 ADIABATIC PROCESS

A process that takes place in such a manner that no heat enters or leaves a system is called an *adiabatic process*. This may be accomplished either by surrounding the system with a thick layer of heat-insulating material (such as cork, asbestos, firebrick, or any light, porous powder) or by performing the process quickly. The flow of heat is a fairly slow process, so that any process performed quickly enough will be practically adiabatic. Applying the first law to an adiabatic process, we get

$$U_2 - U_1 = -W \quad \text{(adiabatic process).}$$

Thus the change in the internal energy of a system, in an adiabatic process, is equal in absolute magnitude to the work. If the work W is negative, as when a system is compressed, then $-W$ is positive, U_2 is greater than U_1, and the internal energy of the system increases. If W is positive, as when a system expands, the internal energy of the system decreases. An increase of internal energy is usually accompanied by a rise in temperature, and a decrease in internal energy by a temperature drop.

The compression of the mixture of gasoline vapor and air that takes place during the compression stroke of a gasoline engine is an example of an approximately adiabatic process involving a temperature rise. The expansion of the combustion products during the power stroke of the engine is an approximately adiabatic process involving a temperature decrease. Adiabatic processes, therefore, play a very important role in mechanical engineering.

19-6 ISOCHORIC PROCESS

If a substance undergoes a process in which the volume remains unchanged, the process is called *isochoric*. The rise of pressure and temperature produced by a flow of heat into a substance contained in a nonexpanding chamber is an example of an isochoric process. If the volume does not change, no work is done and, therefore, from the first law,

$$U_2 - U_1 = Q \qquad \text{(isochoric process)},$$

or all the heat that has been added has served to increase the internal energy. The very sudden increase of temperature and pressure accompanying the explosion of gasoline vapor and air in a gasoline engine may be treated mathematically as though it were an isochoric addition of heat.

19-7 ISOTHERMAL PROCESS

An *isothermal* process takes place at constant temperature. For the temperature of a system to remain strictly constant, the changes in the other coordinates must be carried out slowly, and heat must be transferred. In general none of the quantities Q, W, or $U_2 - U_1$ is zero. There are two ideal materials whose *internal energy depends only on the temperature.* These are an *ideal gas* and an *ideal paramagnetic crystal.* When these two substances undergo an isothermal process, the internal energy does not change, and therefore

$$Q = W \qquad \text{(isothermal process)}.$$

19-8 ISOBARIC PROCESS

A process taking place at constant pressure is called an *isobaric process.* When water enters the boiler of a steam engine and is heated to its boiling point, vaporized, and then the steam is superheated, all these processes take place isobarically. Such processes play an important role in mechanical engineering and also in chemistry.

Consider the change of phase of a mass m of liquid to vapor at constant pressure and temperature. If V_L is the volume of liquid and V_V the volume of vapor, the work done in expanding from V_L to V_V at constant pressure p is

$$W = p(V_V - V_L).$$

The heat absorbed by each unit of mass is the heat of vaporization L. Hence

$$Q = mL.$$

From the first law,

$$U_V - U_L = mL - p(V_V - V_L). \qquad (19\text{-}4)$$

Example. One gram of water (1 cm^3) becomes 1671 cm^3 of steam when boiled at a pressure of 1 atm. The heat of vaporization at this pressure is 539 cal g^{-1}. Compute the external work and the increase in internal energy.

$$W = p(V_V - V_L) = 1.013 \times 10^6 \text{ dyn cm}^{-2}\,(1671 - 1)\text{ cm}^3$$
$$= 1.695 \times 10^9 \text{ ergs} = 169.5 \text{ J} = 41 \text{ cal}.$$

From Eq. (19-4),

$$U_V - U_L = mL - W = 539 - 41 = 498 \text{ cal}.$$

Hence the external work, or the external part of the heat of vaporization, equals 41 cal, and the increase in internal energy, or the internal part of the heat of vaporization, is 498 cal.

19-9 THROTTLING PROCESS

A *throttling process* is one in which a fluid, originally at a constant high pressure, seeps through a porous wall or a narrow opening (needle valve or throttling valve) into a region of constant lower pressure without a transfer of heat taking place. The experiment is sometimes called the *porous plug experiment*. Figure 19-4(a) will help to make the process clear. A fluid is discharged from a pump at a high pressure, then passes through a throttling valve into a pipe which leads directly to the intake or low-pressure side of the pump. Every successive element of fluid undergoes the throttling process in a continuous stream.

Consider any element of fluid enclosed between the piston and throttling valve of Fig. 19-4(b). Suppose this piston to move toward the right and another piston on the other side of the valve to move to the right also at such rates that the pressure on the left remains at a constant high value and that on the right at a constant lower value. After all the fluid has been forced through the valve, the final state is that of Fig. 19-4(c).

The net work done in this process is the difference between the work done in forcing the right-hand piston out and the work done in forcing the left-hand piston in. Let

p_1 = high pressure (on the left),

V_1 = volume of fluid at the high pressure,

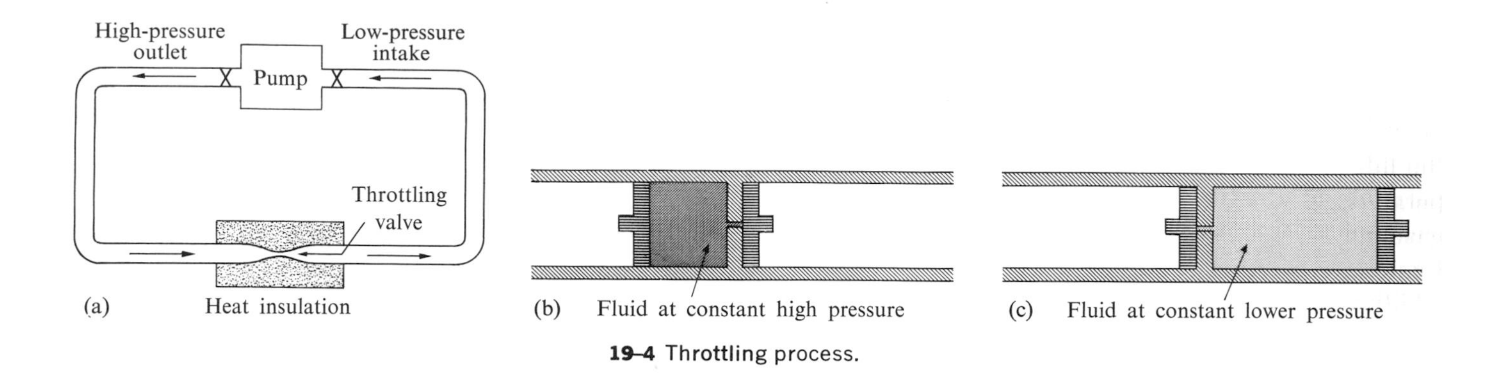

19–4 Throttling process.

p_2 = lower pressure (on the right),

V_2 = volume of fluid at the low pressure.

Since the low-pressure fluid changes in volume from zero to V_2 at the constant pressure p_2, the work is

$$p_2(V_2 - 0),$$

and since the high-pressure fluid changes in volume from V_1 to zero at the constant high pressure p_1, the work is

$$p_1(0 - V_1).$$

The net work W is therefore

$$W = p_2V_2 - p_1V_1.$$

Since the process is adiabatic, $Q = 0$, and hence from the first law,

$$U_2 - U_1 = 0 - (p_2V_2 - p_1V_1)$$

or

$$U_2 + p_2V_2 = U_1 + p_1V_1. \tag{19-5}$$

This result is of great importance in steam engineering and in refrigeration. The sum $U + pV$, called the *enthalpy*, is tabulated for steam and for many refrigerants. The throttling process plays the main role in the action of a refrigerator, since this is the process that gives rise to the drop in temperature needed for refrigeration. Liquids that are about to evaporate (saturated liquids) always undergo a drop in temperature and partial vaporization as a result of a throttling process. Gases, however, may undergo either a temperature rise or drop, depending on the initial temperature and pressure and on the final pressure.

19–10 DIFFERENTIAL FORM OF THE FIRST LAW

Up to this point we have used the first law of thermodynamics only in its finite form,

$$U_2 - U_1 = Q - W.$$

In this form the equation applies to a process in which states 1 and 2 differ in pressure, volume, and temperature by a finite amount. Suppose states 1 and 2 differ only slightly. Then if only a small amount of heat dQ is transferred, and only a small amount of work dW is done, the energy change dU is also very small. In these circumstances, the first law becomes

$$\boxed{dU = dQ - dW.} \tag{19-6}$$

If the system is of such a character that the only work possible is by means of expansion or compression, then $dW = p\,dV$, and

$$\boxed{dU = dQ - p\,dV} \tag{19-7}$$

is the *differential form of the first law*, applicable to solids, liquids, and gases.

19–11 INTERNAL ENERGY OF A GAS

Imagine a thermally insulated vessel with rigid walls, divided into two compartments by a partition. Suppose that there is a gas in one compartment and that the other is empty. If the partition is removed, the gas will undergo what is known as a *free expansion* in which no work is done and no heat is transferred. From the first law, since both Q and W are zero, it follows that *the internal energy remains unchanged during a free expansion.* The question as to whether or not the temperature of a gas changes during a free expansion and, if it does, the magnitude of the temperature change, has engaged the attention of physicists for over a hundred years. Starting with Joule in 1843, many attempts have been made to measure the effect of a free expansion, or, as it is often called, the *Joule effect.*

The reason for noting whether there is a temperature change when a gas undergoes a free expansion is to learn the properties on which the internal energy of a gas depends. For if the temperature should change while the internal energy stays the same, one would have to conclude that the internal energy depends on *both* the temperature and the volume, or both the temperature and the pressure, but certainly not on the temperature alone. If, on the other hand, T remains unchanged during a free expansion in which we know U remains unchanged, then the only conclusion that is admissible is that *U is a function of T only.*

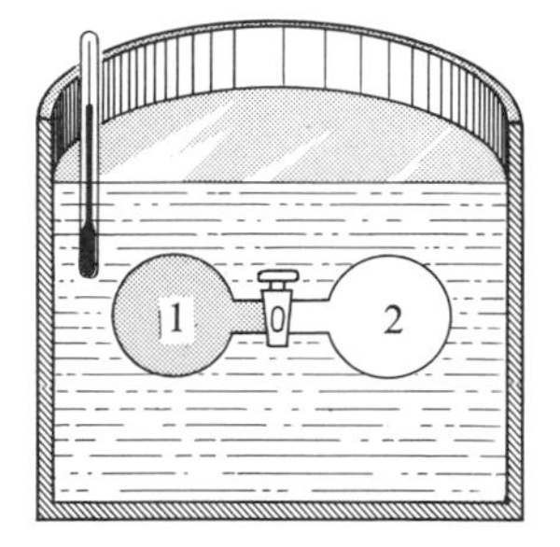

19–5 Free expansion of a gas into an evacuated container.

In the original experiment of Joule, two vessels connected by a short tube and stopcock were immersed in a water bath, as shown in Fig. 19–5. One vessel contained air at high pressure, and the other was evacuated. The temperature of the water was measured before and after the expansion, the idea being to infer the temperature change of the gas from the temperature change of the water. Since the heat capacity of the vessels and the water was approximately 1000 times as large as the heat capacity of the air, Joule was unable to detect any temperature change of the water, although, in the light of our present knowledge, the air must have suffered a temperature change of several degrees.

Experiments performed in recent years have proved that the internal energy of a *real* gas *does* depend on the pressure or volume as well as on the temperature. The volume or pressure dependence, however, is very much smaller than the temperature dependence, so we extend our definition of an *ideal* gas, and say that *the internal energy of an ideal gas depends only on its temperature.*

19–12 HEAT CAPACITIES OF AN IDEAL GAS

The temperature of a substance may be raised under a variety of conditions. The volume may be kept constant, or the pressure may be kept constant, or both may be allowed to vary in some definite manner. In each of these cases, the amount of heat per mole necessary to cause unit rise of temperature is different. In other words, a substance has many different molar heat capacities. Only two, however, are of practical use, namely, those at constant volume and at constant pressure. There is a simple and important relation between these two molar heat capacities of an ideal gas.

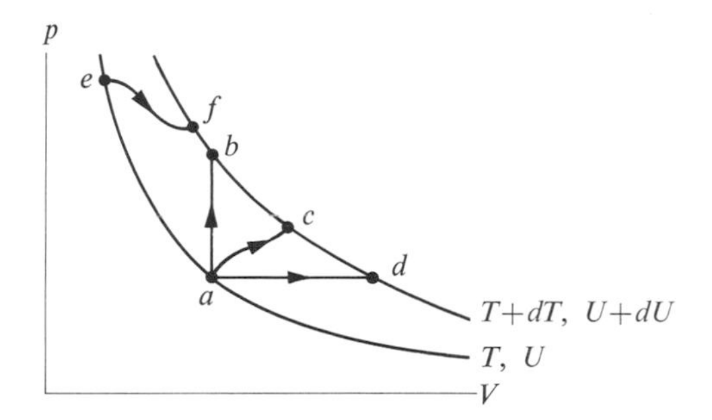

19–6 The change in internal energy of an ideal gas is the same in all processes between two given temperatures.

Figure 19–6 shows two isotherms of an ideal gas at temperatures T and $T + dT$. Since the internal energy of an ideal gas depends only on the temperature, it is constant if the temperature is constant and the isotherms are also curves of *constant internal energy.* The internal energy therefore has a constant value U at all points of the isotherm at temperature T, and a constant value $U + dU$ at all points of the isotherm at $T + dT$. It follows that the *change* in internal energy, dU, is the same in all processes in which the gas is taken from *any* point on one isotherm to *any* point on the other. Thus dU is the same for all the processes *ab*, *ac*, *ad*, and *ef*, in Fig. 19–6.

Consider first the process *ab*, at constant volume. To carry out such a process, the gas at temperature T is enclosed in a rigid container and brought in contact with a body at a slightly higher temperature $T + dT$. There will be a flow of heat dQ into the gas, and by definition of the molar heat capacity at constant volume, C_v,

$$dQ = nC_v\,dT. \qquad (19\text{–}8)$$

The pressure of the gas increases during this process, but no work is done, since the volume is constant. Hence, from the first law (in differential form),

$$dU = dQ - dW,$$

we have

$$dU = nC_v\,dT. \qquad (19\text{–}9)$$

But dU is the same for *all* processes between the isotherms in Fig. 19–6, so Eq. (19–9) gives the change in internal energy in all the processes in Fig. 19–6, *even if they are not at constant volume.*

Now consider the process *ad* in Fig. 19–6, at constant pressure. To carry out such a process the gas might be enclosed in a cylinder with a piston, acted on by a constant external pressure, and brought in contact with a body at a temperature $T + dT$. As heat flows into the gas it expands at constant pressure and does work. By definition of the molar heat capacity at constant pressure, C_p, the heat dQ flowing into the gas is

$$dQ = nC_p\,dT.$$

The work dW is

$$dW = p\,dV.$$

But from the equation of state, since p is constant,

$$pV = nRT, \qquad p\,dV = nR\,dT,$$

so

$$dW = nR\,dT.$$

Then from the first law,

$$dU = dQ - dW,$$
$$nC_v\,dT = nC_p\,dT - nR\,dT,$$

or

$$\boxed{C_p - C_v = R.} \qquad (19\text{–}10)$$

The molar heat capacity of an ideal gas at constant pressure is therefore greater than that at constant volume by the universal gas constant R. Of course, R must be expressed in the same units as C_p and C_v, usually cal mole^{-1} (C°)$^{-1}$. We have shown in Section 18–2 that in these units $R = 1.99$ cal mole^{-1} (C°)$^{-1}$.

Although Eq. (19–10) was derived for an ideal gas, it is very nearly true for real gases at moderate pressures. Measured values of C_p and C_v are given in Table 19–1 for some real gases at low pressures, and the difference is seen to be very nearly 1.99 cal mole^{-1} (C°)$^{-1}$.

TABLE 19–1 MOLAR HEAT CAPACITIES OF GASES AT LOW PRESSURE

Type of gas	Gas	C_p, cal mole^{-1} (C°)$^{-1}$	C_v, cal mole^{-1} (C°)$^{-1}$	$C_p - C_v$	$\gamma = \frac{C_p}{C_v}$
Monatomic	He	4.97	2.98	1.99	1.67
	A	4.97	2.98	1.99	1.67
Diatomic	H_2	6.87	4.88	1.99	1.41
	N_2	6.95	4.96	1.99	1.40
	O_2	7.03	5.04	1.99	1.40
	CO	6.97	4.98	1.99	1.40
Polyatomic	CO_2	8.83	6.80	2.03	1.30
	SO_2	9.65	7.50	2.15	1.29
	H_2S	8.27	6.2	2.1	1.34

In the last column of Table 19–1 are listed the values of the ratio C_p/C_v, denoted by the Greek letter γ (gamma). It is seen that γ is 1.67 for monatomic gases, and is very nearly 1.40 for the so-called permanent diatomic gases. There is no simple regularity for polyatomic gases.

Solids and liquids also expand when heated, if free to do so, and hence perform work. The coefficients of volume expansion of solids and liquids are, however, so much smaller than those of gases that the work is small. The internal energy of a solid or liquid *does* depend on its volume as well as on its temperature, and this must be considered when evaluating the difference between specific heats of solids or liquids. It turns out that here also $C_p > C_v$, but the difference is small and is not expressible as simply as that for a gas. Because of the large stresses set up when solids or liquids are heated and *not* allowed to expand, most heating processes involving them take place at constant pressure, and hence C_p is the quantity usually measured for a solid or liquid.

19–13 ADIABATIC PROCESS OF AN IDEAL GAS

Any process in which there is no flow of heat into or out of a system is called *adiabatic*. To perform a truly adiabatic process it would be necessary that the system be surrounded by a perfect heat insulator, or that the surroundings of the system be kept always at the same temperature as the system. However, if a process such as compression or expansion of a gas is carried out very rapidly, it will be nearly adiabatic, since the flow of heat into or out of the system is slow even under favorable conditions. Thus the compression stroke of a gasoline or diesel engine is approximately adiabatic.

Note that external work may be done on or by a system in an adiabatic process, and that the temperature usually changes in such a process.

Let an ideal gas undergo an infinitesimal adiabatic process. Then $dQ = 0$, $dU = nC_v\,dT$, $dW = p\,dV$, and, from the first law,

$$nC_v\,dT = -p\,dV. \tag{19–11}$$

From the equation of state,

$$p\,dV + V\,dp = nR\,dT.$$

Eliminating dT between these equations and making use of the fact that $C_p - C_v = R$, we obtain the relation

$$\frac{dp}{p} + \frac{C_p}{C_v}\frac{dV}{V} = 0,$$

or, if C_p/C_v is denoted by γ,

$$\frac{dp}{p} + \gamma\frac{dV}{V} = 0.$$

To obtain the relation between p and V in a *finite* adiabatic change we may integrate the preceding equation. This gives

$$\ln p + \gamma \ln V = \ln\,(\text{constant})$$

or

$$pV^{\gamma} = \text{constant}.$$

If subscripts 1 and 2 refer to any two points of the process,

$$p_1V_1^{\gamma} = p_2V_2^{\gamma}. \tag{19–12}$$

It is left as a problem to show that by combining Eq. (19–12) with the equation of state one obtains the alternate forms

$$T_1V_1^{\gamma-1} = T_2V_2^{\gamma-1}, \tag{19–13}$$

$$T_1p_1^{(1-\gamma)/\gamma} = T_2p_2^{(1-\gamma)/\gamma}. \tag{19–14}$$

Values of the specific heat ratio γ are listed in Table 19–1 for some common gases.

An adiabatic expansion or compression of an ideal gas may be represented graphically by a plot of Eq. (19–12), as in Fig. 19–7, in which a number of isothermal curves are shown for comparison. The adia-

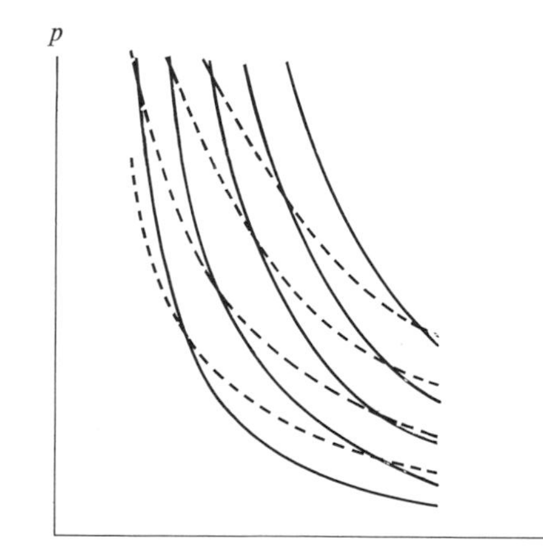

19–7 Adiabatic curves (full lines) versus isothermal curves (dotted lines).

batic curves, at any point, have a somewhat steeper slope than the isothermal curve passing through the same point. That is, as we follow along an adiabatic from right to left (compression process) the curve continually cuts across isotherms of higher and higher temperatures, in agreement with the fact that the temperature generally increases in an adiabatic compression.

The work done by an ideal gas in an adiabatic expansion is computed as follows. We have, from Eq. (19–11),

$$W = \int_{V_1}^{V_2} p\,dV = \int_{T_1}^{T_2} -nC_V\,dT. \qquad (19\text{–}15)$$

Hence the work may be found from either integral. Consider first $\int p\,dV$. Since $pV^\gamma = p_1V_1^\gamma = p_2V_2^\gamma =$ a constant, K, we may write

$$W = \int_{V_1}^{V_2} p\,dV = K\int_{V_1}^{V_2} \frac{dV}{V^\gamma}$$

$$= \frac{1}{1-\gamma}(KV_2^{1-\gamma} - KV_1^{1-\gamma}).$$

In the first term let $K = p_2V_2^\gamma$ and in the second term let $K = p_1V_1^\gamma$. Then

$$W = \frac{p_2V_2 - p_1V_1}{1-\gamma}. \qquad (19\text{–}16)$$

This expresses the work in terms of initial and final pressures and volumes. If the initial and final temperatures are known, we may return to Eq. (19–15) and write

$$W = \int_{T_1}^{T_2} -nC_v\,dT = nC_v(T_1 - T_2). \qquad (19\text{–}17)$$

Example. The compression ratio of a diesel engine, V_1/V_2, is about 15. If the cylinder contains air at 15 lb in^{-2} (absolute) and 60°F (= 520°R) at the start of the compression stroke, compute the pressure and temperature at the end of this stroke. Assume that air behaves like an ideal gas and that the compression is adiabatic. The value of γ for air is 1.40.

From Eq. (19–12),

$$p_2 = p_1\left(\frac{V_1}{V_2}\right)^\gamma$$

or

$$\log p_2 = \log p_1 + \gamma \log\left(\frac{V_1}{V_2}\right) = \log 15 + 1.4 \log 15,$$

$$p_2 = 663 \text{ lb in}^{-2}.$$

The temperature may now be found from Eqs. (19–13), (19–14), or by the ideal gas law. Thus, from Eq. (19–13), we have

$$T_2 = T_1\left(\frac{V_1}{V_2}\right)^{\gamma-1},$$

$$\log T_2 = \log T_1 + (\gamma - 1)\log\left(\frac{V_1}{V_2}\right)$$

$$= \log 520 + (1.4 - 1)\log 15,$$

$$T_2 = 1535°\text{R} = 1075°\text{F}.$$

Or,

$$T_2 = T_1 \times \frac{p_2V_2}{p_1V_1} = 520 \times \frac{663 \times 1}{15 \times 15} = 1535°\text{R}.$$

The work done in the compression stroke is found as follows. Let the initial volume be 60 in^3. From Eq. (19–16),

$$W = \frac{p_2V_2 - p_1V_1}{1-\gamma}.$$

The pressures must be expressed in pounds per square *foot* and the volumes in cubic *feet*.

$$W = \frac{(663 \times 144)(4/1728) - (15 \times 144)(60/1728)}{1 - 1.40}$$

$$= -365 \text{ ft lb}.$$

19–14 THE CONVERSION OF HEAT INTO WORK

The dominating feature of an industrial society is its ability to utilize, whether for wise or unwise ends, sources of energy other than the muscles of men or animals. Except for waterpower, where mechanical energy is directly available, most energy supplies are in the form of potential energy of molecular or nuclear aggregations. In chemical or nuclear reactions some of this potential energy is released and converted to random molecular kinetic energy. In other words, the reaction products are at a relatively high temperature. Heat can be withdrawn from them and utilized for heating buildings, for cooking, or for maintaining a furnace at a high temperature in order to carry out other chemical or physical processes. But to operate a machine, or to propel a vehicle or a projectile, mechanical energy is required, and one of the problems of the mechanical engineer is to withdraw heat from a high-temperature source and convert as large a fraction as possible to mechanical energy.

This transformation always requires the services of a *heat engine*, such as a steam engine, gasoline engine, diesel engine, or jet engine. Since it is only heat and work that are of primary concern in a heat engine, we consider for simplicity an engine in which the so-called "working substance" is carried through a *cyclic* process, that is, a sequence of processes in which it eventually returns to its original state. In the condensing type of steam engine used in marine propulsion, the "working substance," in this case pure water, is actually used over and over again. Water is evaporated in the boilers at high pressure and temperature, does work in expanding against a piston or in a turbine, is condensed by cooling water from the ocean, and pumped back into the boilers. The refrigerant in a household refrigerator also undergoes a cyclic process. Internal combustion engines and steam locomotives do not carry a system through a closed cycle, but they can be analyzed in terms of cyclic processes which approximate their actual operations. All these devices absorb heat from a source at a high temperature, perform some mechanical work, and reject heat at a lower temperature. When a system is carried through a cyclic process, its initial and final internal energies are equal, and from the first law, for any number of complete cycles,

$$U_2 - U_1 = 0 = Q - W,$$
$$Q = W.$$

That is, the **net** *heat flowing into the engine in a cyclic process equals the* **net** *work done by the engine.*

The work is represented by the area enclosed by the curve representing the process in the pV-plane. Thus in Fig. 19–8, for example, where the closed curve shows an arbitrary cyclic process, the area under the upper curve from a to b represents work done *by* the system (positive work) in the expansion from a to b, while the area under the lower curve from b to a represents work done *on* the system (negative work) in the compression from b to a. Since the average pressure during the compression process is less than that during the expansion, the positive work exceeds the negative work and the area bounded by the closed curve is the net positive work done by the system. If the same process were traversed counterclockwise the net work done by the system would be negative.

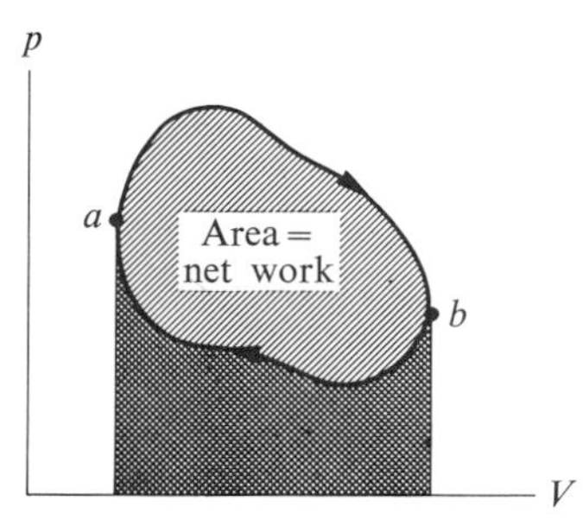

19–8 The area **enclosed** by the curve representing a cyclic process equals the net work.

In the operation of a heat engine or of a refrigerator there are always two bodies capable of either providing or absorbing large quantities of heat without appreciable changes of temperature. Thus the flames and hot gases surrounding the boilers of a marine steam installation can give up large quantities of heat at a high temperature, and constitute therefore what may be called the *hot reservoir*. Heat transferred between the hot reservoir and the working substance in a heat engine will be represented by the symbol Q_H, where it is understood that a positive value of Q_H means heat entering the working substance. The ocean water used to cool the condenser of the marine installation constitutes the *cold reservoir*. (The words "hot" and "cold" are, of course, relative.) The heat transferred between the working substance and the cold reservoir will be denoted by Q_C. A negative value of Q_C means heat rejected by the working substance.

The energy transformations in a heat engine are conveniently represented schematically by the *flow diagram* of Fig. 19–9. The engine itself is represented by the circle. The heat Q_H supplied to the engine by the hot reservoir is proportional to the cross section of the incoming "pipeline" at the top of the diagram. The cross section of the outgoing pipeline at the bottom is

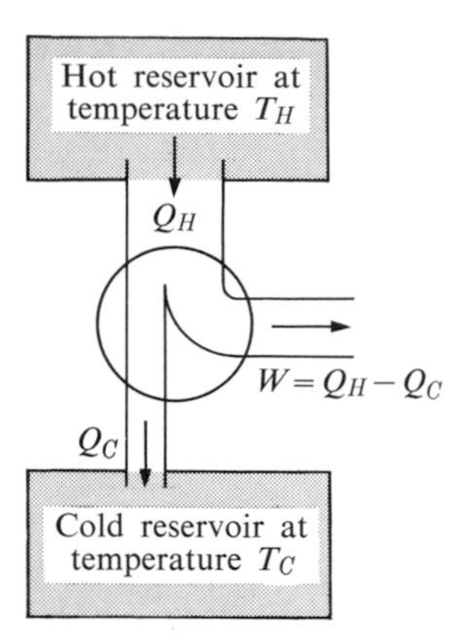

19-9 Schematic flow diagram of a heat engine.

proportional to the heat Q_C which is rejected as heat in the exhaust. The branch line to the right represents that portion of the heat supplied which the engine converts to mechanical work, W.

Consider a heat engine operating in a cycle over and over again and let Q_H and Q_C stand for the heats absorbed and rejected by the working substance *per cycle*. The net heat absorbed is

$$Q = Q_H + Q_C, \tag{19-18}$$

where Q_C is a negative number. The useful output of the engine is the net work W done by the working substance, and from the first law,

$$W = Q = Q_H + Q_C. \tag{19-19}$$

The heat absorbed is usually obtained from the combustion of fuel. The heat rejected ordinarily has no economic value. The *thermal efficiency* of a cycle is defined as the ratio of the useful work to the heat absorbed ("what you get" divided by "what you pay for"):

$$\text{Thermal efficiency} = \frac{W}{Q_H} = \frac{Q_H + Q_C}{Q_H}. \tag{19-20}$$

Because of friction losses, the useful work delivered by an engine is less than the work W, and the overall efficiency is less than the thermal efficiency.

In terms of the flow diagram in Fig. 19-9, the most efficient engine is the one for which the branch pipeline representing the work obtained is as large as possible, and the exhaust pipeline representing the heat rejected is as small as possible, for a given incoming pipeline or quantity of heat supplied.

We shall now consider, without going into the mechanical details of their construction, the gasoline engine, the diesel engine, and the steam engine.

19-15 THE GASOLINE ENGINE

The common gasoline engine is of the four-cycle type, so called because four processes take place in each cycle. Starting with the piston at the top of its stroke, an explosive mixture of air and gasoline vapor is drawn into the cylinder on the downstroke, the inlet valve being open and the exhaust valve closed. This is the *intake* stroke. At the end of this stroke the inlet valve closes and the piston rises, performing an approximately adiabatic compression of the air-gasoline mixture. This is the *compression* stroke. At or near the top of this stroke a spark ignites the mixture of air and gasoline vapor, and combustion takes place very rapidly. The pressure and temperature increase at nearly constant volume.

The piston is now forced down, the burned gases expanding approximately adiabatically. This is the *power stroke* or *working stroke*. At the end of the power stroke the exhaust valve opens. The pressure in the cylinder drops rapidly to atmospheric and the rising piston on the *exhaust stroke* forces out most of the remaining gas. The exhaust valve now closes, the inlet valve opens, and the cycle is repeated.

For purposes of computation, the gasoline engine cycle is approximated by the *Otto* cycle illustrated in Fig. 19-10. Starting at point a, air at atmospheric pressure is compressed adiabatically in a cylinder to point b, heated at constant volume to point c, allowed to expand adiabatically to point d, and cooled at constant volume to point a, after which the cycle is repeated. Line ab corresponds to the compression stroke, bc to the explosion, cd to the working stroke, and da to the exhaust of a gasoline engine. V_1 and V_2 in Fig. 19-10 are respectively the maximum and minimum volumes of the air in the cylinder. The ratio V_1/V_2 is called the *compression ratio*, and is about 10 for a modern internal combustion engine.

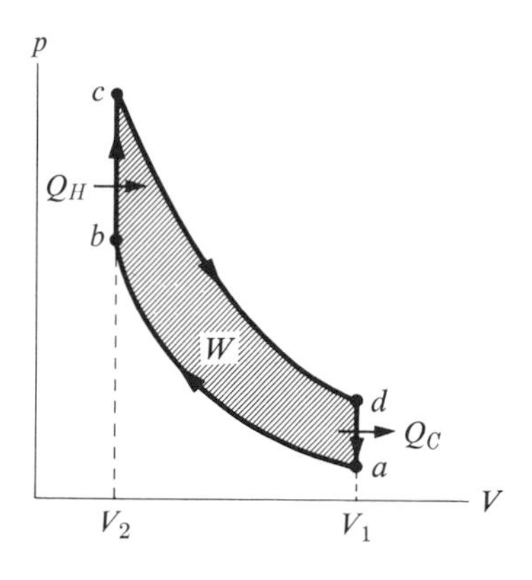

19–10 *pV*-diagram of the Otto cycle.

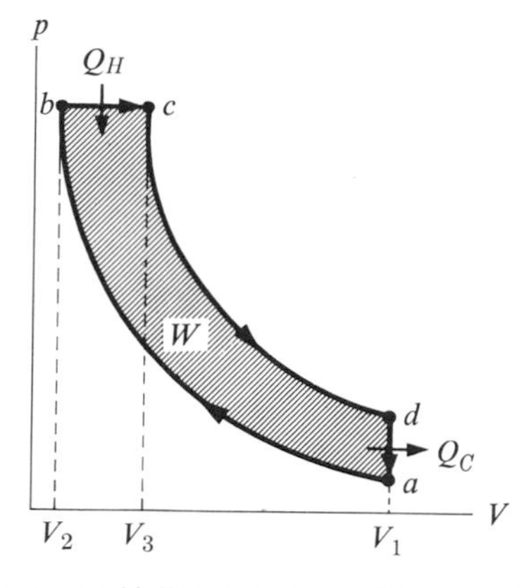

19–11 *pV*-diagram of the Diesel cycle.

The work output in Fig. 19–10 is represented by the shaded area enclosed by the figure *abcd*. The heat *input* is the heat supplied at constant volume along the line *bc*. The exhaust heat is removed along *da*. No heat is supplied or removed in the adiabatic processes *ab* and *cd*.

The heat input and the work output can be computed in terms of the compression ratio, assuming air to behave like an ideal gas. The result is

$$\text{Eff}(\%) = 100\left[1 - \frac{1}{(V_1/V_2)^{\gamma-1}}\right],$$

where γ is the ratio of the specific heat capacity at constant pressure to the specific heat capacity at constant volume, c_p/c_v. For a compression ratio of 10 and a value of $\gamma = 1.4$, the efficiency is about 60%. It will be seen that the higher the compression ratio, the higher the efficiency. Friction effects, turbulence, loss of heat to cylinder walls, etc., have been neglected. All these effects reduce the efficiency of an actual engine below the figure given above.

19–16 THE DIESEL ENGINE

In the Diesel cycle, air is drawn into the cylinder on the intake stroke and compressed adiabatically on the compression stroke to a sufficiently high temperature so that fuel oil injected at the end of this stroke burns in the cylinder without requiring ignition by a spark. The combustion is not as rapid as in the gasoline engine, and the first part of the power stroke proceeds at essentially constant pressure. The remainder of the power stroke is an adiabatic expansion. This is followed by an exhaust stroke which completes the cycle.

The idealized air-Diesel cycle is shown in Fig. 19–11. Starting at point *a*, air is compressed adiabatically to point *b*, heated at constant pressure to point *c*, expanded adiabatically to point *d*, and cooled at constant volume to point *a*.

Since there is no fuel in the cylinder of a diesel engine on the compression stroke, pre-ignition cannot occur and the compression ratio V_1/V_2 may be much higher than that of an internal combustion engine. A value of 15 is typical. The *expansion* ratio V_1/V_3 may be about 5. Using these values, and taking $\gamma = 1.4$, the efficiency of the air-Diesel cycle is about 56%.

19–17 THE STEAM ENGINE

The condensing type of steam engine performs the following sequence of operations. Water is converted to steam in the boiler, and the steam thus formed is superheated above the boiler temperature. Superheated steam is admitted to the cylinder, where it expands against a piston; connection is maintained to the boiler for the first part of the working stroke, which thus takes place at constant pressure. The inlet valve is then closed and the steam expands adiabatically for the rest of the working stroke. The adiabatic cooling causes some of the steam to condense. The mixture of water

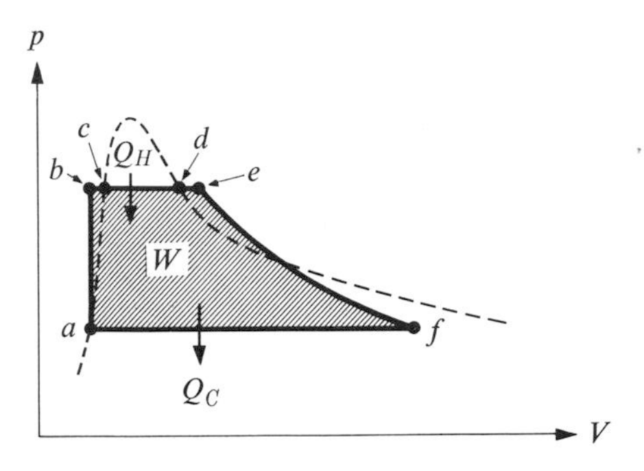

19-12 The Rankine cycle.

droplets and steam (known as "wet" steam) is forced out of the cylinder on the return stroke and into the condenser, where the remaining steam is condensed into water. This water is forced into the boiler by the feed pump, and the cycle is repeated.

An idealized cycle (called the Rankine cycle) which approximates the actual steam cycle is shown in Fig. 19-12. Starting with liquid water at low pressure and temperature (point *a*), the water is compressed adiabatically to point *b* at boiler pressure. It is then heated at constant pressure to its boiling point (line *bc*), converted to steam (line *cd*), superheated (line *de*), expanded adiabatically (line *ef*), and cooled and condensed (along *fa*) to its initial condition.

The efficiency of such a cycle may be computed in the same way as was done in the previous examples, by finding the quantities of heat taken in and rejected along the lines *be* and *fa*. Assuming a boiler temperature of 417°F (corresponding to a pressure of 300 lb in^{-2}), a superheat of 63°F above this temperature (480°F), and a condenser temperature of 102°F, the efficiency of a Rankine cycle is about 32%. Efficiencies of actual steam engines are, of course, considerably lower.

19-18 THE SECOND LAW OF THERMODYNAMICS

Although their efficiencies differ from one another, none of the heat engines described above has a thermal efficiency of 100%. That is, none of them absorbs heat and converts it *completely* into work. There is nothing in the *first* law of thermodynamics that precludes this possibility. The first law requires only that the energy output of an engine, in the form of mechanical work, shall equal the difference between the energies absorbed and rejected in the form of heat. An engine which rejected no heat and which converted all the heat absorbed to mechanical work would therefore be perfectly consistent with the first law.

We know now that there is another principle, independent of the first law and not derivable from it, which determines the maximum fraction of the energy absorbed by an engine as heat that can be converted to mechanical work. The basis of this principle lies in the difference between the natures of internal energy and of mechanical energy. The former is the energy of *random* molecular motion, while the latter represents *ordered* molecular motion. Superposed on their random motion, the molecules of a moving body have an ordered motion in the direction of the velocity of the body. The total molecular kinetic energy associated with the ordered motion is what we call, in mechanics, the kinetic energy of the moving body. The kinetic (and potential) energy associated with the random motion constitutes the internal energy. When the moving body makes an inelastic collision and comes to rest, the *ordered* portion of the molecular kinetic energy becomes converted to *random* motion. Since we cannot control the motions of individual molecules, it is impossible to *completely* reconvert the random motion to ordered motion. We can, however, convert a portion of it and this is what is accomplished by a heat engine.

The science of thermodynamics is concerned only with measurable quantities like heat and work, and the principle known as the *second law of thermodynamics* can be stated entirely apart from any molecular theory. One statement of this law is as follows. *No process is possible whose sole result is the absorption of heat from a reservoir at a single temperature and the conversion of this heat completely into mechanical work.*

If the second law were not true, it would be possible to drive a steamship across the ocean by extracting heat from the ocean, or to run a power plant by extracting heat from the surrounding air. It should be noted again that neither of these "impossibilities" violates the first law of thermodynamics. After all, both the ocean and the surrounding air contain an enormous store of internal energy which, in principle, may be extracted in the form of a flow of heat. The second law, therefore, is not a deduction from the first but stands by itself as a separate law of nature, referring to an aspect of nature different from that contemplated by the first law. The first law denies the possibility of creating or destroying energy; the second denies the possibility of utilizing energy in a particular way.

The fact that work may be dissipated completely into heat, whereas heat may *not* be converted entirely into work, expresses an essential one-sidedness of nature. All natural, spontaneous processes may be studied in the light of the second law, and in all such cases, this peculiar one-sidedness is found. Thus, heat always flows spontaneously from a hotter to a colder body; gases always seep through an opening spontaneously from a region of high pressure to a region of low pressure; gases and liquids left by themselves always tend to mix, not to unmix. Salt dissolves in water but a salt solution does not separate by itself into pure salt and pure water. Rocks weather and crumble; iron rusts; people grow old. These are all examples of *irreversible* processes that take place naturally in only one direction and, by their one-sidedness, express the second law of thermodynamics.

19-19 THE REFRIGERATOR

A refrigerator may be considered to be a heat engine operated in reverse. That is, a heat engine takes in heat from a hot reservoir, converts a part of the heat into mechanical work output, and rejects the difference as heat to a cold reservoir. A refrigerator, however, takes in heat from a cold reservoir, the compressor supplies mechanical work *input*, and heat is rejected to a hot reservoir. With reference to the ordinary home refrigerator, the food and ice cubes constitute the cold reservoir, work is done by the electric motor, and the hot reservoir is the air in the kitchen.

The flow diagram of a refrigerator is given in Fig. 19-13. In one cycle, heat Q_C enters the refrigerator at low temperature T_C, work W is done on the refrigerator and heat Q_H leaves at a higher temperature T_H. Both W and Q_H are negative quantities. It follows from the first law that

$$-Q_H = Q_C - W,$$

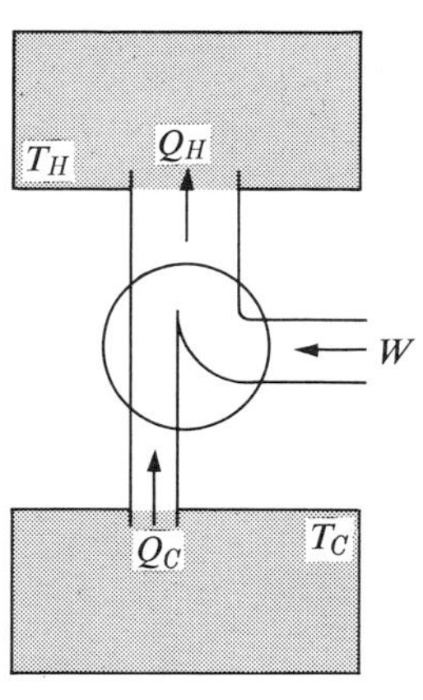

19-13 Schematic flow diagram of a refrigerator.

and the heat rejected to the hot reservoir is the sum of the heat taken from the cold reservoir and the heat equivalent of the work done by the motor.

From an economic point of view, the best refrigeration cycle is one that removes the greatest amount of heat Q_C from the refrigerator, for the least expenditure of mechanical work W. We therefore define the *coefficient of performance* (rather than the efficiency) of a refrigerator as the ratio $-Q_C/W$, and since $W = Q_H + Q_C$,

$$\text{Coefficient of performance} = -\frac{Q_C}{Q_H + Q_C}. \tag{19-21}$$

The principles of the common refrigeration cycle are illustrated schematically in Fig. 19-14. Compressor

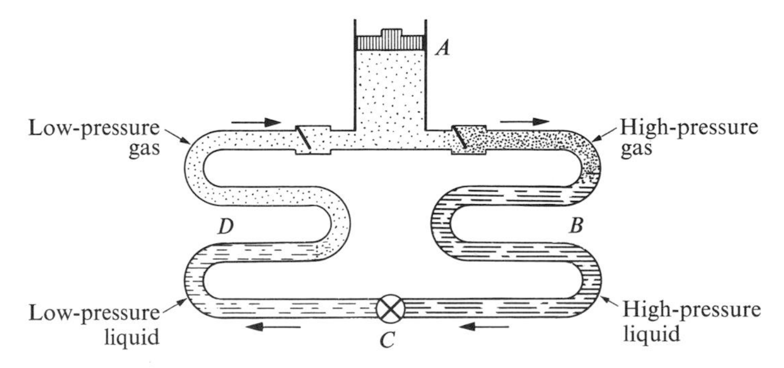

19-14 Principle of the mechanical refrigeration cycle.

A delivers gas (CCl_2F_2, NH_3, etc.) at high temperature and pressure to coils *B*. Heat is removed from the gas in *B* by water or air cooling, resulting in condensation of the gas to a liquid, still under high pressure. The liquid passes through the throttling valve or expansion valve *C*, emerging as a mixture of liquid and vapor at a lower temperature. In coils *D*, heat is supplied that converts the remaining liquid into vapor which enters compressor *A* to repeat the cycle. In a domestic refrigerator, coils *D* are placed in the ice compartment, where they cool the refrigerator directly. In a larger refrigerating plant, these coils are usually immersed in a brine tank and cool the brine, which is then pumped to the refrigerating rooms.

If no work were needed to operate a refrigerator, the coefficient of performance (heat extracted divided by work done) would be infinite. Coefficients of performance of actual refrigerators vary from about 2 to about 6. Experience shows that work is always needed to transfer heat from a colder to a hotter body. This negative statement leads to another statement of the second law of thermodynamics, namely:

No process is possible whose sole result is the transfer of heat from a cooler to a hotter body.

At first sight, this and the previous statement of the second law appear to be quite unconnected, but it can be shown that they are in all respects equivalent. Any device that would violate one statement would violate the other.

19-20 THE CARNOT CYCLE

Although their efficiencies differ from one another, none of the heat engines which have been described has an efficiency of 100%. The question still remains open as to what is the maximum attainable efficiency, given a supply of heat at one temperature and a reservoir at a lower temperature for cooling the exhaust. An idealized engine which can be shown to have the maximum efficiency under these conditions was invented by the French engineer Sadi Carnot in 1824 and is called a *Carnot engine*. The *Carnot cycle*, shown in Fig. 19-15, differs from the Otto and Diesel cycles in that it is bounded by two *isothermals* and two adiabatics. Thus all the heat input is supplied at a *single* high temperature

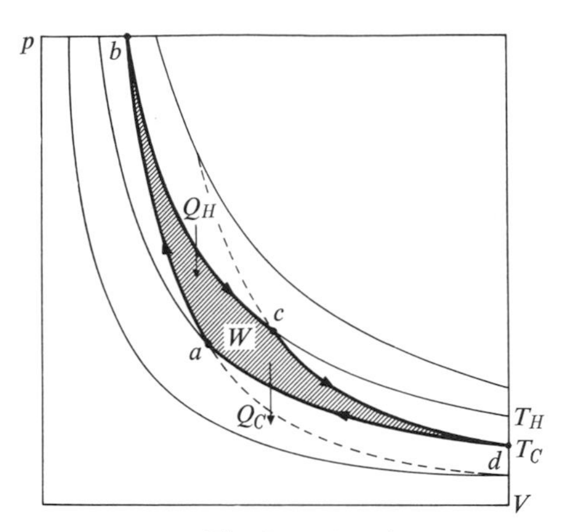

19-15 The Carnot cycle.

and all the heat output is rejected at a *single* lower temperature. (Compare with Figs. 19-10 and 19-11, in which the temperature is different at all points of the lines *bc* and *da*.)

This, however, is not the only feature of the Carnot cycle. There are no "one-way" processes in the Carnot cycle, such as explosions or throttling processes. The isothermal and adiabatic processes of the Carnot cycle are idealizations of actual processes. The direction of either process may be reversed by only a slight change in the external pressure; there is no friction, and the working substance is always very close to equilibrium.

A process that is attended by no equilibrium-disturbing effect (such as an unbalanced force or a finite temperature difference) *and no dissipative effect* (such as friction or electrical resistance) *is said to be reversible.* A reversible process represents the same sort of idealization in thermodynamics that a point mass, a weightless inextensible cord, or a frictionless, moment-of-inertialess pulley represents in mechanics. It may be approximated but not perfectly achieved.

The isothermal and adiabatic processes of the Carnot cycle may be imagined to proceed in either direction. In the direction shown in Fig. 19-15, heat Q_H goes in, heat Q_C goes out, and work *W* is done by the engine. If the arrows in the figure are reversed, the cycle becomes a Carnot refrigeration cycle and, *what is most important, the same amount* of heat Q_H which formerly was taken in from the hot reservoir now goes out, and *the same amount* of work W, which formerly was delivered to the outside is now required from the surroundings, and *the same amount* of heat Q_C which formerly was rejected to the cold reservoir is now taken in. These numerical equalities would *not* exist if any ordinary engine cycle were reversed.

Suppose an engine (not a Carnot engine) were to operate between a source of heat at some temperature and a reservoir of heat at a lower temperature, thereby delivering to the outside an amount of work W. Suppose this work W were used to operate a Carnot refrigerator which extracted heat from the colder reservoir and delivered it to the warmer source. It can be shown that if the first engine were more efficient than the Carnot engine that would result by operating the Carnot refrigerator backward, then the net effect would be a violation of the second law of thermodynamics. Proceeding along these lines, it has been proved that:

No engine operating between two given temperatures can be more efficient than a Carnot engine operating between the same two temperatures,

and also (Carnot's theorem):

All Carnot engines operating between the same two temperatures have the same efficiency, irrespective of the nature of the working substance.

19-21 THE KELVIN TEMPERATURE SCALE

It follows from Carnot's theorem that the efficiency of a Carnot engine operating between reservoirs at two given temperatures *is independent of the nature of the working substance and is a function only of the temperatures.* If we consider a number of Carnot engines using different working substances and absorbing and rejecting heat to the same two reservoirs, the thermal efficiency is the same for all, or

$$\text{Thermal efficiency} = \frac{Q_H + Q_C}{Q_H} = 1 + \frac{Q_C}{Q_H} = \text{constant.}$$

Hence the ratio Q_H/Q_C is a constant for all the engines. Kelvin proposed that the ratio of the temperatures of the reservoirs be *defined* as equal to this constant ratio of the absolute magnitudes of the quantities of heat absorbed and rejected or, since Q_C is a negative quantity, as equal to the negative of the ratio Q_H/Q_C. Thus

$$\frac{T_H}{T_C} = \frac{|Q_H|}{|Q_C|} = -\frac{Q_H}{Q_C}. \tag{19-22}$$

Since the ratio of temperatures does not involve the properties of any particular substance, the Kelvin temperature scale is truly *absolute*.

To complete the definition of the Kelvin scale we proceed, as in Chapter 15, to assign the arbitrary value of 273.16°K to the temperature of the triple point of water, T_3. For a Carnot engine operating between reservoirs at the temperatures T and T_3, we have

$$\frac{|Q|}{|Q_3|} = \frac{T}{T_3}$$

or

$$\boxed{T = 273.16\text{°K}\,\frac{|Q|}{|Q_3|}.} \tag{19-23}$$

Comparing this with the corresponding expression for the ideal gas temperature, namely,

$$273.16\text{°K} \lim_{p_3 \to 0} \left(\frac{p}{p_3}\right)_{\text{constant volume}},$$

it is seen that, in the Kelvin scale, Q plays the role of a "thermometric property." This does not, however, have the objection attached to a coordinate of an arbitrarily chosen thermometer, inasmuch as the behavior of a Carnot engine is independent of the nature of the working substance.

It is a simple matter to show that if an ideal gas is taken around a Carnot cycle, the ratio of the heats absorbed and rejected $|Q_H|/|Q_C|$ is equal to the ratio of the temperatures of the reservoirs *as expressed on the gas scale*, defined in Chapter 15. The proof is somewhat lengthy, and will not be given here. Since, in both scales, the triple point of water is chosen to be 273.16°, it follows that *the Kelvin and the ideal gas scales are identical.*

The efficiency of a Carnot engine rejecting heat Q_C to a reservoir at Kelvin temperature T_C and absorbing heat Q_H from a source at Kelvin temperature T_H is, as usual,

$$\text{Efficiency} = 1 + \frac{Q_C}{Q_H} = 1 - \frac{|Q_C|}{|Q_H|}.$$

But, by definition of the Kelvin scale,

$$\frac{|Q_C|}{|Q_H|} = \frac{T_C}{T_H}.$$

Therefore, the efficiency of a Carnot engine is

$$\boxed{\text{Efficiency (Carnot)} = 1 - \frac{T_C}{T_H}.} \tag{19-24}$$

Equation (19-24) points the way to the conditions which a real engine, such as a steam engine, must fulfill to approach as closely as possible the maximum attainable efficiency. These conditions are that the intake temperature T_H must be made as high as possible and the exhaust temperature T_C as low as possible.

The exhaust temperature cannot be lower than the lowest temperature available for cooling the exhaust. This is usually the temperature of the air, or perhaps of river water if this is available at the plant. The only recourse then is to raise the boiler temperature T_H. Since the vapor pressure of all liquids increases rapidly with increasing temperature, a limit is set by the mechanical strength of the boiler. Another possibility is to use, instead of water, some liquid with a lower vapor pressure. Successful experiments in this direction have been made with mercury vapor replacing steam. At a boiler temperature of 200°C, at which the pressure in a steam boiler would be 225 lb in^{-2}, the pressure in a mercury boiler is only 0.35 lb in^{-2}.

19-22 ABSOLUTE ZERO

It follows from Eq. (19-23) that the heat transferred isothermally between two given adiabatics decreases as the temperature decreases. Conversely, the smaller the value of Q, the lower the corresponding T. The smallest possible value of Q is zero, and the corresponding T is absolute zero. *Thus, if a system undergoes a reversible isothermal process without transfer of heat, the temperature at which this process takes place is called absolute zero.* In other words, at absolute zero, an isotherm and an adiabatic are identical.

It should be noted that the definition of absolute zero holds for all substances and is therefore independent of the peculiar properties of any one arbitrarily chosen substance. Furthermore, the definition is in terms of purely large-scale concepts. No reference is made to molecules or to molecular energy. Whether absolute zero may be achieved experimentally is a question of some interest and importance. To achieve temperatures below 4.2°K, at which ordinary helium (mass number 4) liquefies, it is necessary to lower the vapor pressure by pumping away the vapor as fast as possible. The lowest temperature that has ever been reached in this way is 0.7°K, and this required larger pumps and larger pumping tubes than are usually employed in low-temperature laboratories. With the aid of the light isotope of helium (mass number 3) which liquefies at 3.2°K, vigorous pumping will yield a temperature of about 0.3°K. Still lower temperatures may be achieved magnetically, but it becomes apparent that the closer one approaches absolute zero, the more difficult it is to go further. It is generally accepted as a law of nature that, although absolute zero may be approached as close as we please, it is impossible actually to reach the zero of temperature. This is known as the "*unattainability statement of the third law of thermodynamics.*"

19-23 ENTROPY

The first law of thermodynamics is the law of energy, the second law of thermodynamics is the law of entropy, and every process that takes place in nature, whether it be mechanical, electrical, chemical, or biological, must proceed in conformity with these two laws.

A book of this nature is not the place for a thorough exposition of entropy and the second law of thermodynamics. We shall content ourselves with defining entropy, computing its changes in a few instances, and stating some of its properties.

A re-reading of Section 19-4 will recall to mind that when a system is carried from one state to another it is found by experiment that the difference between the heat added and the work done by the system, $Q - W$, has the same value for all paths. The fact that this difference does have the same value makes it possible to introduce the concept of internal energy, the change in internal energy being defined and measured by the quantity $Q - W$.

Entropy, or rather a change in entropy, may be defined in a similar way. Consider two states of a system and a number of *reversible* paths connecting them. (The restriction to reversible paths need not be made with regard to internal energy changes.) Although the heat added to the system is different along different paths, it may be proved that if the heat added at each point of the path is divided by the absolute temperature of the system at the point, and the resulting ratios summed for the entire path, this sum has the same value for all (reversible) paths between the same endpoints. In mathematical symbols,

$$\int_1^2 \frac{dQ}{T} = \begin{array}{l}\text{constant for all reversible}\\ \text{paths between states 1 and 2.}\end{array}$$

It is therefore *possible* (whether it is of any use or not one can only tell later) to introduce a function whose difference between two states, 1 and 2, is defined by the integral above. This function may be assigned any arbitrary value in some standard reference state, and its value in any other state will be a definite quantity. The function is called the *entropy* of the system, and is denoted by S. We then have

$$S_2 - S_1 = \int_1^2 \frac{dQ}{T} \quad \text{(along any reversible path).} \tag{19-25}$$

If the change is infinitesimal, $dS = dQ/T$. The unit of entropy is 1 cal $(°K)^{-1}$, 1 Btu $(°R)^{-1}$, 1 J $(°K)^{-1}$, etc.

Example 1. One kilogram of ice at 0°C is melted and converted to water at 0°C. Compute its change in entropy.

Since the temperature remains constant at 273°K, T may be taken outside the integral sign. Then

$$S_2 - S_1 = \frac{1}{T}\int dQ = \frac{Q}{T}.$$

But Q is simply the total heat which must be supplied to melt the ice, or 80,000 cal. Hence

$$S_2 - S_1 = \frac{80{,}000}{273} = 293 \text{ cal } °\text{K}^{-1},$$

and the increase in entropy of the system is 293 cal $°K^{-1}$. In any *isothermal* reversible process, the entropy change equals the heat added divided by the absolute temperature.

Example 2. One kilogram of water at 0°C is heated to 100°C. Compute its change in entropy.

The temperature is not constant and dQ and T must be expressed in terms of a single variable in order to carry out the integration. This may readily be done, since

$$dQ = mc\,dT.$$

Hence

$$S_2 - S_1 = \int_{273}^{373} mc\frac{dT}{T} = mc \ln\frac{373}{273} = 312 \text{ cal } °\text{K}^{-1}.$$

Example 3. A gas is allowed to expand adiabatically and reversibly. What is its change in entropy?

In an adiabatic process no heat is allowed to enter or leave the system. Hence $Q = 0$ and there is no change in entropy. It follows that every *reversible* adiabatic process is one of constant entropy, and may be described as *isentropic*.

19-24 THE PRINCIPLE OF THE INCREASE OF ENTROPY

One of the features which distinguishes entropy from such concepts as energy, momentum, and angular momentum is that *there is no principle of conservation of entropy*. In fact, the reverse is true. Entropy *can* be created at will and there is an increase in entropy in every natural process, if all systems taking part in the process are considered.

Consider the process of mixing 1 kg of water at 100°C with 1 kg of water at 0°C. Let us arbitrarily call the entropy of water zero when it is in the liquid state at 0°C. This is the reference state adopted in engineering work. Then from Example 2 in the previous section, the entropy of 1 kg of water at 100°C is 312 cal $°K^{-1}$ and the entropy of 1 kg at 0°C is zero. The entropy of the system, before mixing, is therefore 312 cal $°K^{-1}$.

After the hot and cold water have been mixed, we have 2 kg of water at a temperature of 50°C or 323°K. From the results of Example 2, we see that the entropy of the system is

$$mc \ln\frac{323}{273} = 2000 \times \ln\frac{373}{273} = 336 \text{ cal } °\text{K}^{-1}.$$

There has therefore been an increase in entropy of

$$(336 - 312) \text{ cal } °\text{K}^{-1} = 24 \text{ cal } °\text{K}^{-1}.$$

Physical mixing of the hot and cold water is, of course, not essential in bringing about the final equilibrium state. We might simply have let heat flow by conduction, or be transferred by radiation, from the hot to the cold water. The same increase in entropy would have resulted.

These examples of the mixing of substances at different temperatures, or the flow of heat from a higher to a lower temperature, are illustrative of all natural (i.e., irreversible) processes. When all the entropy changes in the process are included, the increases in entropy are always greater than the decreases. In the special case of a reversible process, the increases and decreases are equal. Hence we can formulate the general principle, which is considered a part of the second law of thermodynamics, that *when all systems taking part in a process are included, the entropy either remains constant or increases.* In other words, *no process is possible in which the entropy decreases*, when all systems taking part in the process are included.

What is the significance of the increase of entropy that accompanies every natural process? The answer, or one answer, is that it represents the extent to which the Universe "runs down" in that process. Consider again the example of the mixing of hot and cold water. We *might* have used the hot and cold water as the high- and low-temperature reservoirs of a heat engine, and in the course of removing heat from the hot water and giving heat to the cold water we could have obtained some mechanical work. But once the hot and cold water have been mixed and have come to a uniform temperature, this opportunity of converting heat to mechanical work is lost and, moreover, it is lost irretrievably. The lukewarm water will never *unmix* itself and separate into a hotter and a colder portion.* Of course, there is no decrease in *energy* when the hot and cold water are mixed, and what has been "lost" in the mixing process is not *energy*, but *opportunity;* the opportunity to convert a portion of the heat flowing out of the hot water to mechanical work. Hence when entropy increases, energy becomes more unavailable, and we say that the Universe has "run down" to that extent. This is the true significance of the term "irreversible."

The tendency of all natural processes such as heat flow, mixing, diffusion, etc., is to bring about a uniformity of temperature, pressure, composition, etc., at all points. One may visualize a distant future in which, as a consequence of these processes, the entire Universe has attained a state of absolute uniformity throughout. When and if such a state is reached, although there would have been no change in the energy of the Universe, all physical, chemical, and presumably biological processes would have to cease. This goal toward which we appear headed has been described as the "heat death" of the Universe.

* The branch of physics called "statistical mechanics" would modify this statement to read, "It is highly improbable that the water will separate spontaneously into a hotter and a colder portion, but it is not impossible."

Problems

19–1 A combustion experiment is performed by burning a mixture of fuel and oxygen in a constant-volume "bomb" surrounded by a water bath. During the experiment the temperature of the water is observed to rise. Regarding the mixture of fuel and oxygen as the system: (a) has heat been transferred? (b) has work been done? (c) what is the sign of ΔU?

19–2 A liquid is irregularly stirred in a well-insulated container and thereby undergoes a rise in temperature. Regarding the liquid as the system: (a) has heat been transferred? (b) has work been done? (c) what is the sign of ΔU?

19–3 A resistor immersed in running water carries an electric current. Consider the resistor as the system under consideration. (a) Is there a flow of heat into the resistor? (b) Is there a flow of heat into the water? (c) Is work done? (d) Assuming the state of the resistor to remain unchanged, apply the first law to this process.

19–4 In a certain process, 500 cal of heat are supplied to a system, and at the same time 100 J of work are done on the system. What is the increase in the internal energy of the system?

19–5 In a certain process, 200 Btu are supplied to a system and at the same time the system expands against a constant external pressure of 100 lb in^{-2}. The internal energy of the system is the same at the beginning and end of the process. Find the increase in volume of the system.

19–6 An inventor claims to have developed an engine which takes in 100,000 Btu from its fuel supply, rejects 25,000 Btu in the exhaust, and delivers 25 kWh of mechanical work. Do you advise investing money to put this engine on the market?

19–7 A vessel with rigid walls and covered with asbestos is divided into two parts by an insulating partition. One part contains a gas at temperature T and pressure P. The other part contains a gas at temperature T' and pressure P'. The partition is removed. What conclusion may be drawn by applying the first law of thermodynamics?

19–8 A mixture of hydrogen and oxygen is enclosed in a rigid insulating container and exploded by a spark. The temperature and pressure both increase considerably. Neglecting the small amount of energy provided by the spark itself, what conclusion may be drawn by applying the first law of thermodynamics?

19–9 When water is boiled under a pressure of 2 atm, the heat of vaporization is 946 Btu lb^{-1} and the boiling point is 250°F. One pound of steam occupies a volume of 14 ft^3, and 1 lb of water a volume of 0.017 ft^3. (a) Compute the external work, in foot-pounds and in Btu when 1 lb of steam is formed at this temperature. (b) Compute the increase in internal energy in Btu.

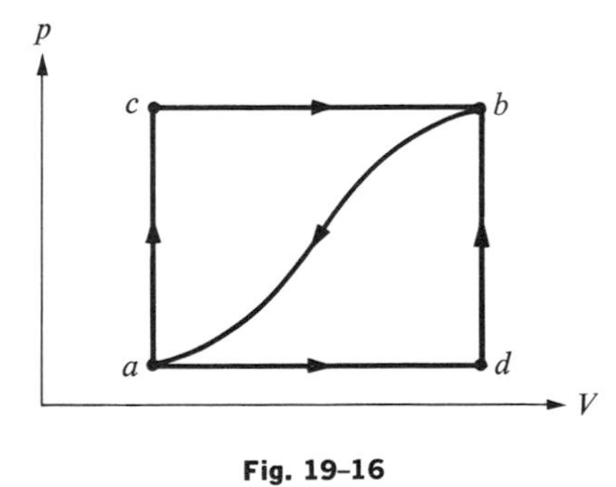

Fig. 19–16

19–10 When a system is taken from state a to state b, in Fig. 19–16, along the path acb, 80 Btu of heat flow into the system, and 30 Btu of work are done. (a) How much heat flows into the system along path adb if the work is 10 Btu? (b) When the system is returned from b to a along the curved path, the work is 20 Btu. Does the system absorb or liberate heat and how much? (c) If $U_a = 0$ and $U_d = 40$ Btu, find the heat absorbed in the processes ad and db.

19–11 A steel cylinder of cross-sectional area 0.1 ft^2 contains 0.4 ft^3 of glycerin. The cylinder is equipped with a tightly fitting piston which supports a load of 6000 lb. The temperature of the system is increased from 60°F to 160°F. Neglect the expansion of the steel cylinder. Find (a) the increase in volume of the glycerin, (b) the mechanical work of the 6000-lb force, (c) the amount of heat added to the glycerin [specific heat of glycerin = 0.58 Btu lb^{-1} (F°)$^{-1}$], (d) the change in internal energy of the glycerin.

19–12 The volume of 1 mole of an ideal gas is increased isothermally (T = constant) from 1 to 20 liters at 0°C. The pressure of the gas at any moment is given by the equation $pV = RT$, where $R = 8.31$ J mole^{-1} (K°)$^{-1}$ and T is the kelvin temperature. How many joules of work are done?

19–13 Calculate the work done when a gas expands from volume V_1 to V_2, the relation between pressure and volume being

$$\left(p + \frac{a}{V^2}\right)(V - b) = K,$$

where a, b, and K are constants.

19–14 (a) The change in internal energy, dU, of a system consisting of n moles of a pure substance, in an infinitesimal process at constant volume, is equal to $nC_v\,dT$. Explain why the internal energy change in a process at constant pressure is *not* equal to $nC_p\,dT$. (b) Explain why the change in internal energy of an ideal gas, in *any* infinitesimal process, is given by $nC_v\,dT$.

19–15 A cylinder contains 1 mole of oxygen gas at a temperature of 27°C. The cylinder is provided with a frictionless piston which maintains a constant pressure of 1 atm on the gas. The gas is heated until its temperature increases to 127°C. (a) Draw a diagram representing the process in the pV-plane. (b) How much work is done by the gas in this process? (c) On what is this work done? (d) What is the change in internal energy of the gas? (e) How much heat was supplied to the gas? (f) How much work would have been done if the pressure had been 0.5 atm?

19–16 (a) Compare the quantity of heat required to raise the temperature of 1 g of hydrogen through 1 C°, at constant pressure, with that required to raise the temperature of 1 g of water by the same amount. (b) Of the substances listed in Table 19–1, which has the largest specific heat capacity in cal g^{-1} (C°)$^{-1}$?

19–17 Ten liters of air at atmospheric pressure are compressed isothermally to a volume of 2 ℓ and is then allowed to expand adiabatically to a volume of 10 ℓ. Show the process in a pV-diagram.

19–18 An ideal gas is contained in a cylinder closed with a movable piston. The initial pressure is 1 atm and the initial volume is 1 ℓ. The gas is heated at constant pressure until the volume is doubled, then heated at constant volume until the pressure is doubled, and finally expanded adiabatically until the temperature drops to its initial value. Show the process in a pV-diagram.

19–19 Compressed air at a gauge pressure of 300 lb in^{-2} is used to drive an air engine which exhausts at a gauge pressure of 15 lb in^{-2}. What must be the temperature of the compressed air in order that there may be no possibility of frost forming in the exhaust ports of the engine? Assume the expansion to be adiabatic. [*Note:* Frost frequently forms in the exhaust ports of an air-driven engine. This happens when the moist air is cooled below 0°C by the expansion which takes place in the engine.]

19–20 Initially at a temperature of 140°F, 10 ft^3 of air expands at a constant gauge pressure of 20 lb in^{-2} to a volume of 50 ft^3, and then expands further adiabatically to a final volume of 80 ft^3 and a final gauge pressure of 3 lb in^{-2}. Sketch the process in the pV-plane and compute the work done by the air.

19–21 The cylinder of a pump compressing air from atmospheric pressure into a very large tank at 60 lb in^{-2} gauge pressure is 10 in. long. (a) At what position in the stroke will air begin to enter the tank? Assume the compression to be adiabatic. (b) If the air is taken into the pump at 27°C, what is the temperature of the compressed air?

19–22 Two moles of helium are initially at a temperature of 27°C and occupy a volume of 20 ℓ. The helium is first expanded at constant pressure until the volume has doubled, and then adiabatically until the temperature returns to its initial value. (a) Draw a diagram of the process in the pV-plane. (b) What is the total heat supplied in the process? (c) What is the total change in internal energy of the helium? (d) What is the total work done by the helium? (e) What is the final volume?

19–23 A heat engine carries 0.1 mole of an ideal gas around the cycle shown in the pV-diagram of Fig. 19–17. Process 1–2 is at constant volume, process 2–3 is adiabatic, and process 3–1 is at a constant pressure of 1 atm. The value of γ for this gas is $\frac{5}{3}$. (a) Find the pressure and volume at points 1, 2, and 3. (b) Find the net work done by the gas in the cycle.

19–24 A cylinder contains oxygen at a pressure of 2 atm. The volume is 3 ℓ and the temperature is 300°K. The oxygen is carried through the following processes: (1) Heated at constant pressure to 500°K. (2) Cooled at constant volume to 250°K. (3) Cooled at constant pressure to 150°K. (4) Heated at constant volume to 300°K. (a) Show the four processes above in a pV-diagram, giving the numerical values of p and V at the end of each process. (b) Calculate the net work done by the oxygen. (c) Find the net heat flowing into the oxygen. (d) What is the efficiency of this device as a heat engine?

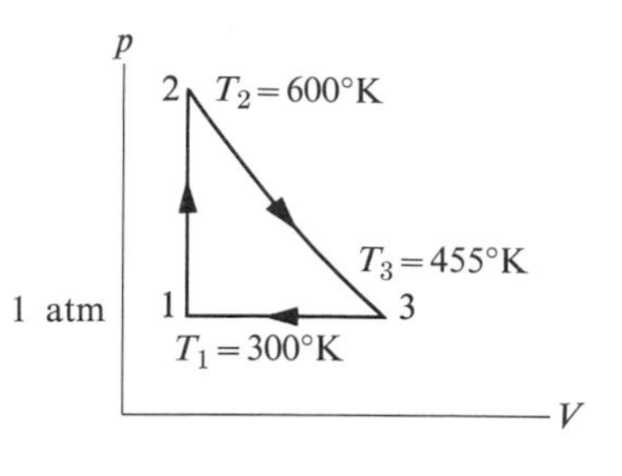

Fig. 19–17

19–25 What is the thermal efficiency of an engine which operates by taking an ideal gas through the following cycle? Let $C_v = 3$ cal mole^{-1} (C°)$^{-1}$.

1) Start with n moles at P_0, V_0, T_0.
2) Change to $2P_0$, V_0, at constant volume.
3) Change to $2P_0$, $2V_0$, at constant pressure.
4) Change to P_0, $2V_0$, at constant volume.
5) Change to P_0, V_0, at constant pressure.

19–26 A Carnot engine whose low-temperature reservoir is at 280°K has an efficiency of 40%. It is desired to increase this to 50%. (a) By how many degrees must the temperature of the high-temperature reservoir be increased if the temperature of the low-temperature reservoir remains constant? (b) By how many degrees must the temperature of the low-temperature reservoir be decreased if that of the high-temperature reservoir remains constant?

19–27 A Carnot engine whose high-temperature reservoir is at 400°K takes in 100 cal of heat at this temperature in each cycle, and gives up 80 cal to the low-temperature reservoir. (a) What is the temperature of the latter reservoir? (b) What is the thermal efficiency of the cycle?

19–28 A Carnot refrigerator takes heat from water at 0°C and rejects heat to a room at 27°C. Suppose that 50 kg of water at 0°C are converted to ice at 0°C. (a) How much heat is rejected to the room? (b) How much energy must be supplied to the refrigerator?

19–29 A Carnot engine is operated between two heat reservoirs at temperatures of 400°K and 300°K. (a) If the engine receives 1200 cal from the reservoir at 400°K in each cycle, how many calories does it reject to the reservoir at 300°K? (b) If the engine is operated in reverse, as a refrigerator, and receives 1200 cal from the reservoir at 300°K, how many calories does it deliver to the reservoir at 400°K? (c) How many calories would be produced if the mechanical work required to operate the refrigerator in part (b) were converted directly to heat?

19–30 Is it possible to cool a closed, insulated room by operating an electric refrigerator in the room, leaving the refrigerator door open?

19–31 Consider the entropy of water to be zero when it is in the liquid phase at 0°C and atmospheric pressure. (a) How much heat must be supplied to 1 kg of water to raise its temperature from 0°C to 100°C? (b) What is the entropy of 1 kg of water at 100°C? (c) 1 kg of water at 0°C is mixed with 1 kg at 100°C. What is the final temperature? (d) Find the entropy of the hot and cold water before mixing, and the entropy of the system after mixing. (This is another example of the increase of entropy in an irreversible process.)

19–32 (a) Draw a graph of a Carnot cycle, plotting kelvin temperature vertically and entropy horizontally (a temperature-entropy diagram). (b) Show that the area under any curve in a temperature-entropy diagram represents the heat absorbed by the system. (c) Derive from your diagram the expression for the thermal efficiency of a Carnot cycle.

20 Molecular Properties of Matter

20–1 MOLECULAR THEORY OF MATTER

The fact that matter is compressible suggests that all substances have a granular or spongy structure with spaces into which the granules can penetrate when the external pressure is increased. The ease with which liquids can flow and gases can diffuse points more to a collection of tiny particles than to a spongelike structure. Thousands of physical and chemical facts support the contention that matter in all phases is composed of tiny particles called *molecules*. The molecules of any one substance are identical. They have the same structure, the same mass, and the same mechanical and electrical properties. Many of the large-scale properties of matter which have been discussed heretofore, such as elasticity, surface tension, condensation, vaporization, etc., can be comprehended with deeper understanding of their significance in terms of the molecular theory. For this purpose we may conceive of a molecule as a rigid sphere, like a small billiard ball, capable of moving, of colliding with other molecules or with a wall, and of exerting attractive or repulsive forces on neighboring molecules. In other parts of physics and chemistry it is important to consider the structure of the molecule, but this is not necessary at this point.

One of the outstanding characteristics of a molecule is the force that exists between it and a neighbor. There is, of course, a force of gravitational attraction between every pair of molecules, but it turns out that this is negligible in comparison with the forces we are now considering. The forces that hold the molecules of a liquid (or solid) together are, in part at least, of electrical origin and do not follow a simple inverse square law. When the separation of the molecules is large, as in a gas, the force is extremely small and is an attraction. The attractive force increases as a gas is compressed and its molecules brought closer together. But since tremendous pressures are needed to compress a liquid, i.e., to force its molecules closer together than their normal spacing in the liquid state, we conclude that at separations only slightly less than the dimensions of a molecule the force is one of repulsion and is relatively large. The force must then vary with separation in somewhat the fashion shown in Fig. 20–1. At large separations the force is one of attraction but is extremely small. As the molecules are brought closer together, the force

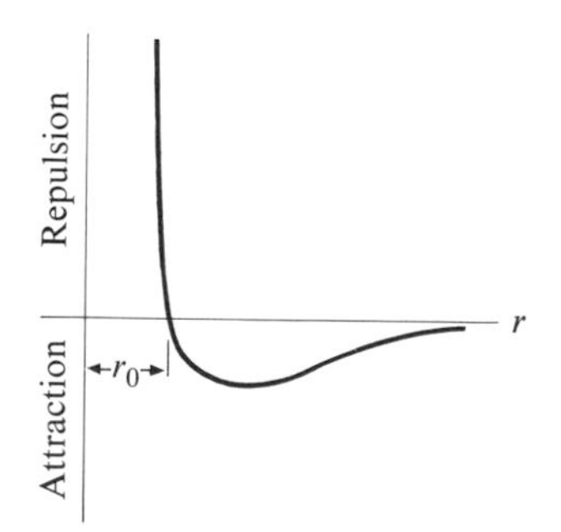

20–1 The force between two molecules changes from an attraction when the separation is large, to a repulsion when the separation is small.

of attraction becomes larger, passes through a maximum, and then decreases to zero at a separation r_0. When the distance between the molecules is less than r_0, the force is one of repulsion.

A single pair of molecules could remain in equilibrium at a center-to-center spacing equal to r_0 in Fig. 20-1. If they were separated slightly, the force between them would be attractive and they would be drawn together. If they were forced closer together than the distance r_0, the force would be one of repulsion and they would spring apart. If they were either pulled apart or pushed together, and then released, they would oscillate about their equilibrium separation r_0.

If the only property of a molecule were the force of attraction between it and its neighbors, all matter would eventually coalesce into the liquid or solid phase. The existence of gases points to another property which enables molecules to stay apart. This is accomplished by molecular motion. The more vigorous the motion the less chance there is for condensation into the liquid or solid phase. In solids the molecules execute vibratory motion about more or less fixed centers. The vibratory motion is relatively weak, and the centers remain fixed at regularly spaced positions which comprise a *space lattice*. This gives rise to the extraordinary regularity and symmetry of crystals. A photograph of the individual, very large molecules of *necrosis virus protein* is shown in Fig. 20-2. It was taken with an electron microscope at a magnification of about 80,000. The molecules look like neatly stacked oranges.

20-2 Crystal of necrosis virus protein. The actual size of the entire crystal is about two thousandths of a millimeter (0.002 mm). (Courtesy of Ralph Wyckoff. Reprinted with the permission of Educational Services, Inc., from *Physics;* D. C. Heath and Co., Boston, 1960.)

In liquids the intermolecular distances are usually only a trifle greater than those of a solid. The molecules execute vibratory motion of greater energy about centers which are free to move but which remain at approximately the same distances from one another. Liquids show a certain regularity of structure only in the immediate neighborhood of a few molecules. This is called *short-range order*, in contrast with the *long-range order* of a solid crystal.

The molecules of a gas have the greatest kinetic energy, so great in fact that they are, on the average, far away from one another, where only very small attractive forces exist. A molecule of a gas therefore moves with linear motion until a collision takes place either with another molecule or with a wall. The molecular definition of an *ideal gas is a gas whose molecules exert no forces of attraction at all*. The mathematical treatment of an assemblage consisting of an enormous number of small, rigid, perfectly elastic spheres executing random linear motions between collisions is called the *kinetic theory of gases*, and was perfected by Clausius, Maxwell, and Boltzmann in the latter part of the nineteenth century. The analysis of an ideal gas given in this chapter is only a simplified version of kinetic theory.

A common substance exists in the solid phase at low temperatures. When the temperature is raised beyond a definite value, the liquid phase results, and when the temperature of the liquid is raised further, the substance exists in the gaseous phase; i.e., from a large-scale or *macroscopic* point of view, the transition from solid to liquid to gas is in the direction of increasing temperature. From a *molecular* point of view this transition is in the direction of increasing molecular kinetic energy. Evidently, *there must be some connection between temperature and molecular kinetic energy.*

TABLE 20-1

Phase	Attractive forces among molecules	Molecular kinetic energy	Temperature
Solid	Strong	Small	Low
Liquid	Moderate	Moderate	Medium
Gas	Weak	Large	High

The preceding paragraphs may be conveniently summarized by means of Table 20-1.

20-2 AVOGADRO'S NUMBER

Surprising as it may seem, it was not until 1811 that the suggestion was first made, by Dalton, that the molecules of a given chemical substance are all alike. Before this, even by those who accepted a molecular theory, it had been assumed that the ultimate particles of the same substance varied in size and shape, just as one might find pebbles of different sizes, all made of the same material.

The first reasonably precise determination of the number of molecules in a mole, now called *Avogadro's number* N_A, was made by the French physicist Jean Perrin in the early 1900's. Before describing Perrin's experiment, we first recall an expression previously derived for the variation of atmospheric pressure with elevation. In Example 3 at the end of Section 18-2 we derived from the laws of hydrostatics and the equation of state of an ideal gas an expression for the ratio of the pressures at two different elevations in an ideal gas at a uniform temperature:

$$\ln \frac{p_2}{p_1} = -\frac{Mg}{RT}(y_2 - y_1). \qquad (20\text{-}1)$$

The molecular "weight" M equals the mass of 1 mole, and can be expressed as the product of Avogadro's number N_A and the mass m of a single molecule:

$$M = N_A m. \qquad (20\text{-}2)$$

Also, the number of moles n in a given sample of a pure substance equals the total number of molecules N divided by Avogadro's number N_A:

$$n = \frac{N}{N_A}.$$

The ideal gas law can therefore be written

$$pV = nRT = \frac{N}{N_A}RT,$$

or

$$\boldsymbol{n} = \frac{N}{V} = p\frac{N_A}{RT}, \qquad (20\text{-}3)$$

where $\boldsymbol{n}$ is the number of molecules per unit volume, N/V. Since N_A and R are universal constants, it follows that at constant temperature T the number of molecules per unit volume is directly proportional to the pressure. Combining Eqs. (20-2) and (20-3) with Eq. (20-1), we get

$$\ln \frac{\boldsymbol{n}_2}{\boldsymbol{n}_1} = -\frac{N_A mg}{RT}(y_2 - y_1), \qquad (20\text{-}4)$$

where $\boldsymbol{n}_2$ and $\boldsymbol{n}_1$ are the numbers of molecules per unit volume at elevations y_2 and y_1.

We might well ask at this point if the earth's atmosphere consists of widely separated molecules, not in contact with one another, and with each one attracted by the earth, why in the course of time do the molecules not all settle down to the earth's surface like drops of rain? The explanation is found in the violent random motions of the molecules, which in the absence of any gravitational force would eventually result in their dispersal throughout interstellar space. There are thus two tendencies acting in opposite directions, the gravitational force drawing all molecules toward the earth and their random motions tending to disperse them. The actual distribution represents the compromise between these two tendencies.

It is well known that extremely small particles, called *colloidal* particles, in a liquid of density less than that of the particles, do not all settle to the bottom but remain permanently suspended. These particles are large enough to be observed with a microscope, and the number per unit volume at different heights in the liquid can be counted. Perrin found that the number per unit volume decreased exponentially with height, following the same law as predicted by Eq. (20-4) for the number of molecules per unit volume in the atmosphere. Furthermore, the particles are observed to be in a state of continuous motion called "Brownian motion" after the English botanist Robert Brown, who first observed this effect in 1809 when studying with a microscope small pollen grains suspended in a liquid.

Perrin reasoned that the colloidal particles might be considered as the molecules of a "gas," molecules which were actually large enough to be seen and counted. If a colloidal suspension could be made in which all the particles were exactly alike, and if the mass m of each could be determined, and if Eq. (20–4) really describes the vertical distribution of particles, then Avogadro's number could be found by counting the number of particles per unit volume at two known heights in the suspension. One minor modification must be made in Eq. (20–4) for particles suspended in a fluid, namely, the "effective" value of g is reduced by the buoyant effect of the fluid, but the correction is readily made if the densities of the particles and the fluid are known.

It would occupy too much space to describe the ingenious methods used by Perrin to obtain uniform particles, to count the number per unit volume, and to measure their masses. A most interesting account will be found in Perrin's *Atoms*, translated by Hammick (1916). Suffice it to say that the difficulties were surmounted and that Perrin finally obtained for N_A a value between 6.5 and 7.2×10^{23} molecules per gram-mole.

The most precise value of N_A to date, obtained by using x-rays to measure the distance between the layers of molecules in a crystal, is

$$N_A = 6.02472 \times 10^{23} \text{ molecules (g-mole)}^{-1}.$$

Once Avogadro's number has been determined, it can be used to compute the mass of a molecule. The mass of 1 mole of atomic hydrogen is 1.008 g. Since by definition the number of molecules in 1 mole is Avogadro's number, it follows that the mass of a single atom of hydrogen is

$$m_H = \frac{1.008}{6.025 \times 10^{23}} = 1.673 \times 10^{-24} \text{ g}.$$

The mass of an atom of atomic weight 1 is $1/N_A$ or 1.660×10^{-24} g. Thus, for an oxygen molecule of molecular weight 32,

$$m_{O_2} = 32 \times 1.660 \times 10^{-24} = 53.12 \times 10^{-24} \text{ g}.$$

At standard conditions, 1 mole of an ideal gas occupies 22,400 cm^3. The number of molecules per cubic centimeter in an ideal gas at standard conditions is therefore

$$\frac{6.025 \times 10^{23}}{22{,}400} = 2.69 \times 10^{19} \text{ molecules cm}^{-3}.$$

This is known as *Loschmidt's number*, and of course it is the same for *all* gases. The number of molecules per unit volume can also be computed from Eq. (20–3), at any pressure and temperature.

The measurement of Avogadro's number therefore gives us the number of molecules per unit volume in a gas, and the mass of an individual molecule. We next turn to a method for computing the average velocities (or the average kinetic energies) of the molecules of a gas.

20–3 EQUATION OF STATE OF AN IDEAL GAS

A molecular theory of matter obviously accomplishes nothing if it simply endows the molecules of a substance with all the properties of that substance. The blue color of copper sulfate is not "explained" by postulating that it consists of molecules, each of which is blue. The fact that a gas is capable of expanding indefinitely is not explained by assuming that it consists of molecules each one of which can expand indefinitely. What one attempts to do is to account for the *complex* properties of matter in bulk as a consequence of *simple* properties ascribed to its molecules. Thus by assuming that a monatomic gas consists of a large number of particles having no properties other than mass and velocity, we can explain the observed equation of state of a gas at low pressure and can show that C_v for the gas should equal 2.99 cal mole^{-1} deg^{-1} (see Table 19–1). If we add the hypothesis that the "particles" are not simply geometrical points but have a finite size, then the general features of the viscosity, thermal conductivity, and coefficient of diffusion of a gas can be understood, as well as the fact that polyatomic gases have larger values of C_v than monatomic gases. By assuming that there are forces between the particles, the equation of state can be brought into better agreement with that of a real gas and the phenomena of liquefaction and solidification at low temperatures can be explained.

The electrical and magnetic properties of matter, and the emission and absorption of light by matter, call

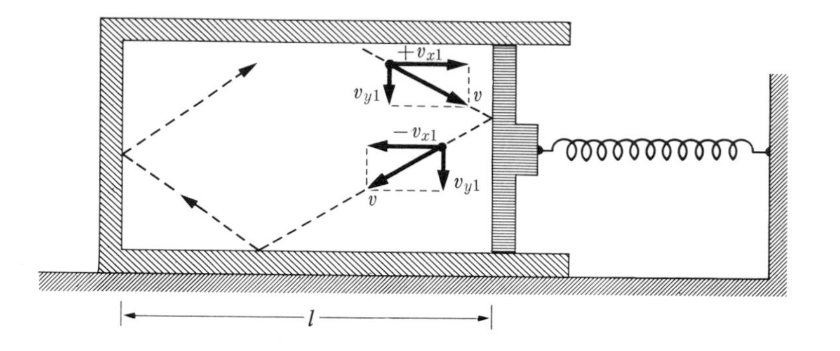

20-3 Collision of a single molecule with a piston.

for a molecular model that is itself an aggregate of subatomic particles. Some of these are electrically charged, and the forces between molecules appear to originate in these electric charges. To pursue this development of molecular theory further would take us too far into the whole problem of atomic and nuclear structure. Let us therefore return to the starting point and see what properties of a gas at low pressure can be explained by the simplest possible molecular model, an aggregate of particles having mass and velocity.

We know that a gas in a cylinder like that in Fig. 20-3 exerts a pressure against the piston. The molecular theory interprets the pressure not as a static push but as the average effect of many tiny impulsive blows resulting from the collisions of the molecules with the piston. Thus if the piston is forced to the left by a spring, we would not expect it to remain absolutely at rest but to oscillate slightly about some average position as if it were being bombarded by a helter-skelter rain of shot fired against it. (It is just these random molecular impacts, which do not exactly balance at all instants, that are responsible for the Brownian motion of suspended particles.)

We assume that the gas is composed of a large number of molecules, all having the same mass m and flying about in the cylinder with randomly directed velocities having magnitudes that differ from one molecule to another. Consider first the collision of a single molecule with the piston. Let the mass of the molecule be m, and let the x-component of its velocity be v_{x1}. Figure 20-3 illustrates the collision process. When all the molecules are considered, the *average* y-components of their velocities are not altered by collisions with the piston, since on the average the gas does not gain or lose any vertical motion as a whole. We shall therefore assume that the y-component of the velocity of each molecule is unchanged in the collision. Also, we identify the internal energy of the gas as the sum total of the kinetic energies of its molecules. Since the internal energy remains constant if the gas is isolated, the average kinetic energy of the molecules is constant. We therefore assume also that the kinetic energy of any one molecule is constant in a collision with the walls, that is, the collisions are perfectly elastic. Therefore the magnitude of the velocity v is the same before and after the collision, and since the y-component is unchanged, the *magnitude* of the x-component is unchanged also, although it is reversed in direction.

The average force exerted on the piston by one molecule can now be computed by setting the average force equal to the average rate of change of momentum. Since the x-velocity of the molecule reverses from $+v_{x1}$ to $-v_{x1}$ in each collision, the magnitude of the change in momentum in each collision is $2mv_{x1}$. To find the average force *during the time of one collision* we would divide the change of momentum by this time. However, after having made one collision with the piston, the molecule must travel to the other end of the cylinder and back before colliding again, and during all this time it exerts no force on the piston. To get the correct time average force we must average over the *total* time be-between collisions. As the molecule travels to the other end of the cylinder and back (see Fig. 20-3), the magnitude of its x-velocity remains constant. If ℓ is the length of the cylinder, the time for the round trip is

$$t = \frac{2\ell}{v_{x1}}$$

and the average force exerted on the piston by one molecule is

$$\text{Average force} = \begin{array}{l}\text{Average rate of change}\\\text{of momentum}\end{array}$$

$$= \frac{2mv_{x1}}{2\ell/v_{x1}} = \frac{mv_{x1}^2}{\ell}. \qquad (20\text{–}5)$$

Now consider other molecules 2, 3, etc., with x-velocities v_{x2}, v_{x3}, etc. The average force exerted by each is given by an expression like that in Eq. (20–5) and the total average force is

$$\text{Average force} = \frac{m}{\ell}(v_{x1}^2 + v_{x2}^2 + v_{x3}^2 + \cdots).$$

The average pressure p equals the total average force divided by the piston area A, so

$$p = \frac{m}{V}(v_{x1}^2 + v_{x2}^2 + v_{x3}^2 + \cdots), \qquad (20\text{–}6)$$

where $V = \ell A$ is the total volume of the cylinder.

Let N be the total number of molecules. The average value of the square of the x-velocity of all molecules is

$$\overline{v_x^2} = \frac{v_{x1}^2 + v_{x2}^2 + v_{x3}^2 + \cdots}{N}. \qquad (20\text{–}7)$$

Hence Eq. (20–6) can be written

$$p = \frac{Nm\overline{v_x^2}}{V}. \qquad (20\text{–}8)$$

The *magnitude* of the resultant velocity v, or the *speed* of any molecule, is given by

$$v^2 = v_x^2 + v_y^2 + v_z^2,$$

and averaging over all molecules,

$$\overline{v^2} = \overline{v_x^2} + \overline{v_y^2} + \overline{v_z^2}.$$

But since the x-, y-, and z-directions are all equivalent,

$$\overline{v_x^2} = \overline{v_y^2} = \overline{v_z^2}.$$

Hence

$$\overline{v^2} = 3\overline{v_x^2},$$

and Eq. (20–8) becomes

$$pV = \tfrac{1}{3}Nm\overline{v^2} = \tfrac{2}{3} \times \tfrac{1}{2}Nm\overline{v^2}. \qquad (20\text{–}9)$$

But $\frac{1}{2}m\overline{v^2}$ is the average kinetic energy of a single molecule, and the product of this and the total number of molecules N equals the total random kinetic energy, or internal energy U. Hence the product pV equals two-thirds of the internal energy:

$$pV = \tfrac{2}{3}U. \qquad (20\text{–}10)$$

By experiment, at low pressures, the equation of state of a gas is

$$pV = nRT.$$

The theoretical and experimental laws will therefore be in complete agreement if we set

$$U = \tfrac{3}{2}nRT. \qquad (20\text{–}11)$$

The average kinetic energy of a single molecule is then

$$\frac{U}{N} = \tfrac{1}{2}m\overline{v^2} = \frac{3}{2}\frac{nRT}{N}.$$

But the number of moles, n, equals the total number of molecules, N, divided by Avogadro's number, N_A, the number of molecules per mole:

$$n = \frac{N}{N_A}, \qquad \frac{n}{N} = \frac{1}{N_A}.$$

Hence

$$\tfrac{1}{2}m\overline{v^2} = \frac{3}{2}\frac{R}{N_A}T.$$

The ratio R/N_A occurs frequently in molecular theory. It is called the *Boltzmann constant, k*:

$$k = \frac{R}{N_A} = \frac{8.31 \times 10^7 \text{ ergs mole}^{-1}\text{ deg}^{-1}}{6.02 \times 10^{23}\text{ molecules mole}^{-1}}$$

$$= 1.38 \times 10^{-16}\text{ erg molecule}^{-1}\text{ deg}^{-1}.$$

Since R and N_A are universal constants, the same is true of k. Then

$$\tfrac{1}{2}m\overline{v^2} = \tfrac{3}{2}kT. \qquad (20\text{–}12)$$

The average kinetic energy per molecule, therefore, *depends only on the temperature* and not on the pressure, volume, or species of molecule.

The average value of the square of the speed is

$$\overline{v^2} = \frac{3kT}{m}$$

and the square root of this, or the root-mean-square speed v_{rms} is

$$v_{rms} = \sqrt{\overline{v^2}} = \sqrt{\frac{3kT}{m}}. \qquad (20\text{–}13)$$

Example 1. What is the average kinetic energy of a molecule of a gas at a temperature of 300°K?

$$\tfrac{1}{2}m\overline{v^2} = \tfrac{3}{2}kT = \tfrac{3}{2} \times 1.38 \times 10^{-16} \times 300 = 6.20 \times 10^{-14} \text{ erg.}$$

Example 2. What is the total random kinetic energy of the molecules in 1 mole of a gas at a temperature of 300°K?

$$U = N_A \times \tfrac{3}{2}kT = \tfrac{3}{2}RT = \tfrac{3}{2} \times 8.3 \times 10^7 \times 300$$
$$= 3750 \times 10^7 \text{ ergs} = 3750 \text{ J} = 900 \text{ cal.}$$

Example 3. What is the root-mean-square speed of a hydrogen molecule at 300°K?

The mass of a hydrogen molecule (see Section 20-2) is $m_{H_2} = 2 \times 1.66 \times 10^{-24}$ g $= 3.32 \times 10^{-24}$ g. Hence

$$v_{rms} = \sqrt{\frac{3kT}{m}} = \sqrt{\frac{3 \times 1.38 \times 10^{-16} \times 300}{3.32 \times 10^{-24}} \text{ cm}^2 \text{ s}^{-2}}$$
$$= 1.95 \times 10^5 \text{ cm s}^{-1}.$$

Example 4. What is the root-mean-square speed of a molecule of mercury vapor at 300°K?

The *kinetic energy* of a mercury molecule (or atom, since mercury vapor is monatomic) is the same as that of a hydrogen molecule at the same temperature. The mass of a mercury atom is

$$m_{Hg} = 201 \times 1.66 \times 10^{-24} \text{ g} = 334 \times 10^{-24} \text{ g.}$$

Therefore

$$v_{rms} = \sqrt{\frac{3 \times 1.38 \times 10^{-16} \times 300}{334 \times 10^{-24}} \text{ cm}^2 \text{ s}^{-2}}$$
$$= 0.19 \times 10^5 \text{ cm s}^{-1}.$$

When a gas expands against a moving piston it does work. The source of this work is the random kinetic energy of the gas molecules. Conversely, when work is done in compressing a gas the random kinetic energy of its molecules increases. But if the collisions with the walls are perfectly elastic, as we have assumed, how can a molecule gain or lose energy in a collision with a piston? To understand this, we must consider the collision of a molecule with a moving wall.

When a molecule makes a collision with a *stationary* wall, it exerts a momentary force on the wall but does no work, since the wall does not move. But if the wall is in motion, work *is* done in a collision. Thus if the piston in Fig. 20-3 is moving to the right, work is done on it by the molecules that strike it, and their velocities (and kinetic energies) after colliding are smaller than they were before a collision. The collision is still completely elastic, since the work done on the moving piston is just equal to the decrease in the kinetic energy of the molecule. Similarly, if the piston is moving toward the left, the kinetic energy of a colliding molecule is *increased*, the increase in kinetic energy being equal to the work done on the molecule.

20-4 MOLAR HEAT CAPACITY OF A GAS

When heat flows into a system in a process at constant volume, no work is done and all of the energy inflow goes into an increase in the internal energy U of the system. From the molecular viewpoint, the internal energy of a system is the sum total of the kinetic and potential energies of its molecules. If this sum can be computed as a function of temperature, then from its rate of change with temperature we can derive a theoretical expression for the molar heat capacity.

The simplest system is a monatomic ideal gas, for which the molecular energy is wholly kinetic and is given by Eq. (20-11):

$$\text{Random kinetic energy} = U = \tfrac{3}{2}nRT.$$

If the temperature is increased by dT, the random kinetic energy increases by

$$dU = \tfrac{3}{2}nR\, dT.$$

In a process in which the temperature increases by dT at constant volume, the energy flowing into a system is, by definition of C_v,

$$dQ = nC_v\, dT = dU.$$

Hence

$$C_v = \tfrac{3}{2}R.$$

But we showed in Section 19-12 and Table 19-1 that the experimental values of C_v for monatomic gases are, in fact, almost exactly equal to $\tfrac{3}{2}R$. This agreement is a striking confirmation of the basic correctness of the kinetic model of a gas and did much to establish this theory at a time when many scientists still refused to accept it.

Since $C_p = C_v + R$, it follows that the theoretical ratio of molar heat capacities for a monatomic ideal gas is

$$\frac{C_p}{C_v} = \gamma = \frac{\frac{3}{2}R + R}{\frac{3}{2}R} = \frac{5}{3} = 1.67.$$

This also is in good agreement with the experimental values in Table 19–1.

Polyatomic gases present a more complicated problem. We might expect that a molecule composed of two or more atoms could have an "internal energy" of its own, in the form of vibratory motion of the atoms relative to each other and of rotation of the molecule about its center of mass. Also, since every atom consists of a positive nucleus and one or more electrons, there may be energy associated with these electric charges.

The *temperature* of a gas depends on the average random *translational* kinetic energy of its molecules. When heat flows into a *monatomic* gas, at constant volume, all of this energy seems to go into an increase in random translational molecular kinetic energy, as evidenced by the agreement between the measured values of C_v and the values computed from the increase in translational kinetic energy. But when heat flows into a *polyatomic* gas, we would expect that a part of the energy would go toward increasing "molecular internal energy." Hence when equal amounts of heat flow into a monatomic and a polyatomic gas containing the same number of molecules, it is to be expected that the temperature rise in the polyatomic gas will be less than in the monatomic, because only a part of the energy is available for increasing *translational* molecular kinetic energy. In other words, to produce *equal* increases of temperature, *more* heat would have to be supplied to the polyatomic gas. This is a qualitative explanation, then, of the larger values of C_v for polyatomic gases listed in Table 19–1.

20–5 EXPERIMENTAL MEASUREMENT OF MOLECULAR SPEEDS

Direct measurements of the distribution of molecular speeds have been made by a number of methods. Figure 20–4 is a diagram of the apparatus used by Zartman and Ko in 1930–1934, a modification of a technique developed by Stern in 1920. Metallic silver is melted and evaporated in the oven O. A beam of silver atoms escapes through a small opening in the oven and passes through the slits S_1 and S_2 into an evacuated region. The cylinder C can be rotated at approximately 6000 rpm about the axis A. If the cylinder is at rest, the molecular beam enters the cylinder through a slit S_3 and strikes a curved glass plate G. The molecules stick to the glass plate, and the number arriving at any portion can be determined by removing the plate and measuring with a recording microphotometer the darkening that has resulted.

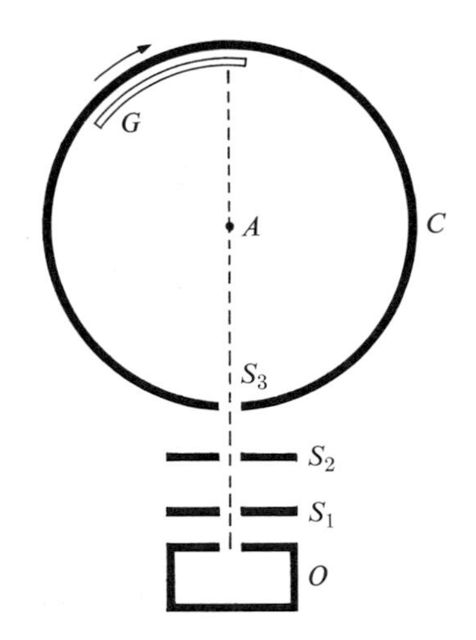

20–4 Apparatus used by Zartman and Ko in studying distribution of velocities.

Now suppose the cylinder is rotated. Molecules can enter it only during the short time intervals during which the slit S_3 crosses the molecular beam. If the rotation is clockwise, as indicated, the glass plate moves toward the right while the molecules cross the diameter of the cylinder. They therefore strike the plate at the left of the point of impact when the cylinder is at rest, and the more slowly they travel, the farther to the left is this point of impact. The blackening of the plate is therefore a measure of the "velocity spectrum" of the molecular beam.

A more recent and more precise measurement, making use of the free fall of the molecules in a beam, was performed by Estermann, Simpson, and Stern, in 1947. A simplified diagram of the apparatus is given in

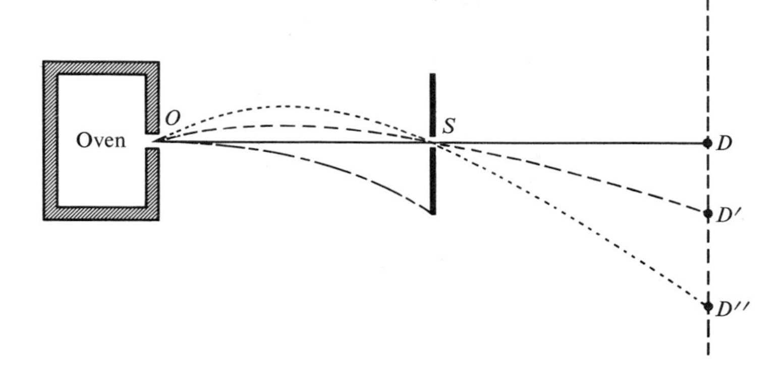

20-5 Schematic diagram of apparatus of Estermann, Simpson, and Stern.

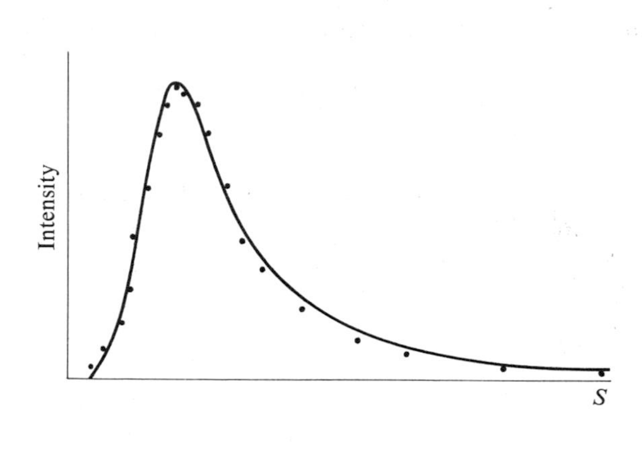

Fig. 20-6

Fig. 20-5. A beam of cesium atoms emerges from the oven slit O, passes through the slit S, and impinges on a hot tungsten wire D. The pressure of the residual gas in the apparatus is of the order of 10^{-8} mm of mercury. Both the slits and the detecting wire are horizontal. The cesium atoms striking the tungsten wire become ionized, re-evaporate, and are collected by a negatively charged cylinder surrounding the wire but not shown in the diagram. The ion current to the collecting cylinder then gives directly the number of cesium atoms impinging on the wire per second.

In the absence of a gravitational field, only those atoms emerging in a horizontal direction would pass through the slit S, and they would all strike the collector in the position D, regardless of their velocities. Actually, the path of each atom is a parabola, and an atom emerging from the slit O in a horizontal direction, as indicated by the dot and dash line (with the vertical scale greatly exaggerated), would not pass through the slit S. The dashed line and the dotted line represent the trajectories of two atoms that can pass through the slit S, the velocity along the dashed trajectory being greater than that along the dotted. Hence as the detector is moved down from the position D, those atoms with velocities corresponding to the dashed trajectory will be collected at D', those with the slower velocity corresponding to the dotted trajectory will be collected at D'', etc. Measurement of the ion current as a function of the vertical height of the collector then gives the velocity distribution.

Figure 20-6 is a graph of the results. The ordinate is proportional to the collector current and the abscissa to the vertical distance S of the collector below the position D in Fig. 20-5. The points are experimental values and the solid line is the theoretical curve computed from the Maxwell velocity distribution. The measured currents at large distances (small velocities) are somewhat smaller than the predicted values, an effect resulting from the collisions of molecules in the beam with one another near the oven slit. The general agreement between theory and experiment is excellent.

20-6 CRYSTALS

If a stream of ordinary illuminating gas is bubbled through benzine and then ignited, a smoky flame will result. When this flame comes in contact with a cool surface, *soot* (also called lampblack) will be deposited. Soot consists of small pieces of solid carbon. The so-called "lead" in a pencil also contains pieces of solid carbon called *graphite*, and a diamond is another form of solid carbon. The remarkably dissimilar mechanical, thermal, electrical, and optical properties of these three forms of carbon are explained in terms of the arrangement of the carbon atoms. In soot the atoms have no regular arrangement and are said to constitute an *amorphous* solid. Graphite and diamond are both crystals, but the orderly arrangement of atoms is different in the

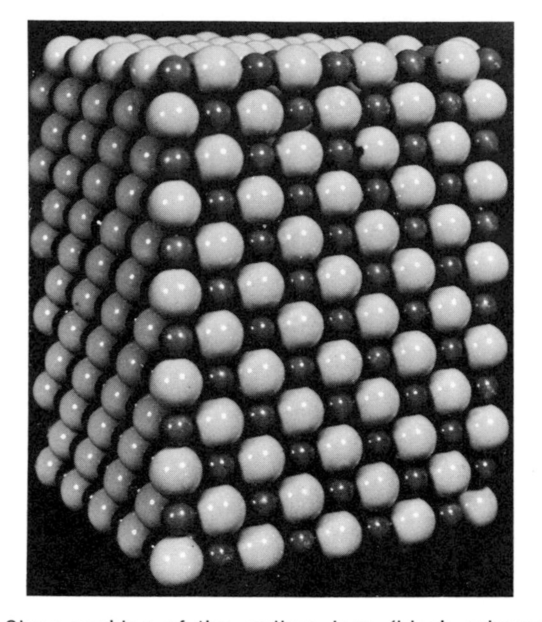

20–7 Close packing of the sodium ions (black spheres) and chlorine ions (white spheres) in a sodium chloride crystal. (Courtesy of Alan Holden; reprinted with permission of Educational Services, Inc., from *Crystals and Crystal Growing,* Doubleday and Co., New York, 1960.)

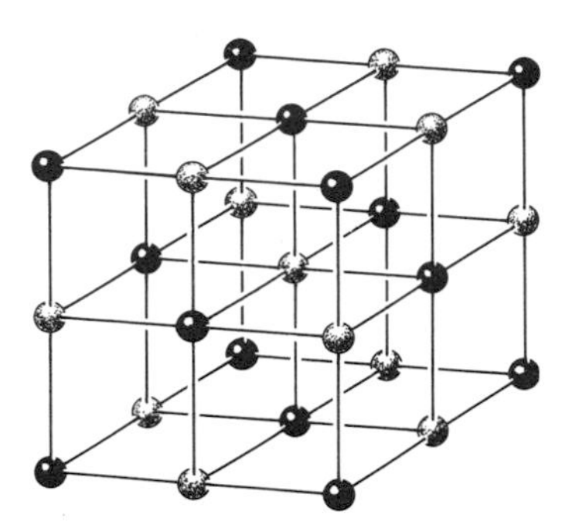

20–8 Symbolic representation of a sodium chloride crystal, with exaggerated distances between ions.

two crystal lattices. The study of the spatial arrangements of atoms (or molecules or ions) in the various types of crystal lattices is a fascinating and complicated part of physics. We shall be concerned in the following pages with only a few of the fundamental ideas needed to understand some of the mechanical and thermal properties of single crystals or solids composed of an aggregate of crystals, that is, *polycrystalline solids.*

The particles comprising a crystal lattice are *closely packed,* as seen in Fig. 20–2 and as suggested by the model in Fig. 20–7, where the sodium and chlorine ions of a sodium chloride crystal are represented as spheres touching one another. Strictly speaking, each ion consists of a nucleus surrounded by electrons in motion, so that the boundary of an ion must be thought of as the average positions of the outermost electrons. It follows that the outer electrons of one particle in a crystal lattice come close to and even at times interpenetrate with the outer electrons of a neighboring particle.

To make crystal structures simpler to comprehend, it is customary to exaggerate the distances between neighboring particles, as shown in Fig. 20–8, where a *face-centered cubic* lattice structure is depicted. Much can be learned by representing other crystal lattices in this way, with the aid of a supply of gum drops and toothpicks.

Suppose that the chemical composition and density of a very small volume element of a substance are measured at many different places. If the composition and density are the same at all points, the substance is said to be *homogeneous.* If the physical properties determining the transport of heat, electricity, light, etc., are the *same in all directions,* the substance is said to be *isotropic.* Cubic crystals are both homogeneous and isotropic, whereas all other crystals, although homogeneous, are not isotropic, i.e., they are *anisotropic.*

The *anisotropy* (accent on the syllable "ot") of noncubic crystals can be understood when the forces between neighboring particles in a lattice are taken into account. A convenient way of visualizing these forces is shown in Fig. 20–9, where there is depicted a large number of spheres, each connected to its neighbors by springs. The force constants of parallel springs are imagined equal, but the force constant of each spring pointing in one direction may be different from that of each spring pointing in a different direction. If each of the particles represented by spheres in Fig. 20–9 is imagined to be vibrating about its equilibrium position, a fairly accurate picture of the dynamic character of a crystal lattice will result. The mathematical calculation

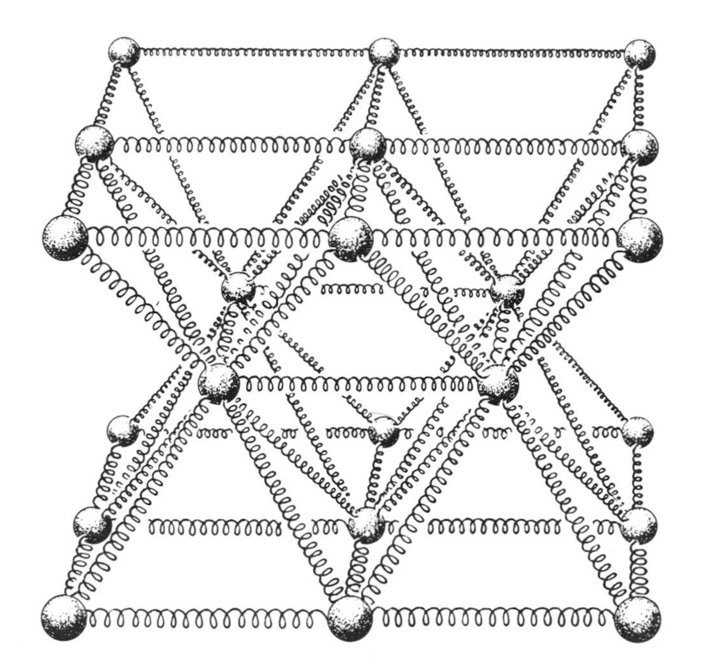

20-9 The forces between neighboring particles in a crystal may be visualized by imagining every particle to be connected to its neighbors by springs. In the case of a cubic crystal, all springs are assumed to have the same spring constant. Anisotrophy is connected with differing spring constants in different directions.

of the vibration frequencies of such a complex mechanical system was first attempted by Born and Von Karman early in the twentieth century. It is, of course, a very laborious and complicated calculation, but it has been carried out in detail for sodium chloride, potassium chloride, diamond, and silver.

The elastic properties of crystals can be partially understood from a diagram such as that in Fig. 20-9. Young's modulus is simply a measure of the stiffness of the springs in the direction in which the crystal is pulled or pushed. The shear modulus depends on the springs which are stretched and those which are compressed when the crystal is twisted. One of the first tests of the correctness of a lattice structure representing a particular material is to compare a calculated value of an elastic modulus with a measured value.

20-7 HEAT CAPACITY OF A CRYSTAL

In Section 20-3 a gas was assumed to consist of N particles in rapid linear motion so far apart on the average that no force of interaction existed except upon collision with one another or with a wall. By applying simple mechanical principles, it was shown that the product of the pressure p and the volume V is equal to

$$pV = \tfrac{2}{3} \times \tfrac{1}{2}Nm\overline{v^2} = \tfrac{2}{3}U,$$

where U is the total energy of the particles, which is entirely translational kinetic energy. In order for this result to agree with the equation of state of an ideal gas,

$$pV = nRT,$$

it is necessary to conclude that the average total energy per particle U/N, under equilibrium conditions, is equal to

$$\frac{U}{N} = \tfrac{3}{2}kT.$$

Since there are three translational degrees of freedom, $\frac{1}{2}kT$ of energy on the average is attributed to each degree of freedom. This can be extended also to rotational degrees of freedom, and the rough statement that to every degree of freedom of motion of a molecule there is attributed an average energy, at equilibrium, equal to $\frac{1}{2}kT$ is termed the *principle of the equipartition of energy.*

The methods of calculation just referred to represent a simplified version of the branch of physics called the *kinetic theory of gases*. There is, however, another way in which the behavior of large numbers of particles may be described and calculated, which, in some respects, is more general and more powerful. This more general method is called *statistical mechanics*. In statistical mechanics it is not necessary to specify whether the particles are molecules of a gas or ions occupying sites in a crystal lattice. The important thing is whether they are almost independent, and if so, whether the weak interaction among the particles is sufficient to enable the particles to exchange energy and finally come to equilibrium. If this is the case, the methods of statistical mechanics enable us to find the equilibrium value of the average energy per particle from a knowledge of the type of energy possessed by each individual particle. If the energy of an individual particle is expressed as the sum of a number of squared terms, like $\frac{1}{2}mv_x^2$ or $\frac{1}{2}I\omega^2$, or $\frac{1}{2}kx^2$, then it can be proved *rigorously* that at equilibrium the average energy per particle *associated with each squared* term is $\frac{1}{2}kT$. This is the rigorous statement of the principle of the equipartition of energy.

If we consider an assemblage of N vibrators, each vibrating along the x-axis with very small amplitude almost independently (but not enough to prevent a weak interaction with neighbors), we may consider the energy of each vibrator to consist of two squared terms,

$$\tfrac{1}{2}mv_x^2 + \tfrac{1}{2}kx^2.$$

The principle of equipartition then gives us for the average energy per particle at equilibrium the sum of two terms each equal to $\frac{1}{2}kT$, or a total of kT.

Now a crystal composed of N particles each of which interacts strongly with its neighbors vibrates in a very complicated way. In spite of the complications, it may be shown quite generally that the actual situation is *equivalent* to $3N$ oscillators, almost independent of one another, each vibrating with its own frequency. It follows therefore that the total energy at equilibrium is

$$U = 3NkT,$$

or, in terms of the number of moles n and the universal gas constant R,

$$U = 3nRT,$$

and finally

$$C_v = \frac{1}{n}\frac{dU}{dT} = 3R = 5.97 \text{ cal mole}^{-1} (\text{K}°)^{-1}.$$

But this is the *Dulong and Petit law* which a solid *approaches* as T increases but which is in complete disagreement with experiment as T approaches zero. The curves in Fig. 16–3 indicating the temperature dependence of C_v of some solids show in *every* case that C_v is *not* constant but approaches zero as T approaches zero.

A very careful analysis shows that there is nothing wrong with the methods of statistical mechanics nor with the equivalence between the actual vibrations of a crystal lattice and $3N$ almost independent oscillators. What is wrong is the assumption that a vibrator can assume *any value* of the energy given by $\frac{1}{2}mv_x^2 + \frac{1}{2}kx^2$. The fundamental idea of the quantum theory is that a vibrator of frequency f may *not* take on *any* value of the energy, but only *discrete energy values* given by whole-number multiples of the product hf:

$$1hf,\quad 2hf,\quad 3hf,\ \ldots,$$

where h is a constant, called *Planck's constant*. When this postulate is used in conjunction with the classical methods of statistical mechanics, we get as the average energy per particle at equilibrium the value

$$\frac{hf}{e^{hf/kT} - 1}.$$

When T gets very large, the exponent hf/kT gets small and the exponential becomes practically equal to $1 + hf/kT$. The average energy therefore reduces to kT, the classical value, and Dulong and Petit's law is approached as T approaches infinity. As T gets smaller, however, the exponent hf/kT gets larger and the average energy *decreases* and, along with it, the heat capacity of the crystal, in agreement with the experimental curves of Fig. 16–3.

Problems

Molecular Data

N_A = Avogadro's number = 6.02×10^{23} molecules mole^{-1},
Mass of a hydrogen atom = 1.67×10^{-24} g,
Mass of a nitrogen molecule = $28 \times 1.66 \times 10^{-24}$ g,
Mass of an oxygen molecule = $32 \times 1.66 \times 10^{-24}$ g.

20–1 Consider an ideal gas at 0°C and at 1 atm pressure. Imagine each molecule to be, on the average, at the center of a small cube. (a) What is the length of an edge of this small cube? (b) How does this distance compare with the diameter of a molecule?

20–2 A mole of liquid water occupies a volume of 18 cm^3. Imagine each molecule to be, on the average, at the center of a small cube. (a) What is the length of an edge of this small cube? (b) How does this distance compare with the diameter of a molecule?

20–3 What is the length of the side of a cube, in a gas at standard conditions, that contains a number of molecules equal to the population of the United States (about 200 million)?

20–4 (a) What is the average translational kinetic energy of a molecule of oxygen at a temperature of 300°K? (b) What is the average value of the square of its speed? (c) What is the root-mean-square speed? (d) What is the momentum of an oxygen molecule traveling at this speed? (e) Suppose a molecule traveling at this speed bounces back and forth between opposite sides of a cubical vessel 10 cm on a side. What is the average force it exerts on the walls of the container? (f) What is the average force per unit area? (g) How many molecules traveling at this speed are necessary to produce an average pressure of 1 atm? (h) Compare with the number of oxygen molecules actually contained in a vessel of this size, at 300°K and atmospheric pressure.

20–5 The speed of propagation of a sound wave in air at 27°C is about 1100 ft sec^{-1}. Compare this with the root-mean-square speed of nitrogen molecules at this temperature.

20–6 What is the total random kinetic energy of the molecules in 1 mole of helium at a temperature of (a) 300°K? (b) 301°K? (c) Compare the difference between these with the change in internal energy of 1 mole of helium when its temperature is increased by 1 K°, as computed from the relation $\Delta U = nC_v\,\Delta T$. (See Table 19–1.)

20–7 A flask contains a mixture of mercury vapor, neon, and helium. Compare (a) the average kinetic energies of the three types of atoms, and (b) the root-mean-square speeds.

20–8 (a) At what temperature is the rms speed of hydrogen molecules equal to the speed of the first earth satellite (about 18,000 mi hr^{-1})? (b) At what temperature is the rms speed equal to the escape speed from the gravitational field of the earth?

20–9 The oven in Fig. 20–4 contains bismuth at a temperature of 840°K, the drum is 10 cm in diameter and rotates at 6000 rev min^{-1}. Find the displacement on the glass plate G, measured from a point directly opposite the slit, of the points of impact of the molecules Bi and Bi_2. Assume that all molecules of each species travel with the rms speed appropriate to that species.

20–10 In the apparatus of Estermann, Simpson, and Stern in Fig. 20–5, the distances *OS* and *SD* are each 1 m. (a) Calculate the distance of the detector below the central position, *D*, for cesium atoms having a speed equal to the rms speed in a beam emerging from an oven at a temperature of 460°K. (b) Calculate also the "angle of departure" of the trajectory. The atomic weight of cesium is 133.

21 Traveling Waves

21-1 INTRODUCTION

Waves on the surface of a body of water, produced by the wind or by some other disturbance, are a familiar sight. A source of sound is heard by means of traveling waves in the intervening atmosphere, and the vibrations of the sound source itself constitute a so-called *stationary* wave. Many of the observed properties of light are best explained by a wave theory, and we believe that light waves are of the same fundamental nature as radio waves, infrared and ultraviolet waves, x-rays, and gamma rays. One of the outstanding developments of twentieth century physics has been the discovery that all matter is endowed with wave properties and that a beam of electrons, for example, is reflected by a crystal in much the same way as a beam of x-rays.

The subject of wave motion is closely related to that of harmonic motion. When a wave travels in a material substance, every particle of the substance oscillates about its equilibrium position, so that we must deal with the vibrations of a large number of particles instead of only one.

When the vibrations of the particles are at right angles to the direction of travel of the wave, the wave is called *transverse*. If the particles oscillate in the direction of propagation, the wave is called *longitudinal*. Suppose that one end of a medium is forced to vibrate periodically, the displacement y (either transverse or longitudinal) varying with the time according to the equation of simple harmonic motion:

$$y = \begin{cases} Y \sin \omega t, \\ \quad \text{or} \\ Y \cos \omega t. \end{cases}$$

During half a cycle, a displacement in one direction is propagated through the medium, and during the other half, a displacement in the opposite direction is caused to proceed. The resulting continuous train of disturbances traveling with a speed depending on the properties of the medium is called a *wave*.

To fix our ideas, suppose that one end of a stretched string is forced to vibrate periodically in a transverse direction with simple harmonic motion of amplitude Y, frequency f, and period $\tau = 1/f$. For the present we shall assume the string to be long enough so that any effects at the far end need not be considered. A *continuous train* of transverse sinusoidal waves then advances along the string. The shape of a portion of the string near the end, at intervals of $\frac{1}{8}$ of a period, is shown in Fig. 21-1 for a total time of one period. The string is assumed to have been vibrating for a sufficiently long time so that the shape of the string is sinusoidal for an indefinite distance from the driven end. It will be seen from the figure that the waveform advances steadily toward the right, as indicated by the short arrow pointing to one particular wave crest, while any one point on the string (see the black dot) oscillates about its equilibrium position with simple harmonic motion. It is important to distinguish between the motion of the *waveform*, which moves with constant velocity c along the string, and the motion of *a particle of the string*, which is simple harmonic and transverse to the string.

The distance between two successive maxima (or between any two successive points in the same phase) is the *wavelength* of the wave and is denoted by λ. Since the waveform, traveling with constant velocity c, advances a distance of one wavelength in a time interval

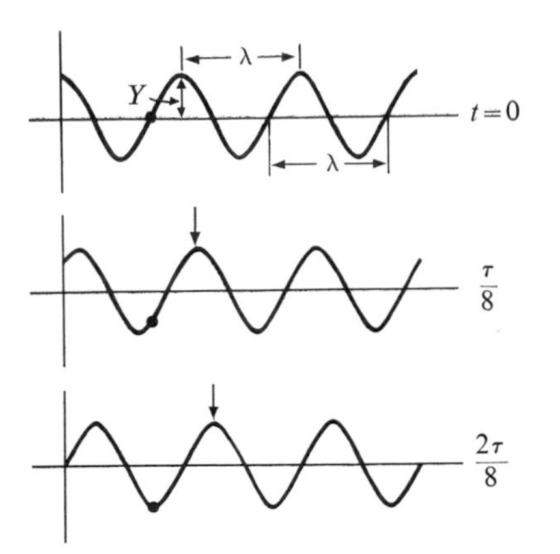

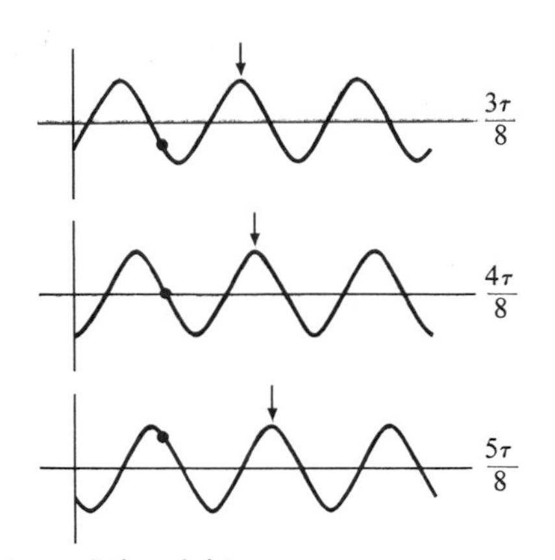

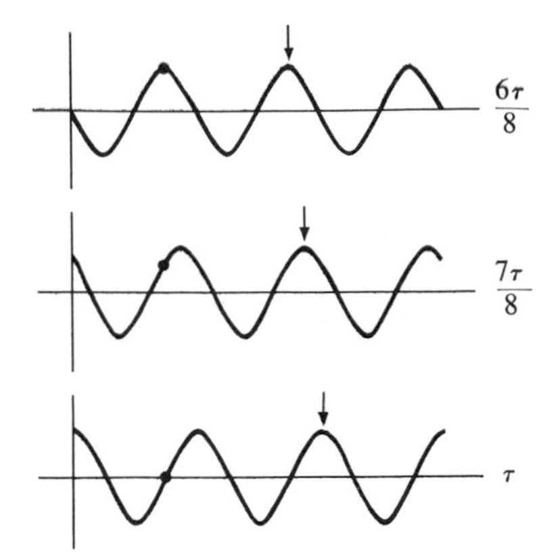

21-1 A sinusoidal transverse wave traveling toward the right, shown at intervals of one-eighth of a period.

of one period, it follows that $c = \lambda/\tau$, or

$$\boxed{c = f\lambda.} \tag{21-1}$$

That is, *the velocity of propagation equals the product of frequency and wavelength.*

To understand the mechanics of a longitudinal wave, consider a long tube filled with a fluid and provided with a plunger at the left end, as shown in Fig. 21-2. The dots represent particles of the fluid. Suppose the plunger is forced to undergo a simple harmonic vibration parallel to the direction of the tube. During a part of each oscillation, a region whose pressure is above the equilibrium pressure is formed. Such a region is called a *condensation* and is represented by closely spaced dots. Following the production of a condensation, a region is formed in which the pressure is lower than the equilibrium value. This is called a *rarefaction* and is represented by widely spaced dots. The condensations and rarefactions move to the right with constant velocity c, as indicated by successive positions of the small vertical arrow. The velocity of longitudinal waves in air (sound) at 20°C is 344 m s^{-1} or 1130 ft s^{-1}. The motion of a single particle of the medium, shown by a heavy black dot, is simple harmonic, parallel to the direction of propagation.

The wavelength is the distance between two successive condensations or two successive rarefactions, and the same fundamental equation, $c = f\lambda$, holds in this, as in all types of waves.

21-2 MATHEMATICAL REPRESENTATION OF A TRAVELING WAVE

Suppose that a wave of any sort travels from left to right in a medium. Let us compare the motion of any one particle of the medium with that of a second particle to the right of the first. We find that the second particle moves in the same manner as the first, but after a lapse of time that is greater the greater the distance of the second particle from the first. Hence if one end of a stretched string oscillates with simple harmonic motion, all other points oscillate with simple harmonic motion of the same amplitude and frequency. The *phase angle* of the motion, however, is different for different points.

Let the displacement of a particle at the origin ($x = 0$) be given by $y = Y \sin \omega t$.

The displacement of a particle at the right of the origin lags that of the particle at the origin by some angle ϕ. That is,

$$y = Y \sin (\omega t - \phi).$$

The lag angle ϕ is proportional to the distance of the particle from the origin, or to its coordinate x:

$$\phi = kx, \tag{21-2}$$

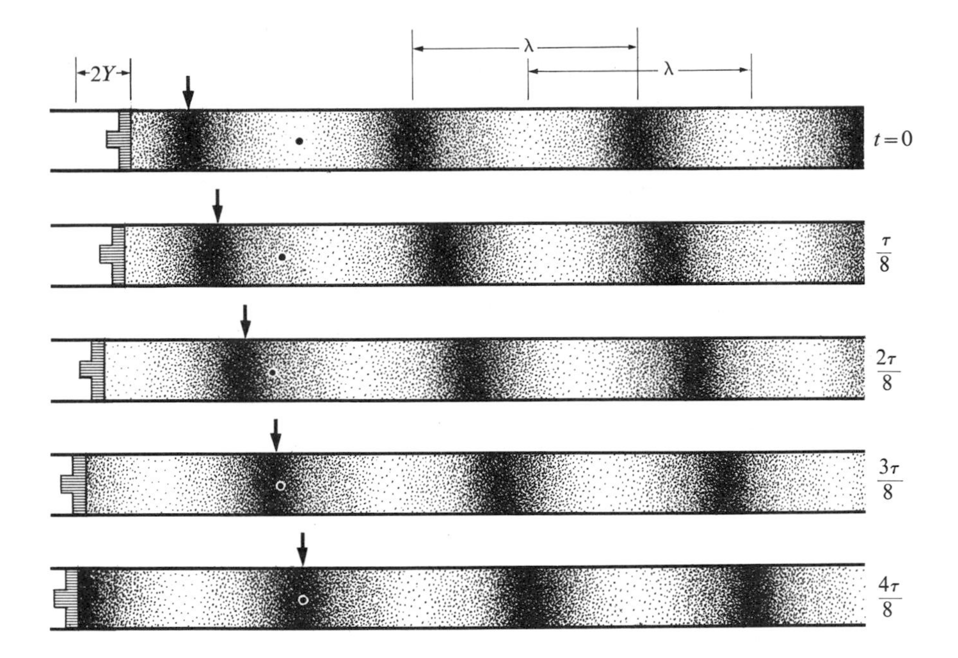

21-2 A sinusoidal longitudinal wave traveling toward the right, shown at intervals of one-eighth of a period.

where k is called the *propagation constant*. Hence for such a particle

$$y = Y \sin(\omega t - kx). \tag{21-3}$$

This equation represents a sine wave traveling to the *right*. If the wave travels to the *left*, particles at the right of the origin *lead* the particle at the origin in phase, and the equation of the wave is

$$y = Y \sin(\omega t + kx). \tag{21-4}$$

A particle at a distance of 1 wavelength from the origin vibrates in step with that at the origin. Hence it lags (or leads) by an angle $\phi = 2\pi$. By setting $x = \lambda$ and $\phi = 2\pi$ in Eq. (21-2), we get

$$2\pi = k\lambda, \qquad k = \frac{2\pi}{\lambda}. \tag{21-5}$$

The propagation constant k is therefore 2π times the reciprocal of the wavelength. Since $\omega = 2\pi f = 2\pi/\tau$, the equation of a traveling wave can also be written as

$$y = Y \sin 2\pi\left(\frac{t}{\tau} \pm \frac{x}{\lambda}\right).$$

At any given *time* t, Eq. (21-3) gives the displacement y of a particle, from its equilibrium position, as a function of the *coordinate* x of the particle. If the wave is a transverse wave in a string, the equation represents the *shape* of the string at that instant, as if we had taken a snapshot of the string. Thus at time $t = 0$,

$$y = Y \sin(-kx) = -Y \sin kx = -Y \sin 2\pi \frac{x}{\lambda}.$$

This curve is plotted in Fig. 21-3.

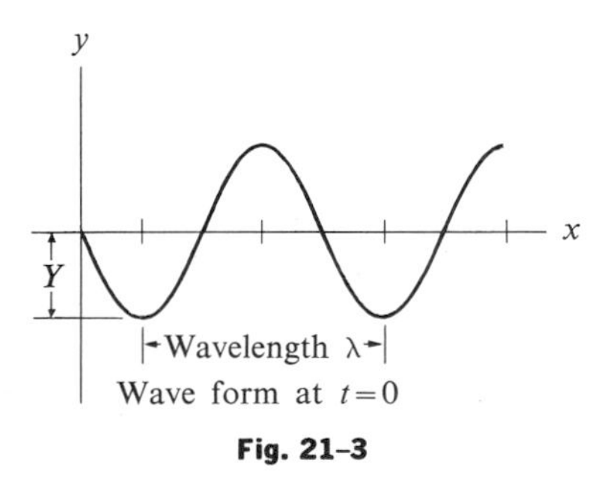

Fig. 21-3

At any given *coordinate* x, Eq. (21-3) gives the displacement y of the particle at that coordinate, as a

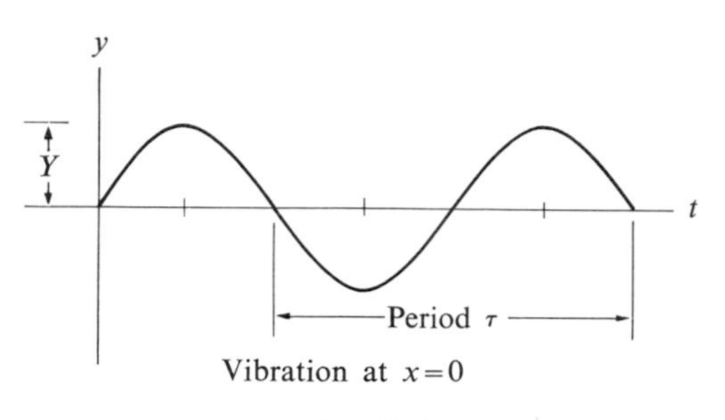

Fig. 21-4

function of *time*. Thus at the coordinate $x = 0$,

$$y = Y \sin \omega t = Y \sin 2\pi \frac{t}{\tau}.$$

This curve is plotted in Fig. 21-4.

We see, therefore, that the displacement y of a particle depends both on the coordinate of the particle and on the time. In other words, *y is a function of two* **independent** *variables, x* and *t*. A graph of Eq. (21-3) would be a *surface* if we plotted x and t along rectangular axes in a horizontal plane, and plotted y vertically above this plane.

It is important to distinguish carefully between the *speed of propagation* of the waveform, c, and the *particle speed*, v, of a particle of the medium in which a wave is traveling. The wave speed c is given by $f\lambda$, and since $\omega = 2\pi f$ and $k = 2\pi/\lambda$, we have

$$c = f\lambda = \frac{\omega}{2\pi} \cdot \frac{2\pi}{k} = \frac{\omega}{k}. \qquad (21\text{-}6)$$

For a transverse traveling sine wave given by Eq. (21-4), namely

$$y = Y \sin (\omega t + kx),$$

the particle displacement y is seen to be a function of *two variables, t* and *x*. The particle speed v at any point where x has a constant value is obtained by taking the derivative of y with respect to t, *holding x constant*. Such a derivative is called a *partial derivative* and is written $\partial y/\partial t$. Thus

$$v = \frac{\partial y}{\partial t} = \omega Y \cos (\omega t + kx). \qquad (21\text{-}7)$$

The acceleration of the particle a is obtained by differentiating partially a second time, or

$$a = \frac{\partial^2 y}{\partial t^2} = -\omega^2 Y \sin (\omega t + kx). \qquad (21\text{-}8)$$

There is, however, another second-order partial derivative of the particle displacement, namely, with respect to x, *holding t constant*. Thus

$$\frac{\partial^2 y}{\partial x^2} = -k^2 Y \sin (\omega t + kx). \qquad (21\text{-}9)$$

It follows from the last two equations that

$$\frac{\partial^2 y/\partial t^2}{\partial^2 y/\partial x^2} = \frac{\omega^2}{k^2} = c^2,$$

since $c = \omega/k$. The *partial differential equation*

$$\frac{\partial^2 y}{\partial t^2} = c^2 \frac{\partial^2 y}{\partial x^2}$$

is one of the most important in all of physics. It is called the *wave equation*, and whenever it occurs, the conclusion is made immediately that y is propagated as a traveling wave along the x-axis with a wave speed c.

21-3 CALCULATION OF THE SPEED OF A TRANSVERSE PULSE

Suppose that a string is stretched between two fixed supports. If the string is struck a transverse blow at some point, or if a small portion is displaced sidewise and released, disturbances will be observed to travel outward in both directions from the displaced portion. Each of these is called a *transverse pulse*, the direction of motion of the particles of the string being at right angles to the direction of travel, or of *propagation*, of the pulse. Each pulse retains its shape as it travels, and each travels with a constant speed which we shall represent by c.

Consider the string depicted in Fig. 21-5, under a tension S and with linear density (mass per unit length) μ. In Fig. 21-5(a) the string is at rest. At time $t = 0$, a constant transverse force F is applied at the left end of the string. As a result, this end moves up with a constant transverse speed v. Figure 21-5(b) shows the shape of the string after a time t has elapsed. All points of the string at the left of the point P are moving with speed v,

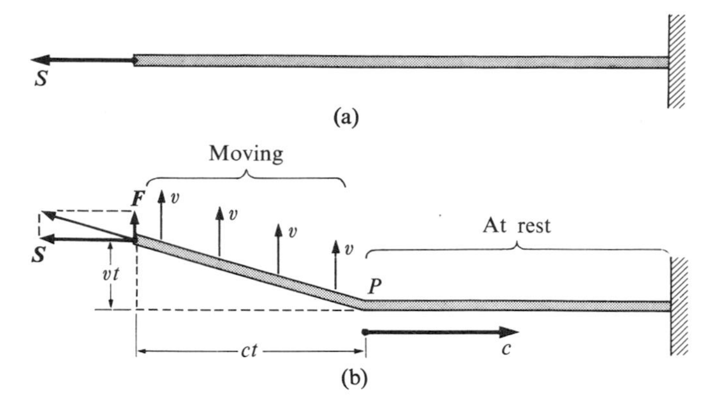

21-5 Propagation of a transverse disturbance in a string.

whereas all points at the right of P are still at rest. The boundary between the moving and the stationary portions is traveling to the right with the *speed of propagation* c. The left end of the string has moved up a distance vt and the boundary point P has advanced a distance ct.

The speed of propagation c can be calculated by setting the transverse impulse (transverse force $\times$ time) equal to the change of transverse momentum of the moving portion (mass $\times$ transverse velocity). The impulse of the transverse force F in time t is Ft. By similar triangles,

$$\frac{F}{S} = \frac{vt}{ct}, \qquad F = S\frac{v}{c}.$$

Hence

$$\text{Transverse impulse} = S\frac{v}{c}t.$$

The mass of the moving portion is the product of the mass per unit length μ and the length ct. Hence

$$\text{Transverse momentum} = \mu ctv.$$

Applying the impulse-momentum theorem, we get

$$S\frac{v}{c}t = \mu ctv,$$

and therefore

$$\boxed{c = \sqrt{S/\mu} \quad \text{(transverse).}} \qquad (21\text{–}10)$$

Thus it is seen that the velocity of propagation of a transverse pulse in a string depends only on the tension and the mass per unit length.

In numerical applications of Eq. (21–6), attention must be paid to the units employed. Appropriate units are shown in Table 21–1.

TABLE 21–1 UNITS FOR EQ. (21–6)

System of units	S	μ	c
mks	newton	kg m^{-1}	m s^{-1}
cgs	dyne	g cm^{-1}	cm s^{-1}
Engineering	pound	slug ft^{-1}	ft s^{-1}

Example 1. One end of a rubber tube is fastened to a fixed support. The other end passes over a pulley at a distance of 8 m from the fixed end, and carries a load of 2 kg. The mass of the tube, between the fixed end and the pulley, is 600 g. What is the speed of a transverse wave in the tube?

The tension in the tube equals the weight of the 2-kg load, or

$$S = 2 \times 9.8 = 19.6 \text{ N}.$$

The mass of the tube, per unit length, is

$$\mu = \frac{m}{L} = \frac{0.60 \text{ kg}}{8 \text{ m}} = 0.075 \text{ kg m}^{-1}.$$

The speed of propagation is therefore

$$c = \sqrt{\frac{S}{\mu}} = \sqrt{\frac{19.6 \text{ N}}{0.075 \text{ kg m}^{-1}}} = 16 \text{ m s}^{-1}.$$

Example 2. Suppose that a sine wave of amplitude $Y = 10$ cm and wavelength $\lambda = 3$ m travels along the tube from left to right. What is the maximum *transverse* speed of a point of the tube?

The equation of the wave is

$$y = Y \sin(\omega t - kx).$$

The transverse speed is

$$v = \frac{\partial y}{\partial t} = \omega Y \cos(\omega t - kx).$$

The maximum transverse speed is

$$v_{max} = \omega Y = 2\pi f Y = 2\pi \frac{c}{\lambda} Y$$

$$= 2\pi \times \frac{16 \text{ m s}^{-1}}{3 \text{ m}} \times 0.10 \text{ m} = 4.9 \text{ m s}^{-1}.$$

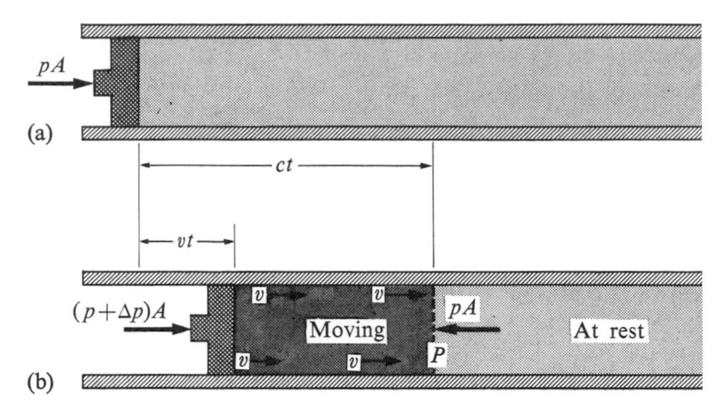

21-6 Propagation of a longitudinal disturbance in a fluid confined in a tube.

21-4 CALCULATION OF THE SPEED OF A LONGITUDINAL PULSE

Figure 21-6 shows a fluid (liquid or gas) of density ρ in a tube of cross-sectional area A and under a pressure p. In Fig. 21-6(a) the fluid is at rest. At time $t = 0$, the piston at the left end of the tube is set in motion toward the right with a speed v. Figure 21-6(b) shows the fluid after a time t has elapsed. All portions of the fluid at the left of point P are moving with speed v, whereas all portions at the right of P are still at rest. The boundary between the moving and the stationary portions travels to the right with the speed of propagation c. The piston has moved a distance vt and the boundary has advanced a distance ct. As for a transverse disturbance in a string, the speed of propagation can be computed from the impulse-momentum theorem.

The quantity of fluid set in motion in time t is that originally occupying a volume of length ct and of cross-sectional area A. The mass of this fluid is therefore ρctA and the longitudinal momentum it has acquired is

$$\text{Longitudinal momentum} = \rho ctAv.$$

We next compute the increase of pressure, Δp, in the moving fluid. The original volume of the moving fluid, Act, has been decreased by an amount Avt. From the definition of bulk modulus B (see Chapter 10),

$$B = \frac{\text{change in pressure}}{\text{fractional change in volume}} = \frac{\Delta p}{Avt/Act}.$$

Therefore

$$\Delta p = B\frac{v}{c}.$$

The pressure in the moving fluid is therefore $p + \Delta p$, and the force exerted on it by the piston is $(p + \Delta p)A$. The *net* force on the moving fluid (see Fig. 21-6b) is ΔpA, and the longitudinal impulse is

$$\text{Longitudinal impulse} = \Delta pAt = B\frac{v}{c}At.$$

Applying the impulse-momentum theorem, we get

$$B\frac{v}{c}At = \rho ctAv,$$

and therefore

$$c = \sqrt{B/\rho} \qquad \text{(longitudinal).} \tag{21-11}$$

The speed of propagation of a longitudinal pulse in a fluid therefore depends only on the bulk modulus and density of the medium.

When a solid bar is struck a blow at one end, the situation is somewhat different from that of a fluid confined in a tube of constant cross section, since the bar will expand slightly sidewise when it is compressed longitudinally. It can be shown by the same type of reasoning as that just given that the velocity of a longitudinal pulse in the bar is given by

$$c = \sqrt{Y/\rho} \qquad \text{(longitudinal),} \tag{21-12}$$

where Y is Young's modulus, defined in Chapter 10.

21-5 ADIABATIC CHARACTER OF A LONGITUDINAL WAVE

It is a familiar fact that compression of a fluid causes a rise in its temperature unless heat is withdrawn in some way. Conversely, an expansion is accompanied by a temperature decrease unless heat is added. As a longitudinal wave advances through a fluid, the regions which are compressed at any instant are slightly warmer than those that are expanded. The condition is present, therefore, for the conduction of heat from a condensation to a rarefaction. The quantity of heat conducted per unit time and per unit area depends on the thermal conductivity of the fluid and upon the distance between a condensation and its adjacent rarefaction (half a wavelength). Now for ordinary frequencies, say from 20 vibrations per second to 20,000 vibrations per second, and for even the best known heat conductors, the wavelength is too large and the thermal conductivity too small for an appreciable amount of heat to flow. The compressions and rarefactions are therefore *adiabatic* rather than isothermal.

In the expression for the speed of a longitudinal wave in a fluid, $c = \sqrt{B/\rho}$, the bulk modulus B is defined by the relation

$$B = \frac{\text{change in pressure}}{\text{fractional change in volume}}.$$

The change in volume produced by a given change of pressure depends upon whether the compression (or expansion) is adiabatic or isothermal. There are therefore two bulk moduli, the adiabatic bulk modulus B_{ad} and the isothermal bulk modulus. The expression for the speed of a longitudinal wave should therefore be written

$$\boxed{c = \sqrt{\frac{B_{ad}}{\rho}}}. \tag{21-13}$$

In the case of an ideal gas, the relation between pressure p and volume V during an adiabatic process is given by

$$pV^{\gamma} = \text{constant}, \tag{21-14}$$

where γ is the ratio of the heat capacity at constant pressure to the heat capacity at constant volume.

The definition of the adiabatic bulk modulus is

$$B_{ad} = -\left(\frac{dp}{dV/V}\right)_{ad} = -V\left(\frac{dp}{dV}\right)_{ad}.$$

To calculate the adiabatic bulk modulus B_{ad}, we must evaluate the derivative $(dp/dV)_{ad}$ with the aid of the adiabatic equation. Thus, taking logarithms of both sides of Eq. (21-14), we get

$$\ln p + \gamma \ln V = \ln \text{constant},$$

and, taking the differential of this equation,

$$\frac{dp}{p} + \gamma\frac{dV}{V} = 0, \qquad \text{whence} \qquad \left(\frac{dp}{dV}\right)_{ad} = -\gamma\frac{p}{V},$$

and

$$B_{ad} = \gamma p. \tag{21-15}$$

Therefore

$$c = \sqrt{\gamma p/\rho} \qquad \text{(ideal gas)}. \tag{21-16}$$

But, for an ideal gas,

$$p/\rho = RT/M,$$

where R is the universal gas constant, M the molecular weight, and T the Kelvin temperature. Therefore

$$c = \sqrt{\gamma RT/M} \qquad \text{(ideal gas)}, \tag{21-17}$$

and since for a given gas, γ, R, and M are constants, we see that the velocity of propagation is proportional to the square root of the Kelvin temperature.

Let us use Eq. (21-17) to compute the velocity of longitudinal waves in air. The mean molecular weight of air is 28.8, $\gamma = 1.40$, and $R = 8.31 \times 10^7$ ergs mole^{-1} deg^{-1}. Let $T = 300$°K. Then

$$c = \sqrt{\frac{1.40\,(8.31 \times 10^7 \text{ dyn cm mole}^{-1}\text{ deg}^{-1})\,(300 \text{ deg})}{28.8 \text{ g mole}^{-1}}}$$
$$= 34{,}800 \text{ cm s}^{-1} = 348 \text{ m s}^{-1} = 1140 \text{ ft s}^{-1}.$$

This is in good agreement with the measured velocity at this temperature.

Longitudinal waves in air give rise to the sensation of sound. The ear is sensitive to a range of sound frequencies from about 20 to about 20,000 hertz (one hertz = 1 Hz = 1 cycle s^{-1}). From the relation $c = f\lambda$, the corresponding wavelength range is from about 56 ft, corresponding to a 20-Hz note, to about 0.056 ft or 5/8 in., corresponding to 20,000 Hz.

The molecular nature of a gas has been ignored in the preceding discussion, and a gas has been treated as though it were a continuous medium. Actually, we know that a gas is composed of molecules in random motion, separated by distances which are large compared with their diameters. The vibrations which constitute a wave in a gas are superposed on the random thermal motion. At atmospheric pressure, the mean free path is about

21–7 Diagram for illustrating longitudinal traveling waves.

10^{-5} cm, while the displacement amplitude of a faint sound may be only one ten-thousandth of this amount. An element of gas in which a sound wave is traveling can be compared to a swarm of gnats, where the swarm as a whole can be seen to oscillate slightly while individual insects move about through the swarm, apparently at random.

Since the *shape* of a fluid does not change when a longitudinal wave passes through it, it is not as easy to visualize the relation between particle motion and wave motion as it is for transverse waves in a string. Figure 21–7 may help in correlating these motions. To use this figure, cut a slit about $\frac{1}{16}$ in. wide and $3\frac{1}{4}$ in. long in a 3-in. $\times$ 5-in. card, place the card over the figure with the slit at the top of the diagram, and move the card downward with constant velocity. The portions of the sine curves that are visible through the slit correspond to a row of particles along which there is traveling a longitudinal, sinusoidal wave. Notice that each particle executes simple harmonic motion about its equilibrium position, with a phase that increases continuously along the slit, while the regions of maximum condensation and rarefaction move from left to right with constant speed. Moving the card upward produces a wave traveling from right to left.

21–6 WAVES IN A CANAL

Probably the most familiar type of wave motion is that observed at the surface of a body of water and produced by the winds or some other disturbance. The oscillations of the water particles in these waves are not confined to the surface, however, but extend with diminishing amplitude to the very bottom. Furthermore, the oscillations have both a longitudinal and a transverse component. For this reason, and also because the motion is governed by the laws of hydrodynamics, a complete derivation of the wave equation calls for mathematical methods beyond the scope of this book. The basic physical approach, however, is perfectly straightforward.

Consider a long canal of rectangular cross section and with frictionless walls, containing a depth h of an ideal incompressible liquid of density ρ and surface tension γ. When a train of waves travels along the canal, each element of the liquid is displaced from its equilibrium position both horizontally and vertically. The restoring force on the element results in part from pressure differences brought about by the variations in depth from point to point, and in part from surface tension effects arising from the curvature of the free surface. The solution of the resulting differential equation of motion must satisfy the boundary conditions that the pressure at the upper surface is constant and equal to atmospheric pressure, and that the vertical displacement at the bottom of the canal is always zero. In addition, the motion of the fluid must satisfy the equation of continuity.

For simple harmonic waves, in which the x- and y-components of displacement are sine or cosine func-

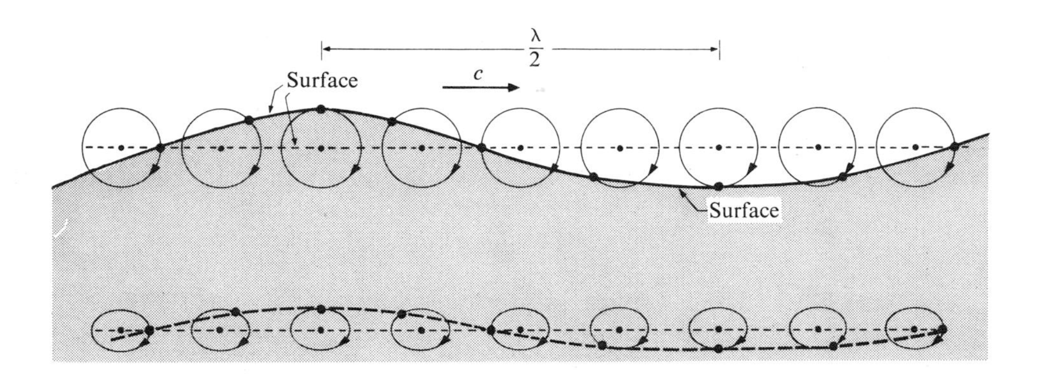

21-8 Particle motion and wave shape for waves in a canal.

tions of $(x - ct)$, the speed c is given by

$$c^2 = \left(\frac{g\lambda}{2\pi} + \frac{2\pi\gamma}{\rho\lambda}\right) \tanh \frac{2\pi h}{\lambda}, \tag{21-18}$$

where g is the acceleration of gravity and λ is the wavelength.

We immediately recognize one very significant difference between this expression and that for the speed of propagation of transverse waves in a string, or compressional waves in a fluid. Here, the speed depends not only on the properties of the medium *but on the wavelength* λ *as well.* When this is the case, the medium is said to be *dispersive.* Such a dependence of speed on wavelength occurs in many other types of waves. For example, a prism separates a parallel beam of white light into a spectrum, with different wavelengths emerging in different directions, because glass is a dispersive medium for light waves. In general, the particles of liquid move in ellipses in vertical planes parallel to the length of the canal, the long axis of the ellipse being horizontal. This motion can be considered as the superposition of two simple harmonic oscillations of the same frequency and different amplitudes, one in the horizontal and one in the vertical direction, and with a phase difference of 90°. The wave can therefore be considered as the superposition of a longitudinal and a transverse wave 90° out of phase with each other and having different amplitudes. If the wavelength is as small or smaller than the depth, the two amplitudes are very nearly equal at the surface and the surface particles move in circles. The amplitudes of both components decrease with increasing depth, but the vertical component decreases more rapidly than does the horizontal. At the bottom, the vertical component becomes zero and the oscillation is wholly longitudinal.

Figure 21-8 illustrates the paths of motion of particles in the surface layer and at some depth below the surface. The upper horizontal dotted line represents the free surface of the liquid at rest. The circles are the paths of particles whose equilibrium position is at the center of the circle. When a train of waves travels from left to right, the particles revolve clockwise in these paths. The full line gives the shape of the surface at the instant shown.

The lower dotted line passes through the equilibrium positions of particles at some depth below the surface. Their paths are elliptical, as indicated, and the dashed line is their locus at the instant when the free surface has the shape shown by the full line.

There are certain properties possessed by all waves. Propagation, reflection, refraction, absorption, interference, and diffraction are the most common. These wave phenomena may be demonstrated effectively by means of a "ripple tank" consisting of a horizontal sheet of glass or plastic forming the bottom of a shallow tank, filled with water about 1 cm deep. A strong source of light beneath the tank sends light through the water waves, which cast a shadow on a translucent screen (tracing cloth) above the tank. A convenient source of water waves is provided by a smooth sphere about the size of a pencil eraser attached to a rod and dipped periodically into the water. At low frequencies the wavelength λ may be made as large as, say, 10 cm.

Then, since the surface tension of water is about 70 dyn cm^{-1} and the density of water is 1 g cm^{-3}, we have

$$\frac{g\lambda}{2\pi} \approx \frac{980 \times 10}{6} \approx 1600, \qquad \frac{2\pi\gamma}{\rho\lambda} \approx \frac{6 \times 70}{10} \approx 40,$$

and

$$\frac{2\pi h}{\lambda} \approx \frac{6 \times 1}{10} \approx 0.6.$$

Under these conditions, the second term in the parentheses of Eq. (21–18) is negligible compared with the first, and also tanh $2\pi h/\lambda$ is very nearly equal to $2\pi h/\lambda$. Then

$$c \approx \sqrt{\frac{g\lambda}{2\pi} \cdot \frac{2\pi h}{\lambda}} \approx \sqrt{gh}. \qquad (21\text{–}19)$$

Thus for long waves the speed depends only on the acceleration due to gravity, g, and the depth h, and is independent of the wavelength (no dispersion). Since, according to Eq. (21–19), the shallower the water, the slower the speed of the waves, it is possible to eliminate reflections from the edges of the ripple tank by beveling the edges of the tank. A wave approaching a "shore" is therefore slowed down to zero speed.

Problems

21–1 A steel wire 6 m long has a mass of 60 g and is stretched with a tension of 1000 N. What is the speed of propagation of a transverse wave in the wire?

21–2 One end of a horizontal string is attached to a prong of an electrically driven tuning fork whose frequency of vibration is 240 Hz. The other end passes over a pulley and supports a weight of 6 lb. The linear weight density of the spring is 0.0133 lb ft^{-1}. (a) What is the speed of a transverse wave in the string? (b) What is the wavelength?

21–3 The equation of a certain traveling transverse wave is

$$y = 2 \sin 2\pi \left(\frac{t}{0.01} - \frac{x}{30}\right),$$

where x and y are in centimeters and t is in seconds. What are (a) the amplitude, (b) the wavelength, (c) the frequency, and (d) the speed of propagation of the wave?

21–4 One end of a rubber tube 50 ft long, weighing 2 lb, is fastened to a fixed support. A cord attached to the other end passes over a pulley and supports a body of weight 20 lb. The tube is struck a transverse blow at one end. Find the time required for the pulse to reach the other end.

21–5 A metal wire has the properties: coefficient of linear expansion = $1.5 \times 10^{-5}(\text{C}°)^{-1}$, Young's modulus = 2.0×10^{12} dyn cm^{-2}, density = 9.0 g cm^{-3}. At each end are rigid supports. If the tension is zero at 20°C, what will be the speed of a transverse wave at 8°C?

21–6 A metal wire of density 20 slugs ft^{-3} with a Young's modulus equal to 15×10^6 lb in^{-2} is stretched between rigid supports. At one temperature the speed of a transverse wave is found to be 657 ft s^{-1}. When the temperature is raised 100 F°, the speed decreases to 536 ft s^{-1}. What is the coefficient of linear expansion?

21–7 A transverse sine wave of amplitude 10 cm and wavelength 200 cm travels from left to right along a long horizontal stretched string with a speed of 100 cm s^{-1}. Take the origin at the left end of the undisturbed string. At time $t = 0$, the left end of the string is at the origin and is moving downward. (a) What is the frequency of the wave? (b) What is the angular frequency? (c) What is the propagation constant? (d) What is the equation of the wave? (e) What is the equation of motion of the left end of the string? (f) What is the equation of motion of a particle 150 cm to the right of the origin? (g) What is the (absolute) maximum transverse velocity of any particle of the string? (h) Find the transverse displacement and the transverse velocity of a particle 150 cm to the right of the origin, at time $t = 3.25$ s. (i) Make a sketch of the shape of the string, for a length of 400 cm, at time $t = 3.25$ s.

21–8 What must be the stress (F/A) in a stretched wire of a material whose Young's modulus is Y, in order that the speed of longitudinal waves shall equal 10 times the speed of transverse waves?

21-9 The speed of longutidinal waves in water is approximately 1450 m s^{-1} at 20°C. Compute the adiabatic compressibility ($1/B_{ad}$) of water and compare with the isothermal compressibility in Table 10-2.

21-10 Provided the amplitude is sufficiently great, the human ear can respond to longitudinal waves over a range of frequencies from about 20 Hz to about 20,000 Hz. Compute the wavelengths corresponding to these frequencies (a) for waves in air, (b) for waves in water. (See Problem 21-9.)

21-11 At a temperature of 27°C, what is the speed of longitudinal waves in (a) argon, (b) hydrogen? (c) Compare with the speed in air at the same temperature.

21-12 What is the difference between the speeds of longitudinal waves in air at −3°C and at 57°C?

21-13 The sound waves from a loudspeaker spread out nearly uniformly in all directions when their wavelength is large compared with the diameter of the speaker. When the wavelength is small compared with the diameter of the speaker, much of the sound energy is concentrated in the forward direction. For a speaker of diameter 10 in., compute the frequency for which the wavelength of the sound waves, in air, is (a) 10 times the diameter of the speaker, (b) equal to the diameter of the speaker, (c) 1/10 the diameter of the speaker.

21-14 A steel pipe 200 ft long is struck at one end. A person at the other end hears two sounds as a result of two longitudinal waves, one in the pipe and the other in the air. What is the time interval between the two sounds? Take Young's modulus of steel to be 30×10^6 lb in^{-2}.

21-15 (a) By how many meters per second, at a temperature of 27°C, does the speed of sound in air increase per centigrade degree rise in temperature? [*Hint:* compute dc in terms of dT, and approximate finite changes by differentials.] (b) Is the rate of change of speed with temperature the same at all temperatures?

21-16 Explain why the breakers on a sloping beach are always parallel to the shore line, whatever may be the direction of the waves far from shore.

21-17 A loose coil of flexible rope of length L and mass M rests on a frictionless table. A force S is applied at one end of the rope and more and more rope is pulled from the coil with a constant velocity v. (a) What is the relation between S and v? (b) A transverse pulse is produced while the rope is being pulled from the coil. How does the pulse behave? (c) What is the work of the force S at the moment when the last bit of rope leaves the coil? (d) What is the kinetic energy of the rope? (e) How come?

21-18 The ends of a flexible rope of length L and mass M are joined so as to form a circular loop of radius R. The loop rotates about an axis through its center on a horizontal frictionless surface with a constant linear speed v. (a) Apply Newton's second law to a short piece of rope and calculate the tension S in the rope. (b) Explain the effect of imparting to the rope, while it is spinning, a small rapid radial displacement.

22 Vibrating Bodies

22-1 BOUNDARY CONDITIONS FOR A STRING

Let us now consider what will happen when a wave pulse or wave train advancing along a stretched string arrives at the end of the string. If fastened to a rigid support, the end must evidently remain at rest. The arriving pulse exerts a force on the support, and the reaction to this force "kicks back" on the string and sets up a *reflected* pulse traveling in the reverse direction. At the opposite extreme from a rigidly fixed end would be one which was perfectly free—a case of no great importance here (it may be realized by a string hanging vertically) but which is of interest since its analog does occur in other types of waves. At a free end the arriving pulse causes the string to "overshoot" and a reflected wave is also set up. The conditions which must be satisfied at the ends of the string are called *boundary conditions.*

The multiflash photograph of Fig. 22-1 shows the reflection of a pulse at a fixed end of a string. (The camera was tipped vertically while the photographs were taken so that successive images lie one under the other. The "string" is a rubber tube and it sags somewhat.) It will be seen that the pulse is reflected with its displacement and its velocity both reversed. When reflection takes place at a free end, the direction of the velocity is reversed but the direction of the displacement is unchanged.

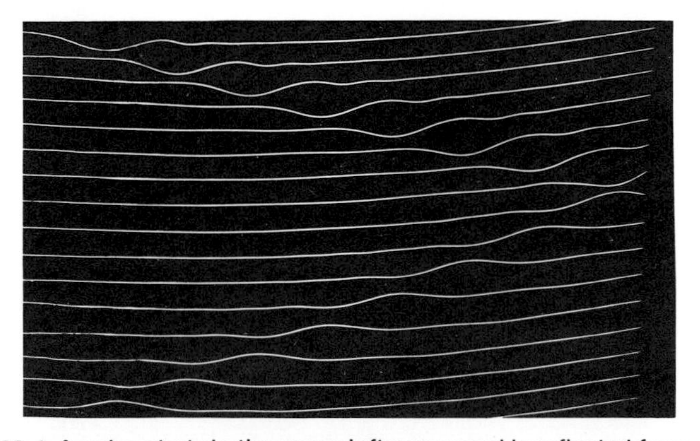

22-1 A pulse starts in the upper left corner and is reflected from the fixed end of the string at the right.

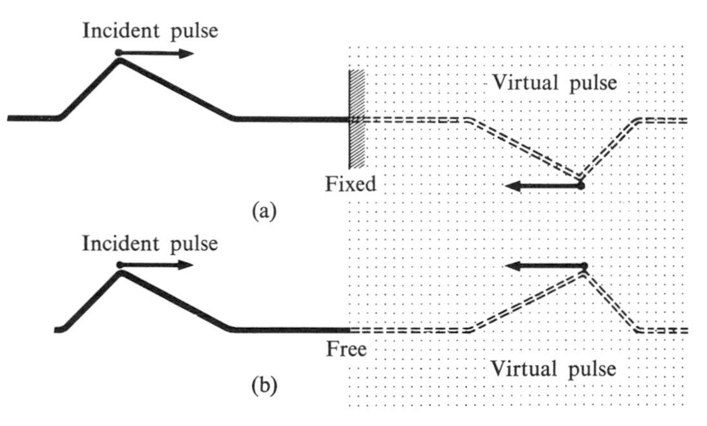

22-2 Description of reflection of a pulse (a) at a fixed end of a string, and (b) at a free end, in terms of an imaginary "virtual" pulse.

It is helpful to think of the process of reflection in the following way. Imagine the string to be extended indefinitely beyond its actual terminus. The actual pulse can be considered to continue on into the imaginary portion as though the support were not there, while at the same time a "virtual" pulse, which has been traveling in the imaginary portion, moves out into the real

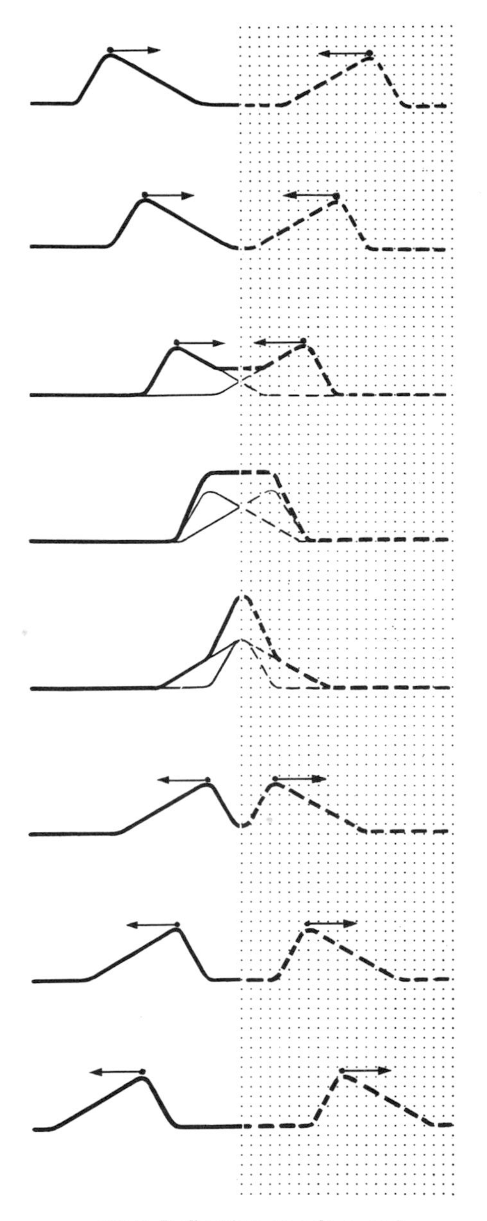

22-3 Reflection at a free end.

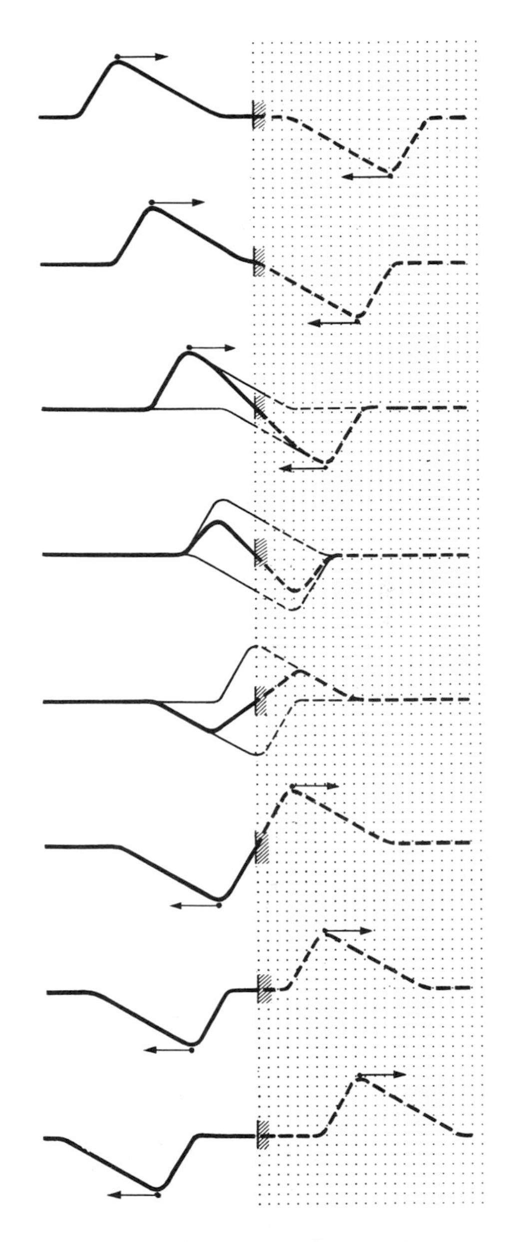

22-4 Reflection at a fixed end.

string and forms the reflected pulse. The nature of the reflected pulse depends on whether the end is fixed or free. The two cases are shown in Fig. 22-2.

The displacement at a point where the actual and virtual pulses cross each other is the algebraic sum of the displacements in the individual pulses. Figures 22-3 and 22-4 show the shape of the end of the string for both types of reflected pulses. It will be seen that Fig. 22-3 corresponds to a free end and Fig. 22-4 to a fixed end. In the latter case, the incident and reflected pulses combine in such a way that the displacement of the end of the string is always zero.

22-2 STATIONARY WAVES IN A STRING

When a continuous train of waves arrives at a fixed end of a string, a continuous train of reflected waves appears to originate at the end and travel in the opposite direction. Provided the elastic limit of the string is not exceeded and the displacements are sufficiently small, the actual displacement of any point of the string is the algebraic sum of the displacements of the individual waves, a fact which is called the *principle of superposition*. This principle is extremely important in all types of wave motion and applies not only to waves in a string but to sound waves in air, to light waves, and, in fact, to wave motion of any sort. The general term *interference* is applied to the effect produced by two (or more) sets of wave trains which are simultaneously passing through a given region.

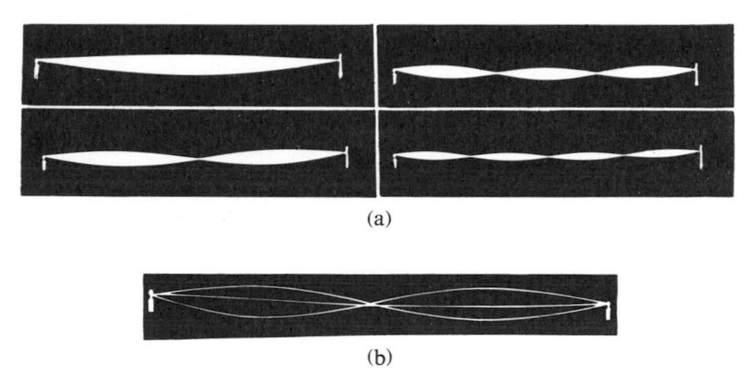

22-5 (a) Standing waves in a stretched string (time exposure). (b) Multiflash photograph of a standing wave, with nodes at the center and at the ends.

The appearance of the string in these circumstances gives no evidence that two waves are traversing it in opposite directions. If the frequency is sufficiently great so that the eye cannot follow the motion, the string appears subdivided into a number of segments, as in the time exposure photograph of Fig. 22-5(a). A multiflash photograph of the same string, in Fig. 22-5(b), indicates a few of the instantaneous shapes of the string. At any instant (except those when the string is straight) its shape is a sine curve, but whereas in a traveling wave the amplitude remains constant while the wave progresses, here the waveform remains fixed in position (longitudinally) while the amplitude fluctuates. Certain points known as the *nodes* remain always at rest. Midway between these points, at the *loops* or *antinodes*, the fluctuations are a maximum. The vibration as a whole is called a *stationary* wave.

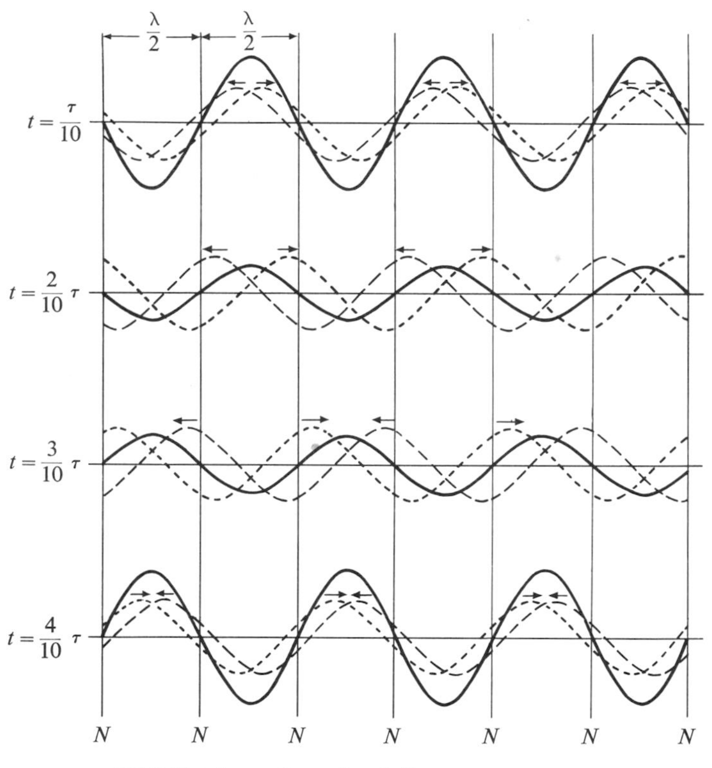

22-6 The formation of a stationary wave.

To understand the formation of a stationary wave, consider the separate graphs of the waveform at four instants $\frac{1}{10}$ of a period apart, shown in Fig. 22-6. The dotted curves represent a wave traveling to the right. The dashed curves represent a wave of the same velocity, same wavelength, and same amplitude traveling to the left. The heavy curves represent the resultant waveform, obtained by applying the principle of superposition, that is, by adding displacements. At those places marked N at the bottom of Fig. 22-6, the resultant displacements are always zero. These are the nodes. Midway between the nodes, the vibrations have the largest amplitude. These are the antinodes. It is evident from the figure that

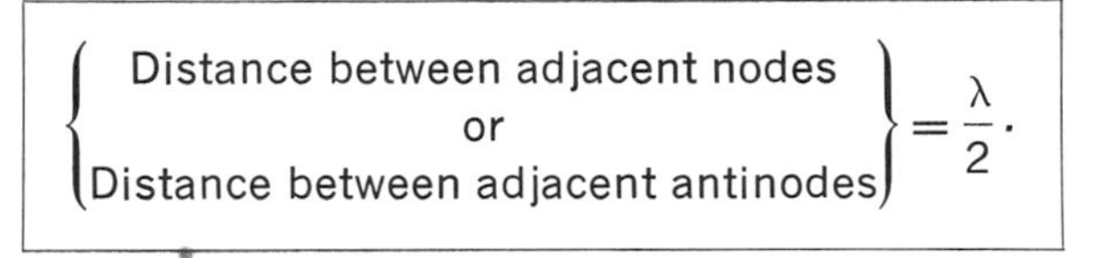

$$\left\{\begin{array}{c}\text{Distance between adjacent nodes}\\ \text{or}\\ \text{Distance between adjacent antinodes}\end{array}\right\} = \frac{\lambda}{2}.$$

The equation of a stationary wave may be obtained by adding the displacements of two waves of equal amplitude, period, and wavelength, but traveling in opposite directions.

Thus if

$$y_1 = A \sin(\omega t - kx) \qquad \text{(positive } x\text{-direction)},$$

$$y_2 = -A \sin(\omega t + kx) \qquad \text{(negative } x\text{-direction)},$$

then

$$y_1 + y_2 = A[\sin(\omega t - kx) - \sin(\omega t + kx)].$$

Introducing the expressions for the sine of the sum and difference of two angles and combining terms, we obtain

$$y_1 + y_2 = -[2A \cos \omega t] \sin kx. \qquad (22\text{-}1)$$

The shape of the string at each instant is, therefore, a sine curve whose amplitude (the expression in brackets) varies with time.

22-3 VIBRATION OF A STRING FIXED AT BOTH ENDS

Thus far we have been discussing a long string fixed at one end and have considered the stationary waves set up near that end by interference between the incident and reflected waves. Let us next consider the more usual case, that of a string fixed at both ends. A continuous train of sine or cosine waves is reflected and re-reflected, and since the string is fixed at both ends, both ends must be nodes. Since the nodes are $\frac{1}{2}$ wavelength apart, the length of the string may be

$$\frac{\lambda}{2}, \quad 2\frac{\lambda}{2}, \quad 3\frac{\lambda}{2},$$

or, in general, any integral number of half-wavelengths. Or, to put it differently, if one considers a particular string of length L, stationary waves may be set up in the string by vibrations of a number of different frequencies, namely, those which give rise to waves of wavelengths

$$2L, \quad \frac{2L}{2}, \quad \frac{2L}{3}, \quad \ldots .$$

From the relation $f = c/\lambda$, and since c is the same for all frequencies, the possible frequencies are

$$\frac{c}{2L}, \quad 2\frac{c}{2L}, \quad 3\frac{c}{2L}, \quad \ldots .$$

The lowest frequency, $c/2L$, is called the *fundamental* frequency f_1 and the others are the *overtones*. The frequencies of the latter are, therefore, $2f_1$, $3f_1$, $4f_1$, and so on. Overtones whose frequencies are integral multiples of the fundamental are said to form a *harmonic series*. The fundamental is the *first harmonic*. The frequency $2f_1$ is the *first overtone* or the *second harmonic*, the frequency $3f_1$ is the *second overtone* or the *third harmonic*, and so on.

We can now see an important difference between a spring-mass system and a vibrating string. The former has but one natural frequency, while the vibrating string has an infinite number of natural frequencies, the fundamental and all of the overtones. If a body suspended from a spring is pulled down and released, only one frequency of vibration will ensue. If a string is initially distorted so that its shape is the same as *any one* of the possible harmonics, it will vibrate, when released, at the frequency of that particular harmonic. But when a piano string is struck, not only the fundamental, but many of the overtones are present in the resulting vibration. The fundamental frequency of the vibrating string is $f_1 = c/2L$, where $c = \sqrt{S/\mu}$. It follows that

$$f_1 = \frac{1}{2L}\sqrt{\frac{S}{\mu}}. \qquad (22\text{-}2)$$

Stringed instruments afford many examples of the implications of this equation. For example, all such instruments are "tuned" by varying the tension S, an increase of tension increasing the frequency or pitch, and vice versa. The inverse dependence of frequency on length L is illustrated by the long strings of the bass

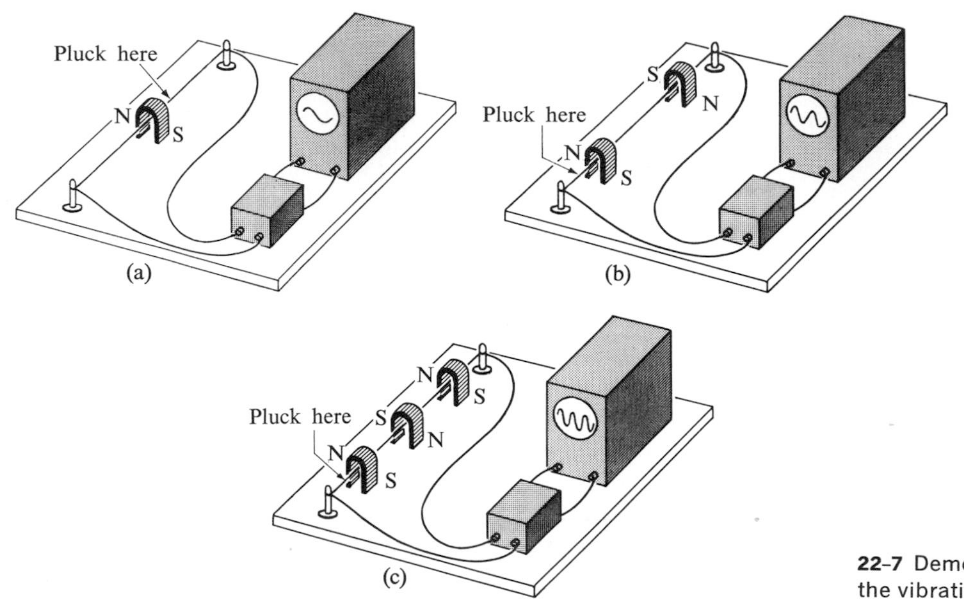

22-7 Demonstration of the harmonics present in the vibration of a plucked string.

section of the piano or the bass viol compared with the shorter strings of the piano treble or the violin. One reason for winding the bass strings of a piano with wire is to increase the mass per unit length μ, so as to obtain the desired low frequency without resorting to a string which is inconveniently long.

22-4 DEMONSTRATION OF THE HARMONIC SERIES IN A VIBRATING STRING

We have seen that a string is capable of vibrating at a number of different frequencies. That it may vibrate with many different frequencies *at the same time* may be demonstrated graphically with the aid of the apparatus depicted in Fig. 22-7. A metal wire is stretched between two metal posts which are in turn connected to a "step-up" transformer. The secondary of the transformer is then connected to those plates of a cathode-ray oscilloscope which impart vertical motion to the electron beam. If the string is made to oscillate in a magnetic field, an alternating current will be set up whose frequency is exactly the same as that of the string. With proper adjustment of the oscilloscope, this alternating current may be caused to give rise to figures on the screen such as those shown.

Suppose that one small magnet is placed over the center of the string (Fig. 22-7a) and the string is plucked near the center so that this part of the string vibrates perpendicular to the magnetic lines of force. The figure on the oscilloscope shows the fundamental frequency.

By placing one magnet $\frac{1}{4}$ of the way along the string and another magnet *with its polarity reversed* at the $\frac{3}{4}$ mark (Fig. 22-7b), and plucking the string near the $\frac{1}{4}$ point, the second harmonic may be obtained. If one of the magnets is quickly reversed while the string is sounding, so as to set the two magnetic fields in the same direction, the fundamental will occur again, showing that the fundamental and second harmonic exist at the same time.

We now place three magnets on the string at the $\frac{1}{6}$, $\frac{1}{2}$, $\frac{5}{6}$ points, with the polarity shown in Fig. 22-7(c). By plucking the string near the $\frac{1}{6}$ point, the third harmonic is obtained. While the string is vibrating, if the middle magnet is reversed so as to make all the magnets point in the same direction, the fundamental will appear. Thus the fundamental and third harmonic exist at the same time.

Proceeding in this manner, we may pick up higher harmonics and demonstrate that a string can vibrate with all of these frequencies at the same time.

22-5 RESONANCE

In general, whenever a body capable of oscillating is acted on by a periodic series of impulses having a frequency equal to one of the natural frequencies of oscillation of the body, the body is set into vibration with a relatively large amplitude. This phenomenon is called *resonance*, and the body is said to *resonate* with the applied impulses.

A common example of mechanical resonance is provided by pushing a swing. The swing is a pendulum with a single natural frequency depending on its length. If a series of regularly spaced pushes is given to the swing, with a frequency equal to that of the swing, the motion may be made quite large. If the frequency of the pushes differs from the natural frequency of the swing, or if the pushes occur at irregular intervals, the swing will hardly execute a vibration at all.

Unlike a simple pendulum, which has only one natural frequency, a stretched string (and other systems to be discussed later in this chapter) has a large number of natural frequencies. Suppose that one end of a stretched string is fixed while the other is moved back and forth in a transverse direction. The amplitude at the driven end is fixed by the driving mechanism. Stationary waves will be set up in the string, whatever the value of the frequency f. If the frequency is not equal to one of the natural frequencies of the string, the amplitude at the antinodes will be fairly small. However, if the frequency is equal to *any one* of the natural frequencies, the string is in resonance and the amplitude at the antinodes will be very much larger than that at the driven end. In other words, although the driven end is not a node, it lies much closer to a node than to an antinode when the string is in resonance. In Fig. 22-5(a), the right end of the string was fixed and the left end was forced to oscillate vertically with small amplitude. Stationary waves of relatively large amplitude resulted when the frequency of oscillation of the left end was equal to the fundamental frequency or to any of the first three overtones.

A bridge or, for that matter, any structure, is capable of vibrating with certain natural frequencies. If the regular footsteps of a column of soldiers were to have a frequency equal to one of the natural frequencies of a bridge which the soldiers are crossing, a vibration of dangerously large amplitude might result. Therefore, in crossing a bridge, a column of soldiers is ordered to break step.

Tuning a radio is an example of electrical resonance. By turning a dial, the natural frequency of an alternating current in the receiving circuit is made equal to the frequency of the waves broadcast by the desired station. Optical resonance may also take place between atoms in a gas at low pressure and light waves from a lamp containing the same atoms. Thus light from a sodium lamp may cause the sodium atoms in a glass bulb to glow with characteristic yellow sodium light.

The phenomenon of resonance may be demonstrated with the aid of the longitudinal waves set up in air by a vibrating plate or tuning fork. If two identical tuning forks are placed some distance apart and one is struck, the other will be heard when the first is suddenly damped. Should a small piece of wax or modeling clay be put on one of the forks, the frequency of that fork will be altered enough to destroy the resonance.

22-6 INTERFERENCE OF LONGITUDINAL WAVES

The phenomenon of interference between two longitudinal waves in air may be demonstrated with the aid of the apparatus depicted in Fig. 22-8. A wave emitted by a source S is sent into a metal tube, where it divides into two waves, one following the constant path SAR, the other the path SBR, which may be varied by sliding the tube B to the right. Suppose the frequency of the source is 1100 Hz. Then the wavelength $\lambda = c/f = 1$ ft. If both paths are of equal length, the two waves will arrive at R at the same time and the vibrations set up by both waves will be in phase. The resulting vibration will have an

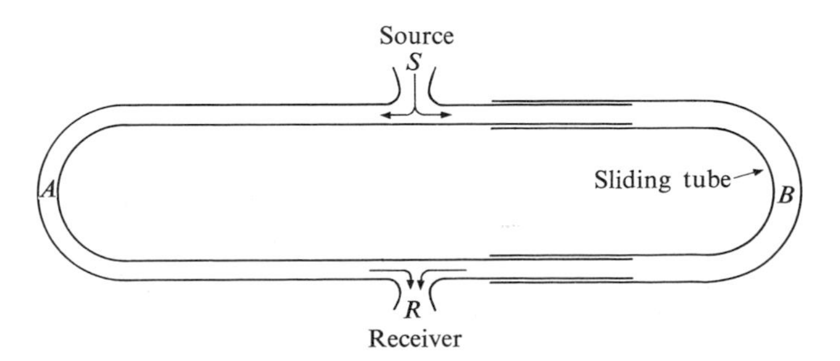

22-8 Apparatus for demonstrating interference of longitudinal waves.

amplitude equal to the sum of the two individual amplitudes and the phenomenon of *reinforcement* may be detected either with the ear at R or with the aid of a microphone, amplifier, and loudspeaker.

Now suppose the tube B is moved out a distance of 3 in., thereby making the path SBR 6 in. longer than the path SAR. The right-hand wave will have traveled a distance $\lambda/2$ greater than the left-hand wave and the vibration set up at R by the right-hand wave will therefore be in opposite phase to that set up by the left-hand wave. The consequent interference is shown by the marked reduction in sound at R.

If the tube B is now pulled out another 3 in., so that the *path difference*, SBR minus SAR, is 1 ft (1 wavelength), the two vibrations at R will again reinforce each other. Thus

$$\left\{\begin{matrix}\text{Reinforcement takes place}\\ \text{when the path difference}\end{matrix}\right\} = 0, \lambda, 2\lambda, \text{etc.}$$

$$\left\{\begin{matrix}\text{Interference takes place}\\ \text{when the path difference}\end{matrix}\right\} = \frac{\lambda}{2}, \frac{3\lambda}{2}, \frac{5\lambda}{2}, \text{etc.}$$

An acoustical interferometer of this sort is of value only in demonstrating the phenomenon of interference. Optical interferometers, however, whose principles of operation are the same, have many practical uses in physical optics.

22-7 STATIONARY LONGITUDINAL WAVES

Longitudinal waves traveling along a tube of finite length are reflected at the ends of the tube in much the same way that transverse waves in a string are reflected at its ends. Interference between the waves traveling in opposite directions gives rise to stationary waves.

If reflection takes place at a closed end, the displacement of the particles there must always be zero. Hence a closed end is a *displacement node*. If the end of the tube is open, the nature of the reflection is more complex and depends on whether the tube is wide or narrow compared with the wavelength. If the tube is narrow compared with the wavelength, which is the case in most musical instruments, the reflection is such that the open end is a *displacement antinode*. Thus the longitudinal waves in a column of fluid are reflected at the closed and open ends of a tube in the same way that transverse waves in a string are reflected at fixed and free ends, respectively.

The reflections at the openings where the instrument is blown are found to be such that an antinode is located at or near the opening. The effective length of the air column of a wind instrument is thus less definite than the length of a string fixed at its ends.

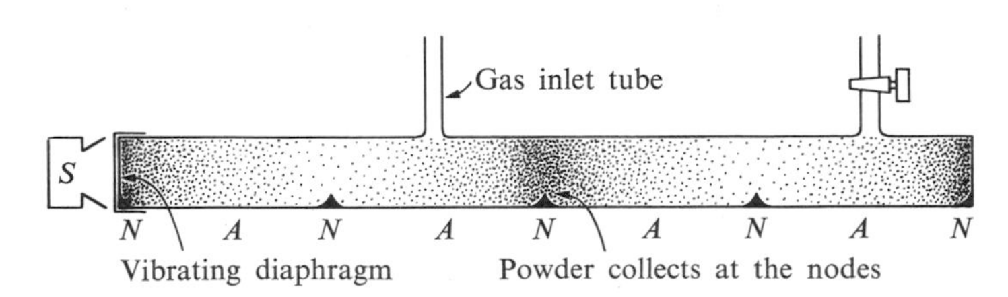

22-9 Kundt's tube for determining the velocity of sound in a gas. The dots represent the density of the gas molecules at an instant when the pressure at the displacement nodes is a maximum or a minimum.

Stationary longitudinal waves in a column of gas may be demonstrated conveniently with the aid of the apparatus shown in Fig. 22-9, known as Kundt's tube. A glass tube a few feet long is closed at one end with glass and at the other with a flexible diaphragm. The gas to be studied is admitted to the tube at a known temperature and at atmospheric pressure. A powerful source of longitudinal waves, S, whose frequency may be varied, causes vibration of the flexible diaphragm. A small amount of light powder or cork dust is sprinkled uniformly along the tube.

When a frequency is found at which the air column is in resonance, the amplitude of the stationary waves becomes large enough for the gas particles to sweep the cork dust along the tube, at all points where the gas is in motion. The powder therefore collects at the displacement nodes, where the gas remains at rest. Sometimes a wire, running along the axis of the tube, is maintained at a dull red heat by an electric current and the nodes show themselves as hot points, compared with the antinodes.

With careful manipulation and with a good variable frequency source, a fair determination of the velocity of the wave may be obtained with Kundt's tube. Since, in a stationary wave, the distance between two adjacent nodes is $\frac{1}{2}$ wavelength, the wavelength λ is obtained by measuring the distance between alternate clumps of powder. When the frequency f is known, the velocity c is then

$$c = f\lambda.$$

A constant frequency source may be used if the vibrating element is a piston which may be moved along the tube until resonance is obtained.

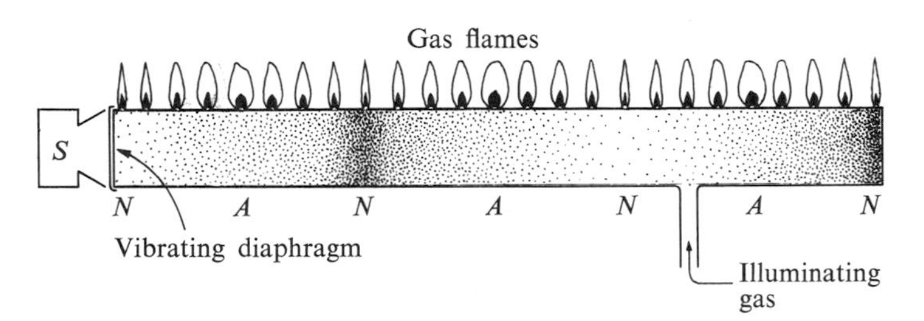

22-10 The variations in gas pressure are greatest at the displacement nodes. The dots represent the density of the gas molecules at an instant when the pressure at the displacement nodes is a maximum or a minimum.

At a displacement node, the pressure variations above and below the average are maximum, whereas at a displacement antinode, pressure does not vary. This may be understood easily when it is realized that two small masses of gas on opposite sides of a displacement node vibrate in *opposite phase.* When they approach each other, the pressure at the node is maximum, and when they recede from each other, the pressure at the node is minimum. Two small masses of gas, however, on opposite sides of a displacement antinode vibrate *in phase,* and hence cause no pressure variations at the antinode. This may be demonstrated in the case of illuminating gas with the aid of the apparatus shown in Fig. 22-10, where the amplitude of the harmonic variations of gas pressure determines the shape and color of the gas flames.

Figure 22-11 may be helpful in visualizing a longitudinal stationary wave. To use this figure, cut a slit about $\frac{1}{16}$ in. wide and $3\frac{1}{2}$ in. long in a card. Place the card over the diagram with the slit horizontal and move it vertically with constant velocity. The portions of the curves that appear in the slit will correspond to the oscillations of the particles in a longitudinal stationary wave.

Fig. 22-11

22-8 VIBRATIONS OF ORGAN PIPES

If one end of a pipe is open and a stream of air is directed against an edge, vibrations are set up and the tube resonates at its natural frequencies. As in the case of a plucked string, the fundamental and overtones exist at the same time. In the case of an open pipe, the fundamental frequency f_1 corresponds to an antinode at each end and a node in the middle, as shown at the top of

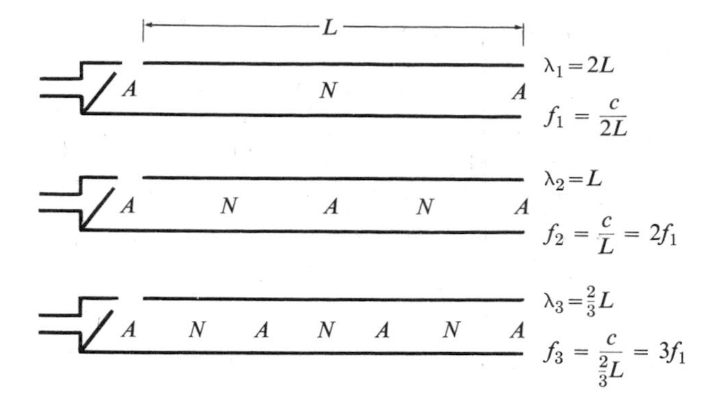

22-12 Modes of vibration of an open organ pipe.

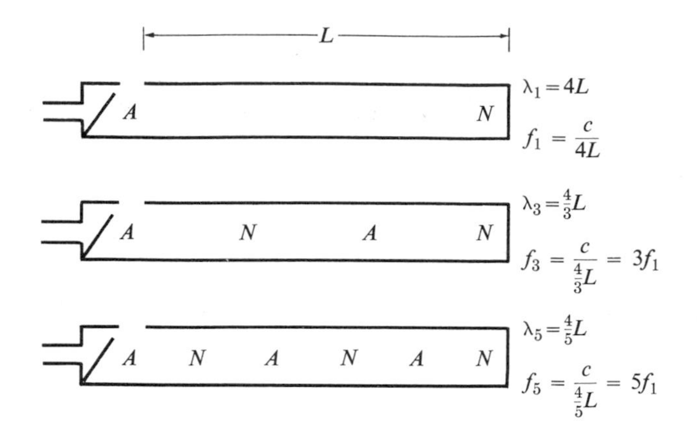

22-13 Modes of vibration of a closed organ pipe.

Fig. 22-12. Succeeding diagrams of Fig. 22-12 show two of the overtones, which are seen to be the second and third harmonics. *In an open pipe the fundamental frequency is $c/2L$ and all harmonics are present.*

The properties of a closed pipe are shown in the diagrams of Fig. 22-13. The fundamental frequency is seen to be $c/4L$, which is one-half that of an open pipe of the same length. In the language of music, the pitch of a closed pipe is one octave lower than that of an open pipe of equal length. From the remaining diagrams of Fig. 22-13, it may be seen that the second, fourth, etc., harmonics are missing. Hence, *in a closed pipe, the fundamental frequency is $c/4L$ and only the odd harmonics are present.*

22-9 VIBRATIONS OF RODS AND PLATES

A rod may be set in longitudinal vibration by clamping it at some point and stroking it with a chamois skin that has been sprinkled with rosin. In Fig. 22-14(a) the rod is clamped in the middle; consequently, when it is stroked near the end, a stationary wave is set up with a node in the middle and antinodes at each end, exactly the same as the fundamental mode of an open organ pipe. The fundamental frequency of the rod is then $c/2L$, where c is the velocity of a longitudinal wave in the rod. Since the velocity of a longitudinal wave in a solid is much greater than that in air, a rod has a higher fundamental frequency than an open organ pipe of the same length.

By clamping the rod at a point $\frac{1}{4}$ of its length from one end, as shown in Fig. 22-14(b), the second harmonic may be produced.

If a stretched flexible membrane, such as a drumhead, is struck a blow, a two-dimensional pulse travels outward from the struck point and is reflected and re-reflected at the boundary of the membrane. If some point of the membrane is forced to vibrate periodically, continuous trains of waves travel along the membrane. Just as with the stretched string, stationary waves can be set up in the membrane and each of these waves has a certain natural frequency. The lowest frequency is the fundamental and the others are overtones. In general, when the membrane is vibrating, a number of overtones are present.

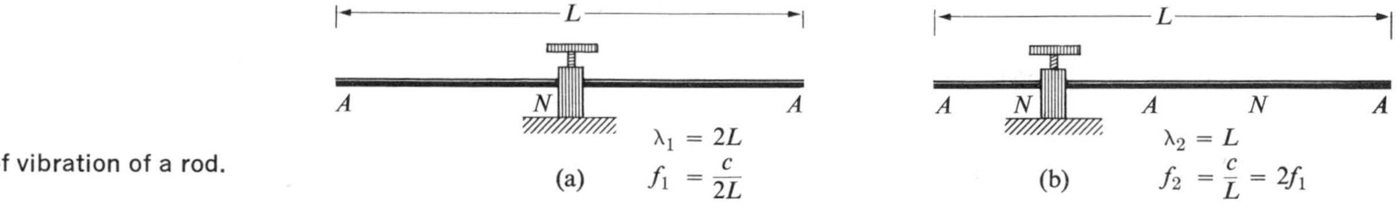

22-14 Modes of vibration of a rod.

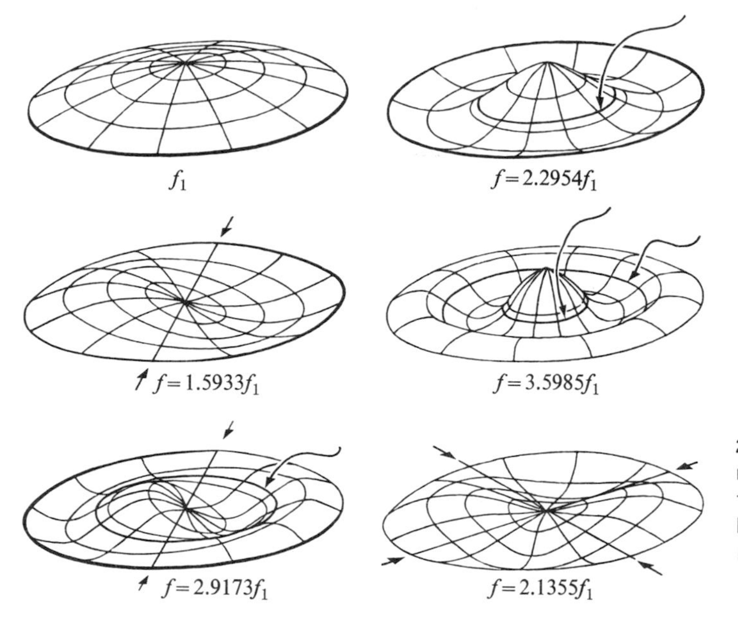

22-15 Possible modes of vibration of a membrane, showing nodal lines. The frequency of each mode is given in terms of fundamental frequency, f_1. (Adapted from *Vibration and Sound,* by Philip M. Morse, 2nd Edition, McGraw-Hill Book Company, Inc., 1948. By permission of the publishers.)

The nodes of a vibrating membrane are lines (nodal lines) rather than points. The boundary of the membrane is evidently one such line. Some of the other possible nodal lines of a circular membrane are indicated by arrows in Fig. 22-15. The natural frequency of each mode is given in terms of the fundamental f_1. It will be noted that the frequencies of the overtones are *not* integral multiples of f_1. That is, they are not harmonics.

The restoring force in a vibrating flexible membrane arises from the tension with which it is stretched. A metal plate, if sufficiently thick, will vibrate in a similar way, the restoring force being produced by bending stresses in the plate. The study of vibrations of membranes and plates is of importance in connection with the design of loudspeaker diaphragms and the diaphragms of telephone receivers and microphones.

Problems

22-1 A steel piano wire 50 cm long, of mass 5 g, is stretched with a tension of 400 N. (a) What is the frequency of its fundamental mode of vibration? (b) What is the number of the highest overtone that could be heard by a person who is capable of hearing frequencies up to 10,000 Hz?

22-2 A steel wire of length $L = 100$ cm and density $\rho = 8$ g cm^{-3} is stretched tightly between two rigid supports. Vibrating in its fundamental mode, the frequency is $f = 200$ Hz. (a) What is the speed of transverse waves on this wire? (b) What is the longitudinal stress in the wire (in dynes per square centimeter)? (c) If the maximum acceleration at the midpoint of the wire is 80,000 cm s^{-2}, what is the amplitude of vibration at the midpoint?

22-3 A stretched string is observed to vibrate with a frequency of 30 Hz in its fundamental mode when the supports are 60 cm apart. The amplitude at the antinode is 3 cm. The string has a mass of 30 g. (a) What is the speed of propagation of a transverse wave in the string? (b) Compute the tension in the string.

22–4 The fundamental frequency of the A-string on a cello is 220 Hz. The vibrating portion of the string is 68 cm long and has a mass of 1.29 g. With what tension, in pounds, must it be stretched?

22–5 A standing wave of frequency 1100 Hz in a column of methane at 20°C produces nodes that are 20 cm apart. What is the ratio of the heat capacity at constant pressure to that at constant volume?

22–6 An aluminum object is hung from a steel wire. The fundamental frequency for transverse stationary waves on the wire is 300 Hz. The object is then immersed in water so that one-half of its volume is submerged. What is the new fundamental frequency?

22–7 Stationary waves are set up in a Kundt's tube by the longitudinal vibration of an iron rod 1 m long, clamped at the center. If the frequency of the iron rod is 2480 Hz and the powder heaps within the tube are 6.9 cm apart, what is the speed of the waves (a) in the iron rod, and (b) in the gas?

22–8 The speed of a longitudinal wave in a mixture of helium and neon at 300°K was found to be 758 m s^{-1}. What is the composition of the mixture?

22–9 The atomic weight of iodine is 127. A stationary wave in iodine vapor at 400°K produces nodes that are 6.77 cm apart when the frequency is 1000 Hz. Is iodine vapor monatomic or diatomic?

22–10 A copper rod 1 m long, clamped at the $\frac{1}{4}$ point, is set in longitudinal vibration and is used to produce stationary waves in a Kundt's tube containing air at 300°K. Heaps of cork dust within the tube are found to be 4.95 cm apart. What is the speed of the longitudinal waves in copper?

22–11 Find the fundamental frequency and the first four overtones of a 6-in. pipe (a) if the pipe is open at both ends, (b) if the pipe is closed at one end. (c) How many overtones may be heard by a person having normal hearing for each of the above cases?

22–12 A long tube contains air at a pressure of 1 atm and temperature 77°C. The tube is open at one end and closed at the other by a movable piston. A tuning fork near the open end is vibrating with a frequency of 500 Hz. Resonance is produced when the piston is at distances 18.0, 55.5, and 93.0 cm from the open end. (a) From these measurements, what is the speed of sound in air at 77°C? (b) From the above result, what is the ratio of the specific heats γ for air?

22–13 An organ pipe *A* of length 2 ft, closed at one end, is vibrating in the first overtone. Another organ pipe *B* of length 1.35 ft, open at both ends, is vibrating in its fundamental mode. Take the speed of sound in air as 1120 ft s^{-1}. Neglect end corrections. (a) What is the frequency of the tone from *A*? (b) What is the frequency of the tone from *B*?

22–14 A plate cut from a quartz crystal is often used to control the frequency of an oscillating electrical circuit. Longitudinal standing waves are set up in the plate with displacement antinodes at opposite faces. The fundamental frequency of vibration is given by the equation

$$f_1 = \frac{2.87 \times 10^5}{s},$$

where f_1 is in Hertz and s is the thickness of the plate in centimeters. (a) Compute Young's modulus for the quartz plate. (b) Compute the thickness of plate required for a frequency of 1200 k Hz. (1 kilocycle = 1000 cycles.) The density of quartz is 2.66 g cm^{-3}.

23 Acoustical Phenomena

23-1 PRESSURE VARIATIONS IN A SOUND WAVE

We shall limit ourselves in this chapter to the consideration of longitudinal waves only, and in particular to those which, when striking the ear, give rise to the sensation of sound. Such waves, within the frequency range from about 20 to about 20,000 Hz, are called, for simplicity, *sound waves.*

The reception of a sound wave by the ear gives rise to a vibration of the air particles at the eardrum with a definite frequency and a definite amplitude. This vibration may also be described in terms of the variation of air pressure at the same point. The air pressure rises above atmospheric pressure and then sinks below atmospheric pressure with simple harmonic motion of the same frequency as that of an air particle. The maximum amount by which the pressure differs from atmospheric pressure is called the *pressure amplitude.* It can be proved that the pressure amplitude is proportional to the displacement amplitude.

Measurements of sound waves show that the maximum pressure variations in the loudest sounds which the ear can tolerate are of the order of magnitude of 280 dyn cm^{-2} (above and below atmospheric pressure, which is about 1,000,000 dyn cm^{-2}). The corresponding maximum displacement for a frequency of 1000 Hz is about 0.001 cm. The displacement amplitudes, even in the loudest sounds, are therefore extremely small.

The maximum pressure variations in the *faintest* sound of frequency 1000 Hz are only about 2×10^{-4} dyn cm^{-2}. The corresponding displacement amplitude is about 10^{-9} cm. By way of comparison, the wavelength of yellow light is 6×10^{-5} cm, and the diameter of a molecule is about 10^{-8} cm. It will be appreciated that the ear is an extremely sensitive organ.

23-2 INTENSITY

From a purely geometrical point of view, that which is propagated by a traveling wave is the *waveform.* From a physical viewpoint, however, something else is propagated by a wave, namely, *energy.* The most outstanding example, of course, is the energy supply of the earth, which reaches us from the sun via electromagnetic waves. The *intensity* I of a traveling wave is defined as *the time average rate at which energy is transported by the wave per unit area* across a surface perpendicular to the direction of propagation. More briefly, the intensity is the average power transported per unit area.

We have seen that the power developed by a force equals the product of force and velocity. Hence the power per unit area in a sound wave equals the product of the excess pressure (force per unit area) and the *particle* velocity. Averaging over 1 cycle, it can be proved that

$$I = \frac{P^2}{2\rho c}, \tag{23-1}$$

where P is the pressure amplitude, ρ is the average density of the air, and c is the velocity of the sound wave. It will be noted that the *intensity* is proportional to the *square of the amplitude,* a result which is true for any sort of wave motion.

The intensity of a sound wave of pressure amplitude $P = 280$ dyn cm^{-2} (roughly, the loudest tolerable sound) is

$$I = \frac{(280 \text{ dyn cm}^{-2})^2}{2 \times 1.22 \times 10^{-3} \text{ g cm}^{-3} \times 3.46 \times 10^4 \text{ cm s}^{-1}}$$
$$= 940 \text{ ergs s}^{-1} \text{ cm}^{-2} = 94 \times 10^{-6} \text{ W cm}^{-2}.^*$$

* The "W cm^{-2}" is a hybrid unit, neither cgs nor mks. We shall retain it to conform with general usage in acoustics.

The pressure amplitude of the faintest sound wave which can be heard is about 0.0003 dyn cm^{-2} and the corresponding intensity is about 10^{-16} W cm^{-2}.

The total power carried across a surface by a sound wave equals the product of the intensity at the surface and the surface area, if the intensity over the surface is uniform. The average power developed as sound waves by a person speaking in an ordinary conversational tone is about 10^{-5} W, while a loud shout corresponds to about 3×10^{-2} W. Since the population of the city of New York is about eight million persons, the acoustical power developed if all were to speak at the same time would be about 60 W, or enough to operate a moderate-sized electric light. On the other hand, the power required to fill a large auditorium with loud sound is considerable. Suppose the intensity over the surface of a hemisphere 20 m in radius is 10^{-4} W cm^{-2}. The area of the surface is about 25×10^6 cm^2. Hence the acoustic power output of a speaker at the center of the sphere would have to be

$$10^{-4} \times 25 \times 10^6 = 2500 \text{ W},$$

or 2.5 kW. The electrical power input to the speaker would need to be considerably larger, since the efficiency of such devices is not very high.

23-3 INTENSITY LEVEL AND LOUDNESS

Because of the large range of intensities over which the ear is sensitive, a logarithmic rather than an arithmetic intensity scale is convenient. Accordingly, the *intensity level* β of a sound wave is defined by the equation

$$\beta = 10 \log \frac{I}{I_0}, \qquad (23\text{-}2)$$

where I_0 is an arbitrary reference intensity which is taken as 10^{-16} W cm^{-2}, corresponding roughly to the faintest sound which can be heard. Intensity levels are expressed in *decibels*, abbreviated db.*

If the intensity of a sound wave equals I_0 or 10^{-16} W cm^{-2}, its intensity level is zero. The maximum intensity which the ear can tolerate, about 10^{-4} W cm^{-2}, corresponds to an intensity level of 120 db. Table 23-1 gives the intensity levels in decibels of a number of familiar noises. It is taken from a survey made by the New York City Noise Abatement Commission.

TABLE 23-1 NOISE LEVELS DUE TO VARIOUS SOURCES (representative values)

Source or description of noise	Noise level, db
Threshold of pain	120
Riveter	95
Elevated train	90
Busy street traffic	70
Ordinary conversation	65
Quiet automobile	50
Quiet radio in home	40
Average whisper	20
Rustle of leaves	10
Threshold of hearing	0

The range of frequencies and intensities to which the ear is sensitive is conveniently represented by a diagram like that of Fig. 23-1, which is a graph of the *auditory area* of a person of good hearing. The lower curve represents the intensity level of the faintest pure tone which can be heard. It will be seen from the diagram that the ear is most sensitive to frequencies between 2000 and 3000 Hz, where the *threshold of hearing*, as it is called, is about −5 db. At intensities above those corresponding to the upper curve, which is called the *threshold of feeling or pain*, the sensation changes from one of hearing to discomfort or even pain. The height of the upper curve is approximately constant at a level of about 120 db for all frequencies. Every pure tone which can be heard may be represented by a point lying somewhere in the area between these two curves.

Only about 1% of the population has a threshold of hearing as low as the bottom curve in Fig. 23-1; 50% of the population can hear pure tones of a frequency of 2500 Hz when the intensity level is about 8 db, and 90% when the level is 20 db.

For a loud tone of intensity level 80 db, the frequency range of the hearing mechanism is from about 20 to about 20,000 Hz, but at a level of 20 db it is only from about 200 to about 15,000 Hz. At a frequency of 1000 Hz the range of intensity level is from about 3 db

* Originally, a scale of intensity levels in *bels* was defined by the relation

$$\text{Intensity level} = \log I/I_0.$$

This unit proved rather large and hence the decibel, one-tenth of a bel, has come into general use. The unit is named in honor of Alexander Graham Bell.

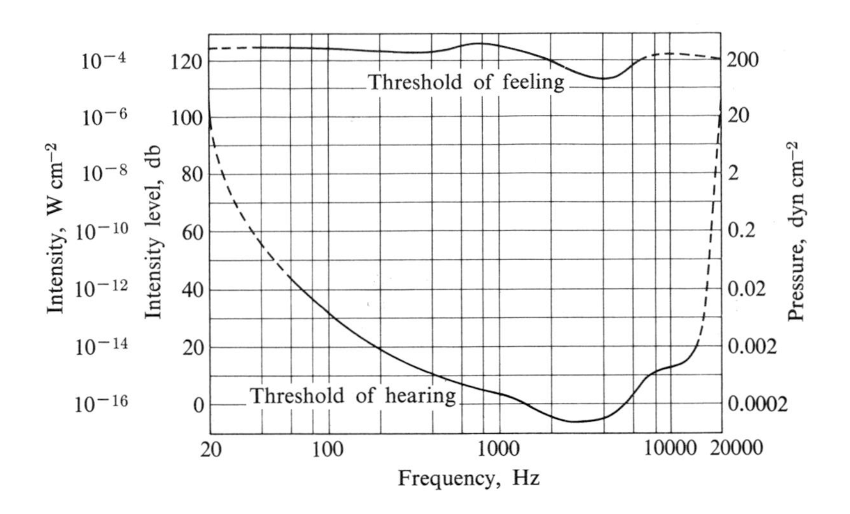

23–1 Auditory area between threshold of hearing and threshold of feeling. (Courtesy of Dr. Harvey Fletcher.)

to about 120 db, whereas at 100 Hz it is only from 30 db to 120 db.

The term *loudness* refers to a sensation in the consciousness of a human observer. It is purely subjective, as contrasted with the objective quantity *intensity*, and is not directly measurable with instruments. Loudness increases with intensity but there is no simple linear relationship. Pure tones of the same intensity but different frequencies do not necessarily produce sensations of equal loudness. Thus for a listener whose auditory area is represented in Fig. 23–1, a pure tone of intensity level 30 db and frequency 60 Hz is completely inaudible, while one of the same intensity level but of frequency 1000 Hz is well above the threshold of audibility. For the first tone to appear as loud as the second, its intensity level would have to be raised to about 65 db.

23–4 QUALITY AND PITCH

A string that has been plucked or a plate that has been struck, if allowed to vibrate freely, will vibrate with many frequencies at the same time. It is a rare occurrence for a body to vibrate with only one frequency. A carefully made tuning fork struck lightly on a rubber block may vibrate with only one frequency, but in the case of musical instruments, the fundamental and many harmonics are usually present at the same time. The impulses that are sent from the ear to the brain give rise to one net effect which is characteristic of the instrument. Suppose, for example, the sound spectrum of a tone consisted of a fundamental of 200 Hz and harmonics 2, 3, 4, and 5, all of different intensity, whereas the sound spectrum of another tone consisted of exactly the same frequencies but with a different intensity distribution. The two tones would sound different; they are said to differ in *quality*.

Adjectives used to describe the quality of musical tones are purely subjective in character, such as reedy, golden, round, mellow, tinny, etc. *The quality of a sound is determined by the number of overtones present and their respective intensity-versus-time curves.* The sound spectra of various musical instruments are shown in Fig. 23–2.

The term *pitch* refers to the attribute of a sound sensation that enables one to classify a note as "high" or "low." Like loudness, it is a subjective quantity and cannot be measured with instruments. Pitch is related to the objective quantity *frequency*, but there is no one-to-one correspondence. For a pure tone of constant intensity, the pitch becomes higher as the frequency is increased, but the pitch of a pure tone of constant frequency becomes lower as the intensity level is raised.

Many of the notes played on musical instruments are rich in harmonics, some of which may be more prominent than the fundamental. Presented with an array of frequencies constituting a harmonic series, the ear will still assign a characteristic pitch to the combination, this pitch being that associated with the fundamental frequency of the series. So definite is this pitch sensation that it is possible to eliminate the

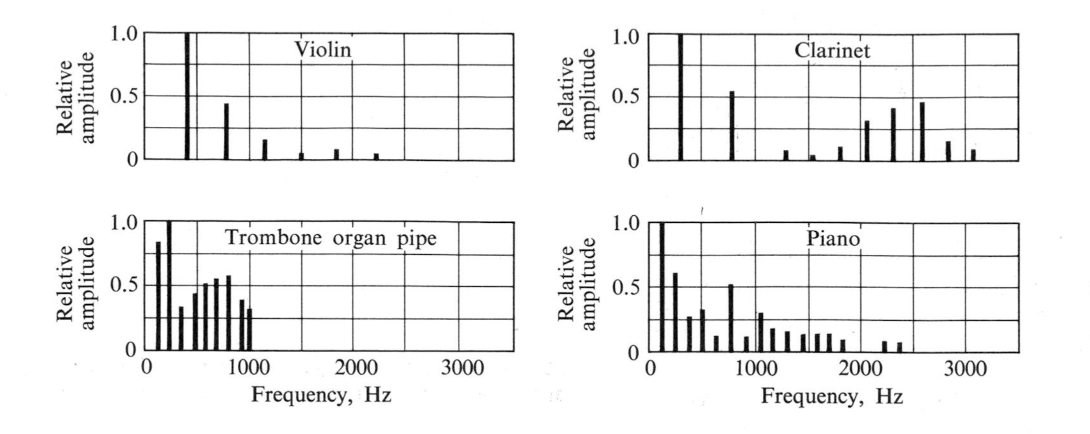

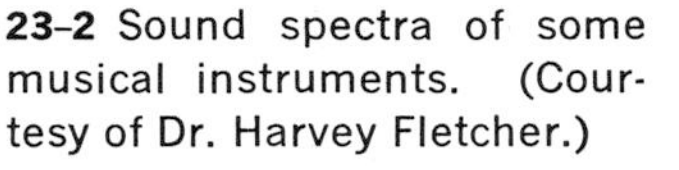
23-2 Sound spectra of some musical instruments. (Courtesy of Dr. Harvey Fletcher.)

fundamental frequency entirely, by means of filters, without any observable effect upon the pitch! The ear apparently will supply the fundamental, provided the correct harmonics are present. It is this rather surprising property of the ear that enables a small loudspeaker which does not radiate low frequencies well to give nevertheless the impression of good radiation in the low-frequency region. Because the speaker is a fairly efficient radiator for the frequencies of the harmonics, the listener believes he is actually hearing the low frequencies, when instead he is hearing only multiples of these frequencies and his ear is supplying the fundamental. It is possible, by deliberate distortion of the harmonics associated with low musical notes, to make a very small radio set, totally inadequate in the low-frequency range, sound somewhat like a larger, acoustically superior console set. Such synthetic bass is, to the critical ear, inferior to true bass reproduction, where the harmonic content is closer to that of the original sound.

23-5 SPHERICAL WAVES

In the preceding chapters we have discussed the propagation of plane waves in a fluid contained in a tube of constant cross section. In the absence of frictional effects, a plane wave in the tube remains plane and its amplitude does not change as the wave advances. On the other hand, the waves originating at a vibrating body in the air, or under water, spread out in all directions from the source. We shall consider only the simplest type of source, namely, a sphere whose surface performs radial oscillations as would the surface of a rubber balloon if air were alternately forced into it and withdrawn from it. Such a source is called a *pulsing sphere* and the waves it emits are spherical, concentric with the source. Since the energy transported by the waves spreads out over spheres of radius r and area $4\pi r^2$, the intensity of the waves (energy per unit time per unit area) must vary inversely as the square of the distance from the source. Also, since the intensity varies as the square of the amplitude, according to Eq. (23-1), it follows that the amplitude of the wave varies inversely as the first power of the distance. At sufficiently large distances from the source, however, the waves can be considered plane, with the pressure variations in phase with the particle velocity, so that there is always a flow of energy outward from the source.

At relatively small distances from the source, the situation is more complicated. The pressure is neither exactly in phase with the particle velocity, as in a traveling wave, nor is it exactly 90° out of phase, as in a stationary wave, but the phase difference is somewhere between 0° and 90°. However, the pressure can be resolved into two components, one of which is in phase with the particle velocity while the other is 90° out of phase. The former component gives rise to a flow of energy that is always outward, while the latter is associated with energy surging periodically in and out from the source. Energy which leaves the source permanently is said to be *radiated* by the source, and of course the radiated energy is the same across all surfaces surrounding the source.

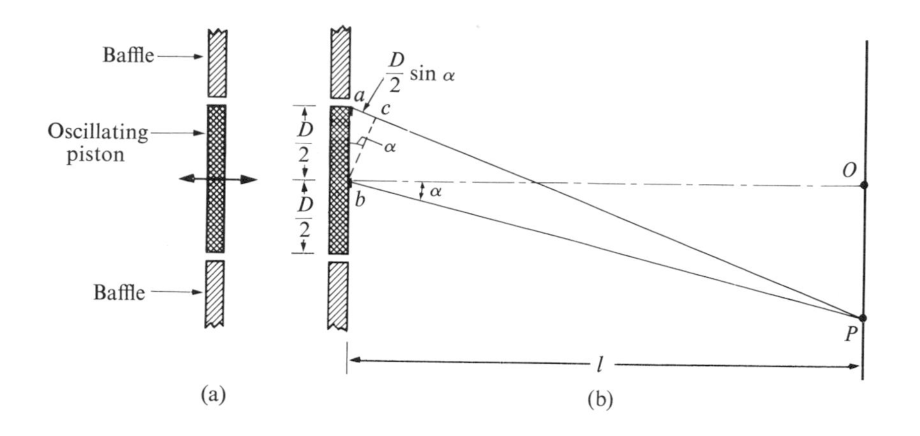

23-3 (a) An oscillating piston set in a baffle. (b) Construction for locating diffraction minima in radiation pattern from a long rectangular source.

The fact that energy is radiated from a sound source means that energy must be supplied to the source to keep it vibrating; therefore the radiation process is analogous to a frictional resistance to the motion of the source. If the source is started vibrating and left to itself, its motion is damped out; we speak of the effect as *radiation damping*. The greater the damping, the more effective is the source as a radiator.

23-6 RADIATION FROM A PISTON. DIFFRACTION

An actual sound source, such as a loudspeaker diaphragm, can be better approximated by a plane vibrating normally to its surface than by a pulsing sphere. A single surface oscillating in this way radiates from both sides. To simplify the problem, we shall assume that we have an oscillating piston fitting closely in a large wall or *baffle*, as in Fig. 23-3(a). Then only the waves radiated from the right side of the piston need be considered in the space at the right of the baffle.

The distribution of radiated energy in direction depends on the shape of the piston and on its dimensions relative to the wavelength of the emitted waves. The mathematics becomes impossibly complicated for any shapes except circles and rectangles. The simplest case is that of a very long rectangle like a long plank, and Fig. 23-3(b) is an end view of such a piston. Imagine the right face of the piston, of width D, to be subdivided into a large number of very narrow strips parallel to the length of the piston and perpendicular to the plane of the diagram. Each of these strips can be treated as a *line source*, sending out waves having *cylindrical* wave surfaces coaxial with the source. We now apply the principle of superposition to these cylindrical waves.

Point P is a point in space at a distance ℓ large compared with the width D of the source, and with the wavelength λ. Consider the two elementary line sources at a and b, one at the top edge of the piston and the other just below its centerline. With P as a center and Pb as radius, strike the arc shown by the dotted line, intersecting the line Pa at c. The distance ac is then the path difference between the waves reaching P from a and b. Since ℓ is large compared with D, the arc bc is very nearly a straight line and abc is very nearly a right triangle with the angle α equal to the angle between Pb and the normal Ob. The path difference ac is then

$$ac = \frac{D}{2}\sin\alpha. \tag{23-3}$$

If the angle α has such a value that ac equals $\frac{1}{2}$ wavelength, the waves from a and b will reach P 180° out of phase and will very nearly cancel one another. The cancellation will not be complete because (1) the waves have slightly different distances to travel, and the amplitude decreases with distance, and (2) the directions to P are not exactly the same for both sources, and a line source, unlike a spherical source, does not radiate uniformly in all directions. However, both of these effects will be small if ℓ is large compared with D.

Consider next the pair of line sources just below a and b. Except for second-order differences, the same diagram as in Fig. 23-3(b) can be constructed for them and the waves from these sources will also cancel at P. Proceeding in this way over the entire surface of the piston, we see that very nearly complete cancellation results in a direction making an angle α with the normal, provided that

$$ac = \frac{\lambda}{2},$$

or, from Eq. (23-3), that

$$\sin\alpha = \frac{\lambda}{D}. \qquad (23\text{-}4)$$

As an example, if $D = 12$ in. and $\lambda = 6$ in. (corresponding to a frequency of about 2000 Hz),

$$\sin\alpha = \tfrac{6}{16} = 0.50,$$
$$\alpha = 30°,$$

and no energy is radiated at an angle of 30° on either side of the normal.

Other minima will occur in directions for which $\sin\alpha = 2\lambda/D$, $3\lambda/D$, etc. This can be shown by dividing the surface of the piston into quarters, sixths, etc., and pairing off one element against another, as in Fig. 23-3(b).

The relatively simple discussion above, while it gives the angular positions of the *minima*, does not give those of the *maxima* nor does it give the relative intensities in various directions. The complete analysis is too lengthy to give here and we shall only state the results. There is a maximum intensity of radiated energy along the normal to the piston ($\alpha = 0$) and in other directions very nearly halfway between the minima. However, by far the greatest amount of energy is concentrated in the region between the first two minima on either side of the normal, and for most purposes the energy radiated in other directions can be neglected.

The analysis of the radiation pattern from a *circular* piston is carried out in the same way as that for a long rectangle. The piston is subdivided into narrow circular zones instead of long strips, and the effect of the waves from all the zones is summed at a distant point. There is a maximum of intensity along the axis of the piston, as would be expected. The angle α at which the first minimum occurs is given by

$$\sin\alpha = 1.22\,\frac{\lambda}{D}, \qquad (23\text{-}5)$$

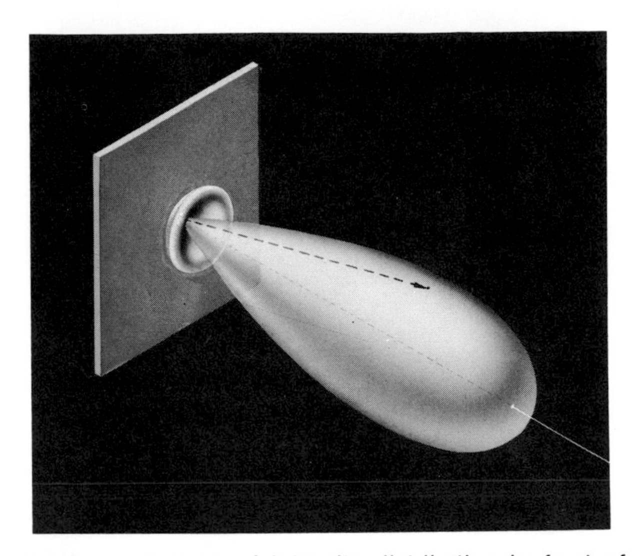

23-4 Space diagram of intensity distribution in front of an oscillating circular piston set in a baffle. A vector drawn from the center of the piston to any point on the surface has a length proportional to the sound intensity in that direction, as observed at a large fixed distance from the piston.

where D is now the piston diameter, and about 85% of the radiated energy is concentrated within a cone of this half-angle. If $D = 12$ in., $\lambda = 6$ in.,

$$\sin\alpha = 1.22 \times \tfrac{6}{12} = 0.61,$$
$$\alpha \approx 37°.$$

Other minima, and maxima of rapidly decreasing intensity, surround the central maximum.

Figure 23-4 is a space diagram of the intensity distribution in front of an oscillating circular piston set in a baffle.

This *directivity* of the sound radiated by a piston has a number of important applications. In a motion picture theater, for example, we wish the sound waves radiated by a speaker behind the screen to spread out over a large angle. If a loudspeaker 12 in. in diameter, approximated by a circular piston, is the sound source, then at a frequency of 2000 Hz the directly radiated wave is concentrated mainly in a "beam" of half-angle 37°, centered on the speaker axis. For a frequency of 10,000 Hz, or a wavelength of about 0.1 ft (about the upper frequency limit for such sound systems), the half-angle is

$$\sin \alpha = 1.22 \frac{0.1}{1.0} = 0.122, \qquad \alpha \approx 7°,$$

while at a frequency of 1200 Hz, or a wavelength of about 0.8 ft,

$$\sin \alpha = 1.22 \frac{0.8}{1.0} \approx 1.0, \qquad \alpha \approx 90°.$$

This means that for frequencies of 1200 Hz or less the radiated energy spreads out fairly uniformly, while at 10,000 Hz it is concentrated in a narrow beam, like a searchlight beam.

Since the intelligibility of speech depends largely on the high-frequency components, these are usually channeled into a number of speakers directed toward different parts of the auditorium, while the low frequencies can be handled by a single speaker, since the low-frequency radiation pattern has a much wider angular spread. On the other hand, it is sometimes desirable, as in underwater sound signaling or when leading the cheering section, to produce a beam having only a small angular divergence. To accomplish this, the diameter of the source (approximating it by a circular piston) must be large compared with the wavelength of the radiated sound.

If instead of a vibrating piston set in a wall there is merely an aperture in the wall and a train of plane waves is incident from the left, the waves transmitted through the aperture will propagate beyond it in the same way as waves originating at a piston set in the aperture. If the wavelength is small compared with the dimensions of the aperture, the spreading of the waves is small, while if the wavelength is large, the waves spread out in all directions.

When there is an obstacle in the path of a train of waves, the resultant effect at the far side of the obstacle is due to those portions of the advancing wave surface that are *not* obstructed. In very general terms, if the wavelength is relatively small, the spreading is small and the obstacle casts a sharp "shadow." The larger the wavelength, the greater the spreading or bending of the waves. Thus one *can* hear around the corner of a wall but the effect is greater for waves of long wavelength (or low frequency) than for waves of short wavelength (or high frequency). The general term for the phenomena described above, in which one is concerned with the resultant effect of a large number of waves from different parts of a source, is *diffraction*.

23-7 RADIATING EFFICIENCY OF A SOUND SOURCE

The directivity of a sound source is not the only factor to be considered in designing such a source. When an oscillating surface is large compared with the wavelength of the waves it emits, most of the sound energy is propagated directly ahead of the source in a fairly well-defined beam of plane waves. The phase relations between pressure and particle velocity in a plane wave are such that energy always travels away from the source. If, on the other hand, the surface is small compared with the wavelength, the phase relations near the source are such that much of the energy in the sound field simply surges alternately away from and back to the source and only a small amount is radiated.

Consider a sound wave having a frequency of 250 Hz and a wavelength of about 4 ft. A loudspeaker 6 in. in diameter (small compared with the wavelength) will radiate waves of this frequency quite uniformly in all directions, but the *power* it can radiate is relatively small unless it is driven with an impossibly large amplitude. A speaker 8 ft in diameter would be a much more efficient radiator at this frequency, but its disadvantages are obvious. A horn, such as a megaphone, the horn of an early phonograph, or the horns often used in loudspeakers, is a device which makes it possible for a source of small dimensions to radiate low frequencies just as efficiently as a much larger source. At the same time, of course, the directivity is increased, which may or may not be desirable.

In very general terms, what a horn does is to control the way in which the waves from a source spread out immediately after leaving the source. Instead of being free to diverge in all directions they are restricted by the walls of the horn. The result is to modify the phase relations in the wave so that a large fraction of the sound energy reaching the mouth of the horn is radiated and only a small fraction is reflected back to the source. Of course, if the horn is itself a musical instrument, like a bugle, trumpet, or French horn, some energy must be reflected at its mouth in order to establish a system of stationary waves in the horn and cause the air column to resonate at the desired frequency.

Because of the necessarily large bulk of a horn with a wide mouth, horns are not widely used in sound reproducing systems in the home. Sufficiently effective radiation of low frequencies can be accomplished with the conventional cone speaker if the amplifier which precedes it has a relatively greater "gain" for low frequencies than for high. Another common practice is to utilize, at low frequencies, the waves radiated from the back of the speaker cone. This is accomplished by mounting the speaker in an enclosure in which there is an opening. If the volume of the enclosure and the area of the opening have the proper values, the enclosure resonates at low frequencies in such a way that waves radiated by the opening are in phase with those from the front of the speaker. The characteristic of the hearing mechanism mentioned in Section 23-4, namely, that the *pitch* associated with a harmonic series is that of the fundamental even if this is absent, also makes the output of a small speaker at least acceptable even if the low frequencies present in the original sound are not reproduced with their original relative intensities.

23-8 BEATS

Stationary waves in an air column have been cited as one example of interference. They arise when two wave trains of the same amplitude and frequency are traveling through the same region in opposite directions. We now wish to consider another type of interference, which results when two wave trains of equal amplitude but slightly different frequency travel through the same region. Such a condition exists when two tuning forks of slightly different frequency are sounded simultaneously or when two piano wires struck by the same key are slightly "out of tune."

Let us consider some one point of space through which the waves are simultaneously passing. The displacements due to the two waves separately are plotted as a function of the time on graph (a) in Fig. 23-5. If the total extent of the time axis represents 1 s, the graphs correspond to frequencies of 16 Hz and of 18 Hz. Applying the principle of superposition to find the resultant vibration, we get graph (b), where it is seen that the amplitude varies with the time. These variations of amplitude give rise to variations of loudness which are called *beats*. Two strings may be tuned to the same frequency by tightening one of them while sounding both until the beats disappear.

The production of beats may be treated mathematically as follows. The displacements due to the two waves passing simultaneously through some one point of space may be written

$$y_1 = Y \cos 2\pi f_1 t, \qquad y_2 = Y \cos 2\pi f_2 t$$

(the amplitudes are assumed equal).

By the principle of superposition, the resultant displacement is

$$y = y_1 + y_2 = Y(\cos 2\pi f_1 t + \cos 2\pi f_2 t),$$

and, since

$$\cos a + \cos b = 2 \cos \frac{a+b}{2} \cos \frac{a-b}{2},$$

this may be written

$$y = \left[2Y \cos 2\pi \left(\frac{f_1 - f_2}{2}\right) t\right] \cos 2\pi \frac{f_1 + f_2}{2} t. \quad (23\text{-}6)$$

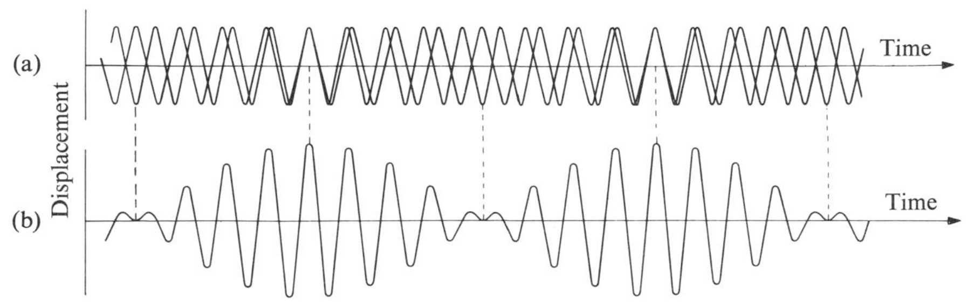

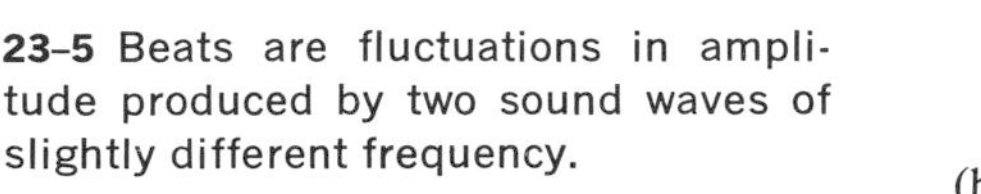
23-5 Beats are fluctuations in amplitude produced by two sound waves of slightly different frequency.

The resulting vibration can then be considered to be of frequency $(f_1 + f_2)/2$, or the average frequency of the two tones, and of amplitude given by the expression in brackets. The amplitude therefore varies with time at a frequency $(f_1 - f_2)/2$. If f_1 and f_2 are nearly equal, this term is small and the amplitude fluctuates very slowly. When the amplitude is large the sound is loud, and vice versa. A beat, or a maximum of amplitude, will occur when $\cos 2\pi[(f_1 - f_2)/2]t$ equals 1 or -1. If each of these values occurs once in each cycle, the number of beats per second is twice the frequency $(f_1 - f_2)/2$, *or the number of beats per second equals the difference of the frequencies.*

23-9 THE DOPPLER EFFECT

When a source of sound, or a listener, or both, are in motion relative to the air, the pitch of the sound, as heard by the listener, is in general not the same as when source and listener are at rest. The most common example is the sudden drop in pitch of the sound from an automobile horn as one meets and passes a car proceeding in the opposite direction. This phenomenon is called the *Doppler effect.*

Let v_L and v_S represent the velocities of a listener and a source, relative to the air. We shall consider only the special case in which the velocities lie along the line joining listener and source. Since these velocities may be in the same or opposite directions, and the listener may be either ahead of or behind the source, a convention of signs is required. We shall take the positive directions of v_L and v_S as that *from* the position of the listener to the position of the source. The velocity of propagation of sound waves, c, will always be considered positive.

In Fig. 23-6, a listener L is at the left of a source S. The positive direction is then from left to right and both v_L and v_S are positive in the diagram. The sound source is at point a at time $t = 0$ and at point b at time t. The outer circle represents the wave surface emitted at time $t = 0$. This surface (in free space) is a sphere with center at a, and is traveling radially outward at all points with speed c. (The fact that the wave originated at a *moving* source does not affect its speed after leaving the source. The wave speed c is a property of the *medium* only; the waves forget about the source as soon as they leave it.) The radius of this sphere (the distance ea or ad) is therefore ct. The distance ab equals v_St, so

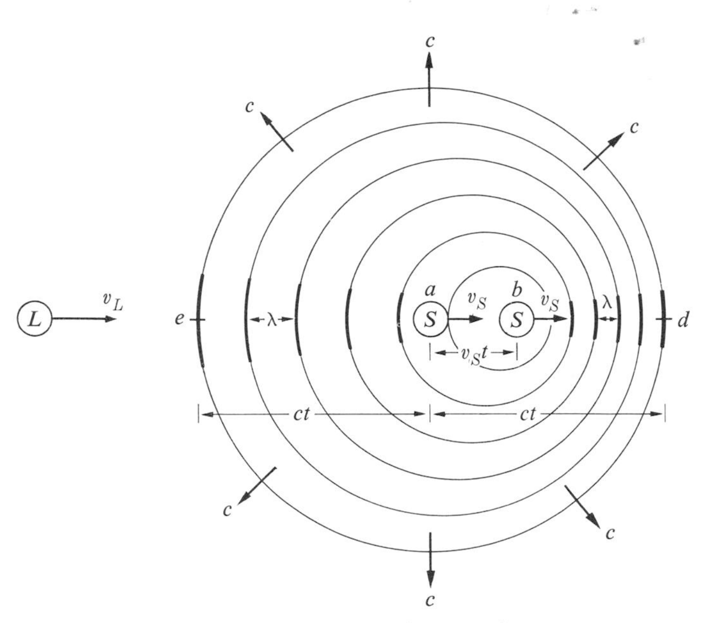

23-6 Wave surfaces emitted by a moving source.

$$eb = (c + v_S)t, \qquad bd = (c - v_S)t.$$

In the time interval between $t = 0$ and $t = t$, the number of waves emitted by the source is f_St, where f_S is the frequency of the source. In front of the source these waves are crowded into the distance bd, while behind the source they are spread out over the distance eb. The wavelength in front of the source is therefore

$$\lambda = \frac{(c - v_S)t}{f_St} = \frac{c - v_S}{f_S},$$

while the wavelength behind the source is

$$\lambda = \frac{(c + v_S)t}{f_St} = \frac{c + v_S}{f_S}.$$

The waves approaching the moving listener L have a speed of propagation relative to him, given by $c + v_L$. The frequency f_L at which the listener encounters these waves is

$$f_L = \frac{c + v_L}{\lambda} = \frac{c + v_L}{(c + v_S)/f_S},$$

or

$$\frac{f_L}{c + v_L} = \frac{f_S}{c + v_S}, \tag{23-7}$$

which expresses the frequency f_L as heard by the listener in terms of the frequency f_S of the source. It is unnecessary to derive equations for other special cases if consistent use is made of the sign convention given above. See the following examples.

Examples. Let $f_S = 1000$ Hz, and $c = 1000$ ft s^{-1}. The wavelength of the waves emitted by a stationary source is then $c/f_S = 1.00$ ft.

a) What are the wavelengths ahead of and behind the moving source in Fig. 23-6 if its velocity is 100 ft s^{-1}?

In front of the source,

$$\lambda = \frac{c - v_S}{f_S} = \frac{1000 - 100}{1000} = 0.90 \text{ ft.}$$

Behind the source,

$$\lambda = \frac{c + v_S}{f_S} = \frac{1000 + 100}{1000} = 1.10 \text{ ft.}$$

b) If the listener L in Fig. 23-6 is at rest and the source is moving away from him at 100 ft s^{-1}, what is the frequency as heard by the listener?

Since

$$v_L = 0 \quad \text{and} \quad v_S = 100 \text{ ft s}^{-1},$$

we have

$$f_L = f_S \frac{c}{c + v_S} = 1000 \frac{1000}{1000 + 100} = 909 \text{ Hz.}$$

c) If the source in Fig. 23-6 is at rest and the listener is moving toward the left at 100 ft s^{-1}, what is the frequency as heard by the listener?

The positive direction (from listener to source) is still from left to right, so

$$v_L = -100 \text{ ft s}^{-1}, \quad v_S = 0,$$

$$f_L = f_S \frac{c + v_L}{c} = 1000 \frac{1000 - 100}{1000} = 900 \text{ Hz.}$$

Thus while the frequency f_L as heard by the listener is less than the frequency f_S both when the source moves away from the listener and when the listener moves away from the source, the decrease in frequency is not the same for the same speed of recession.

In the preceding equations, the velocities v_L, v_S and c are all *relative to the air,* or more generally, to the medium in which the waves are traveling. The Doppler effect exists also for electromagnetic waves in empty space, such as light waves or radio waves. In this case, there is no "medium" relative to which a velocity can be defined, and we can speak only of the *relative* velocity *v of source and receiver.*

The wave velocity c is the velocity of light and is the same for both source and receiver. Let f_S be the frequency of an *electromagnetic* wave emitted by a source, and f_R the frequency observed by a receiver. These frequencies are related by the equations:

$$\frac{f_R}{\sqrt{c + v}} = \frac{f_S}{\sqrt{c - v}} \quad \begin{pmatrix}\text{source and receiver} \\ \text{approaching}\end{pmatrix}, \tag{23-8}$$

$$\frac{f_R}{\sqrt{c - v}} = \frac{f_S}{\sqrt{c + v}} \quad \begin{pmatrix}\text{source and receiver} \\ \text{receding}\end{pmatrix}. \tag{23-9}$$

The quantity v is the *absolute magnitude* of the relative velocity and is considered *positive* always. Although these equations do not have the same form as those for sound waves in air, it will be seen that the received frequency f_R is greater than the source frequency f_S when source and receiver are approaching, and that f_R is smaller than f_S when source and receiver are receding.

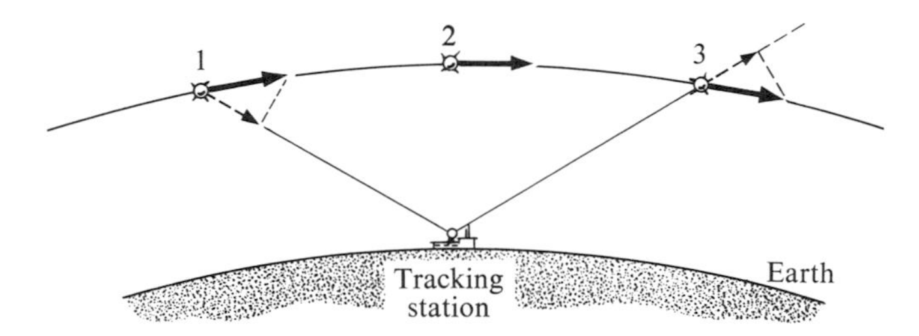

23-7 Change of velocity component along the line of sight of a satellite passing a tracking station.

The Doppler effect provides a convenient means of tracking an artificial satellite which is emitting a radio signal of constant frequency f_S. The frequency f_R of the signal that is received on the earth decreases as the satellite is passing, since the velocity component *toward* the earth decreases from position 1 to position 2 in

Fig. 23–7, and then points *away* from the earth from 2 to 3. If the received signal is combined with a constant signal generated in the receiver to give rise to *beats,* then the beat frequency may be such as to produce an audible note whose pitch decreases as the satellite passes overhead.

Problems

23–1 (a) If the pressure amplitude in a sound wave is tripled, by how many times is the intensity of the wave increased? (b) By how many times must the pressure amplitude of a sound wave be increased in order to increase the intensity by a factor of 16 times?

23–2 (a) Two sound waves of the same frequency, one in air and one in water, are equal in intensity. What is the ratio of the pressure amplitude of the wave in water to that of the wave in air? If the pressure amplitudes of the waves are equal, (b) what is the ratio of their intensities? (c) What is the difference between their intensity levels? The speed of sound in water may be taken as 1490 m s^{-1}.

23–3 (a) Relative to the arbitrary reference intensity of 10^{-16} W cm^{-2}, what is the intensity level in decibels of a sound wave whose intensity is 10^{-10} W cm^{-2}? (b) What is the intensity level of a sound wave in air whose pressure amplitude is 2 dyn cm^{-2}?

23–4 The intensity due to a number of independent sound sources is the sum of the individual intensities. How many decibels greater is the intensity level when all five quintuplets cry simultaneously than when a single one cries? How many more crying babies would be required to produce a further increase in the intensity level of the same number of decibels?

23–5 A window whose area is 1 m^2 opens on a street where the street noises result in an intensity level, at the window, of 60 db. How much "acoustic power" enters the window via the sound waves?

23–6 (a) What are the upper and lower limits of intensity level of a person whose auditory area is represented by the graph of Fig. 23–1? (b) What are the highest and lowest frequencies he can hear when the intensity level is 40 db?

23–7 Two loudspeakers, *A* and *B*, radiate sound uniformly in all directions. The output of acoustic power from *A* is 8×10^{-4} W, and from *B* it is 13.5×10^{-4} W. Both loudspeakers are vibrating in phase at a frequency of 173 Hz. (a) Determine the difference in phase of the two signals at a point *C* along the line joining *A* and *B*, 3 m from *B* and 4 m from *A*. (b) Determine the intensity at *C* from speaker *A* if speaker *B* is turned off, and the intensity at *C* from speaker *B* if speaker *A* is turned off. (c) With both speakers on, what is the intensity and intensity level at *C*?

23–8 What should be the diameter of a sound source in the form of a circular piston set in a wall, if the central lobe of the diffraction pattern is to have a half-angle of 45°, for a frequency of 10,000 Hz?

23–9 The sound source of a sonar system operates at a frequency of 50,000 Hz. Approximate the source by a circular disk set in the hull of a destroyer. The velocity of sound in water can be taken as 1450 m s^{-1}. (a) What is the wavelength of the waves emitted by the source? (b) What must be the diameter of the source if the half-angular divergence of the main beam is not to be more than 10°? (c) What is the difference in frequency between the directly radiated waves and the waves reflected from the hull of a submarine traveling directly away from the destroyer at 15 mi hr^{-1}?

23–10 Two identical piano wires when stretched with the same tension have a fundamental frequency of 400 Hz. By what fractional amount must the tension in one wire be increased in order that 4 beats s^{-1} shall occur when both wires vibrate simultaneously? (Approximate finite changes by differentials.)

23–11 The frequency ratio of a half-tone interval on the diatonic scale is $\frac{16}{15}$. Find the velocity of an automobile passing a listener at rest in still air, if the pitch of the car's horn drops a half tone between the times when the car is coming directly toward him and when it is moving directly away from him.

23–12 (a) Refer to Fig. 23–6 and the examples in Section 23–9. Suppose that a wind of velocity 50 ft s^{-1} is blowing in the same direction as that in which the source is moving. Find the wavelengths ahead of and behind the source. (b) Find the frequency heard by a listener at rest when the source is moving away from him.

23–13 A railroad train is traveling at 100 ft s^{-1} in still air. The frequency of the note emitted by the locomotive whistle is 500 Hz. What is the wavelength of the sound waves (a) in front of and (b) behind the locomotive? What would be the frequency of the sound heard by a stationary listener (c) in front of and (d) behind the locomotive? What frequency would be heard by a passenger on a train traveling at 50 ft s^{-1} and (e) approaching the first, (f) receding from the first? (g) How would each of the preceding answers be altered if a wind of velocity 30 ft s^{-1} were blowing in the same direction as that in which the locomotive was traveling?

23–14 A train of plane sound waves of frequency f_0 and wave length λ_0 travels horizontally toward the right. It strikes and is reflected from a large, rigid, vertical plane surface, perpendicular to the direction of propagation of the wave train and moving toward the left with a velocity v. (a) How many waves strike the surface in a time interval t? (b) At the end of this time interval, how far to the left of the surface is the wave that was reflected at the beginning of the time interval? (c) What is the wavelength of the reflected waves, in terms of λ_0? (d) What is the frequency, in terms of f_0? (e) A listener is at rest at the left of the moving surface. Describe the sensation of sound which he hears as a result of the combined effect of the incident and reflected wave trains.

23–15 Two whistles, A and B, each have a frequency of 500 Hz. A is stationary and B is moving toward the right (away from A) at a velocity of 200 ft s^{-1}. An observer is between the two whistles, moving toward the right with a velocity of 100 ft s^{-1}. Take the velocity of sound in air as 1100 ft s^{-1}. (a) What is the frequency from A as heard by the observer? (b) What is the frequency from B as heard by the observer? (c) What is the beat frequency heard by the observer?

23–16 A man stands at rest in front of a large smooth wall. Directly in front of him, between him and the wall, he holds a vibrating tuning fork of frequency 400 Hz. He now moves the fork toward the wall with a velocity of 4 ft s^{-1}. How many beats per second will he hear between the sound waves reaching him directly from the fork, and those reaching him after being reflected from the wall?

23–17 A source of sound waves, S, emitting waves of frequency 1000 Hz, is traveling toward the right in still air with a velocity of 100 ft s^{-1}. At the right of the source is a large smooth reflecting surface moving toward the left with a velocity of 400 ft s^{-1}. (a) How far does an emitted wave travel in 0.01 s? (b) What is the wavelength of the emitted waves in front of (i.e., at the right of) the source? (c) How many waves strike the reflecting surface in 0.01 s? (d) What is the velocity of the reflected waves? (e) What is the wavelength of the reflected waves?

23–18 (a) Show that Eq. (23–8) can be written

$$f_R = f_S\left(1+\frac{v}{c}\right)^{1/2}\left(1-\frac{v}{c}\right)^{-1/2}.$$

b) Use the binomial theorem to show that if $v \ll c$, then, to a good approximation,

$$f_R = f_S\left(1+\frac{v}{c}\right).$$

c) An earth satellite emits a radio signal of frequency 10^8 Hz. An observer on the ground produces beats between the received signal and a local signal also of frequency 10^8 Hz. At a particular moment, the beat frequency is 2400 Hz. What is the component of the satellite's velocity directed toward the earth at this moment?

Answers to Odd-Numbered Problems

Answers to Odd-Numbered Problems

Chapter 1

1–3 a) 19.3 lb in a direction midway between the 10-lb forces
b) 8.46 lb in a direction midway between the 10-lb forces
1–5 15 lb, 53° above the x-axis
26 lb, 28° below the x-axis
1–7 25.7 lb, 30.6 lb
1–9 a) 18.5 lb b) 9.2 lb
1–11 308 lb, 25° above the x-axis
1–13 a) 7 lb, 2.9 lb b) 7.6 lb c) 11 lb
1–15 $F_x = 380$ lb, W; $F_y = 760$ lb, N

Chapter 2

2–3 a) 10 lb b) 20 lb
2–5 a) 150 lb in A, 180 lb in B, 200 lb in C
b) 200 lb in A, 280 lb in B, 200 lb in C
c) 550 lb in A, 670 lb in B, 200 lb in C
d) 167 lb in A, 58 lb in B, 125 lb in C
2–7 a) Parts (b) and (c) can be solved
b) In part (a) another side or angle is needed
2–9 630 lb
2–11 a) 20 lb b) 30 lb
2–13 a) $w/(2 \sin \theta)$ b) $w/(2 \tan \theta)$
2–17 22 lb
2–19 a) 76 lb b) 24 lb c) from 15.4 to 84.6 lb
2–21 a) Held back b) 144 lb
2–23 a) 3 lb b) 4 lb c) 5 lb
2–25 b) 10 lb c) 30 lb

Chapter 3

3–3 b) 5 ft c) 5.75 ft
3–5 25 lb, 20 lb to the right, 5 lb up
3–7 a) −6 lb, 10 lb b) $\frac{5}{3}$ c) 12 lb d) 2 ft from right end
3–9 a) 54 lb b) 24 lb
3–11 a) 65 lb, 75 lb b) 45 lb c) 47 lb
3–13 a) arctan 0.3 b) 24 cm c) 0.53
3–17 a) 43 lb b) 37 lb c) 59 lb
3–19 a) 20 lb b) 0.2 ft to right of c.g. c) 1.25 ft
3–23 About 3000 mi from the center of the earth

Chapter 4

4–1 a) 12.5 cm s^{-1} b) 7840 cm
4–3 a) 300 ft b) 60 ft s^{-1}
4–5 a) 24 ft s^{-1}, 29 ft s^{-1}, 34 ft s^{-1} b) 2.5 ft s^{-2}
c) 21.5 ft s^{-1} d) 8.6 s e) 23 ft f) 1 s g) 24 ft s^{-1}
4–7 a) 32 ft s^{-1} b) 2.83 s c) 16 ft d) 22.6 ft s^{-1}
e) 16 ft s^{-1}
4–9 a) 8.7 s b) 75 ft c) $v_A = 52$ ft s^{-1}, $v_T = 35$ ft s^{-1}
4–11 a) 94 ft s^{-1} b) 124 ft c) 53 ft s^{-1} d) 150 ft s^{-2}
e) 1.96 s f) 93 ft s^{-1}
4–13 39.4 ft
4–15 a) 48 ft s^{-1} b) 36 ft c) Zero d) 32 ft s^{-2}, downward
e) 80 ft s^{-1}
4–17 a) 32 ft b) 13 ft s^{-1}, −32 ft s^{-2}
c) −51 ft s^{-1}, −32 ft s^{-2} d) 37 ft s^{-1}
4–19 a) 36 ft b) 0.5 s
c) 32 ft s^{-1}, −32 ft s^{-2}, −16 ft s^{-1}, −32 ft s^{-2}
4–21 a) 4 m s^{-2} b) 6 m s^{-1} c) 4.5 m
4–25 a) $t \approx 1$ min b) $y \approx 100$ mi
4–27 d) $k = 3.33 \times 10^{-3}$ ft^{-1} e) −1.33 ft s^{-2}
4–31 6.55 knots
4–33 a) 50 mi hr^{-1}, 53° S of W b) 30° N of W
4–35 a) 5 mi hr^{-1}, 53° E of N b) 0.75 mi c) 15 min
4–37 11.6 mi hr^{-1}, 59° N of E

Chapter 5

5-1 a) 0.102 kg b) 0.00102 g c) 0.0312 slug
5-5 a) 0.5 slug b) 500 ft
5-7 a) 1.62×10^{-10} dyn b) 3.33×10^{-9} s
c) 1.8×10^{17} cm s^{-2}
5-9 6.2×10^{27} g
5-11 6.32 ft s^{-2}
5-13 a) 59.0 N b) const. velocity c) 6.25 m s^{-1}
5-15 a) 40 lb b) 4 ft s^{-2} downward c) zero
5-17 240 ft
5-19 $W = 2wa/(g + a)$
5-21 a) 37° b) 6.4 ft s^{-2} c) 2.5 s
5-23 a) 6 lb b) 8 ft s^{-2}
5-25 a) 2 lb b) 1.9 lb
5-27 a) 196 cm s^{-2} b) 314,000 dyn
5-29 a) to left b) 2.13 ft s^{-2} c) 43.3 lb
5-31 1.33 lb
5-33

	a_1	a_2
a)	0	0
b)	0	0
c)	0	16 ft s^{-2}
d)	4 ft s^{-2}	28 ft s^{-2}
e)	16 ft s^{-2}	48 ft s^{-2}

5-35 a) 4.4 ft s^{-2} b) 1.2 lb
5-37 127.5 lb
5-39 g/μ
5-43 a) 133 lb b) 33 lb c) 33 lb
5-45 c) $v_T = \sqrt{ma_0/k}$ d) $t_R\sqrt{m/ka_0}$ e) $v = v_T \dfrac{e^{2t/t_R} - 1}{e^{2t/t_R} + 1}$

Chapter 6

6-1 a) 0.5 s b) 12 ft s^{-1}
c) 20 ft s^{-1}, 53° below horizontal
6-3 a) 20 s b) 6000 ft
c) $v_x = 300$ ft s^{-1} $v_y = 640$ ft s^{-1}
6-5 a) 100 ft b) 200 ft
c) $v_x = 80$ ft s^{-1}, $v_y = 80$ ft s^{-1}, $v = 113$ ft s^{-1}, 45° below horizontal
6-7 a) 40 s b) 55°
6-9 a) 8 ft b) 19°, 71°
6-11 100 ft
6-13 Yes. Ball clears fence by 10 ft
6-15 1.98 s, 157 s
6-17 1300 ft
6-19 $R = 2v_0^2 \sin(\theta_0 - \alpha) \cos\theta_0/g \cos^2\alpha$
6-23 a) About 25° above horizontal b) About 650 mi
c) About 10,000 mi hr^{-1} d) About 875 mi
e) About 5 min
6-25 a) 6.5 cm s^{-1}, 25° E of N
b) 3.25 cm s^{-2}, 25° E of N
6-27 a) 66,700 mi hr^{-1} b) 0.0193 ft s^{-2}
6-29 22 m s^{-1}
6-31 a) 0.43 b) To the very end
6-33 a) 6.4 s b) No
6-35 a) 52° b) No
c) The bead would remain at the bottom
6-37 a) 33° b) 1.45 mi c) 0.65
6-39 a) 1530 ft b) 1440 lb
6-41 a) 0.37 b) 4.3 s c) Twice his actual weight
d) He would strike the ground at a point 30 ft from the center of the wheel
6-45 36,000 km above the earth
6-47 a) 98 min b) 26.5 ft s^{-2}
6-49 9.8 oz

Chapter 7

7-1 63.4×10^6 ft lb
7-3 9850 ft lb
7-5 a) 6 lb b) 48 lb c) 22.5 ft lb
7-7 a) 55,000 ft lb b) 4 times
7-9 4.5×10^{-10} erg
7-11 9.8 J
7-13 a) 5 lb, 10 lb, 20 lb b) 1.25 ft lb, 5 ft lb, 20 ft lb
7-15 a) 33 ft lb b) about 13 ft lb
7-17 a) 160 ft lb b) 160 ft lb
7-19 a) 3300 ft lb b) 1300 ft lb c) 1500 ft lb
d) 500 ft lb Goes into heat. e) $b + c + d = a$
7-21 a) 42 lb b) 105 ft lb
7-23 a) 40 ft s^{-1} b) 19 ft s^{-1}
7-25 10 cm
7-27 a) 0.25 b) 3.5 ft lb
7-31 1.81 m s^{-1}
7-33 a) 24.6 ft s^{-1} b) 27.2 ft s^{-1} c) 22.5 lb
d) 10 ft s^{-1}
7-39 154 hp
7-41 $1.29
7-43 a) 250 lb b) 5 hp, 80 hp
7-45 a) 8200 J b) 8200 J c) 0.55 hp
7-47 a) 6300 lb b) 30 ft s^{-1} c) 68,000 ft lb
d) 220,000 ft lb e) 250 hp
7-49 a) 250 lb b) 39 hp c) 16 hp d) 10.5% grade
7-51 a) 1.8×10^8 MJ b) 1.8×10^{14} MW c) 2.5×10^{10} lb or 4×10^8 ft^3

Chapter 8

8–1 a) 28,000 slug ft s^{-1} b) 60 mi hr^{-1} c) 43 mi hr^{-1}
8–3 a) 8×10^5 m s^{-2} b) 4×10^4 N c) 5×10^{-4} s
d) 20 N · s
8–7 a) 1 ft s^{-1} b) $\frac{2}{3}$ of original E_k c) 1.5 ft s^{-1}
8–9 780 ft s^{-1}
8–11 30 cm
8–13 a) 0.39 ft b) 1950 ft lb c) 3.9 ft lb
8–15 3.3 cm
8–17 17 mi hr^{-1} 53° E of S
8–19 a) 10 cm s^{-1} b) 0.14 J c) −70 cm s^{-1}, 80 cm s^{-1}
8–21 5 cm s^{-1}, −25 cm s^{-1}
8–23 a) 50 cm s^{-1} 53° below the x-axis in the fourth quadrant
b) 380,000 ergs
8–27 a) $v_A = 58.5$ ft s^{-1}, $v_B = 41.5$ ft s^{-1}
8–29 a) 58,700 ft b) 18×10^4 m ft lb
8–31 a) 7.20 ft s^{-1} b) 375
8–33 a) 18 ft s^{-1} b) 125
8–35 1.82 ft s^{-1}
8–37 a) 149° from direction of electron
b) 10.6×10^{-16} g cm s^{-1} c) 14.5×10^{-10} erg
8–39 4.22 lb
8–41 a) 25×10^6 dyn, 250 N

Chapter 9

9–1 16, 31, 47 ft s^{-1}; 600, 1800, 5400 rev min^{-1}
9–3 20 rad s^{-2}
9–7 a) At the bottom of the wheel b) 178 ft s^{-2}
9–9 a) 135° b) 2.36 ft s^{-2}, 11.1 ft s^{-2}
9–11 164,000 rev min^{-1}
9–15 a) A smallest b) D largest
9–17 a) 2.56×10^{30} slug mi^2 b) 2530 mi
9–19 a) $mb^2/12$ b) $mb^2/3$
9–21 0.47
9–23 a) 19.1 kg m^2 b) 2 N m c) 91.6 rev
9–25 a) 63 m b) 38 N c) 185 N
9–27 a) 0.0625 slug ft^2 b) 12 lb, 4 lb
9–29 a) 2 rad s^{-2} b) 8 rad s^{-1} c) 240 ft lb
d) 7.5 slug ft^2
9–31 a) 4400 lb ft b) 5900 lb c) 190 ft s^{-1}
9–33 69 mr^2
9–35 1.1 slug ft^2
9–37 a) 2×10^7 J b) 18 min
9–39 a) 4 rad s^{-2} b) 14 rad s^{-1} c) 32,000 ft lb
9–41 1.45 J
9–43 a) 24 slug ft^2 s^{-1} b) 6 rad s^{-1} c) 24 ft lb, 72 ft lb
9–45 1 ft
9–47 a) 6 rev min^{-1} b) 6 rev min^{-1}
9–49 4500 ft lb
9–51 0.15 rad s^{-1} in same direction
9–53 a) 1.8 hr b) 130,000 lb ft
9–55 2.5 rad s^{-1}
9–57 a) 5 lb upward b) 7.4 lb up, 2.6 lb up
c) 16.5 lb up, 6.5 lb down d) 0.087 rev s^{-1}

Chapter 10

10–1 25×10^6 lb in^{-2}
10–3 b) 14×10^6 lb in^{-2} c) 0.016×10^6 lb in^{-2}
10–5 a) 27.5 lb b) 0.028 ft
10–7 0.0253 in.
10–9 a) 1.8 ft
b) 30,000 lb in^{-2} in steel, 12,000 lb in^{-2} in copper
c) 0.001 in steel, 0.0006 in copper
10–13 2000 lb
10–15 Steel, 0.64×10^{-6} atm^{-1}
Water, 50×10^{-6} atm^{-1}
Water is 78 times more compressible
10–17 5.1 in.

Chapter 11

11–3 a) 9470 cm s^{-2}, 377 cm s^{-1}
b) 5680 cm s^{-2}, 301 cm s^{-1} c) 0.0368 s
11–5 a) 2400 π^2 ft s^{-2} b) 740 lb c) 43 mi hr^{-1}
11–7 23.2 lb
11–9 a) 9 Hz b) 20×10^6 lb in^{-2}
11–11 a) 4 lb b) 2.59 in. below equilibrium, moving upward
c) 5.2 lb
11–13 a) 31 cm s^{-1} b) 49 cm s^{-2} c) 0.33 s d) 100 cm
11–15 0.79 s
11–17 a) 5.6 lb b) 13.6 lb, 8 lb, 2.4 lb
c) 0.62 ft lb, 0.077 ft lb
11–19 a) $k = k_1 + k_2$ b) $k = k_1 + k_2$ c) $\frac{1}{k} = \frac{1}{k_1} + \frac{1}{k_2}$
d) $\sqrt{2}$
11–21 a) 1.4 s, 3.5 cm
11–23 0.817 ft
11–25 a) 97 cm b) -2×10^{-4}
11–27 a) 6.9 in. b) 0.70 rad s^{-1}
11–29 a) 81 ft
11–31 67 cm

Chapter 12

12–1 21 lb in^{-2}
12–5 a) 1.077×10^6 dyn cm^{-2} b) 1.037×10^6 dyn cm^{-2} c) 1.037×10^6 dyn cm^{-2} d) 5 cm of mercury e) 68 cm of water
12–7 a) 9.80×10^6 dyn b) 1.95×10^5 dyn
12–9 34.8 ft^2
12–11 270,000 ft^3, 9.3 tons lift, using helium
12–13 a) 5 cm b) 4900 dyn cm^{-2}
12–15 a) 100 lb ft^{-3} b) E will read 5 lb, D will read 15 lb
12–19 a) 4500 lb b) 10,000 lb c) 230 lb
12–21 100.87 g
12–23 b) 4 lb c) 1 ft^3
12–25 0.781 g cm^{-3}
12–29 6.25×10^4 lb ft
12–31 a) 4.17×10^7 lb ft b) 5×10^7 lb ft
12–33 a) $\ell a/g$ b) $\omega^2\ell^2/2g$

Chapter 13

13–3 13,720 dyn cm^{-2}
13–5 a) 70.7 cm of Hg b) 71.2 cm of Hg c) 11 cm
13–7 4.3 cm

Chapter 14

14–1 a) 36 ft s^{-1} b) 0.2 ft^3 s^{-1}
14–3 a) 10 cm b) 11 s
14–5 39.6 ft s^{-1}
14–7 a) 0.056 ft^3 s^{+1} b) 3 ft
14–9 a) 16 ft s^{-1} b) 0.79 ft^3 s^{-1}
14–11 56.7 lb in^{-2}
14–13 a) 12 lb in^{-2} b) 12 ft^3 s^{-1}
14–15 0.91 ft^3 min^{-1}
14–17 12 ft^3 min^{-1}
14–19 26.8 ft^3 s^{-1}
14–21 3.88×10^3 cm s^{-1}
14–25 a) 0.77 cm s^{-1} b) 1.89 cm s^{-1}
14–27 325 ft s^{-1}
14–29 a) Turbulent flow b) 5.3 ℓ s^{-1}
14–31 a) Yes b) No

Chapter 15

15–1 600.45°K
15–5 About 7%
15–7 235 cm
15–9 20 cm, 10 cm
15–11 About 250 ft
15–13 a) About 1 C°
15–15 270,000 lb
15–17 9.35×10^8 dyn cm^{-2}
15–19 a) 3×10^{-4} b) 9000 lb in^{-2} c) 450,000 lb d) 21,000 lb in^{-2} e) 105°F
15–21 a) 893 atm b) 36.2°C
15–23 494 atm

Chapter 16

16–1 a) Yes b) No
16–3 4.0 Btu
16–5 370 years
16–7 a) 64.7
16–9 a) 1:0.093:0.031 b) 1:0.83:0.35
16–11 0.1 Btu lb^{-1} (F°)$^{-1}$
16–13 a) 0.092 cal g^{-1} (C°)$^{-1}$
16–15 a) 72°C
16–17 a) Higher c) 0.72 cal g^{-1} (C°)$^{-1}$
16–19 35,300 cal
16–21 0.023 cal mole^{-1} (K°)$^{-1}$, 0.555 cal mole^{-1} (K°)$^{-1}$
16–23 106 g
16–25 0°C with 0.2 g of ice left
16–27 539 cal g^{-1}
16–29 24°C
16–31 40°C
16–33 1.84 kg

Chapter 17

17–1 d) 31.4 C° cm^{-1} e) 29 cal s^{-1} f) zero h) 10 C° s^{-1}
17–3 a) 1.8 cal sec^{-1} b) 20 cm
17–5 0.20 Btu in. hr^{-1} ft^{-2} (F°)$^{-1}$
17–7 110°C
17–9 a) $\sqrt{R_1R_2}$
17–11 84.2 C°
17–13 a) 7.32×10^5 cal (vertical) b) 8.05×10^5 cal (horizontal)
17–15 a) 460 W m^{-2} b) 4600 kW m^{-2}
17–17 1.78 W
17–19 20.03°K

Chapter 18

18–1 3.6 lb in^{-2}
18–3 a) 0.75 atm b) 2 g
18–5 $\rho = pM/RT$
18–7 3.25

18–9 a) 82 cm^3 b) 0.33 g
18–11 When the piston has descended 13.12 in.
18–13 0.0023 g
18–15 b) 600°K c) 4 atm
18–19 a) Meniscus goes down b) Meniscus goes up
c) Meniscus stays approximately in the same place
18–23 16%
18–25 a) 12°C b) 10 g m^{-3}
18–27 1.3 lb hr^{-1}

Chapter 19

19–1 a) Yes b) no c) negative
19–3 a) No b) yes c) yes
d) Work done on the resistor equals the heat transferred to the water.
19–5 10.8 ft^3
19–7 $U_1 = U_2$
19–9 a) 59,300 ft lb, 76 Btu b) 870 Btu
19–11 a) 1.1×10^{-2} ft^3 b) 650 ft lb c) 1800 Btu
d) 1.4×10^6 ft lb
19–13 $K \ln (V_2 - b)/(V_1 - b) - (a/V_1 - a/V_2)$
19–15 b) 830 J c) on the atmosphere d) 2110 J
e) 2940 J f) Same as (b)
19–19 267°C
19–21 a) When piston is 3.13 in. from bottom b) 477°K
19–23 a) $p_1 = 1$ atm, $V_1 = 2.46\ \ell$
$p_2 = 2$ atm, $V_2 = 2.46\ \ell$
$p_3 = 1$ atm, $V_3 = 3.74\ \ell$
b) 49.4 J
19–25 18%
19–27 a) 320°K b) 20%
19–29 a) 900 cal b) 1600 cal c) 400 cal
19–31 a) 10^5 cal b) 312 cal $(K°)^{-1}$ c) 50°C
d) 312 cal $(K°)^{-1}$ before, 165 cal $(K°)^{-1}$ after.

Chapter 20

20–1 a) About 3×10^{-7} cm b) About 10 times as great
20–3 6.32×10^{-12} cm^3
20–5 1700 ft s^{-1}
20–9 Bi, 9.9 mm; Bi_2, 14 mm

Chapter 21

21–1 320 m s^{-1}
21–3 a) 2 cm b) 30 cm c) 100 Hz d) 3000 cm s^{-1}
21–5 6320 cm s^{-1}
21–7 a) 0.5 Hz b) π rad s^{-1} c) $\pi/100$ cm^{-1}
d) $y = 10 \sin [(\pi x/100) - \pi t]$ e) $y = -10 \sin \pi t$
f) $y = 10 \sin [(3\pi/2) - \pi t]$ g) 10π cm s^{-1}
h) $y = 5\sqrt{2}$ cm, $v = -5\pi\sqrt{2}$ cm s^{-1}
21–9 48×10^{-6} atm^{-1}
21–11 Argon: 322 m s^{-1}, hydrogen: 1320 m s^{-1}
21–13 a) 132 Hz b) 1320 Hz c) 13,200 Hz
21–15 a) 0.58 m s^{-1} b) No

Chapter 22

22–1 a) 200 Hz b) 49th overtone
22–3 a) 3600 cm s^{-1} b) 6.5×10^6 dyn
22–5 1.28
22–7 a) 5000 m s^{-1} b) 340 m s^{-1}
22–9 Diatomic
22–11 a) 1140, 2280, 3420, 4560, 5700 Hz
b) 570, 1710, 2850, 3990, 5130 Hz c) 16, 17
22–13 a) 420 Hz b) 415 Hz

Chapter 23

23–1 a) 9 times b) 4 times
23–3 a) 60 db b) 77 db
23–5 10^{-6} W
23–7 a) π rad
b) $I_A = 4 \times 10^{-10}$ W cm^{-2}, $I_B = 12 \times 10^{-10}$ W cm^{-2}
c) 2.1×10^{-10} W cm^{-2}, 63.2 db
23–9 a) 0.029 m b) 21 cm c) 455 Hz
23–11 24 mi hr^{-1}
23–13 a) 2.06 ft b) 2.46 ft c) 548 Hz d) 460 Hz
e) 572 Hz f) 439 Hz
g) 2.12 ft, 2.40 ft, 547, 458, 572, and 437 Hz
23–15 a) 454 Hz b) 462 Hz c) 8 beats s^{-1}
23–17 a) 11 ft b) 1 ft c) about 15 waves d) 1100 ft s^{-1}
e) 0.49 ft

NATURAL TRIGONOMETRIC FUNCTIONS

Angle				
Degree	Radian	Sine	Cosine	Tangent
0°	.000	0.000	1.000	0.000
1°	.017	.017	1.000	.017
2°	.035	.035	0.999	.035
3°	.052	.052	.999	.052
4°	.070	.070	.998	.070
5°	.087	.087	.996	.087
6°	.105	.104	.994	.105
7°	.122	.122	.992	.123
8°	.140	.139	.990	.140
9°	.157	.156	.988	.158
10°	.174	.174	.985	.176
11°	.192	.191	.982	.194
12°	.209	.208	.978	.212
13°	.227	.225	.974	.231
14°	.244	.242	.970	.249
15°	.262	.259	.966	.268
16°	.279	.276	.961	.287
17°	.297	.292	.956	.306
18°	.314	.309	.951	.325
19°	.332	.326	.946	.344
20°	.349	.342	.940	.364
21°	.366	.358	.934	.384
22°	.384	.375	.927	.404
23°	.401	.391	.920	.424
24°	.419	.407	.914	.445
25°	.436	.423	.906	.466
26°	.454	.438	.899	.488
27°	.471	.454	.891	.510
28°	.489	.470	.883	.532
29°	.506	.485	.875	.554
30°	.524	.500	.866	.577
31°	.541	.515	.857	.601
32°	.558	.530	.848	.625
33°	.576	.545	.839	.649
34°	.593	.559	.829	.674
35°	.611	.574	.819	.700
36°	.628	.588	.809	.726
37°	.646	.602	.799	.754
38°	.663	.616	.788	.781
39°	.681	.629	.777	.810
40°	.698	.643	.766	.839
41°	.716	.656	.755	.869
42°	.733	.669	.743	.900
43°	.750	.682	.731	.933
44°	.768	.695	.719	.966
45°	.785	.707	.707	1.000

Angle				
Degree	Radian	Sine	Cosine	Tangent
46°	0.803	0.719	0.695	1.036
47°	.820	.731	.682	1.072
48°	.838	.743	.669	1.111
49°	.855	.755	.656	1.150
50°	.873	.766	.643	1.192
51°	.890	.777	.629	1.235
52°	.908	.788	.616	1.280
53°	.925	.799	.602	1.327
54°	.942	.809	.588	1.376
55°	.960	.819	.574	1.428
56°	.977	.829	.559	1.483
57°	.995	.839	.545	1.540
58°	1.012	.848	.530	1.600
59°	1.030	.857	.515	1.664
60°	1.047	.866	.500	1.732
61°	1.065	.875	.485	1.804
62°	1.082	.883	.470	1.881
63°	1.100	.891	.454	1.963
64°	1.117	.899	.438	2.050
65°	1.134	.906	.423	2.145
66°	1.152	.914	.407	2.246
67°	1.169	.920	.391	2.356
68°	1.187	.927	.375	2.475
69°	1.204	.934	.358	2.605
70°	1.222	.940	.342	2.747
71°	1.239	.946	.326	2.904
72°	1.257	.951	.309	3.078
73°	1.274	.956	.292	3.271
74°	1.292	.961	.276	3.487
75°	1.309	.966	.259	3.732
76°	1.326	.970	.242	4.011
77°	1.344	.974	.225	4.331
78°	1.361	.978	.208	4.705
79°	1.379	.982	.191	5.145
80°	1.396	.985	.174	5.671
81°	1.414	.988	.156	6.314
82°	1.431	.990	.139	7.115
83°	1.449	.992	.122	8.144
84°	1.466	.994	.104	9.514
85°	1.484	.996	.087	11.43
86°	1.501	.998	.070	14.30
87°	1.518	.999	.052	19.08
88°	1.536	.999	.035	28.64
89°	1.553	1.000	.017	57.29
90°	1.571	1.000	.000	∞

COMMON LOGARITHMS

N	0	1	2	3	4	5	6	7	8	9
0	....	0000	3010	4771	6021	6990	7782	8451	9031	9542
1	0000	0414	0792	1139	1461	1761	2041	2304	2553	2788
2	3010	3222	3424	3617	3802	3979	4150	4314	4472	4624
3	4771	4914	5051	5185	5315	5441	5563	5682	5798	5911
4	6021	6128	6232	6335	6435	6532	6628	6721	6812	6902
5	6990	7076	7160	7243	7324	7404	7482	7559	7634	7709
6	7782	7853	7924	7993	8062	8129	8195	8261	8325	8388
7	8451	8513	8573	8633	8692	8751	8808	8865	8921	8976
8	9031	9085	9138	9191	9243	9294	9345	9395	9445	9494
9	9542	9590	9638	9685	9731	9777	9823	9868	9912	9956
10	0000	0043	0086	0128	0170	0212	0253	0294	0334	0374
11	0414	0453	0492	0531	0569	0607	0645	0682	0719	0755
12	0792	0828	0864	0899	0934	0969	1004	1038	1072	1106
13	1139	1173	1206	1239	1271	1303	1335	1367	1399	1430
14	1461	1492	1523	1553	1584	1614	1644	1673	1703	1732
15	1761	1790	1818	1847	1875	1903	1931	1959	1987	2014
16	2041	2068	2095	2122	2148	2175	2201	2227	2253	2279
17	2304	2330	2355	2380	2405	2430	2455	2480	2504	2529
18	2553	2577	2601	2625	2648	2672	2695	2718	2742	2765
19	2788	2810	2833	2856	2878	2900	2923	2945	2967	2989
20	3010	3032	3054	3075	3096	3118	3139	3160	3181	3201
21	3222	3243	3263	3284	3304	3324	3345	3365	3385	3404
22	3424	3444	3464	3483	3502	3522	3541	3560	3579	3598
23	3617	3636	3655	3674	3692	3711	3729	3747	3766	3784
24	3802	3820	3838	3856	3874	3892	3909	3927	3945	3962
25	3979	3997	4014	4031	4048	4065	4082	4099	4116	4133
26	4150	4166	4183	4200	4216	4232	4249	4265	4281	4298
27	4314	4330	4346	4362	4378	4393	4409	4425	4440	4456
28	4472	4487	4502	4518	4533	4548	4564	4579	4594	4609
29	4624	4639	4654	4669	4683	4698	4713	4728	4742	4757
30	4771	4786	4800	4814	4829	4843	4857	4871	4886	4900
31	4914	4928	4942	4955	4969	4983	4997	5011	5024	5038
32	5051	5065	5079	5092	5105	5119	5132	5145	5159	5172
33	5185	5198	5211	5224	5237	5250	5263	5276	5289	5302
34	5315	5328	5340	5353	5366	5378	5391	5403	5416	5428
35	5441	5453	5465	5478	5490	5502	5514	5527	5539	5551
36	5563	5575	5587	5599	5611	5623	5635	5647	5658	5670
37	5682	5694	5705	5717	5729	5740	5752	5763	5775	5786
38	5798	5809	5821	5832	5843	5855	5866	5877	5888	5899
39	5911	5922	5933	5944	5955	5966	5977	5988	5999	6010
40	6021	6031	6042	6053	6064	6075	6085	6096	6107	6117
41	6128	6138	6149	6160	6170	6180	6191	6201	6212	6222
42	6232	6243	6253	6263	6274	6284	6294	6304	6314	6325
43	6335	6345	6355	6365	6375	6385	6395	6405	6415	6425
44	6435	6444	6454	6464	6474	6484	6493	6503	6513	6522
45	6532	6542	6551	6561	6571	6580	6590	6599	6609	6618
46	6628	6637	6646	6656	6665	6675	6684	6693	6702	6712
47	6721	6730	6739	6749	6758	6767	6776	6785	6794	6803
48	6812	6821	6830	6839	6848	6857	6866	6875	6884	6893
49	6902	6911	6920	6928	6937	6946	6955	6964	6972	6981
50	6990	6998	7007	7016	7024	7033	7042	7050	7059	7067
N	0	1	2	3	4	5	6	7	8	9

N	0	1	2	3	4	5	6	7	8	9
50	6990	6998	7007	7016	7024	7033	7042	7050	7059	7067
51	7076	7084	7093	7101	7110	7118	7126	7135	7143	7152
52	7160	7168	7177	7185	7193	7202	7210	7218	7226	7235
53	7243	7251	7259	7267	7275	7284	7292	7300	7308	7316
54	7324	7332	7340	7348	7356	7364	7372	7380	7388	7396
55	7404	7412	7419	7427	7435	7443	7451	7459	7466	7474
56	7482	7490	7497	7505	7513	7520	7528	7536	7543	7551
57	7559	7566	7574	7582	7589	7597	7604	7612	7619	7627
58	7634	7642	7649	7657	7664	7672	7679	7686	7694	7701
59	7709	7716	7723	7731	7738	7745	7752	7760	7767	7774
60	7782	7789	7796	7803	7810	7818	7825	7832	7839	7846
61	7853	7860	7868	7875	7882	7889	7896	7903	7910	7917
62	7924	7931	7938	7945	7952	7959	7966	7973	7980	7987
63	7993	8000	8007	8014	8021	8028	8035	8041	8048	8055
64	8062	8069	8075	8082	8089	8096	8102	8109	8116	8122
65	8129	8136	8142	8149	8156	8162	8169	8176	8182	8189
66	8195	8202	8209	8215	8222	8228	8235	8241	8248	8254
67	8261	8267	8274	8280	8287	8293	8299	8306	8312	8319
68	8325	8331	8338	8344	8351	8357	8363	8370	8376	8382
69	8388	8395	8401	8407	8414	8420	8426	8432	8439	8445
70	8451	8457	8463	8470	8476	8482	8488	8494	8500	8506
71	8513	8519	8525	8531	8537	8543	8549	8555	8561	8567
72	8573	8579	8585	8591	8597	8603	8609	8615	8621	8627
73	8633	8639	8645	8651	8657	8663	8669	8675	8681	8686
74	8692	8698	8704	8710	8716	8722	8727	8733	8739	8745
75	8751	8756	8762	8768	8774	8779	8785	8791	8797	8802
76	8808	8814	8820	8825	8831	8837	8842	8848	8854	8859
77	8865	8871	8876	8882	8887	8893	8899	8904	8910	8915
78	8921	8927	8932	8938	8943	8949	8954	8960	8965	8971
79	8976	8982	8987	8993	8998	9004	9009	9015	9020	9025
80	9031	9036	9042	9047	9053	9058	9063	9069	9074	9079
81	9085	9090	9096	9101	9106	9112	9117	9122	9128	9133
82	9138	9143	9149	9154	9159	9165	9170	9175	9180	9186
83	9191	9196	9201	9206	9212	9217	9222	9227	9232	9238
84	9243	9248	9253	9258	9263	9269	9274	9279	9284	9289
85	9294	9299	9304	9309	9315	9320	9325	9330	9335	9340
86	9345	9350	9355	9360	9365	9370	9375	9380	9385	9390
87	9395	9400	9405	9410	9415	9420	9425	9430	9435	9440
88	9445	9450	9455	9460	9465	9469	9474	9479	9484	9489
89	9494	9499	9504	9509	9513	9518	9523	9528	9533	9538
90	9542	9547	9552	9557	9562	9566	9571	9576	9581	9586
91	9590	9595	9600	9605	9609	9614	9619	9624	9628	9633
92	9638	9643	9647	9652	9657	9661	9666	9671	9675	9680
93	9685	8689	9694	9699	9703	9708	9713	9717	9722	9727
94	9731	9736	9741	9745	9750	9754	9759	9763	9768	9773
95	9777	9782	9786	9791	9795	9800	9805	9809	9814	9818
96	9823	9827	9832	9836	9841	9845	9850	9854	9859	9863
97	9868	9872	9877	9881	9886	9890	9894	9899	9903	9908
98	9912	9917	9921	9926	9930	9934	9939	9943	9948	9952
99	9956	9961	9965	9969	9974	9978	9983	9987	9991	9996
100	0000	0004	0009	0013	0017	0022	0026	0030	0035	0039
N	0	1	2	3	4	5	6	7	8	9

PERIODIC TABLE OF THE ELEMENTS

The atomic weights, based on the exact number 12 as the assigned atomic mass of the principal isotope of carbon, are the most recent (1961) values adopted by the International Union of Pure and Applied Chemistry. (For artificially produced elements, the approximate atomic weight of the most stable isotope is given in brackets.)

Group→		I	II	III	IV	V	VI	VII	VIII			O
Period	Series											
1	1	1 H 1.00797										2 He 4.0026
2	2	3 Li 6.939	4 Be 9.0122	5 B 10.811	6 C 12.01115	7 N 14.0067	8 O 15.9994	9 F 18.9984				10 Ne 20.183
3	3	11 Na 22.9898	12 Mg 24.312	13 Al 26.9815	14 Si 28.086	15 P 30.9738	16 S 32.064	17 Cl 35.453				18 A 39.948
4	4	19 K 39.102	20 Ca 40.08	21 Sc 44.956	22 Ti 47.90	23 V 50.942	24 Cr 51.996	25 Mn 54.9380	26 Fe 55.847	27 Co 58.9332	28 Ni 58.71	
	5	29 Cu 63.54	30 Zn 65.37	31 Ga 69.72	32 Ge 72.59	33 As 74.9216	34 Se 78.96	35 Br 79.909				36 Kr 83.80
5	6	37 Rb 85.47	38 Sr 87.62	39 Y 88.905	40 Zr 91.22	41 Nb 92.906	42 Mo 95.94	43 Tc [99]	44 Ru 101.07	45 Rh 102.905	46 Pd 106.4	
	7	47 Ag 107.870	48 Cd 112.40	49 In 114.82	50 Sn 118.69	51 Sb 121.75	52 Te 127.60	53 I 126.9044				54 Xe 131.30
6	8	55 Cs 132.905	56 Ba 137.34	57–71 Lanthanide series*	72 Hf 178.49	73 Ta 180.948	74 W 183.85	75 Re 186.2	76 Os 190.2	77 Ir 192.2	78 Pt 195.09	
	9	79 Au 196.967	80 Hg 200.59	81 Tl 204.37	82 Pb 207.19	83 Bi 208.980	84 Po [210]	85 At [210]				86 Rn [222]
7	10	87 Fr [223]	88 Ra [226.05]	89–Actinide series**								

*Lanthanide series:	57 La 138.91	58 Ce 140.12	59 Pr 140.907	60 Nd 144.24	61 Pm [147]	62 Sm 150.35	63 Eu 151.96	64 Gd 157.25	65 Tb 158.924	66 Dy 162.50	67 Ho 164.930	68 Er 167.26	69 Tm 168.934	70 Yb 173.04	71 Lu 174.97
**Actinide series:	89 Ac [227]	90 Th 232.038	91 Pa [231]	92 U 238.03	93 Np [237]	94 Pu [242]	95 Am [243]	96 Cm [245]	97 Bk [249]	98 Cf [249]	99 Es [253]	100 Fm [255]	101 Md [256]	102 No	103

FUNDAMENTAL CONSTANTS

Largely based on values in the *American Institute of Physics Handbook* (1957). The probable error for each value has been omitted here; it should properly be considered part of the datum.

Name of Quantity	Symbol	Value
Velocity of light in vacuum	c	2.9979×10^8 m s^{-1}
Charge of electron	q_e	-1.602×10^{-19} C = -4.803×10^{-10} stC
Rest mass of electron	m_e	9.108×10^{-31} kg
Ratio of charge to mass of electron	q_e/m_e	1.759×10^{11} C kg^{-1} = 5.273×10^{17} stC g^{-1}
Planck's constant	h	6.625×10^{-34} J s
Boltzmann's constant	k	1.380×10^{-23} J °K^{-1}
Avogadro's number (chemical scale)	N_0	6.023×10^{23} molecules mole^{-1}
Universal gas constant (chemical scale)	R	8.314 J mole^{-1} °K^{-1}
Mechanical equivalent of heat	J	4.185×10^3 J kcal^{-1}
Standard atmospheric pressure	1 atm	1.013×10^5 N m^{-2}
Volume of ideal gas at 0°C and 1 atm (chemical scale)		22.415 liters mole^{-1}
Absolute zero of temperature	0°K	−273.16°C
Acceleration due to gravity (sea level, at equator)		9.78049 m s^{-2}
Universal gravitational constant	G	6.673×10^{-11} N · m^2 kg^{-2}
Mass of earth	m_E	5.975×10^{24} kg
Mean radius of earth		6.371×10^6 m = 3959 mi
Equatorial radius of earth		6.378×10^6 m = 3963 mi
Mean distance from earth to sun	1 AU	1.49×10^{11} m = 9.29×10^7 mi
Eccentricity of earth's orbit		0.0167
Mean distance from earth to moon		3.84×10^8 m ≐ 60 earth radii
Diameter of sun		1.39×10^9 m = 8.64×10^5 mi
Mass of sun	m_S	1.99×10^{30} kg = 333,000 × mass of earth
Coulomb's law constant	C	8.98×10^9 N · m^2 C^{-2}
Faraday's constant (1 faraday)	F	96,500 C mole^{-1}
Mass of neutral hydrogen atom	m_{H^1}	1.008142 amu
Mass of proton	m_p	1.007593 amu
Mass of neutron	m_n	1.008982 amu
Mass of electron	m_e	5.488×10^{-4} amu
Ratio of mass of proton to mass of electron	m_p/m_e	1836.12
Rydberg constant for nucleus of infinite mass	R_∞	109,737 cm^{-1}
Rydberg constant for hydrogen	R_H	109,678 cm^{-1}
Wien displacement law constant		0.2898 cm °K^{-1}

Numerical constants: $\pi = 3.142$; $e = 2.718$; $\sqrt{2} = 1.414$; $\sqrt{3} = 1.732$

CONVERSION FACTORS

Length:

$1 \text{ m} = 100 \text{ cm} = 1000 \text{ mm}$
$1 \text{ km} = 1000 \text{ m} = 0.6214 \text{ mi}$
$1 \text{ m} = 39.37 \text{ in.}; 1 \text{ cm} = 0.3937 \text{ in.}$
$1 \text{ ft} = 30.48 \text{ cm}; 1 \text{ in.} = 2.540 \text{ cm}$
$1 \text{ mi} = 5280 \text{ ft} = 1.609 \text{ km}$
$1 \text{ A} = 10^{-8} \text{ cm}; 1\ \mu \text{ (micrometer)} = 10^{-4} \text{ cm}$

Area:

$1 \text{ cm}^2 = 0.155 \text{ in}^2; 1 \text{ m}^2 = 10^4 \text{ cm}^2 = 10.76 \text{ ft}^2$
$1 \text{ in}^2 = 6.452 \text{ cm}^2; 1 \text{ ft}^2 = 144 \text{ in}^2 = 0.0929 \text{ m}^2$

Volume:

$1 \text{ liter} = 1000 \text{ cm}^2 = 10^{-3} \text{ m}^3 = 0.0351 \text{ ft}^3 = 61 \text{ in}^3$
$1 \text{ ft}^3 = 0.0283 \text{ m}^3 = 28.32 \text{ liters} = 7.5 \text{ gal}$

Velocity:

$1 \text{ cm s}^{-1} = 0.03281 \text{ ft s}^{-1}; 1 \text{ ft s}^{-1} = 30.48 \text{ cm s}^{-1}$
$1 \text{ mi min}^{-1} = 60 \text{ mi hr}^{-1} = 88 \text{ ft s}^{-1}$

Acceleration:

$1 \text{ cm s}^{-2} = 0.03281 \text{ ft s}^{-2} = 0.01 \text{ m s}^{-2}$
$30.48 \text{ cm s}^{-2} = 1 \text{ ft s}^{-2} = 0.3048 \text{ m s}^{-2}$
$100 \text{ cm s}^{-2} = 3.281 \text{ ft s}^{-2} = 1 \text{ m s}^{-2}$

Force:

$1 \text{ dyn} = 2.247 \times 10^{-6} \text{ lb} = 10^{-5} \text{ N}$
$1.383 \times 10^4 \text{ dyn} = 0.0311 \text{ lb} = 0.1383 \text{ N}$
$4.45 \times 10^5 \text{ dyn} = 1 \text{ lb} = 4.45 \text{ N}$
$10^5 \text{ dyn} = 0.2247 \text{ lb} = 1 \text{ N}$

Mass:

$1 \text{ g} = 6.85 \times 10^{-5} \text{ slug} = 10^{-3} \text{ kg}$
$453.6 \text{ g} = 0.0311 \text{ slug} = 0.4536 \text{ kg}$
$1.459 \times 10^4 \text{ g} = 1 \text{ slug} = 14.59 \text{ kg}$
$10^3 \text{ g} = 0.0685 \text{ slug} = 1 \text{ kg}$

Pressure:

$1 \text{ atm} = 14.7 \text{ lb in}^{-2} = 1.013 \times 10^6 \text{ dyn cm}^{-2}$

Energy:

$1 \text{ J} = 10^7 \text{ ergs} = 0.239 \text{ cal}; 1 \text{ cal} = 4.18 \text{ J}$
$1 \text{ eV} = 10^{-6} \text{ MeV} = 1.60 \times 10^{-12} \text{ erg} = 1.07 \times 10^{-9} \text{ amu}$
$1 \text{ amu} = 1.66 \times 10^{-24} \text{ g} = 1.49 \times 10^{-3} \text{ erg} = 931 \text{ MeV}$

Index

Index

ABCDE79876543210